Fluid Mechanics and Its Applications

Volume 106

Series editor

André Thess, Ilmenau, Germany

Founding Editor

René Moreau, Ecole Nationale Supérieure d'Hydraulique de Grenoble, Saint Martin d'Hères Cedex, France

For further volumes:
http://www.springer.com/series/5980

Aims and Scope of the Series

The purpose of this series is to focus on subjects in which fluid mechanics plays a fundamental role.

As well as the more traditional applications of aeronautics, hydraulics, heat and mass transfer etc., books will be published dealing with topics which are currently in a state of rapid development, such as turbulence, suspensions and multiphase fluids, super and hypersonic flows and numerical modeling techniques.

It is a widely held view that it is the interdisciplinary subjects that will receive intense scientific attention, bringing them to the forefront of technological advancement. Fluids have the ability to transport matter and its properties as well as to transmit force, therefore fluid mechanics is a subject that is particularly open to cross fertilization with other sciences and disciplines of engineering. The subject of fluid mechanics will be highly relevant in domains such as chemical, metallurgical, biological and ecological engineering. This series is particularly open to such new multidisciplinary domains.

The median level of presentation is the first year graduate student. Some texts are monographs defining the current state of a field; others are accessible to final year undergraduates; but essentially the emphasis is on readability and clarity.

Ki-Han Kim · Georges Chahine
Jean-Pierre Franc · Ayat Karimi
Editors

Advanced Experimental and Numerical Techniques for Cavitation Erosion Prediction

Springer

Editors
Ki-Han Kim
Office of Naval Research
Arlington, VA
USA

Georges Chahine
DYNAFLOW, INC.
Jessup, MD
USA

Jean-Pierre Franc
Laboratoire des Ecoulements Géophysiques et Industriels (LEGI)
Grenoble
France

Ayat Karimi
Swiss Federal Institute of Technology (EPFL)
Lausanne
Switzerland

ISSN 0926-5112 ISSN 2215-0056 (electronic)
ISBN 978-94-017-8538-9 ISBN 978-94-017-8539-6 (eBook)
DOI 10.1007/978-94-017-8539-6
Springer Dordrecht Heidelberg New York London

Library of Congress Control Number: 2013957427

Printed on acid-free paper

Springer is part of Springer Science+Business Media (www.springer.com)

Foreword

Cavitation is a problem that has continuously plagued engineers in a variety of disciplines, ranging from the aerospace engineer designing rocket pumps to the civil engineer concerned with the service life of spillway structures and energy dissipaters. Cavitation is known to occur in a variety of fluid machinery including pumps, turbines, and marine propellers. Performance breakdown, noise, vibration, and erosion are difficult problems that are often encountered in practice. Euler, in his 1754 memoir on the theory of hydraulic machines, was probably the first person to conjecture that cavitation might be a problem. As early as 1873, Reynolds had carried out a fundamental study of cavitation in tubular constrictions. However, the real impetus for research in this area came from the introduction of the marine propeller. In fact, R. E. Froude, a naval architect, coined the term "cavitation" in about 1895.

The focus of this book is cavitation erosion. Although Rayleigh was the first to quantify the erosive potential of a single bubble collapse in 1917, progress since that time in actually quantifying the erosion process has been slow. The actual erosion process is far more complex than the simple Rayleigh model implies. Previous work covered several different aspects of the problem such as the inception of cavitation, the formation of complex cavitation patterns, the formation of bubble clouds and the collective collapse of a collection of bubbles, material properties and resistance to cavitation, but a complete description of the erosion process has not been available.

This book is a welcome addition to the literature in that it provides a cohesive blending of sophisticated experimentation and numerical modeling with material properties in order to provide the tools for obtaining a quantitative description of the erosion process. The book is divided into two parts. In Part I, the authors draw upon their broad experience in numerical modeling and experimentation to provide a very comprehensive and detailed account of the physics of the problem. Part II is a selection of nine papers by recognized experts presented at the International Workshop on *Advanced Experimental and Numerical Techniques for Cavitation Erosion Prediction* held in Grenoble, France in 2011. These papers have been selected to complement and expand on the principles outlined in Part I.

There are many features of this new text that are quite different from other cavitation texts. One unique feature is the focus exclusively on cavitation erosion. The introduction contains an elegant summary of cavitation in general followed by the physical principles of the erosion process. A nice overview of experimental procedures is followed by three chapters that provide a detailed interpretation of the data, including signal processing. Another unique feature of this book is the very detailed chapter on numerical methods to determine impulsive loads due to bubble collapse using a fully coupled fluid–structure interaction approach. These ideas are expanded in a separate chapter devoted to the modeling of cavitating propellers and bubble clouds in Part II. Finally, Part I closes with a prediction method of erosion rate based on cavitation loading and material response.

It is interesting to note that this book is motivated by research related to naval applications, just as the early research in this area more than a century ago was motivated by the introduction of the marine propeller. However, the information contained herein is applicable to a broad range of engineering problems. This book will be very useful to both newcomers to the field as well as to experienced researchers.

Roger E. A. Arndt
Prof. Emeritus
University of Minnesota

Preface

This book is devoted to cavitation erosion. It provides a comprehensive review of the phenomenon and the state of the art of the research in cavitation erosion.

In brief, cavitation is the formation of vapor bubbles in a liquid flow due to local low pressures. After an initial growth phase in the low pressure (or high velocity) regions, cavitation bubbles collapse in the regions of pressure recovery. When the collapses occur repeatedly close to a solid boundary, they may induce wear and erosion. After some incubation period, material loss can increase steadily with the exposure time.

Cavitation erosion involves both liquid flow and material properties. On the liquid side, cavitation erosion depends upon the "aggressiveness" of the cavitating flow, defined in terms of the frequency and intensity of the collapses. On the material side, it depends upon the material properties which govern the response of the boundary to the cavitating flow. The actual damage will be the result of the competition between the cavitation intensity and the material strength.

Material strength may be characterized by conventional properties such as hardness, strain energy, or ultimate resilience. Correlations between cavitation erosion (typically mass loss) and material properties are unfortunately far from being universal and are generally valid only within a given class of materials and cavitation intensities. This is the reason why researchers have recently attempted to develop analytical techniques as opposed to correlative techniques.

Thanks to their built-in physical information, analytical techniques open the way for numerical predictions of cavitation damage with limited recourse to erosion tests. These techniques are made increasingly more possible due to the recent increase in computational power and resources. An overview of the founding principles of analytical methods of cavitation erosion prediction is provided in this book.

In principle, analytical techniques in cavitation erosion are very similar to conventional approaches in structural mechanics which consist of identifying the loads applied to a structure and computing the behavior of the structure under these mechanical loads using the material properties. However, there are many challenges in making this ideal approach effective in cavitation erosion.

The loading conditions in a cavitating flow are quite complex. Even for a single collapsing bubble, the impact load is still largely uncertain because of its extreme features in amplitude, frequency, and size. The situation is further complicated in real cavitating flows by a large spectrum of bubbles of various characteristics possibly interacting with each other. The response of the material to the distribution of impact loads on the boundary is also a major source of difficulties since complex mechanical and metallurgical phenomena are involved within the material, including fatigue and fracture. In spite of all the difficulties, significant progress has recently been made in the development of analytical techniques which offer the most promising option for cavitation erosion prediction.

In an attempt to address these challenges more effectively, an international collaborative project was initiated with the support of the U.S. Office of Naval Research (ONR) and ONR Global Naval International Cooperative Science and Technology Opportunity Program (NICOP program) during 2009–2012. The participating organizations were the Laboratory of Geophysical and Industrial Flows (LEGI, Grenoble, France), the Swiss Federal Institute of Technology (EPFL, Lausanne, Switzerland), DYNAFLOW, INC. (USA), the Naval Surface Warfare Center, Carderock Division (USA), and the Naval Research Laboratory (USA).

As part of this project, the International Workshop on Advanced Experimental and Numerical Techniques for Cavitation Erosion Prediction was held in Grenoble during 1–2 March 2011, supported by ONR Global. The objectives of the Workshop were to provide an informal forum to researchers currently involved in cavitation erosion research, to share the knowledge, and identify unresolved issues in understanding the phenomena, thus providing insights into analytical approaches. More than 40 experts in cavitation erosion participated and 17 technical papers were presented.

This book consists of two parts. Part I (Chaps. 1–7) was jointly written by the NICOP project team members; G. L. Chahine (DYNAFLOW, INC.), J.-P. Franc (LEGI) and A. Karimi (EPFL), summarizing the major accomplishments of the project. Part II (Chaps. 8–16) consists of selected papers presented at the International Workshop in Grenoble.

The fundamental principles of analytical techniques are presented in Part I of this book. The guiding thread throughout Part I is the concept of impact load defined as the elementary loading conditions resulting from the collapse of a cavitation bubble.

After a general introduction to cavitation and cavitation erosion (Chap. 1) and a description of typical laboratory testing methods of cavitation erosion (Chap. 2), the early stage of erosion (the incubation period) is analyzed in detail in Chap. 3. This chapter is focused on the elementary damage—a cavitation erosion pit—that results from a single cavitation impact load and provides a technique for estimating the impact load amplitude from the material properties. A statistical analysis of pits and associated impact loads is also provided for a typical cavitating flow.

Chapter 4 is devoted to the measurement of impact loads by means of pressure sensors. Distribution of impact loads in terms of amplitude and time duration are discussed as well as scaling laws followed by the impact load spectra.

The advanced stages of erosion are presented in Chap. 5. They are characterized by material removal and mass loss which result from the repetitive loading of the material surface.

Chapter 6 offers an insight into the potential of numerical methods to compute the impact load resulting from the collapse of a cavitation bubble and the formation of a permanent pit in the case of an elastic–plastic material using a fully coupled fluid–structure interaction approach.

Finally, Chap. 7 addresses the modeling of the advanced stages of erosion. A simple model of the response of a ductile material to successive impact loads is provided which makes it possible to estimate the erosion rate.

Part II is devoted to a selection of papers (each presented in a chapter) representative of the state-of-the-art research in cavitation erosion. Chapters 8–12 deal with experimental aspects, whereas Chaps. 13–16 are devoted to numerical aspects.

Chapter 8 by G. Bark and R. E. Bensow (Chalmers University of Technology, Sweden) presents various concepts of the specific hydrodynamic processes controlling cavitation erosion on marine propellers with a special emphasis on the concept of focusing of collapse energy.

The dynamics of sheet and cloud cavitation in a Venturi-type test section (converging–diverging 2D nozzle) is more specifically studied using a copper layer placed in the diverging section by P. F. Pelz, T. Keil, and G. Ludwig (Technical University of Darmstadt, Germany) in Chap. 9, whereas Chap. 10 by M. Dular (University of Ljubljana, Slovenia) confirms the strong correlation between the dynamics of cavitation structures and the erosion damage on copper-coated hydrofoils.

In Chap. 11, recent investigations in cavitation erosion conducted at the University of Fukui (Japan) are presented by S. Hattori. They include the effect of temperature, erosion in liquid metals, and prediction of erosion based on impact load measurements.

Chapter 12 is devoted to erosion by a submerged cavitating jet with a special emphasis on the effect of the nozzle geometry on the flow aggressiveness. This chapter was prepared by S. Nishimura, O. Takakuwa, and H. Soyama from Tohoku University (Japan).

The next Chaps. 13–16 address modeling and simulation issues in cavitation erosion in pumps, propellers, and hydrofoils. Using a bubble flow model, M. Fukaya (Hitachi, Japan) was able to numerically assess the cavitation intensity in a centrifugal pump and also the compressive residual stresses in a sample exposed to a cavitating jet in Chap. 13.

A compressible Euler model including shock developed by S. J. Schmidt, M. S. Mihatsch, M. Thalhamer, and N. A. Adams (Technical University of Munich, Germany) is presented in Chap. 14 to predict bubble cloud collapses and resulting peak pressures responsible for cavitation erosion. This chapter showed promising results indicating a possibility of practical applications of the tool to predict cavitation erosion susceptibility.

Modeling and simulation of bubble cloud dynamics is also the subject of Chap. 15 by G. L. Chahine, C.-T. Hsiao, and R. Raju (DYNAFLOW, INC., USA). The first part of this chapter deals with cavitation dynamics on a rotating propeller using a Eulerian–Lagrangian approach with special emphasis on scaling of pressure loading. The second part focuses on the modeling of bubble cloud and shows the significance of tuning between the bubble cloud characteristics and the pressure field that is essential to generating very high cavitation impulsive load.

Part II ends with a presentation of numerical simulations of shock emission by bubble collapse near a rigid surface in Chap. 16 by E. Johnsen (University of Michigan, USA) with potential applications to shock wave lithotripsy and the spallation neutron source.

We would like to express our sincere gratitude to all the authors who graciously agreed to contribute to this book. We are also especially grateful to Martin Donnelly (Naval Surface Warfare Center, Carderock Division, USA); Eric Johnsen (University of Michigan, USA); Farrel Martin, and Erik Knudsen (Naval Research Laboratory, USA) for their contributions in reviewing the book manuscript. This book would not have materialized without the commitment and continuous support from the Office of Naval Research. We are very grateful to Richard Vogelsong (ONR Global, London) who supported the Grenoble Workshop that initiated this book project. We would like to thank other contributors to Part I, including Michel Riondet (LEGI); Davide Carnelli (EPFL); Jin-Keun Choi, Chao-Tsung Hsiao, Arvind Jayaprakash, and Sowmitra Singh (DYNAFLOW, INC.). We are also very grateful to René Moreau and André Thess, editors of the Springer series "Fluid Mechanics and Its Applications" who encouraged the publication of our efforts and the referee who kindly agreed to review our manuscript and contributed valuable suggestions for improvement.

Ki-Han Kim
Georges Chahine
Jean-Pierre Franc
Ayat Karimi

Contents

Part I
Cavitation and Cavitation Erosion: Computational and Experimental Approaches

Chapter 1
Cavitation and Cavitation Erosion

Georges L. Chahine, Jean-Pierre Franc and Ayat Karimi

Abstract In this chapter, an introduction to cavitation and cavitation erosion is presented. Cavitation involves the development of various types of vapor structures (such as attached cavities, travelling bubbles, vortical cavities, bubble clouds) in liquid flow due to a drop in the local pressure below a critical value usually close to the vapor pressure. These structures generally originate from cavitation nuclei, typically gaseous microbubbles contained in the liquid. The critical pressure of a nucleus is defined as the particular value of the pressure below which no equilibrium is possible. If a nucleus is subject to pressure lower than its critical pressure, it will explosively grow into a macroscopic cavitation bubble. The bubble will collapse when transported by the liquid flow into regions of pressure recovery. If the collapse occurs near a wall, the resulting high amplitude and small duration impulsive loads may cause local damage. Repeated impulsive loads may cause increasing cavitation erosion damage. The response of the material to cavitation impulsive loads is discussed and material properties most relevant to cavitation erosion, such as sensitivity to strain rate, are presented.

G. L. Chahine (✉)
DYNAFLOW, INC., 10621-J Iron Bridge Road, Jessup, MD, USA
e-mail: glchahine@dynaflow-inc.com

J.-P. Franc (✉)
LEGI, Grenoble, France
e-mail: jean-pierre.franc@legi.grenoble-inp.fr

A. Karimi (✉)
EPFL, Lausanne, Switzerland
e-mail: ayat.karimi@epfl.ch

K.-H. Kim et al. (eds.), *Advanced Experimental and Numerical Techniques for Cavitation Erosion Prediction*, Fluid Mechanics and Its Applications 106,
DOI: 10.1007/978-94-017-8539-6_1, © Springer Science+Business Media Dordrecht 2014

1.1 Introduction to Cavitation

Cavitation is the explosive growth and intense collapse of bubble nuclei in a liquid when exposed to large pressure variations. Practically, cavitation is identified as either the visual appearance of cavities in the absence of air injection and heat input, or the acoustical emission of repeated high-pressure high-frequency recognizable sound produced by these cavities. Sound emission usually occurs earlier than visual identification (i.e. at lower speeds and/or higher local pressures) because microscopic bubbles, invisible to the naked eye, can undergo very strong dynamics and emit pressure impulses before they can be seen with further reduction of the pressure or increase in the velocity [1–7].

The relationship between pressure and velocity in cavitation study is traditionally defined by the *cavitation number*:

$$\sigma = \frac{P_\infty - P_v}{\frac{1}{2}\,\rho\, V_\infty^2}, \tag{1.1}$$

where P_∞ is the reference pressure, P_v is the liquid vapor pressure, ρ the liquid density, and V_∞ the reference velocity. From Eq. (1.1), the effect of decreasing reference pressure on the cavitation number is equivalent to increasing the reference velocity. σ is commonly used as an indicator of the development or extent of cavitation in a liquid flow. At high σ values, cavitation is absent, while cavitation is increasingly more developed as σ is reduced.

Cavitation can take a variety of forms depending upon geometry, hydrodynamic conditions and liquid quality. As an example, different cavitation patterns are visible in the ship model picture of Fig. 1.1. An attached cavity can be seen on the rudder. This type of cavitation is generally associated with flow separation near the leading edge. Cavity closure is usually much more unstable than cavity detachment. Fluctuations at closure are accompanied by the shedding of vapor clouds as shown in Fig. 1.1. Cloud cavitation is among the most damaging types of cavitation. Vortex cavitation is another common type of cavitation. It is due to the pressure drop in the vortex core caused by centrifugal forces associated with liquid rotation. A typical example is tip vortex cavitation which develops from the tip of each blade of a propeller, as observed in Fig. 1.1. In shear flows, such as in submerged jets or in the wakes of bluff bodies, shear cavitation may also develop, as shown in Fig. 2.7 for cavitating jets. This type of cavitation is connected to the turbulent structure of shear flows where various types of coherent rotational structures, whose cores are regions of relatively low pressure, develop. Travelling bubble cavitation may also occur in various hydraulic machines. This cavitation pattern depends primarily upon the water quality in terms of nuclei content (see Sect. 1.1.2). Each macroscopic cavitation bubble originates from a microscopic nucleus that is carried by the liquid flow and grows in low pressure regions before collapsing in regions of pressure recovery, with a potential risk of erosion on neighboring walls. Whatever the type of cavitation maybe, its occurrence is strongly dependent upon the minimum pressure in the flow.

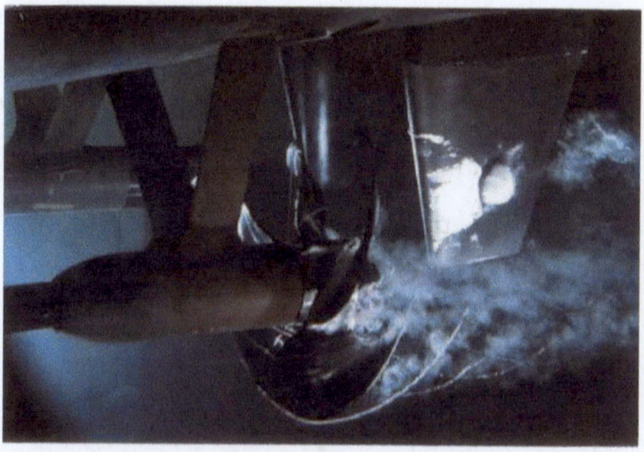

Fig. 1.1 Various types of cavitation on a naval ship model in the U.S. Navy's Large Cavitation Channel (LCC), Memphis, Tennessee. (*Courtesy* U.S. Naval Surface Warfare Center, Carderock Division)

1.1.1 Engineering Definition of Cavitation

It is often assumed that cavitation occurs when the local pressure in the liquid drops below its vapor pressure at the given temperature. Such a definition is based on the phase diagram of a substance, i.e. the curves which separate solid, liquid, and vapor phases of the substance at different temperatures (see Fig. 1.2). Any process that raises the temperature or reduces the pressure by a sufficient amount will result in a phase change from liquid to vapor. Conventionally, boiling is defined as the phase change resulting from raising the temperature at ambient pressure above the boiling temperature, while cavitation is the process resulting in phase change at ambient temperature through a pressure drop below the vapor pressure P_v [2, 3, 5].

Even though this definition has allowed significant progress in practical cavitation studies and design work, it is also often responsible for lack of further advances in design to accommodate cavitation since it ignores bubble dynamics effects and implies that understanding the liquid phase flow only is sufficient to predict cavitation.

1.1.2 Cavitation Nuclei

The above engineering definition of cavitation inception is only true in static conditions when the liquid is in a container and is in contact with its vapor through a flat liquid/vapor free surface. For the more common conditions of a flowing liquid around a solid object such as a rotating machine, liquid vaporization can only occur through the presence of microbubbles, also called "cavitation nuclei". This is a fundamental concept for understanding cavitation and its effects.

Indeed, a pure liquid free of nuclei can sustain very large tensions, measured in the hundreds of atmospheres (tens of MPa), before a cavity can be generated

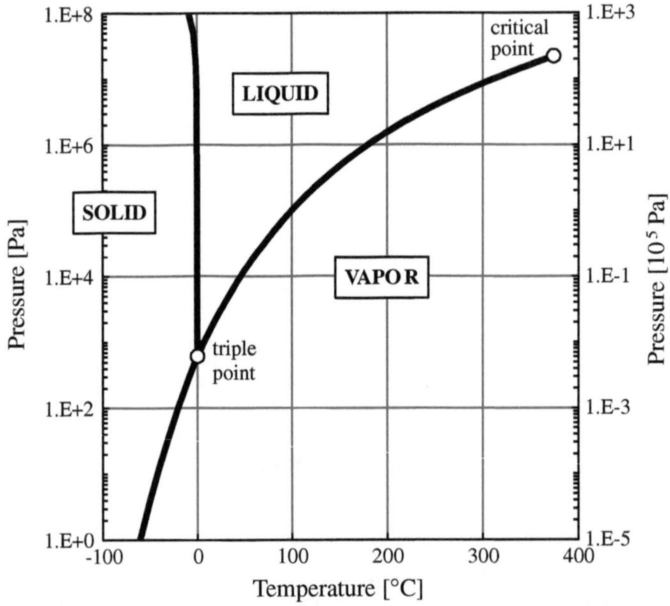

Fig. 1.2 Phase diagram of water. *Vertical axis* (pressure) is logarithmic whereas *horizontal axis* (temperature) is linear. The solid/liquid/vapor triple point for water is at 0.01 °C and 611 Pa and the critical point is at 374 °C and 22.1 MPa. The *curve* which joins the triple point to the critical point is the *vapor pressure curve*. Vapor pressure is 2337 Pa at 20 °C and 1.01×10^5 Pa at 100 °C

through separation of the liquid molecules [1–4]. Cavitation initiates at weak spots of the liquid or nuclei. These weak spots are very small microscopic bubbles either in suspension or trapped in the crevices or discontinuities of solid surfaces or particles [1].

Several techniques have been used to measure nuclei distributions both in the ocean and in laboratory cavitation channels. These include Coulter counter, holography, light scattering methods, cavitation susceptibility meters, and acoustic methods [8–12]. The measurements have indicated that the number density distribution, $n(R)$, whose unit is m^{-4}, follows $n(R) \approx R^{-4}$ where R is the nuclei radius [5, 10].

Therefore, any fundamental analysis of cavitation inception has to start from the observation that a liquid contains nuclei which, when subjected to variations in the local ambient pressure, will respond dynamically by oscillating and eventually growing and collapsing violently (i.e. cavitating).

Cavitation inception can appear in several forms, such as:

- Explosive growth of individual travelling bubbles,
- Sudden appearance of transient cavities or "flashes" on boundaries,
- Development of attached partial cavities, or sheet cavities,

- Appearance, growth, and collapse of bubble clouds behind attached cavities or on a vibrating surface,
- Sudden appearance of cavitating rotating filaments, or vortex cavitation.

Figure 1.1 presents a typical example of cavitation on a high-speed naval ship propeller and rudder. Upon further analysis, all these forms are initiated by the dynamics of the pre-existing nuclei responding to various forms of local pressure fluctuations. These are either imposed pressure variations, pressure drops due to local liquid accelerations, or strongly non-uniform pressure fields due to stream-wise or transverse large vortical structures. These nuclei or weak spots in the liquid are therefore essential in cavitation dynamics. As we will see later, the collapse of a large distribution of elementary nuclei-like cavities is responsible for cavitation erosion.

1.1.3 Nuclei Critical Pressure

A first correction to the engineering definition of cavitation inception is based on the consideration of the static equilibrium of a bubble nucleus. The nucleus is assumed to be spherical and to contain non-condensable gas of partial pressure P_g and vapor of the liquid of partial pressure P_v. Therefore, at the bubble surface, the balance among the internal pressure, the liquid pressure, and surface tension can be written:

$$P_L = P_v + P_g - \frac{2\gamma}{R}, \tag{1.2}$$

where P_L is the pressure in the liquid, γ is the surface tension parameter, and R is the radius of the bubble.

If the liquid ambient pressure changes very slowly, the bubble radius will change accordingly to adapt to the new balance. This is accompanied with a modification of the pressure inside the bubble. The vaporization of the liquid at the bubble–liquid interface occurs very fast relative to the time scale of the bubble dynamics in hydrodynamic cavitation applications, so that the liquid and the vapor can be considered in equilibrium throughout, and the partial pressure of the vapor in the bubble remains constant. On the other hand, gas diffusion occurs at a much longer time scale, so that the amount of gas inside the bubble remains almost constant. This results in a gas partial pressure which varies with the bubble volume. For quasi-steady equilibrium, the gas follows an isothermal compression law, and P_g is related to the initial values, P_{g0}, R_0, and to the new bubble radius R through:

$$P_g = P_{g0}\left(\frac{R_0}{R}\right)^3. \tag{1.3}$$

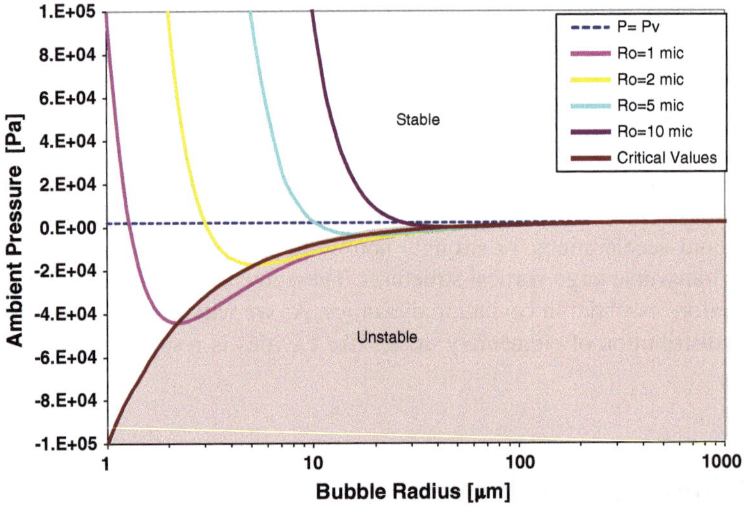

Fig. 1.3 *Static equilibrium curves* of spherical bubbles and definition of *critical curves*. Example given for bubbles of 1, 2, 5, and 10 μm initially at equilibrium at a pressure of 1 atmosphere

The balance of pressures at the bubble wall becomes:

$$P_L(R) = P_v + P_{g0}\left(\frac{R_0}{R}\right)^3 - \frac{2\gamma}{R}, \tag{1.4}$$

where the notation $P_L(R)$ is meant to associate the liquid pressure, P_L, with the bubble radius, R. The bubble static equilibrium can be understood by considering the curve $P_L(R)$. As illustrated in Fig. 1.3, this curve has a minimum value below which there is no equilibrium bubble radius. Only the branch of the curve on the left side of the minimum corresponds to a stable equilibrium.

Solving for the minimum value of $P_L(R)$ using Eq. (1.4) provides the values of the critical pressure P_c, and the corresponding critical radius R_c:

$$\begin{cases} P_c = P_v - \dfrac{4\gamma}{3R_c}, \\ R_c = R_0\sqrt{\dfrac{3P_{g0}}{2\gamma/R_0}}. \end{cases} \tag{1.5}$$

If the pressure in the flow field drops below the critical pressure, an explosive bubble growth, i.e. cavitation, is created. This provides an improved definition for cavitation inception: a liquid flow experiences cavitation if the local pressure drops below the critical pressure, P_c. Since the critical pressure varies with nucleus size, one requires knowledge of the nuclei size distribution in the liquid, which is an essential information in cavitation scaling study.

Equation (1.5) illustrates the fact that the critical pressures are always lower than the vapor pressure. For large initial nuclei sizes, however, P_c is close to P_v. This explains why the engineering definition is acceptable for liquids containing a significant number of large nuclei. This is not the case for many Navy, industrial, and engineering conditions, however, and accurate cavitation inception prediction requires knowledge of the bubble population in the liquid.

1.1.4 Need for Numerical Modeling of Dynamic Effects

Often numerical modeling of cavitation, including nuclei tracking and bubble dynamics computation, is needed to predict cavitation accurately. When the variations in the pressure field to which the bubble is subjected are not slow compared to the bubble response time, the nuclei cannot instantaneously adapt to the pressure field due to inertia effects. One then needs to consider bubble dynamics equations. This is the case for nuclei travelling through the flow field of a rotating machine such as a propeller. The bubbles then act as resonators excited by temporal and spatial variations of the flow field. In the case of a vortical flow field the bubble size can be of the same order as the vortex core. The bubble then experiences strong spatial pressure gradients and time variations, which couple with its motion and dynamics. This makes such a case much more complex than what occurs for a bubble travelling along a foil where the position of the bubble is less coupled to its dynamics.

The flow field pressure fluctuations have various time scales: e.g. relatively long for cavitation bubbles travelling over a blade or captured in a vortical flow region, or very short for cavitation in strongly sheared turbulent flow regions. The amplitudes of these fluctuations and the relationship between the various characteristic times determine the potential for cavitation inception. In all these cases, numerical simulations with correct modeling would accurately predict the dynamic behavior and scaling of cavitation.

Figure 1.4 illustrates bubble dynamic behavior and the relative importance of the minimum pressure, P_{min}, of the imposed pressure field and the critical pressure of the nuclei. As shown in Fig. 1.4a, when $P_{min} < P_c$ the bubble grows explosively (red curve), its radius grows to a maximum that significantly exceeds the initial radius, and it then acquires large potential energy making the implosion very violent. When $P_{min} > P_c$, bubble radius changes only very slightly (blue curve). Figure 1.4b shows the corresponding emitted pressures. For the explosive behavior, large pressure fluctuations are seen during the whole dynamics with large pressure impulses at each bubble collapse (red curve), while the non-explosive case shows much smaller amplitude oscillations during the whole history (blue curve).

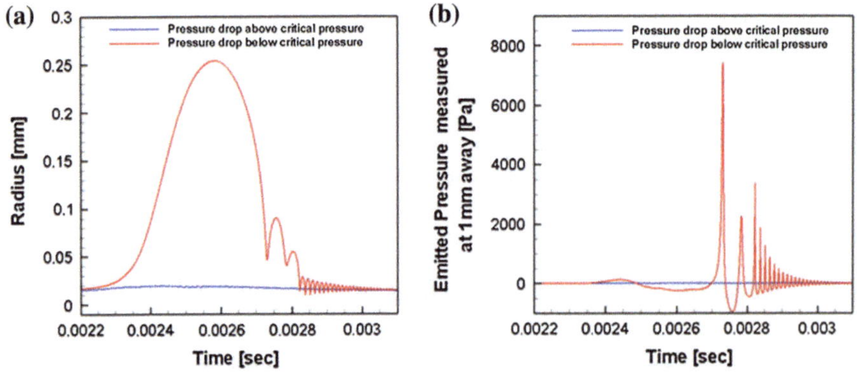

Fig. 1.4 Illustration of bubble dynamic behavior when the excitation pressure field drops below P_c (*red curves*) or remains above P_c (*blue curves*). The nuclei critical pressure P_c is −1,319 Pa (initial nuclei size 10 μm, ambient pressure 0.158 MPa). The minimum pressures encountered by the bubble are −2,572 Pa for the *red curves* and 6,398 Pa for the *blue curves*. **a.** Bubble radius versus time in a propeller flow field. **b.** Acoustic pressure computed at 1 mm away from the bubble wall [6, 7, 13]

1.2 Cavitation Erosion Overview

Cavitation is triggered in the low pressure region as an explosive growth of microscopic nuclei into much larger bubbles. These are then transported by the liquid toward regions of higher pressure where they collapse. This collapse is generally very violent. This is because, at their maximum volume, the pressure inside these bubbles is quite low and becomes close to the vapor pressure. As an example, vapor pressure of water at ambient temperature is as small as 2,300 Pa. As a consequence, the difference between the liquid pressure and the bubble inner pressure may be quite large and able to provide a large acceleration to the bubble wall.

The bubble collapse is accompanied with compression of its content and the subsequent emission of a large amplitude pressure wave (see Fig. 1.4b). In addition, if the bubble collapses close enough to a wall (at a distance less than about twice its radius), a re-entrant jet forms, crosses the bubble and impacts the wall and a high impulsive pressure results (see Fig. 1.5). If the impulsive pressure resulting either from the impact of the micro-jet or from the impact of the shock waves exceeds some appropriate material threshold, such as its yield stress or its ultimate strength, local damage is induced. Figure 1.6 shows an example of cavitation erosion damage on a rudder. Further details on the cavitation erosion process will be given in various chapters of this book.

Material damage depends upon the duration of exposure to cavitation. For metallic surfaces, there is generally an incubation period during which no measurable mass loss occurs (see Chap. 3). Damage remains quite limited throughout the incubation period and takes the form of small permanent plastic deformations

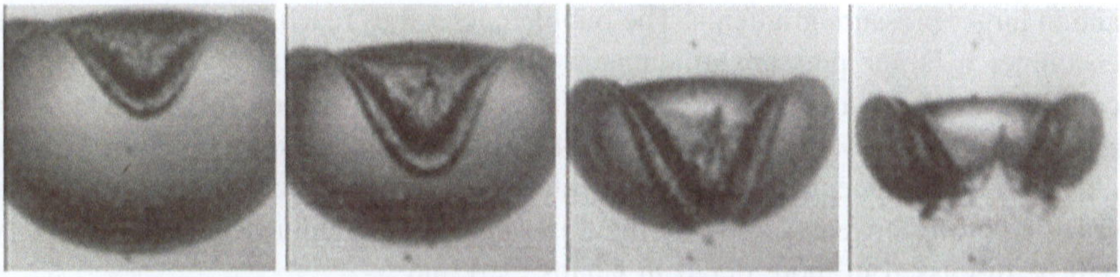

Fig. 1.5 Illustration of the development of a re-entrant jet during the collapse of a bubble near a flat rigid boundary. A pressure impulse is generated when the jet impacts the other side of the bubble (liquid–liquid impact, 3rd image). A second impulse is generated when the remaining toroidal bubble in the 4th image completes its collapse. Spark-generated bubble using a 16 kV condenser discharge in a low pressure tank at DYNAFLOW: $P_{amb} = 0.1$ atm, image size: 2 cm × 2 cm. [14, 15]

Fig. 1.6 Cavitation damage on the rudder of a full-scale naval combatant. (Courtesy: U.S. Naval Surface Warfare Center, Carderock Division)

or pits (typically in the range of a few microns to a few hundreds of microns in diameter) which are due to bubble collapses whose intensities exceed the elastic limit of the material. Pits are first isolated on the material surface and progressively overlap as the exposure time increases.

In general for metallic materials, when a material surface element has been subjected to a large enough number of superimposed cavitation impacts, failure and material removal occur. This is the mass loss period (presented in Chap. 5) which follows incubation. This applies to most materials used in hydrodynamic applications but cannot be generalized to all materials. For instance, for brittle materials, such as glass, coatings, or composites, failure can occur before pits overlap.

Cavitation erosion is strongly dependent upon the flow velocity for two major reasons. One is that the rate of supply of bubble nuclei in the cavitation erosion region increases with the flow velocity. The second is that the impulsive pressure pulses induced by collapsing bubbles also increase with the flow velocity. The latter effect is due to two factors: a larger velocity results in a larger pressure drop and in larger pressure gradients. The bubble nuclei thus reach larger sizes and then encounter a larger pressure gradient at collapse resulting in an increase of the collapse intensity and pressure pulses. The combination of these phenomena results in a strongly non-linear effect of the flow velocity on cavitation erosion damage. Measurements generally indicate that cavitation erosion damage increases as a power between 4 and 9 of the flow velocity. The influence of flow velocity will be discussed in more detail in Chaps. 3, 4 and 5.

1.3 Cavitation Impact Loads

A cavitation flow field is characterized by the omnipresence of a large space and time distribution of very sharp (high frequency) and strong (high amplitude) distinct pressure pulses, which produce a distinguishable cavitation sound. The pulse spectrum results from a statistical distribution of bubbles with violent dynamic behavior including explosive growth, collapse, fission and rebound.

Figure 1.7 shows an illustrative example of a pressure signal detected by a pressure transducer in a cavitating flow field. Such typical pressure-versus-time traces are observed universally in all applications which involve cavitation erosion. This indicates that an obstacle or boundary in the cavitation field will be subjected to repeated impulsive loads which are statistically distributed in both time and space.

The concept of elementary impulsive loads resulting from the pressure pulses or discrete spikes visible in Fig. 1.7 is a key concept in cavitation erosion since the pioneering work of Hammitt [16]. Typical impulsive loads measured using polyvinylidene fluoride (PVDF) pressure sensors are presented in Fig. 1.8. They have a short duration usually measured in microseconds [17, 18].

Another important feature of the cavitation impulsive loads is the small size of the affected area, which for impacting microjets is only a portion (approximately 1/10) of the maximum bubble size [19]. The characteristic size of the affected zone is of the same order of magnitude as the size of the cavitation erosion pits. A typical range is a few micrometers to several hundreds of micrometers.

Cavitation impulsive loads also have a high amplitude. Values on the order of tens to hundreds of Newtons have been reported in the literature [17, 21, 22]. Assuming that the load is applied on a typical surface area of the order of a pit area, the deduced amplitude of the pressure pulse can be as high as the order of GPa and can thus exceed the conventional yield stress of usual materials. This is the reason why cavitation impulsive loads can induce plastic deformations, if not failure.

These extreme values in duration, size, and amplitude make it difficult to measure impact loads accurately. Two main techniques are generally used. One is based on the use of small size, high frequency response, pressure sensors (see Chap. 4) which need to be resistant to cavitation erosion. Another is based on controlled pitting tests carried out on reference materials. In this method, the material itself is used as a sensor (limited to loads exceeding the material elastic

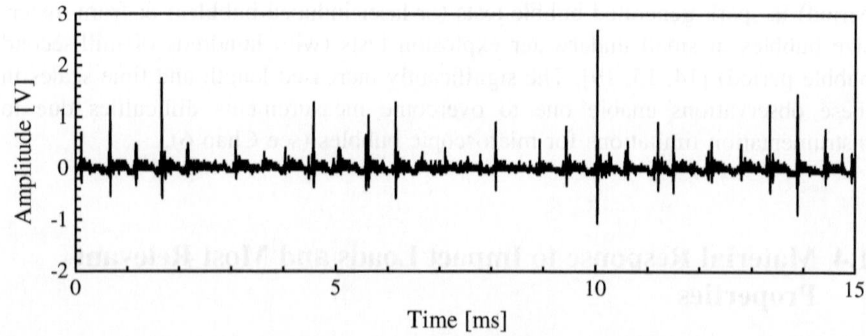

Fig. 1.7 Typical pressure signal in a cavitation flow [20]. The signal exhibits successive pulses of various amplitudes whose highest ones are likely responsible for erosion damage (piezoelectric pressure sensor, natural frequency 250 kHz, sensitive surface diameter 3.6 mm, transducer sensitivity 137 N/V, sampling frequency 2 M samples/s, flow velocity 89 m/s)

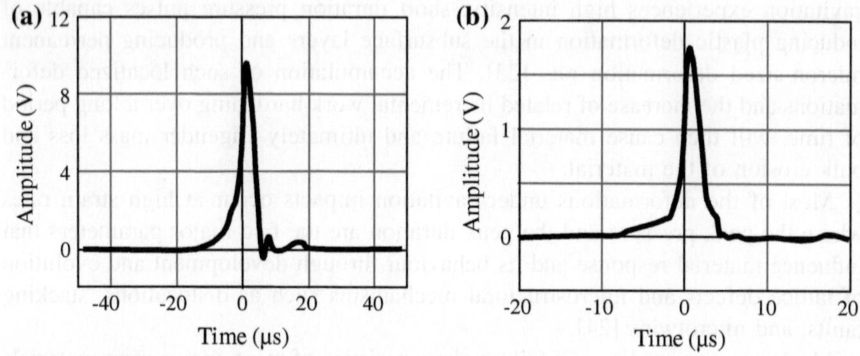

Fig. 1.8 Typical examples of cavitation erosion impulsive loads measured using PVDF pressure sensors. **a.** Example of an impulsive load measured in a cavitating jet apparatus. The PVDF film was 110 µm thick and protected against cavitation by polyamide tapes. Sensitivity estimated from both the pencil lead breaking technique and the ball dropping technique is of the order of 10 N/V (adapted from [17], with permission from ASME) **b.** Example of an impulsive load measured in a cavitating mercury loop with a 40 µm thick PVDF film. The flow velocity in mercury was smaller than 9 m/s. Sensitivity determined from the ball dropping test technique is of the order of 30 N/V [18]

limit) and each erosion pit is considered as the signature of an impact. The method consists of estimating the impact load which is at the origin of each pit from the geometric features of the pit and the properties of the material, which may be deduced from nano-indentation tests (see Chap. 3).

A third and less conventional method uses fundamental single bubble dynamics experimental and numerical investigations at a larger scale than the cavitation bubbles and transposes the results using knowledge of scaling rules. Such studies deduce results from centimeter-size bubbles (with tens of millisecond bubble period) in spark-generated bubble tests (or laser induced bubbles) or from meter-size bubbles in small underwater explosion tests (with hundreds of millisecond bubble period) [14, 15, 19]. The significantly increased length and time scales in these observations enable one to overcome measurements difficulties due to instrumentation limitations for microscopic bubbles (see Chap 6).

1.4 Material Response to Impact Loads and Most Relevant Properties

As discussed above and in further detail in Chaps. 3, 4 and 5, cavitation exerts on nearby solid surfaces a mechanical loading characterized by repeated and randomly distributed impulsive stresses due to emitted shock waves and/or impacting microjets generated by the collapsing bubbles. Independent of the loading mechanisms (pressure waves, shock waves, or microjets), the material exposed to cavitation experiences high intensity, short duration pressure pulses capable of inducing plastic deformation in the subsurface layers and producing permanent micron-sized deformation pits [23]. The accumulation of such localized deformations and the increase of related incremental work hardening over a long period of time, will then cause material failure and ultimately engender mass loss and bulk erosion of the material.

Most of the deformations under cavitation impacts occur at high strain rates where the peak pressure and the peak duration are the two major parameters that influence material response and its behaviour through development and evolution of lattice defects and microstructural mechanisms such as dislocations, stacking faults, and microtwins [24].

Mechanical properties and failure characteristics of most engineering materials vary with the strain rate, $\dot{\varepsilon}$ [25]. Figure 1.9 depicts examples of the influence of strain rate on the flow stress of different alloys. The flow stress is defined as the instantaneous stress to sustain plastic deformation at a particular strain. In a tensile curve, the flow stress at the strain of 0.2 % corresponds to the yield stress, and at the fracture limit refers to the rupture stress. As shown in Fig. 1.9, two regions of sensitivity of the stress to the strain rate can be distinguished over a wide range of strain rates:

- In the range, $\dot{\varepsilon} < 10^3 s^{-1}$, the strain rate has only a slight influence on the flow stress and the dependency can be assumed to be logarithmic.
- In contrast, for $\dot{\varepsilon} > 10^3 s^{-1}$, the flow stress increases more rapidly with strain rate and shows a strong upturn indicative of transition in hardening mechanisms and emergence of additional limitations to plastic deformation.

It is thus important to determine the range of strain rates applicable for a given cavitation impulsive load configuration in order to take into account such a marked effect of strain rate sensitivity in material response. Consideration of the strain rate

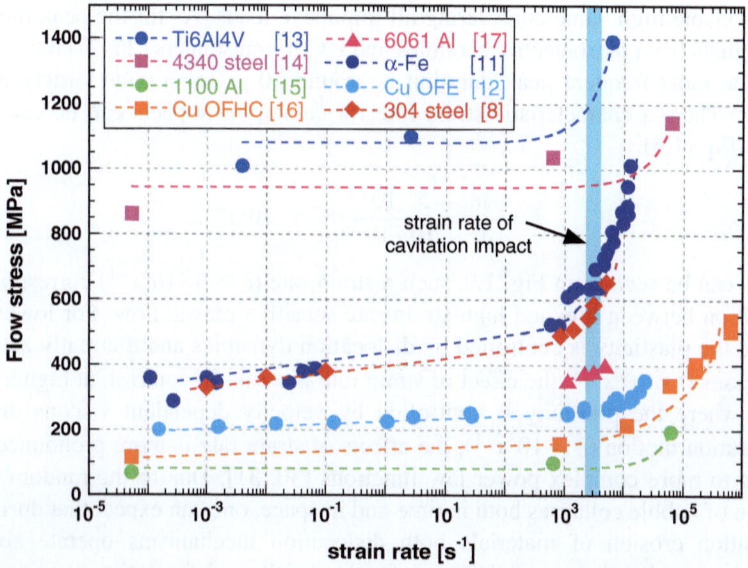

Fig. 1.9 Influence of strain rate on flow stress of different alloys. Two regions of strain rate sensitivity can be distinguished

effects allows not only a better understanding of the erosion behavior of different materials, but also helps to elaborate more accurate predictive models for long-term erosion.

The strain rate of cavitation impulsive loads can be estimated either from analysis of the imploding cavities, or from the strain dynamic concept of material deformations. In both cases, by definition, the strain rate can be expressed using the following equation:

$$\dot{\varepsilon} = \frac{\Delta\varepsilon}{\Delta t} = \frac{\Delta l}{l} \times \frac{1}{\Delta t} = \frac{\Delta l}{\Delta t} \times \frac{1}{l} = \frac{V}{l}, \quad (1.6)$$

where $\Delta\varepsilon$ is the mean strain caused by an impulsive load, Δt is the peak duration, V is the displacement velocity of the impacted surface and l is the length of the plastic zone resulting from an impulsive load. The parameters V and Δt can be evaluated experimentally using, for example, pressure sensitive transducers [26–28]. In contrast, $\Delta\varepsilon$ and l are difficult to measure, but can be estimated from a plastic deformation model of the material as discussed in Sect. 3.4.1. From Eq. (1. 6), it is clear that the strain rate of the impulsive loads for a given cavitation is not a unique value but varies over an order of magnitude depending on the intensity of the impulsive loads expressed by $\Delta\varepsilon$ and its duration denoted by Δt.

Analysis of the stress–strain relationship for a variety of cavitation pits [29] revealed that although strains up to 15–16 % can be reached for strong single impulsive loads, the majority of the impulsive loads generates a deformation around $\varepsilon = 3 - 4\%$ (cf. Sect. 3.4.1). Therefore, a deformation of $\varepsilon = 5\%$ would be a reasonable high value considering all impulsive loads. As for the peak duration, the signals from piezoelectric ceramics and PVDF transducers [26, 27] have shown that the most frequent peak duration is around 10 μs for a wide variety of collapses. Then, a characteristic strain rate for cavitation impacts can be calculated using Eq. (1.6):

$$\dot{\varepsilon} = \frac{\Delta\varepsilon}{\Delta t} = \frac{5 \cdot 10^{-2}}{10 \cdot 10^{-6}} = 5 \cdot 10^{3} s^{-1}. \tag{1.7}$$

As can be seen from Fig. 1.9, such a strain rate ($\dot{\varepsilon} \cong 5 \cdot 10^{3} s^{-1}$) corresponds to transition between low and high strain rate sensitive plastic flow. For lower rates where the plasticity is controlled by dislocation dynamics and thermally activated processes ($\dot{\varepsilon} < 10^{3} s^{-1}$), the effect of strain rate is weak. In contrast, at higher strain rates where the plasticity is controlled by velocity dependent viscous drag on dislocation motion ($\dot{\varepsilon} > 10^{3} s^{-1}$), the effects of strain rate is more pronounced and obeys to more complex power law functions [30, 31]. Due to the random distribution of bubble collapses both in time and in space, one can expect that during the cavitation erosion of materials both dislocation mechanisms operate simultaneously, but that the preponderance of either of these deformation processes will depend on flow characteristics and cavitation types. For example, for ultrasonic cavitation, plasticity is controlled more by thermally activated dislocation dynamics, while for hydrodynamic cavitation plasticity can be described more by viscous drag mechanisms. This could explain why the vibratory cavitation erosion results do not often match with the results of hydrodynamic cavitation erosion.

The high strain rate characteristics of cavitation induced strains can be recognized also from the configuration of deformation microstructures in the subsurface layers. From the microstructural point of view, plastic deformation is directly related to nucleation and arrangement of deformation substructures usually made of dislocations, stacking faults, twins, and point defects. High strain rates induce higher density of dislocations [32], larger vacancy concentrations [33], and higher tendency for deformation twinning as compared to low strain rate [34]. The contribution of each microscopic component to the total deformation depends not only on loading conditions but also on the stacking fault energy of the material [33].

To provide a simple description of stacking fault energy and its influence on material deformation, it is to be noted that in crystalline materials like cubic and hexagonal structures, the constituent atoms are located in well-defined positions within the lattice that creates particular sequences of the atomic planes. The interruption of the stacking sequences stores a certain amount of energy called stacking fault energy (SFE). This energy influences the nature and arrangement of lattice defects and thereby affects mechanisms of the plastic deformation. When the SFE is low, the mobility of dislocations is reduced making the occurrence of cross-slip and climb more difficult. In contrast at high SFE, materials deform mainly by dislocation glide, which favours deviation of dislocations from one plane to another, and results in the formation of 3D structures. Based on the above description, low SFE materials (for example, stainless steels) yield a significant

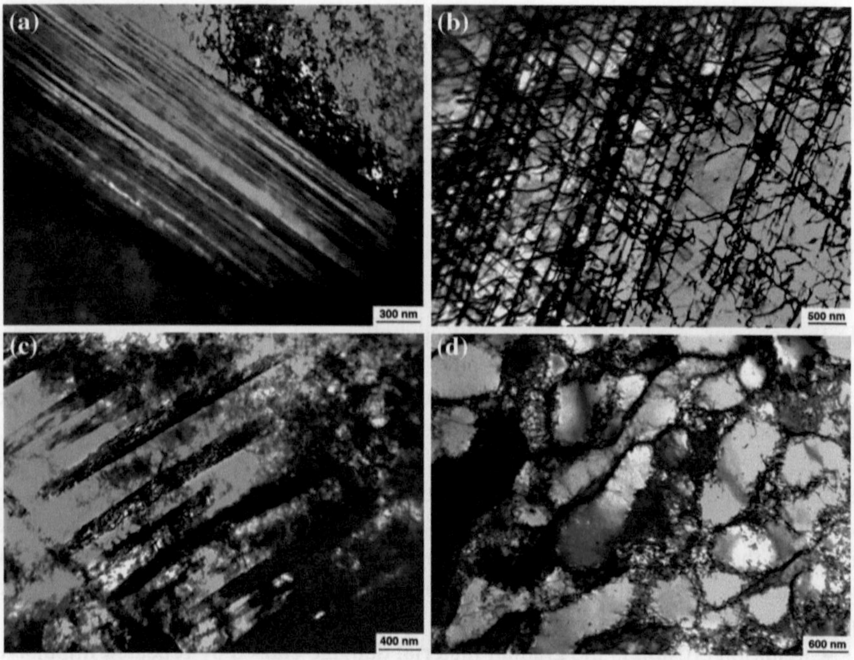

Fig. 1.10 Cross sectional TEM micrographs showing microstructural arrangements at subsurface layers of eroded samples, **a.** stainless steel 18–12, micro twins and dislocation loops at depth of about 10 μm from surface (reproduced from Acta Metallurgica 37(1989) 1079, with permission from Elsevier), **b.** stainless steel 18–12, dislocation patterns at depth of about 100 μm, **c.** deformation twins in copper at depth of 10 μm, **d.** dislocation cells in copper at depth of about 500 μm. The strain rates for these pictures correspond to cavitation, which is highlighted in Fig. 1.9 by a *blue line*

number of planar defects, while high SFE materials (such as copper and aluminum) develop dislocation tangles and cell structures [35]. Therefore, changing strain rate sensitivity and level of strain reflects microscopic transformations in the process of plastic deformation as illustrated by the cross sectional Transmission Electron Microscopy (TEM) micrographs in Fig. 1.10.

In the eroded stainless steel 18–12, the subsurface deformation at a depth of ~10 μm occurs mostly by microtwins (Fig. 1.10a) surrounded by dislocation debris and dislocation loops. In deeper layers corresponding to about 100 μm from the surface, deformation appears as arrays of planar dislocations and loosely tangled dislocation cells. Stainless steel 18–12 is an alloy with low stacking fault energy and thus tends to undergo planar dislocation arrangements. In comparison, copper has high stacking fault energy and deforms by the formation of well-defined dislocation cells as shown in Fig. 1.10d which corresponds to the microstructure of a region at about 500 μm from the eroded surface. Meanwhile, close to the surface at depth of about 10 μm, the deformation process consists mainly of microtwins and dislocation loops (Fig. 1.10c).

Because of the gradient of strain distribution beneath the impacts, a wide variety of deformation substructures were observed as a function of distance from the eroded surface. The TEM micrographs in Fig. 1.10 are only a few examples to illustrate microscopic mechanisms involved in cavitation damage. More detailed analysis of deformation processes under cavitation erosion can be found in [24, 36].

1.5 Summary

In this chapter, a brief introduction of cavitation and cavitation erosion was presented. From an engineering perspective, it is commonly accepted that cavitation occurs when the local pressure in the liquid drops below the liquid vapor pressure at the operating temperature. This engineering definition of cavitation in terms of the vapor pressure is a good approximation for liquids containing large nuclei. For a more accurate definition, cavitation nuclei distribution must be accounted for, and a critical pressure that is defined based on the static equilibrium of bubble nuclei in the liquid should be used. However, in real applications, such as the flow generated by a propeller, bubble dynamics must also be included for accurate prediction of cavitation.

Cavitation erosion is the gradual material damage caused by repeated impulsive loads in the form of shock waves and/or microjet, which are created by the collapsing bubbles. In hydrodynamic cavitation, erosion is strongly dependent upon the liquid velocity. The impulsive loads can be measured or estimated using direct and indirect methods, including pressure sensors, pitting and nano-indentation testing methods (using the material itself as a sensor) and a combined computational and experimental approach for larger-size bubble dynamics utilizing scaling laws. Cavitation erosion occurs in various stages, starting from an initial incubation period where the material surface undergoes plastic deformation, progressively moving towards material failure and progressive weight loss due to repeated cavitation impact loads. These stages are described in detail in the subsequent chapters.

Cavitation loads induce high strain rate in material deformation. Consideration of the strain rate effects not only allows a better understanding of the erosion behavior of different materials, but also helps to elaborate more accurate predictive models for long-term erosion.

References

1. Knapp RT, Daily JW, Hammitt FG (1970) Cavitation. McGraw Hill Book Co., New York
2. Hammitt FG (1980) Cavitation and multiphase flow phenomena. McGraw-Hill International Book Co., New York
3. Young FR (1989) Cavitation. McGraw Hill Book Co., New York

4. Franc J-P, Michel J-M (2004) Fundamentals of cavitation. R Moreau (ed) Fluid mechanics and its applications. Springer, Dordrecht
5. Brennen CE (1995) Cavitation and bubble dynamics. Oxford engineering sciences series 44. Oxford University Press, New York
6. Hsiao C-T, Chahine GL (2004) Prediction of vortex cavitation inception using coupled spherical and non-spherical models and navier-stokes computations. J Mar Sci Technol 8(3):99–108
7. Hsiao C-T, Chahine GL (2005) Scaling of tip vortex cavitation inception noise with a bubble dynamics model accounting for nuclei size distribution. J Fluids Eng 127(1):55–65
8. MacIntyre F (1986) On reconciling optical and acoustical bubble spectra in the mixed layer. In: Monahan EC, Niocaill GM (eds) Oceanic whitecaps. D. Reidel Publishing Company, New York, pp 75–94
9. Oldenziel DM (1982) A new instrument in cavitation research: the cavitation susceptibility meter. J Fluids Eng 104(2):136–141
10. Billet ML (1985) Cavitation nuclei measurements—a review. Paper presented at the ASME cavitation and multiphase flow forum FED vol 23, June 1985
11. Breitz N, Medwin H (1989) Instrumentation for in situ acoustical measurements of bubble spectra under breaking waves. J Acoust Soc Am 86:739–743
12. Duraiswami R, Prabhukumar S, Chahine GL (1998) Bubble counting using an inverse acoustic scattering method. J Acoust Soc Am 105 (5)
13. Hsiao CT, Chahine GL (2008) Numerical study of cavitation inception due to vortex/vortex interaction in a ducted propulsor. J Ship Res 52(2):114–123
14. Jayaprakash A, Hsiao C-T, Chahine G (2012) Numerical and experimental study of the interaction of a spark-generated bubble and a vertical wall. J Fluids Eng 134(3):031301. doi:10.1115/1.4005688
15. Chahine GL, Frederick GS, Lambrecht CJ, Harris GS, Mair HU (1995) Spark generated bubbles as laboratory-scale models of underwater explosions and their use for validation of simulation tools. In: 66th Shock and vibration symposium, Biloxi, MS, November 1995. pp 265–276
16. De MK, Hammitt FG (1982) New method for monitoring and correlating cavitation noise to erosion capability. Trans ASME J Fluids Eng 104(4):434–442
17. Soyama H, Lichtarowicz A, Momma T, Williams EJ (1998) A new calibration method for dynamically loaded transducers and its application to cavitation impact measurement. J Fluids Eng 120(4):712–718
18. Nguyen Trong H (1993) Développement et validation d'une méthode analytique de prévision de l'érosion de cavitation. PhD, Institut National Polytechnique de Grenoble, Grenoble
19. Chahine GL, Annasami R, Hsiao CT, Harris G (2006) Scaling rules for the prediction on UNDEX bubble re-entering jet parameters. SAVIAC Crit Technol Shock Vib 4(1):1–12 (Ed Walter Pilkey)
20. Franc J-P, Riondet M, Karimi A, Chahine GL (2011) Impact load measurements in an erosive cavitating flow. J Fluids Eng 133(12):121301–121308
21. Hattori S, Hirose T, Sugiyama K (2009) Prediction of cavitation erosion based on the measurement of bubble collapse impact loads. Paper presented at the 7th International symposium on cavitation, Ann Arbor, Michigan, USA, August 17–22, 2009
22. Franc J-P, Michel J-M (1997) Cavitation erosion research in France: the state of the art. J Mar Sci Technol 2:233–244
23. Franc J-P (2009) Incubation time and cavitation erosion rate of work-hardening materials. J Fluids Eng 131(2):021303
24. Karimi A, Maamouri M, Martin JL (1989) Cavitation-erosion-induced microstructures in copper single crystals. Mater Sci Eng A 113:287–296. doi:10.1016/0921-5093(89)90317-1
25. Armstrong RW, Arnold W, Zerilli FJ (2009) Dislocation mechanisms of copper and iron in high rate deformation tests. J Appl Phys 105:1–7

26. Konno A, Kato H, Yamaguchi H, Maeda M (1999) Observation of cavitation bubble collapse by high-speed video. Proc, 5th Asian symposium on visualization, Bedugul Bali, Indonesia, March 8–11, 1999
27. Soyama H, Sekine Y, Saito K (2011) Evaluation of the enhanced cavitation impact energy using a PVDF transducer with an acrylic resin backing. Meas 44:1279–1283
28. Hattori S, Hirose T, Sugiyama K (2010) Prediction method for cavitation erosion based on measurement of bubble collapse impact loads. Wear 269(7–8):507–514. doi:10.1016/j.wear.2010.05.015
29. Carnelli D, Karimi A, Franc J-P (2012) Application of spherical nanoindentation to determine the pressure of cavitation impacts from pitting tests. J Mater Res 27(1):91–99. doi:10.1557/jmr.2011.259
30. Lee WS, Lin CF (2001) Impact properties and microstructure evolution of 304L stainless steel. Mater Sci Eng A 308:124–135
31. Ferreira PJ, Sande JBV, Fortes MA, Kyrolainen A (2004) Microstrucrure development during high velocity deformation. Metall Mater Trans 35A:3091–3101
32. Andrade U, Meyers MA, Vecchio KS, Chokshi AH (1994) Dynamic recrystallization in high-strain, high-strain-rate plastic deformation of copper. Acta Metall Mater 42(9):3183–3195. doi:10.1016/0956-7151(94)90417-0
33. Johnson KA, Murr LE, Staudhammer KP (1985) Comparison of residual microstructures for 304 stainless steel shock loaded in plane and cylindrical geometries: implications for dynamic compaction and forming. Acta Metall 33(4):677–684. doi:10.1016/0001-6160(85)90031-8
34. Christian JW, Majahan S (1995) Deformation twinning. Prog Mater Sci 39(1–2):1–157
35. Altynova M, Hu XY, Daehn GS (1996) Increased ductility in high velocity electromagnetic ring expansion. Metall Mater Trans A 27A:1837–1844
36. Karimi A (1989) Cavitation erosion of austenitic stainless steel and effect of boron and nitrogen ion implantation. Acta Metall 37(4):1079–1088. doi:10.1016/0001-6160(89)90104-1

Chapter 2
Laboratory Testing Methods of Cavitation Erosion

Georges L. Chahine, Jean-Pierre Franc and Ayat Karimi

Abstract This chapter presents in detail several cavitation erosion testing methods commonly used in the laboratory. The vibratory cavitation apparatus (G32) is described with its two variants, the direct method using a specimen attached to the vibrating tip of the ultrasonic horn and the alternative method using a fixed specimen facing the horn tip. In the cavitating jet apparatus (G134 and its variants), a jet is discharged at high pressure and velocity in a cell whose pressure may be controlled to adjust the cavitation number. This results in a shear type cavitation whose aggressiveness may be enhanced by a proper design of the nozzle shape and piping assembly. A high-speed cavitation tunnel equipped with a radial divergent test section is also presented. This particular test section generates an unsteady cavity attached to the nozzle exit with cavitation erosion damage concentrated in the cavity closure region. Usual testing procedures together with typical erosion patterns and mass loss results obtained in such facilities are also presented.

2.1 Introduction

Proper evaluation of new materials for their resistance to cavitation erosion requires a comprehensive effort addressing both the intensity of the cavitation field and the resistance of the material. In the absence of historical data on the

G. L. Chahine (✉)
DYNAFLOW, INC., 10621-J Iron Bridge Road, Jessup, MD, USA
e-mail: glchahine@dynaflow-inc.com

J.-P. Franc (✉)
LEGI, Grenoble, France
e-mail: jean-pierre.franc@legi.grenoble-inp.fr

A. Karimi (✉)
EPFL, Lausanne, Switzerland
e-mail: ayat.karimi@epfl.ch

K.-H. Kim et al. (eds.), *Advanced Experimental and Numerical Techniques for Cavitation Erosion Prediction*, Fluid Mechanics and Its Applications 106,
DOI: 10.1007/978-94-017-8539-6_2, © Springer Science+Business Media Dordrecht 2014

performance of a new material in the target cavitating flow fields, experimental studies in the laboratory offer a convenient means of assessing the cavitation erosion performance.

Field erosion studies have been conducted for hydraulic turbines and pumps (e.g. [1–5]), but for marine applications, small scale laboratory tests are more common. These laboratory experimental studies aim at obtaining within the required short time periods an evaluation of the cavitation resistance of the new material, whereas in the real field cavitation erosion may occur after a long duration of exposure.

Such accelerated erosion test techniques include the utilization of ultrasonic vibration devices to generate the cavitation [6–8], cavitation flow loops with strong flow separation, vortex or venturi effects [9–11], rotating discs and submerged cavitating jets [12–15], and other methods. There are also attempts to test model propellers in water tunnels [16].

Some of these techniques are standardized and follow the American Society for Testing and Materials (ASTM) Standards [17]. The ultrasonic technique and the liquid jet technique are the two most popular laboratory techniques for testing cavitation erosion characteristics of materials.

In this chapter, three different laboratory testing methods and equipments are presented in detail, and are used to generate the erosion data presented in Part 1 of this book. They are vibratory devices (ASTM G32), cavitating liquid jets (ASTM G134), and a high-speed cavitation tunnel.

2.2 Vibratory Cavitation Apparatus (ASTM G32)

In ultrasonic cavitation tests, the cavitation is generated by a vibratory device employing a magnetostrictive ultrasonic horn (Fig. 2.1). The high frequency oscillations of the horn, typically tens of kilohertz, induce cyclic formation of very high and very low pressures, which generate high negative tension in the liquid.

This can be understood easily if one considers the acoustic field generated by the imposed amplitude motion of the tip of the horn given by:

$$X(t) = A\cos(2\pi f t), \tag{2.1}$$

where $X(t)$ is the vertical position of the tip of the horn at instant t, A the amplitude and f the frequency of the tip vibratory oscillations.

The resulting acoustic pressure is given by:

$$p = \rho_l c_l \dot{X} = -2\pi f \rho_l c_l A \sin(2\pi f t), \tag{2.2}$$

where ρ_l is the liquid density and c_l is the sound speed in the liquid.

Typically, the vibratory device operates at 20 kHz and the amplitude of the horn tip motion, A, is maintained at 25 μm with the help of a bifilar microscope. This gives for water:

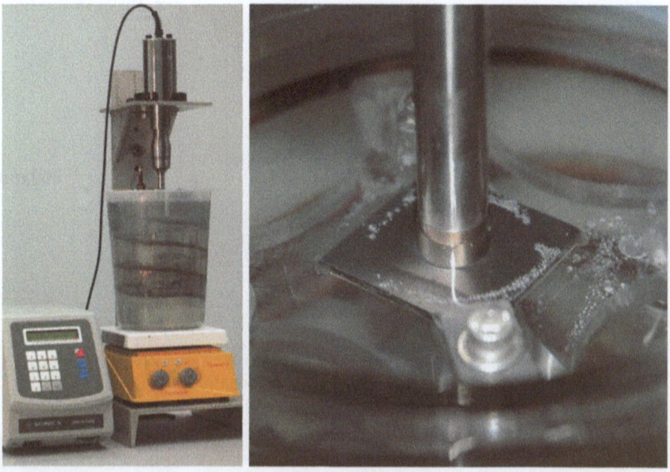

Fig. 2.1 Ultrasonic cavitation erosion test setup at DYNAFLOW. The ultrasonic horn tip vibrates at 20 kHz and generates cavitation bubbles around the tip. The *right picture* shows the alternative G32 configuration. The sample is placed in the square support plate below the cylindrical horn. The *reddish tip* is the Titanium "button". Cavitation under the horn is difficult to see as it is limited to the gap between the "button" and the sample. The *white spots* are bubbles generated by the vibrations at the free surface of the container and at the periphery of the sample holder

$$p = -4.7 \times 10^6 \sin(2\pi f t) \text{ Pascals.} \tag{2.3}$$

Since the amplitude of the pressure oscillations is much larger than the ambient pressure (actually 47 atmospheres), this results in pressure drops during the negative pulse cycle much below the critical pressure of most liquids (see Sect. 1.1.3).

A sample "button" of the material being tested is affixed to the end of the horn and is subjected to the cavitation resulting from the vibration of the horn. A hemispherical cavitation cloud forms at the exposed face of the sample and executes severe dynamics resulting in bubble cloud growth and collapse. The ASTM G32-09 [17, 18] specifies the sample diameter, 16 mm, the vibration frequency, 20 kHz, and amplitude, 50 μm peak-to-peak, and the shape and size of the container in order to minimize variations among different tests and laboratories due to acoustic interaction between transducer and container. A 2,000 ml beaker filled with distilled water and with the tip of the horn submerged 8 mm beneath the free surface is required. In addition, the temperature is controlled by immersing the beaker in a water bath maintained at 25 ± 2 °C.

In an "alternative" G32 test configuration [12, 19] (also known as the stationary specimen method), a stationary material sample is placed at a small distance, typically 0.5 mm, below the vibrating horn tip made of a cavitation resistant button (e.g. Titanium). Deviations from the ASTM G32 method have to be documented. The cavitation erosion tests presented in Chap. 5 used a sample diameter of 12.7 mm instead of 16 mm recommended by the ASTM for both the direct and alternative methods. The alternative G32 method is especially useful for testing

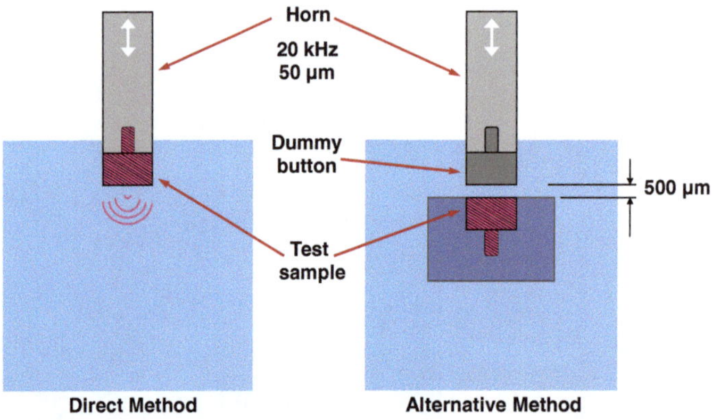

Fig. 2.2 Sketches of the test setups for the ultrasonic cavitation ASTM G32 direct method (*left*) and the alternative method (*right*). In the direct method a hemispherical microbubble cloud is formed under the sample and collapses quasi spherically onto the sample. In the alternative method, the cloud is cylindrical and is confined between the sample and a dummy button, and collapses quasi-cylindrically

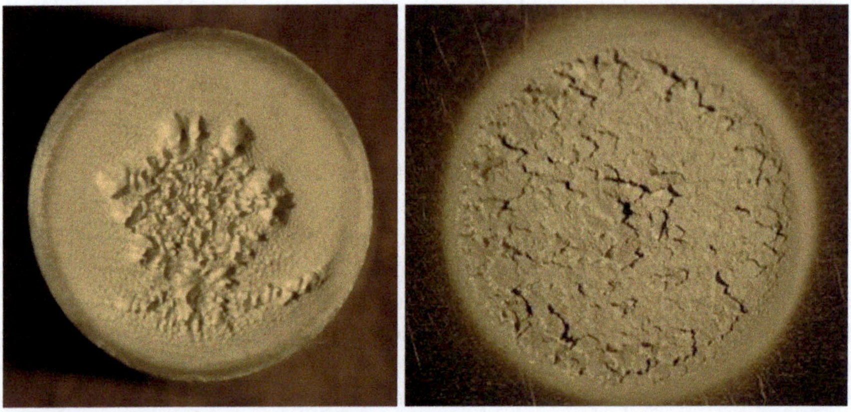

Fig. 2.3 Aluminum alloy Al 7075 samples tested at DYNAFLOW by ASTM G32 direct method (*left*) and alternative method (*right*). Both pictures are shown for 900 min of exposure to cavitation. (Button samples diameter: 12.7 mm)

materials difficult to be made into threaded buttons. Sketches of both setups are shown in Fig. 2.2.

Figure 2.3 shows samples with typical patterns of advanced erosion and mass loss tested by the two ultrasonic cavitation methods. The erosion patterns are significantly different, with the direct method showing a large eroded area concentrated mainly in the central part of the sample, while the alternative method shows a more spread erosion pattern. This is because the shape of the bubble cloud is different between the two schemes. In the direct G32 method the cavitation

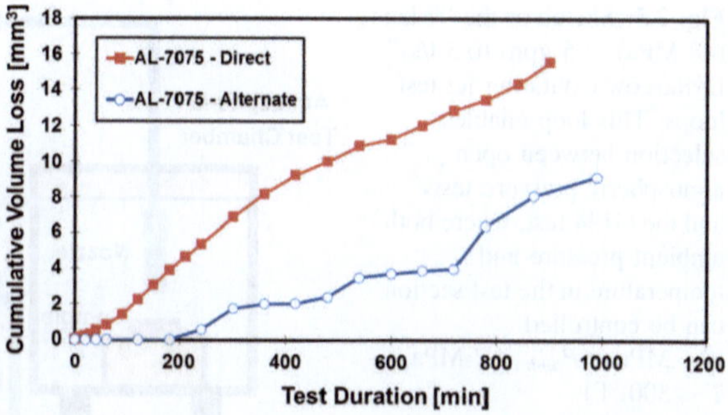

Fig. 2.4 Comparison of material erosion progression on Al 7075 samples tested at DYNAFLOW using both the ASTM G32 direct method and the alternative method. Erosion in the alternative method progresses much slower than with the direct method

cloud collapses in a hemispherical way towards the tested sample (see Fig. 2.2 left), while in the alternative method, the cavitation bubble cloud collapses in a cylindrical way (see Fig. 2.2 right). Cavitation clouds collapsing cylindrically were found to be much less erosive than the hemi-spherically collapsing cavitation clouds [12, 19].

Mass loss versus time curves on the same material (aluminum alloy Al 7075) for the two methods are presented in Fig. 2.4, which illustrates a mass loss rate by the direct method being almost twice that of the alternative method.

The conventional test procedure using the ultrasonic vibrating horn method is to expose the sample to cavitation for a selected period of time, interrupt the test, remove the sample, and record weight to enable calculation of weight loss as a function of time. The sample is then returned to the exact same position on the horn for additional time intervals of erosion. Other erosion characteristics such as volume of erosion imprint, maximum width and depth can also be recorded, together with photographs of the evolution of the eroded region as a function of time.

Erosion tests using ultrasonic cavitation provide reproducible cavitation within a laboratory environment, but the cavitation thus generated is different from that on a propeller or a rudder in a number of ways. The cavitation bubbles are of nearly uniform sizes and are excited by the horn at a fixed frequency, while real cavitation fields have a distribution of bubble nuclei sizes and cavitation forms and vastly different exciting frequencies. The ultrasonic test does not include the effects of bubble nuclei captured by turbulent vortex filaments, break-up of cavities, and presence of liquid flow that interacts with the bubbles. The most important discrepancy is the presence in the ultrasonic method of a cavitation bubble cloud always at the same location.

2.3 Cavitating Liquid Jets (ASTM G134 and Variants)

Cavitating jets have been used extensively for materials testing because of the flexibility these jets provide to control and dial the cavitation intensity. The American Society for Testing and Materials (ASTM) established a standard

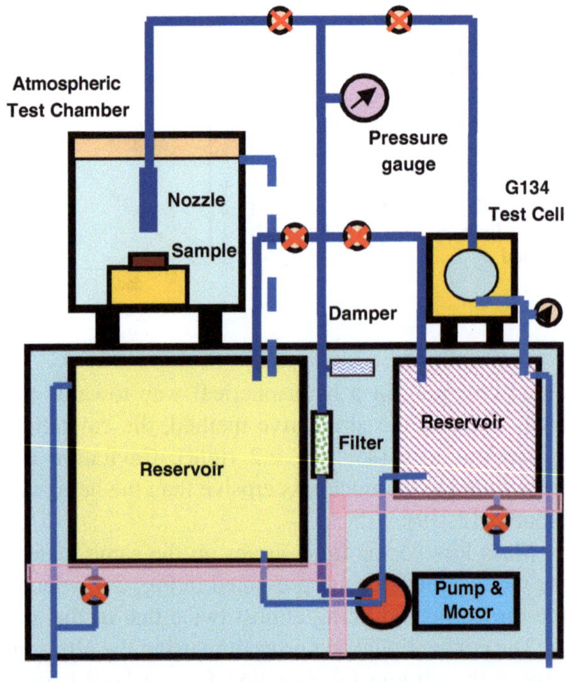

Fig. 2.5 Sketch of the "7 ksi (48 MPa) – 5 gpm (0.3 l/s)" DYNAFLOW cavitating jet test loops. This loop enables selection between open atmospheric pressure tests and the G134 test, where both ambient pressure and temperature in the test section can be controlled (0.1 MPa $< P_{amb} <$ 2 MPa, $T <$ 300 °F)

method using specific conditions and orifice type under the G134 in 1995 [20]. Cavitation intensity produced by cavitating jets can be varied in a wide range through adjustment of the type of the jet, the jet velocity, the jet diameter, the jet angle, the standoff distance, and the ambient pressure in which they are discharged [14]. The jet pressure can be as high as 300 MPa for some applications. This flexibility makes a cavitating jet a useful research and testing tool to study parametrically the effect of cavitation intensity on material behavior.

Compared to the ultrasonic horn testing (G32), the cavitation generated by a cavitating jet provides more realistic cavitation bubble clouds than that by ultrasonic horn, with distribution of various size micro bubbles, shear flows with vortices, and dense bubble clouds, which collapse on the sample. With the control of the operating pressure, the jet angle, and the standoff, the testing time can be controlled to provide either quick erosion for an initial screening or accelerated erosion more relevant to the real flows.

The cavitating jet erosion test setup used in the studies presented in Chaps. 3, 4 and 5 is sketched in Fig. 2.5. The test facility has two testing loops sharing one pump, i.e. only one loop is used at a time by shutting down the other loop using valves. The first flow loop that circulates water through the left side of the setup consists of a cavitating jet nozzle (CAVIJET®), a sample holder, an atmospheric test chamber, a water reservoir, and a pump. A sample holder is used to ensure that the sample can be taken out for measurements and then placed back precisely at the same location to continue testing.

Fig. 2.6 Picture of the G134 "7 ksi (48 MPa) – 5 gpm (0.3 l/s)" test chamber loop

The second flow loop that circulates test liquid through the right half of the setup consists of a cavitating nozzle conforming to G134 specification (0.4 mm orifice diameter), a sample holder, a pressurized test cell, a water reservoir, and a pump. When the cavitation number needs to be controlled or maintained for different jet pressures, the jet is discharged in a pressure controlled cell, where the ambient pressure can be increased (see Fig. 2.6). This is the case for the G134 test cell setup shown in Fig. 2.5.

Different types of jets can be tested for their effect on cavitation erosion. In a conventional submerged jet (see Fig. 2.7 left), cavitation is generated in the turbulent shear layer between the high speed jet and the surrounding liquid. This results in a random distribution of elongated cavitation bubbles with some tendency to organize [21]. This tendency can be harnessed and passive acoustic enhancement can be achieved by proper design of the nozzle shape and piping assembly to result in much more erosive structured cavitating jet (see Fig. 2.7 right) [22, 23]. In this case, vorticity is collected in toroidal vortical structures, whose collapse is intense [24]. Unstructured conventional cavitating jets were used in the studies presented in Chaps. 3, 4 and 5.

For conventional materials erosion testing, where relative performance between samples is assessed, the jet and the sample are submerged in a water tank open to the atmosphere and relative erosion testing is conducted. Under these conditions, the cavitation number is very low and can be defined for cavitating jet as:

$$\sigma_{jet} = \frac{P_{tank} - P_v}{P_{jet} - P_{tank}} \ll 1, \tag{2.4}$$

where P_{tank} is the pressure in the test tank where the sample is located and P_{jet} is the pressure upstream of the nozzle orifice.

A photograph showing a typical setup of the jet nozzle and the sample in its holder is shown in Fig. 2.8. The overall test procedure is similar to that used in the G32 tests other than using a cavitating jet. A normal test procedure for a sample is as follows: (a) the sample is exposed to the cavitating jet for a predetermined

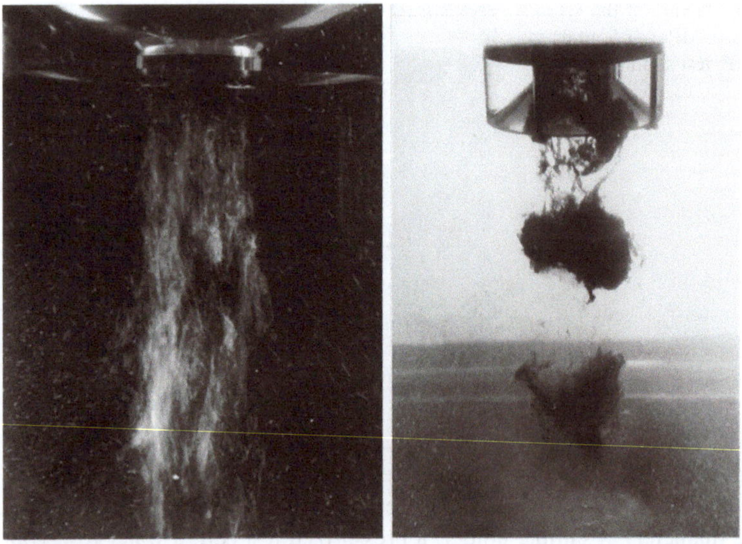

Fig. 2.7 Conventional CaviJet® cavitating jet (*left*) and structured cavitating jet generated by a StratoJet® (*right*). Both pictures were taken using large scale nozzles geometrically scaled up while conserving cavitation number and Strouhal number. The *left* CaviJet® nozzle had an orifice diameter of 2.5 cm, while the *right* StratoJet® orifice had a diameter of 1 cm. The cavitating vortex rings in the StratoJet® were emitted with a frequency corresponding to a Strouhal number of 0.3 at the cavitation number of 0.5

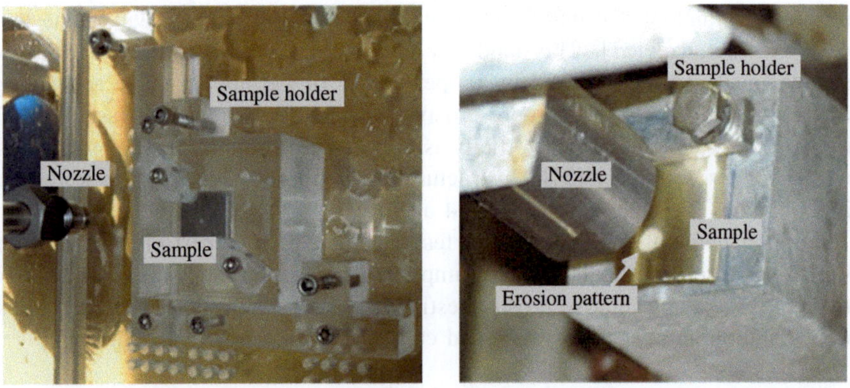

Fig. 2.8 A typical cavitating jet erosion test setup at Dynaflow (*left*): the sample is 2.5 cm × 2.0 cm × 2.5 cm. The nozzle diameter is about 2 mm and the standoff distance is about 2.5 cm. The *right picture* shows more specialized testing; here a cylindrically shaped sample is placed under the nozzle. The whole rod piece can be held in place under the jet. The jet and the samples shown in the pictures are submerged in water during the test

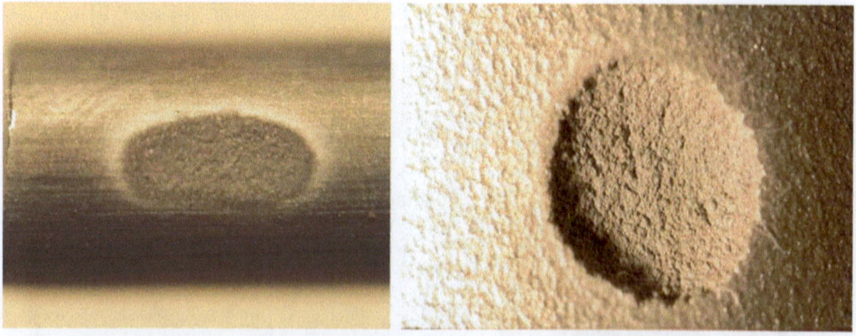

Fig. 2.9 Cavitation erosion pattern on metals created by a CaviJet® cavitating jet. The *left figure* shows the erosion pattern on the rod sample shown in Fig. 2.8 (*right*), which explains the elliptical shape of the eroded area. The *right picture* shows a more typical erosion pattern on a flat sample. In both pictures the erosion areas had typical size of the order of a centimeter, the samples were surface treated proprietary stainless steels and the jet pressure was about 40 MPa

period of time, (b) the test is interrupted, (c) the sample is taken out from its holder for examination, and (d) the erosion is characterized by weight and depth measurement. Photographs of the progression of the erosion patterns such as shown in Fig. 2.9 are taken at selected times. The sample is then returned for additional testing, and the process is repeated. The time intervals are appropriately selected to capture a cumulative weight loss curve displaying as much as possible the characteristic S-curve (see Chap. 5).

2.4 High-speed Cavitation Tunnels

Cavitation erosion tests can also be conducted in high-speed cavitation tunnels. In order to be able to characterize the resistance to cavitation erosion of hard materials within reasonable exposure times, cavitating flows of sufficiently high aggressiveness are required. As aggressiveness increases with flow velocity, cavitation erosion tunnels are often designed for high velocities and consequently high pressures.

Figure 2.10 presents a typical example of such a facility. The whole facility is designed for a maximum pressure of 4 MPa (40 bar) corresponding to a maximum velocity of about 90 m/s. The facility is equipped with a 80 kW centrifugal pump, which can provide a flow rate of up to 11 l/s. A heat exchanger of 80 kW limits the increase in temperature during long duration tests.

The facility comprises a downstream tank of 1 m^3 pressurized with nitrogen by means of a pressurization vessel. The small section of the pressurization vessel limits the dissolution of nitrogen into water so that the dissolved gas content is expected to be almost independent of the pressurization level. Pressurization is required to control the cavitation number which, in turn, controls the extent of

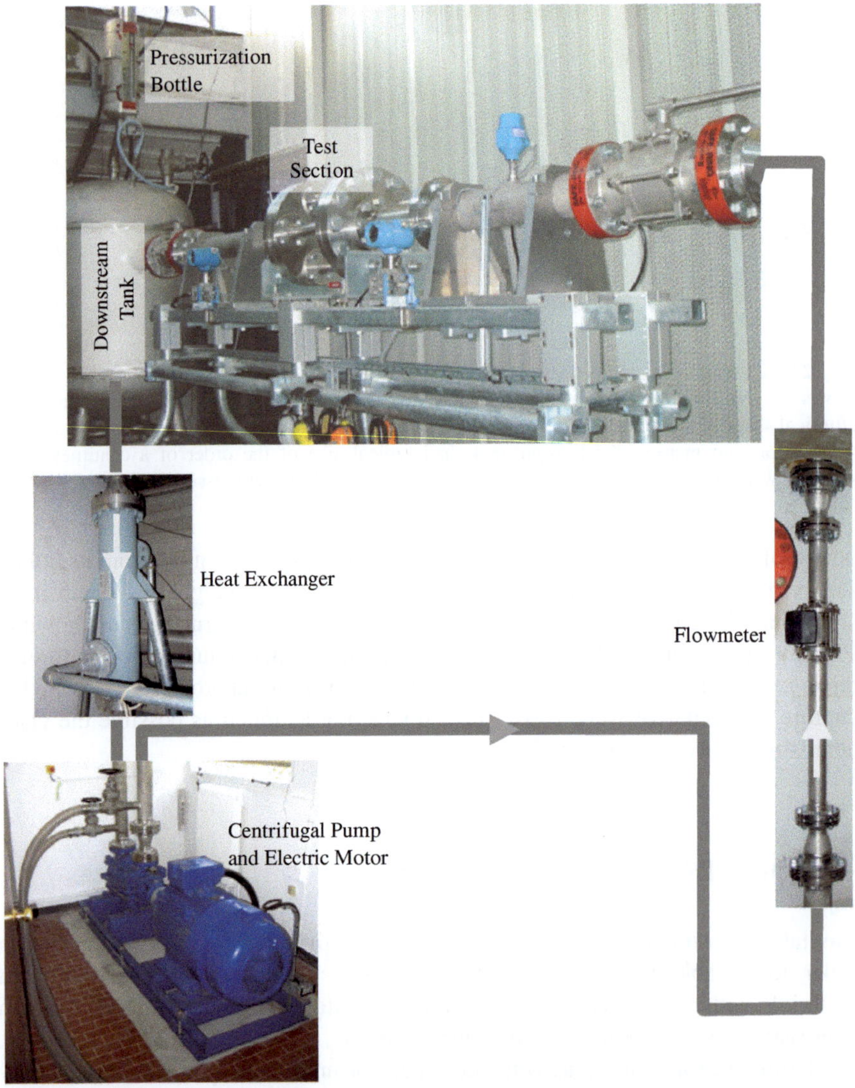

Fig. 2.10 The high-speed cavitation tunnel of the LEGI laboratory (University of Grenoble, France) used for cavitation erosion tests. The tunnel, made of stainless steel, was designed for a maximum operating pressure of 4 MPa (40 bar) corresponding to a maximum flow velocity of 90 m/s. Adapted from [36], with permission from ASME

cavitation and the location of erosion on the sample. In addition, pressurization makes it possible to keep the cavitation number constant when the flow velocity is changed. A similar extent of cavitation is then guaranteed and the effect of flow velocity is separated from the effect of cavitation number or cavity length increase.

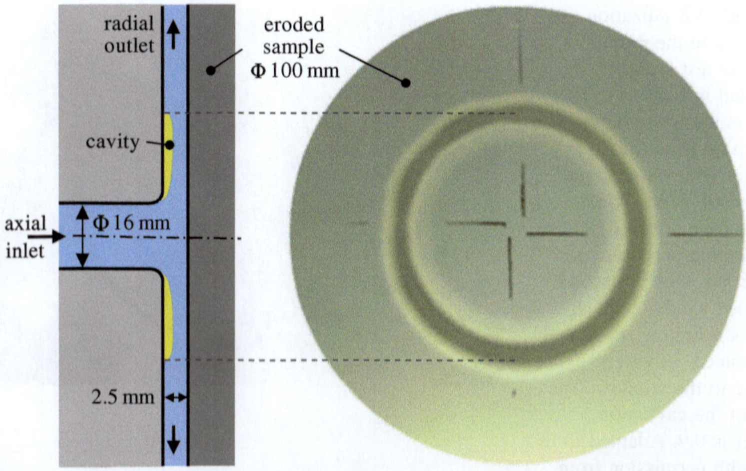

Fig. 2.11 Schematic view of the radial divergent test section used at the LEGI laboratory (University of Grenoble, France) and typical example of an eroded sample. A cavity (*in yellow*) develops at the exit of the 16 mm diameter nozzle, opposite to the sample to be eroded. Erosion is concentrated in the closure region of the cavity and takes the form of a ring due to the axial symmetry of the test section. The mean diameter of the ring is of the order of 45 mm for a value of the cavitation number of 0.9 [37]. Adapted from [36], with permission from ASME

Several pressure sensors are used to control the operating point. A flow meter measures the flow rate Q in the test section and two pressure sensors give the upstream and downstream pressures P_u and P_d respectively. They are located far upstream and downstream of the test section in the inlet and outlet ducts of large diameter (90 mm) with respect to that of the nozzle (16 mm).

The cavitation number is defined by:

$$\sigma = \frac{P_d - P_v}{P_u - P_d}, \tag{2.5}$$

where P_v is the liquid vapor pressure. A temperature sensor is also used to check that the temperature rise during long erosion tests remains limited to typically a few degrees Celsius.

Different types of test sections have been used to investigate cavitation erosion in high-speed tunnels such as a Venturi with or without a central body [25–27], slot cavitator [28–34], cylindrical specimen spanning the tunnel [35] or radial divergent [36].

As an example, the radial divergent test section used in the LEGI ("Laboratoire des Écoulements Géophysiques et Industriels", Grenoble, France) facility is presented in more detail in Figs. 2.11 and 2.12. The inlet flow is axial whereas the outlet flow is radial. Cavitation develops from the nozzle exit and extends into the radial diverging channel. The sample to be eroded faces the nozzle and is located at a distance of 2.5 mm. Cavitation erosion has an annular shape similar to the shape of the closure region of the cavity.

Fig. 2.12 Visualization of the cavity in the radial divergent test section presented in Fig. 2.11. The cavity is the *white* region developing from the 16 mm diameter nozzle exit (*small black circle in the middle*). Flow is from left to right. For visualization purposes, the sample has been replaced by a perspex window, which requires operating the tunnel at a reduced velocity to avoid damage to the window. The value of the cavitation number is 0.9. Adapted from [37], with permission from ASME

For the erosion tests conducted at LEGI, the tunnel is usually operated at a cavitation number around 0.9. With this value of σ, the cavity closure point is located at a radial distance of the order of 22.5 mm from the axis (see Fig. 2.13). Using the definition (2.5) of the cavitation number, the pressure drop through the test section is:

$$P_u - P_d = \frac{P_u - P_v}{1 + \sigma}. \quad (2.6)$$

In this equation, the vapor pressure P_v is generally negligible with respect to the upstream pressure. Since the cavitation number is around 1, Eq. (2.6) shows that the downstream pressure, P_d, in the cavitating test section, and the pressure drop across the nozzle, $P_u - P_d$, are each about half the upstream pressure.

Using Bernoulli equation, a typical velocity on the cavity can be derived:

$$V_c \cong \sqrt{\frac{2P_u}{\rho}}, \quad (2.7)$$

where ρ is the liquid density. Equation (2.7) assumes that the pressure on the cavity surface (which is expected to be close to the vapor pressure) is negligible with respect to the upstream pressure and that the velocity in the inlet duct of large diameter (90 mm) is negligible with respect to the velocity in the test section. As an example, for an upstream pressure of 4 MPa, the velocity on the cavity is $V_c \cong 90\ m/s$. For this typical operating point, the measured flowrate is 8.2 l/s. The equivalent flow velocity in the minimum section area corresponding to the cylindrical section of diameter 16 mm and thickness 2.5 mm at the exit of the

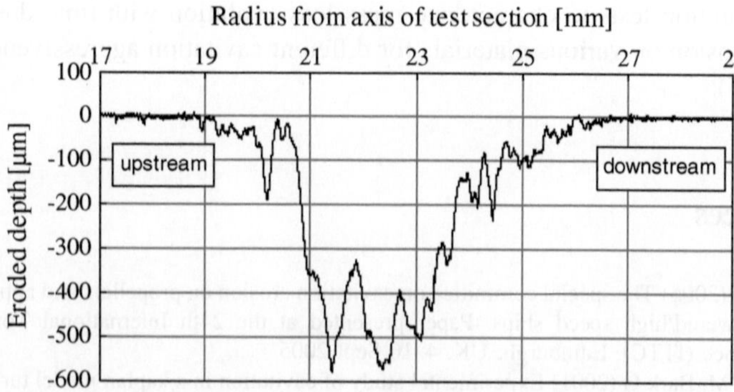

Fig. 2.13 Typical example of a profile of a sample eroded in the cavitation tunnel of the LEGI laboratory. The *horizontal axis* is the radius measured from the center of the test section. The *vertical axis* is the depth of penetration of the damage. Level 0 corresponds to the original non-eroded material surface. Erosion is concentrated in a ring of mean approximate radius 22.5 mm (stainless steel A2205, exposure time: 161 h, upstream pressure: 40 bar, downstream pressure: 18.9 bar, flow rate: 8.2 l/s)

nozzle (see Fig. 2.11) is 65 m/s. This estimate assumes that the flow in this section is purely liquid.

Figure 2.13 presents a typical example of an eroded sample profile along the radial direction. Damage is concentrated in an annular region extending roughly between radius 20 mm and radius 26 mm. The radial location of this region is controlled by the value of the cavitation number. This region moves downstream when the cavitation number is decreased and follows the increase in cavity length.

2.5 Summary

In this chapter, a detailed description of cavitation erosion facilities and equipments was given together with typical measurement results. The facilities include an ultrasonic vibratory horn (G32), a cavitating jet (G134 and its variants) and a high-speed cavitation tunnel with a radial divergent section. They were systematically used to investigate cavitation erosion and the test results are presented in Part I of this book. They include:

- Relatively short duration tests to investigate the cavitation incubation period, various materials pitting, and to deduce impulsive loads on the material (Chap. 3).
- Measurements of the cavitation pressure loads on transducers in order to characterize the amplitude and frequency distribution of cavitation impulsive pressures (Chap. 4).

- Long duration tests to characterize mass loss evolution with time due to cavitation erosion on various materials for different cavitation aggressiveness levels (Chap. 5).

References

1. Billet M (2005) The special committee on cavitation erosion on propellers and appendages on high powered/high speed ships. Paper presented at the 24th International Towing Tank Conference (ITTC), Edinburgh, UK, 4–10 Sept 2005
2. Grekula M, Bark G (2001) Experimental study of cavitation in a kaplan model turbine. Paper presented at the 4th International Symposium on Cavitation (CAV2001), Pasadena, CA, 20–23 Jun 2001
3. Farhat M, Bourdon P (1998) Extending repair intervals of hydro turbines by mitigating cavitation erosion. Paper presented at the CEA Electricity'98 Conference and Exposition, Toronto, 26–29 Apr 1998
4. Farhat M, Bourdon P, Lavigne P, Simoneau R (1997) The hydrodynamic aggressiveness of cavitating flows in hydro turbines. Paper presented at the ASME Fluids Engineering Division Summer Meeting, Vancouver, BC, 22–26 Jun 1997
5. Turbomachinery Society of Japan (2010) Guideline for prediction and evaluation of cavitation erosion in pumps. Tokyo
6. Hammitt FG, Chao C, Kling CL, Mitchell TM, Rogers DO (1970) Round-Robin test with vibratory cavitation and liquid impact facilities of 6061–T 6511 aluminum alloy, 316 stainless steel and commercially pure nickel. Mater Res Stand 10:16–36
7. Chao C, Hammitt FG, Kling CL (1968) ASTM round-robin test with vibratory cavitation and liquid impact facilities of 6061–T6 aluminum alloy, 316 stainless steel, commercially pure nickel, vol 84. The University of Michigan Report MMPP-344-3-T/01357-4-T, Ann Arbor
8. Light KH (2005) Development of a cavitation erosion resistant advanced material system. Master of science thesis, The University of Maine, Orono, ME
9. Dominguez-Cortazar MA, Franc J-P, Michel J-M (1997) The erosive axial collapse of a cavitating vortex: an experimental study. J Fluids Eng 119(3):686–691
10. Hammitt FG (1966) Damage to solids caused by cavitation. Philos Trans R Soc of Lond Ser A Math Phys Sci 260(1110):245–255
11. Escaler X, Avellan F, Egusquiza E (2001) Cavitation erosion prediction from inferred forces using material resistance data. Paper presented at the 4th international symposium on cavitation, Pasadena, California, 20–23 Jun 2001
12. March PA (1987) Evaluating the relative resistance of materials to cavitation erosion: a comparison of cavitating jet results and vibratory results. Paper presented at the ASME cavitation and multiphase flow forum, Cincinnati, 14–17 Jun 1987
13. Momma T, Lichtarowicz A (1995) A study of pressures and erosion produced by collapsing cavitation. Wear 186–187(Part 2):425–436. doi:10.1016/0043-1648(95)07144-x
14. Chahine GL, Courbière P (1987) Noise and erosion of self-resonating cavitating jets. J Fluids Eng 109(4):429–435
15. Lee MK, Kim WW, Rhee CK, Lee WJ (1999) Liquid impact erosion mechanism and theoretical impact stress analysis in tin-coated stream turbine blade materials. Metall Mater Trans A 30A:961–968
16. Pfitsch W, Gowing S, Fry D, Donnelly M, Jessup S (2009) Development of measurement techniques for studying propeller erosion damage in severe wake fields. Paper presented at the 7th international symposium on cavitation (CAV2009), Ann Arbor, Michigan, 17–22 Aug 2009

17. Annual Book of ASTM Standards (2010) Section 3: metals test methods and analytical procedures, vol 03.02. Annual Book of ASTM Standards, West Conshohocken
18. Designation: G 32-09: Standard Test Method for Cavitation Erosion Using Vibratory Apparatus. Annual Book of ASTM Standards (2010) Section 3: Metals test methods and analytical procedures, vol 03.02. West Conshohocken, pp 94–109
19. Hansson I, Mørch KA (1980) The dynamics of cavity clusters in ultrasonic (vibratory) cavitation erosion. J Appl Phys 51(9):4651–4658
20. Designation: G134-95: standard test method for erosion of solid materials by a cavitating liquid Jet. Annual Book of ASTM Standards (2010) Section 3: metals test methods and analytical procedures, vol 03.02, West Conshohocken, pp 558–571
21. Crow SC, Champagne FH (1971) Orderly structure in jet turbulence. J Fluid Mech 48(03):547–591. doi:10.1017/S0022112071001745
22. Johnson VE, Chahine GL, Lindenmuth WT, Conn AF, Frederick GS, Giacchino GJ (1984) Cavitating and structured jets for mechanical bits to increase drilling rate. Part I: theory and concepts. J Energy Res Technol 106(2):282–288
23. Johnson VE, Chahine GL, Lindenmuth WT, Conn AF, Frederick GS, Giacchino GJ (1984) Cavitating and structured jets for mechanical bits to increase drilling rate. Part II: experimental results. J Energy Res Technol 106(2):289–294
24. Chahine GL, Genoux PF (1983) Collapse of a cavitating vortex ring. J Fluids Eng 105(4):400–405
25. Hattori S, Sun B-H, Hammitt FG, Okada T (1985) An application of bubble collapse pulse height spectra to venturi cavitation erosion of 1100-0 aluminum. Wear 103(2):119–131. doi:10.1016/0043-1648(85)90128-0
26. Franc J-P, Michel J-M (1997) Cavitation erosion research in France: the state of the art. J Mar Sci Technol 2:233–244
27. Okada T, Hammitt FG (1981) Cavitation erosion in vibratory and venturi facilities. Wear 69(1):55–69
28. Steller J, Krella A, Koronowicz J, Janicki W (2005) Towards quantitative assessment of material resistance to cavitation erosion. Wear 258(1–4):604–613. doi:10.1016/j.wear.2004.02.015
29. Krella A (2011) An experimental parameter of cavitation erosion resistance for tin coatings. Wear 270(3–4):252–257. doi:10.1016/j.wear.2010.10.065
30. Krella A (2005) Influence of cavitation intensity on X6CrNiTi18-10 stainless steel performance in the incubation period. Wear 258(11–12):1723–1731. doi:10.1016/j.wear.2004.11.025
31. Krella A, Czyzniewski A (2006) Cavitation erosion resistance of Cr–N coating deposited on stainless steel. Wear 260(11–12):1324–1332. doi:10.1016/j.wear.2005.09.018
32. Krella A, Czyzniewski A (2008) Cavitation erosion resistance of nanocrystalline tin coating deposited on stainless steel. Wear 265(7–8):963–970. doi:10.1016/j.wear.2008.02.004
33. Krella A, Czyzniewski A (2009) Cavitation resistance of Cr–N coatings deposited on austenitic stainless steel at various temperatures. Wear 266(7–8):800–809. doi:10.1016/j.wear.2008.11.002
34. Krella A, Czyzniewski A (2007) Influence of the substrate hardness on the cavitation erosion resistance of tin coating. Wear 263(1–6):395–401. doi:10.1016/j.wear.2007.02.003
35. Coleman SL, Scott VD, McEnaney B, Angell B, Stokes KR (1995) Comparison of tunnel and jet methods for cavitation erosion testing. Wear 184(1):73–81. doi:10.1016/0043-1648(94)06563-2
36. Franc J-P (2009) Incubation time and cavitation erosion rate of work-hardening materials. J Fluids Eng 131(2):021303
37. Franc J-P, Riondet M, Karimi A, Chahine GL (2011) Impact load measurements in an erosive cavitating flow. J Fluids Eng 133(12):121301–121308

Chapter 3
Pitting and Incubation Period

Jean-Pierre Franc, Georges L. Chahine and Ayat Karimi

Abstract This chapter is devoted to the initial stage of cavitation erosion known as the incubation period. During this period, failure and material removal are negligible and the damage consists in isolated pits that result from plastic deformation for metallic samples. Typical distributions of pits with respect to their diameter are presented and the effects of various parameters such as flow velocity and material properties are discussed. The characteristic time of the erosion process is defined as the time needed for the pits to fully cover the eroded area without any overlap, and the characteristic pit size is defined as the size of those pits whose contribution to the covered surface is the largest. A new technique for estimating impulsive pressures from pitting tests is suggested. It is based on the similarity between a cavitation erosion pit and a conventional spherical nanoindentation imprint. This technique allows for the determination of the load spectrum in a wide range of pressure amplitudes by combining pitting tests on different materials of various cavitation resistances. The impulsive load associated with a pit can be deduced and correlated with pit volume.

J.-P. Franc (✉)
LEGI, Grenoble, France
e-mail: jean-pierre.franc@legi.grenoble-inp.fr

G. L. Chahine (✉)
DYNAFLOW, INC., 10621-J Iron Bridge Road, Jessup, MD, USA
e-mail: glchahine@dynaflow-inc.com

A. Karimi (✉)
EPFL, Lausanne, Switzerland
e-mail: ayat.karimi@epfl.ch

K.-H. Kim et al. (eds.), *Advanced Experimental and Numerical Techniques for Cavitation Erosion Prediction*, Fluid Mechanics and Its Applications 106,
DOI: 10.1007/978-94-017-8539-6_3,

3.1 Introduction

During the initial stage of erosion, the incubation period, mass loss is negligible and erosion damage is essentially characterized by distributed local indentations or pits as shown in Fig. 3.1. Pits are shallow plastic deformations, roughly circular, which are visible only if the original surface is finely polished since they can be of similar elevation as surface roughness.

When the exposure time to cavitation is small, pits are scattered. As the exposure time is increased, the number of collapse events increases, erosion pits progressively overlap and the material surface becomes increasingly deformed and

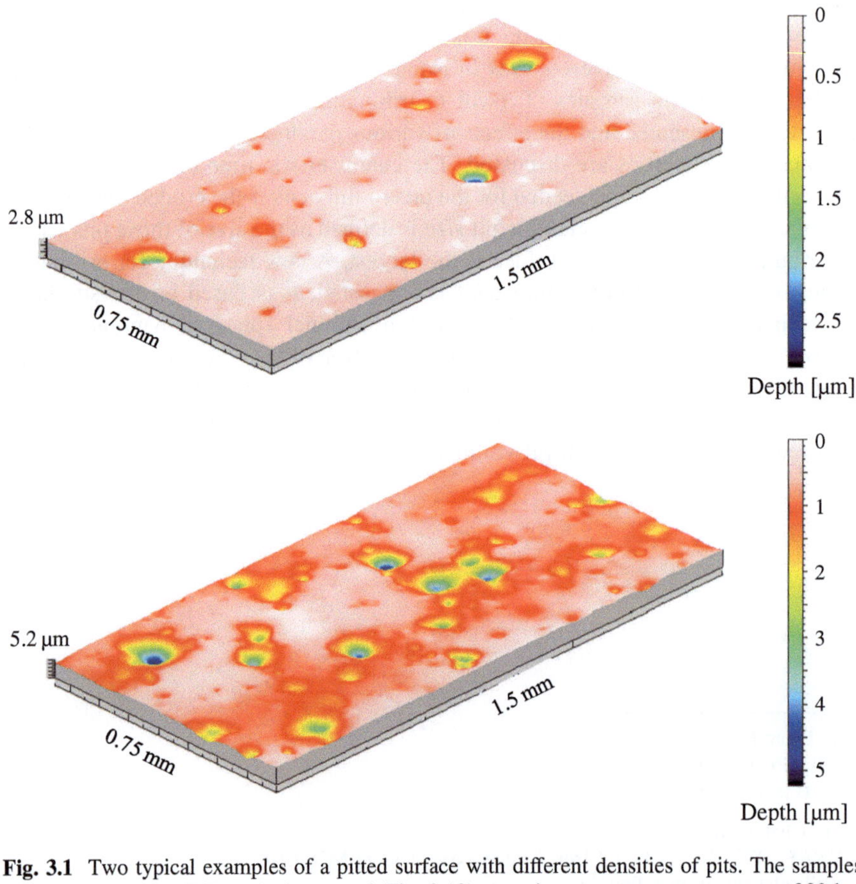

Fig. 3.1 Two typical examples of a pitted surface with different densities of pits. The samples were eroded in the LEGI cavitation tunnel (Fig. 2.10) operating at an upstream pressure of 20 bar (2 MPa), a downstream pressure of 9.5 bar (0.95 MPa) and a flow rate of 5.6 l/s. The material was polished aluminum alloy Al 7075 T651. Exposure time was 5 min. The two different images were obtained from the same sample but at two different radial locations (Fig. 2.11). Pit densities are different because of the non-uniform radial distribution of damage. The bottom picture was taken close to the point of maximum damage

hardens. For much larger exposure times, when the material surface has been covered several times by cavitation impacts, more severe damage including material failure occurs. This is the mass loss period (addressed in Chap. 5) which follows the incubation period considered in the present chapter.

Pitting tests were introduced by Knapp in the 1950s as a way to characterize the intensity or aggressiveness of a cavitating flow [1–4]. The pitting rate (defined as the number of pits per unit surface area and unit exposure time) for a given material was initially used to assess the aggressiveness of the flow. Recent progress in surface analysis techniques allows a more advanced quantitative analysis of pitting tests including the determination of the size, depth and volume of the pits in addition to the pitting rate. Typical results on statistical distributions of pits are presented and discussed in Sects. 3.2 and 3.3.

Each pit is due to an impulsive load generated by the collapse of a vapor cavity. In order to assess the intensity of cavitation, it is essential to be able to determine the characteristics of these loads. Two main techniques can be used.

One is based on pitting tests and regards each pit as the signature of a hydrodynamic loading. In this approach, the material itself is used as a sensor. In Sect. 3.4, a method is presented indicating that the detailed measurement of pit geometry combined with the nano-indentation technique makes it possible to assess the impulsive load responsible for each erosion pit, thus offering the possibility of estimating the cavitation intensity in terms of the distribution of impact loads. Finite element analysis combined with bubble collapse modeling presented in Chap. 6 further discusses the relationship between the actually imposed impulsive load and the load deduced from the computed permanent deformation.

A second technique for estimating impact loads is the direct measurement using pressure sensors. This technique together with typical results is presented in Chap. 4 .

3.2 Erosion Pits

3.2.1 Pitting Test Analysis

A statistical analysis of cavitation erosion pits is presented in the following sections on the basis of pitting tests carried out in the facilities presented in Chap. 2. Pitted samples were analyzed using a contact profilometer at LEGI [5] and optical profilometry at the Naval Research Laboratory for the DYNAFLOW tests [6]. Other surface analysis techniques can be used such as laser profilometry [7], optical interferometry [8, 9] or scanning electron microscopy [10].

Pitting test results obviously depend upon the material since a less resistant material will detect more pits than a highly resistant one. A pit will be formed only

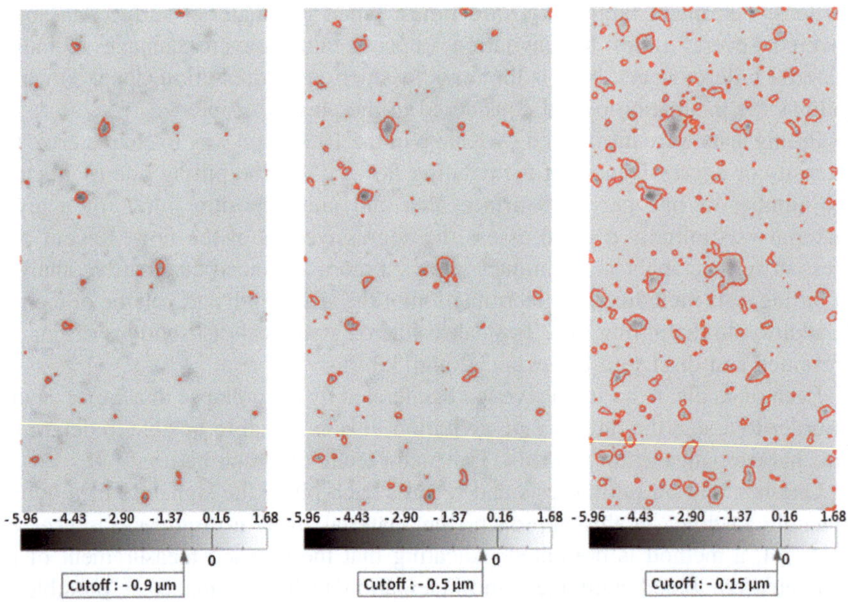

Fig. 3.2 Influence of cutoff depth on pit identification. The gray level measures the local depth between −5.96 (maximum depth) and 1.68 µm (maximum elevation). Level 0 corresponds to the mean level of the whole image. The red contours are the cross-sections of the surface at a given cutoff depth. Three values of cutoff depth are shown from *left* to *right*: 0.9, 0.5 and 0.15 µm. Pitting tests in the LEGI tunnel on stainless steel A2205, upstream pressure: 40 bar (4 MPa), cavitation number: $\sigma = 0.9$, exposure time: 2 min, image size: 4 mm × 2 mm

if the amplitude of the impulsive load exceeds some material threshold. The conventional yield stress is often considered as an appropriate threshold since, by definition, if the applied stress is smaller than the yield stress the material will return elastically to its original non-deformed state after unloading. Even though pitting test results depend upon the material, pitting tests conducted on different materials exhibit similar trends [5] discussed in the following sections.

The identification of pits requires choosing a cutoff depth in order to be able to define the contour of the pits. Figure 3.2 shows the influence of the cutoff depth on pit detection. Ideally, the threshold should be the smallest possible, but larger than the roughness in order to avoid any confusion between pits and roughness elements or material defects. On the other hand, if the cutoff depth is reduced, the risk of merging between two neighboring pits increases as shown in Fig. 3.2 (right). The choice of an appropriate cutoff depth should then be the result of a compromise in order to avoid pit merging and simultaneously include as many shallow pits as possible.

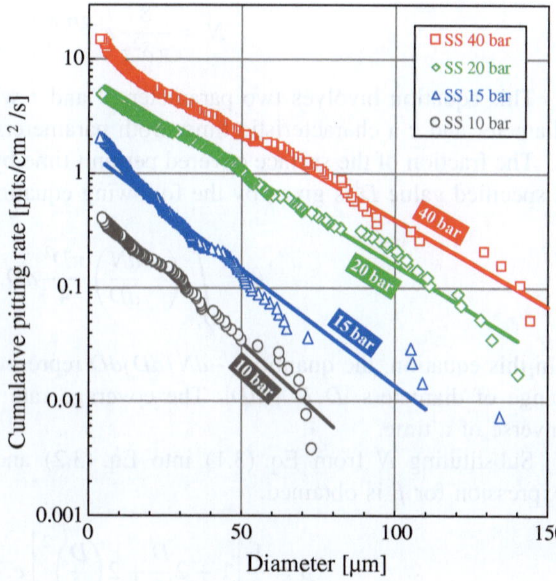

Fig. 3.3 Typical example of cumulative histograms of pitting rates versus pit size. The count for the cumulative number of pits is started from the largest pit. Histograms are approximated by straight lines in semi-logarithmic scales. Pitting tests carried out in the LEGI cavitation erosion facility on stainless steel A2205 at different upstream pressures between 10 bar (1 MPa) and 40 bar (4 MPa) and at a constant cavitation number $\sigma = 0.9$; cutoff depth: 0.5 µm. Adapted from [5], with permission from Elsevier

3.2.2 *Histograms of Pit Size*

A pitted surface exhibits a large variety of pits with various equivalent diameters (based on the contour areas) and maximum depths. The statistical distribution of pits with diameter can be characterized by histograms of pit size. Figure 3.3 shows typical examples of cumulative histograms which count the number of pits whose diameter exceeds a specified value plotted along the horizontal axis. The vertical axis is the pitting rate that is the number of pits per unit exposure time and unit surface area. By differentiating a cumulative histogram relative to the pit diameter, the probability density function is obtained and gives the distribution of pits with respect to diameter.

As shown by the linear fit in the log-lin representation in Fig. 3.3, cumulative pitting rates can be approximated by an exponential function in a wide range of diameters. If needed, the basic exponential law can be refined in order to better account for pits of large size whose distribution may depart from the exponential law (see Sect. 3.2.4). Probability density functions also follow an exponential law. This expresses that the smaller the pit size, the larger the pitting rate and reflects the fact that a cavitating field generates a very large number of small pits while the larger ones are rarer. A detailed analysis of pitting tests makes it possible to assess which ones are the most effective in terms of surface coverage and then the most damaging during the incubation period. This point is discussed below.

If histograms of pit size follow an exponential law, the cumulative pitting rate N can be expressed as a function of pit diameter D by the following equation:

$$N = \frac{8}{\pi \delta^2 \tau} e^{-2D/\delta}. \tag{3.1}$$

This equation involves two parameters δ and τ where δ is a characteristic pit diameter and τ a characteristic time. Both parameters are interpreted below.

The fraction of the surface covered per unit time by all pits whose size exceeds a specified value D is given by the following equation:

$$\beta = \int_D^\infty \left(-\frac{dN}{dD} \right) \frac{\pi D^2}{4} dD. \tag{3.2}$$

In this equation, the quantity $(-dN/dD)dD$ represents the number of pits in the range of diameters $[D, D + dD]$. The coverage rate β has the dimension of the inverse of a time.

Substituting N from Eq. (3.1) into Eq. (3.2) and integrating, the following expression for β is obtained:

$$\beta = \frac{1}{\tau} \left[1 + 2\frac{D}{\delta} + 2\left(\frac{D}{\delta}\right)^2 \right] e^{-2D/\delta}. \tag{3.3}$$

When D approaches zero, that is if all pit sizes are taken into account to compute the covered surface, the coverage rate β tends to $1/\tau$. As a result, time τ is interpreted as the coverage time, i.e. the time required for the surface to be fully covered once by the pits. It is a fundamental characteristic time of the cavitation erosion process. As shown in Chap. 7, it is a relevant time for making the incubation time, the acceleration time and the erosion rate non-dimensional.

As for the length parameter δ, it is interpreted on the basis of the probability density function of the coverage rate. From Eq. (3.3), we get:

$$-\frac{d\beta}{dD} = \frac{4}{\delta^3 \tau} D^2 e^{-2D/\delta}. \tag{3.4}$$

This probability density function gives the contribution of pits to the covered surface as a function of their diameter. As shown in Fig. 3.4, the probability density function exhibits a maximum for a given diameter. It can easily be shown by differentiating Eq. (3.4) that the maximum occurs for a pit diameter equal to δ. Thus, the parameter δ is actually the diameter of the pits which contribute most to the coverage of the surface. The contribution of much smaller and much larger pits is weaker. Although their density is very high, small pits do not contribute significantly because of their small size. As for large pits, they contribute less to the coverage because of their small probability of occurrence which is not compensated by their large area. However, the large loads associated to large pits could be responsible for more severe damage in terms of material failure. It is then difficult to generalize this result to the more advanced stages of erosion when damage is no longer measured by plastic deformations but by failure and mass loss.

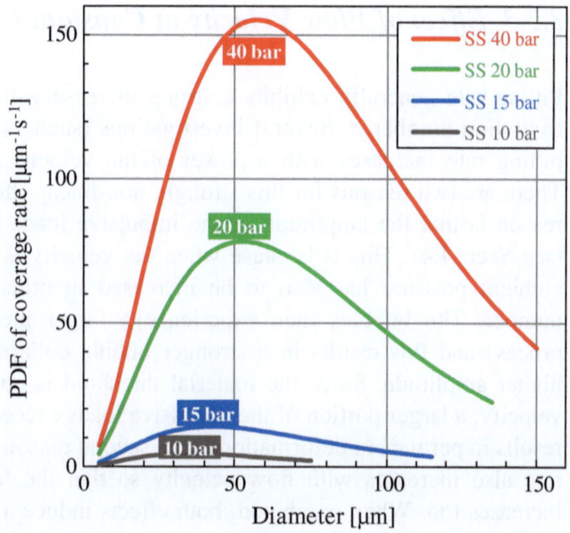

Fig. 3.4 Typical probability density functions of the coverage rate. The plotted curves correspond to the fitted exponential laws presented in Fig. 3.3 and exhibit a maximum for a characteristic value of pit diameter identified as parameter δ. Pits of diameter around this characteristic value δ contribute most to the coverage of the surface. Adapted from [5], with permission from Elsevier

Within the present approach, any histogram of pit size is characterized by only two parameters δ and τ. On the basis of these parameters, a non-dimensional pit diameter and a non-dimensional pitting rate can be introduced:

$$\overline{D} = \frac{D}{\delta}, \tag{3.5}$$

$$\overline{N} = \delta^2 \tau N. \tag{3.6}$$

Just as parameter δ is a reference pit diameter, the group of parameters $1/(\delta^2\tau)$ can be considered as a reference pitting rate used to make the measured pitting rate non-dimensional. Any histogram of pit size can then be reduced to the same non-dimensional histogram defined by:

$$\overline{N} = \frac{8}{\pi} e^{-2\overline{D}}. \tag{3.7}$$

As a consequence, any histogram obtained for given operating conditions can be transposed to other operating conditions by using the normalizations (3.5) and (3.6) provided that the values of the parameters δ and τ are known. The scaling law (3.7) can also account for a possible change of material in addition to a change of operating conditions [5]. They are applicable when histograms follow an exponential law and for geometrically similar cavitating flows, which requires a constant cavitation number. Typical values of δ are given. As for the coverage time, it can vary with several orders of magnitude depending on the material and the flow velocity as explained in the following.

3.2.3 Effect of Flow Velocity at Constant Cavitation Number

Pitting rate generally exhibits a strong increase with flow velocity for the same cavitation number σ. Several investigations (such as [5, 9, 11]) have shown that pitting rate increases with a power of the velocity generally between 5 and 6. There are two reasons for this strongly non-linear effect of the flow velocity. One reason is that the amplitude of the impulsive loads increases with flow velocity (see Sect. 4.4). This is because when the velocity is increased at constant σ, the ambient pressure has also to be increased in order to conserve the cavitation number. The bubbles then experience a larger pressure during their dynamic process and this results in a stronger bubble collapse and an impulsive load of higher amplitude. Since the material threshold is obviously independent of flow velocity, a larger portion of the impulsive loads exceeds the material threshold and results in permanent deformations. The second reason is that the bubble production rate also increases with flow velocity so that the frequency of pressure pulses increases too. When combined, both effects induce a rapid increase of the pitting rate with flow velocity.

The effect of flow velocity on pitting rate is illustrated in Fig. 3.5 where the reference pitting rate, $1/(\delta^2\tau)$, defined in Sect. 3.2.2 is plotted versus flow velocity. These results show an increase of pitting rate with approximately the fifth power of the flow velocity.

An increase in flow velocity also induces an increase in pit size. Figure 3.6 shows that this increase is moderate since the reference pit size δ increases like $V^{0.75}$. By combining both trends presented in Figs. 3.5 and 3.6, the influence of flow velocity on the coverage time τ can be derived. Since $\delta^2\tau \propto V^{-5}$ and $\delta \propto V^{0.75}$, it is concluded that $\tau \propto V^{-6.5}$. Thus, the coverage time decreases very rapidly with flow velocity. This is due to the combined effects of flow velocity on pitting rate and pit size. Since a strong correlation exists between the incubation time and the coverage time (see Chap. 7), it can be expected that the incubation time should also strongly decrease with the flow velocity. In a similar way, since the erosion rate is correlated to the inverse of the coverage time, the erosion rate is expected to strongly increase with flow velocity. These trends are discussed quantitatively in Chap. 5.

3.2.4 Effect of Maximum Pit Size

In the previous sections, it was assumed that the distribution of pits with respect to their diameter extends up to infinity. Even though the exponential law gives a fast drop in the probability of occurrence as pit diameter increases, it may be more realistic to assume that a maximum pit diameter D_{max} exists and that all pits have actually a diameter smaller than D_{max} (see Fig. 3.6).

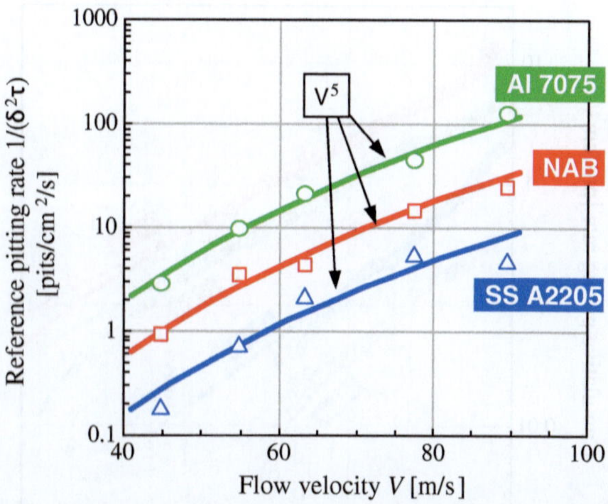

Fig. 3.5 Effect of flow velocity on the reference pitting rate $1/(\delta^2\tau)$ for three different materials (aluminum alloy Al 7075, nickel aluminum bronze NAB and stainless steel A2205). All three curves follow approximately the same power law V^5 with the flow velocity V. Results obtained in the LEGI facility at a constant cavitation number $\sigma = 0.9$. Adapted from [5], with permission from Elsevier

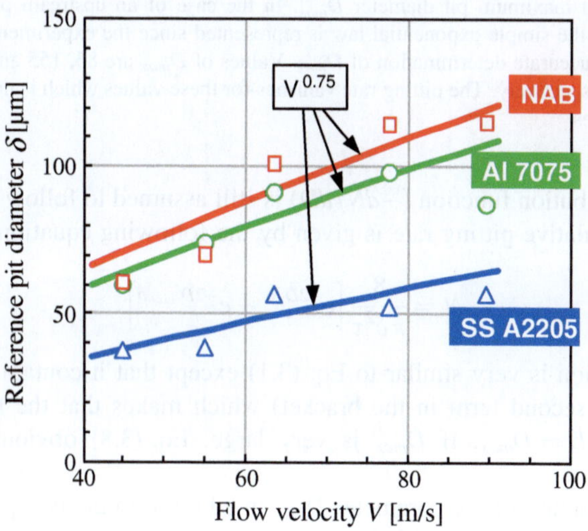

Fig. 3.6 Effect of flow velocity on the reference pit diameter δ for three different materials (aluminum alloy Al 7075, nickel aluminum bronze NAB and stainless steel A2205). All three curves follow approximately the same power law $V^{0.75}$ with flow velocity V. Pit size is similar for aluminum and nickel aluminum bronze alloys but smaller in the case of stainless steel. Results obtained in the LEGI facility at a constant cavitation number $\sigma = 0.9$. Adapted from [5], with permission from Elsevier

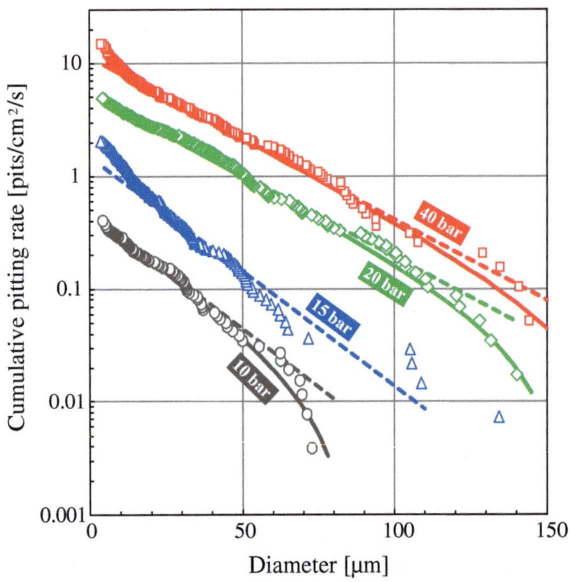

Fig. 3.7 Effect of the introduction of a maximum pit diameter D_{max} in the fitting of cumulative histograms. The data are the same as in Fig. 3.3. The dotted lines correspond to a purely exponential fit given by Eq. (3.1), whereas the continuous lines correspond to Eq. (3.8) with the introduction of a maximum pit diameter D_{max}. In the case of an upstream pressure of 15 bar (1.5 MPa), only the simple exponential law is represented since the experimental points are too scattered for an accurate determination of D_{max}. Values of D_{max} are 85, 155 and 175 μm for 10, 20 and 40 bar respectively. The pitting rate vanishes for these values which is not visible in semi-logarithmic scales

If the distribution function $(-dN/dD)$ is still assumed to follow an exponential law, the cumulative pitting rate is given by the following equation:

$$N = \frac{8}{\pi \delta^2 \tau} \left[e^{-2D/\delta} - e^{-2D_{max}/\delta} \right]. \tag{3.8}$$

This equation is very similar to Eq. (3.1) except that it contains an additional constant (the second term in the bracket) which makes that the number of pits vanishes for $D = D_{max}$. If D_{max} is very large, Eq. (3.8) obviously reduces to Eq. (3.1).

The introduction of the parameter D_{max} in addition to the two previous ones, δ and τ, improves the fitting of the cumulative histograms as shown in Fig. 3.7 particularly for large pit sizes. However, it does not change the conclusions of the previous section since the probability density function which is the derivative of Eq. (3.8) keeps the same exponential form. In practice D_{max} can be adjusted for best fit of experimental data.

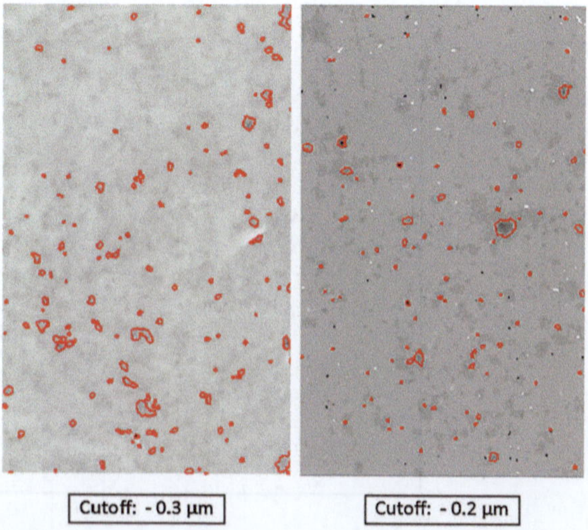

Fig. 3.8 Pitting patterns on aluminum alloy Al 7075 observed under an ultrasonic horn (*left*) and a cavitating jet (*right*). Image size: 1.2 mm × 1.7 mm. Notice strong similarities between the two and with the hydrodynamic pitting in Fig. 3.2. The *gray* levels correspond to the local depths. The *red* contours are the cross-sections of the surface at the cutoff depths shown under each picture. Dark *gray* pixels not surrounded by *red* contours are pits shallower than the selected cutoff depths. *Left*, Ultrasonic G32 settings, exposure time: 1 min. *Right*, Jet pressure: 70 bar (7 MPa), cavitation number: $\sigma = 0.014$, exposure time: 1 min. Testing at DYNAFLOW

3.3 Scaling of Pit Distribution for Cavitating Jets

In this section, results of the pitting tests using cavitating jets and ultrasonic horns are presented. Analysis of the pitting results is done in the same manner as in Sect. 3.2 and compared with the hydrodynamic cavitation tunnel test results presented in Sect. 3.2.

3.3.1 Pit Distribution

Pitting tests conducted using cavitating jets and ultrasonic horns show remarkably similar pit shapes and pitting distributions (Fig. 3.8) as in the hydrodynamic tunnel tests shown in the previous section.

The same materials tested in the tunnel were also exposed to cavitation generated by the submerged cavitating jets and the ultrasonic device described in Chap. 2. In the cavitating jet tests, 25.4 mm × 25.4 mm × 6.35 mm mirror-polished material samples were exposed to the cavitating jet for 1 min for various jet velocities and the results were compared. The central jet impact area where maximum pitting occurred was chosen for detailed analysis.

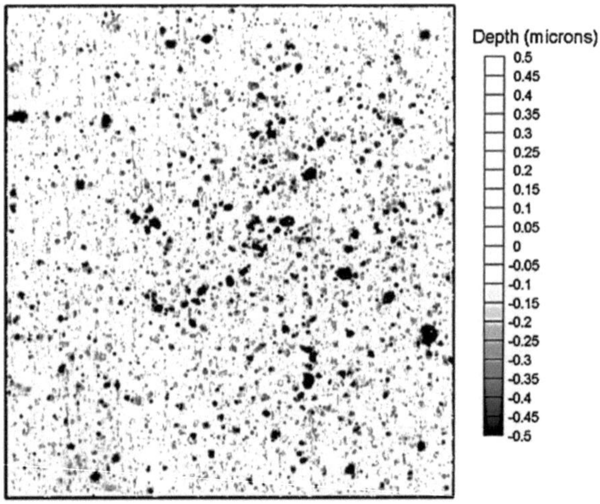

Fig. 3.9 Typical data obtained from optical scanning of a pitted sample at the U.S. Naval Research Laboratory. Shown here is the actual 1.587 mm × 1.719 mm scan of an Al 7075 sample pitted under a 80 bars (8 MPa) DYNAJETS® for a duration of 2 min. Testing conducted at DYNAFLOW. Based on [6], reprinted with permission from Elsevier

Optical scanning profiles of the surface of the pitted sample were obtained using an Alicona InfiniteFocus G4 scanner at the US Naval Research Laboratory. The spatial resolution of the instrument could be set to a few nanometers and an interrogation mesh size of 1.5 μm × 1.5 μm was selected. Figure 3.9 shows an example of the contour plots of the surface data obtained from the optical scan of the pitted surface. Very little pitting overlap occurred, individual pits could be well identified, and the geometric characteristics of each individual pit could be accurately measured and used in the statistical analysis. Results from several cut-off depths were compared and the results were very similar to Fig. 3.2. The cut-off depth of 0.3 μm was selected for the results shown below.

From the scanned profilometer measurements, an average pitting rate was calculated as the ratio of the cumulative number of pits per unit area to the exposure time. As in the previous section, the counted pits were classified according to their equivalent diameter to define the cumulative diameter distribution, $N(D)$, where N is the number of pits per unit area per unit time with a diameter larger than the value D. Using the analysis in the previous sections or also from simple direct dimensional analysis, $N(D)$ can be made non-dimensional through the definition of a characteristic number of pits, N^*, and a characteristic pit diameter, D^*. The cavitating jet pitting phenomenon follows a universal scaling law, as the pitting measured in the high-speed water tunnel does in Sect. 3.2, if all distributions are represented by the same function $\overline{N}(\overline{D})$ where $\overline{N} = N/N^*$ and $\overline{D} = D/D^*$.

Analysis of all cavitating jet pitting data has shown that $\overline{N}(\overline{D})$ can be fitted well by a Weibull distribution [12], expressed by the following simple mathematical function:

$$\begin{aligned} \overline{N} &= e^{-\overline{D}^k}, \\ N &= N^* e^{-(D/D^*)^k}, \end{aligned} \tag{3.9}$$

where k is a shape parameter. The main difference between this scaling law (3.9) and (3.7) is the shape parameter which is the exponent on the diameter. This parameter was selected to be 1 in Eq. (3.7). In the data shown below, the three parameters, N^*, D^*, k, were determined through best fit to the measured data by the least square method.

3.3.2 Effect of Jet Velocity on Pitting

Pitting tests were carried out on three materials (Al 7075, SS A2205, NAB) for jet pressures varying from ~70 bars (7 MPa) to ~480 bars (48 MPa) or jet speeds from 115 to 215 m/s [6]. In all these tests, the jets were discharged into a container where the local pressure was the atmospheric pressure (see Fig. 2.5). Therefore all results were for the same local pressure and this resulted in differences in the cavitation number between runs. However, in all cases the cavitation numbers were very small (between 0.002 and 0.007), i.e. much smaller than the cavitation inception number ($\sigma_i \sim 0.5$). This difference in the procedure is responsible for some difference with the results obtained at constant cavitation number in the previous section. The effect of the jet pressure (or the cavitation intensity) on the pitting characterization parameters and, as a consequence, on the distribution function of pitting rate versus pit diameter is discussed in the following paragraphs.

Figure 3.10 shows the pit distributions at different jet speeds for the A2205 stainless steel. In all cases, Weibull fits using the shape parameter, $k = 0.7$, were able to cover the full range of pit sizes including the larger diameter values, with the exception of higher pressures where the curve fits deviate for the larger pit diameters. The larger pits are of particular interest as they correspond to the rare high intensity events, which would result in micro-fracture and later weight loss.

Equation (3.1) is equivalent to Eq. (3.9) with the shape factor $k = 1$ instead of 0.7. Different curve fits using Eq. (3.1) are presented in Fig. 3.11.

3.3.3 Effect of Jet Velocity on Characteristic Parameters

Figures 3.12 and 3.13 show the variation of the characteristic pit diameter, D^*, and the characteristic pitting rate, N^*, for $k = 0.7$ as functions of the jet pressure for stainless steel A2205. It is clear that the characteristic pitting rate and pit size increase with the jet pressure or jet speed, with approximately 0.88 power of the jet pressure (or 1.76 power of the jet velocity) for D^*, and 1.56 power of the jet pressure (or 3.12 power of the jet velocity) for N^*.

Although the qualitative trends are the same for jet cavitation and tunnel cavitation, the exponents with respect to the flow velocity for both the

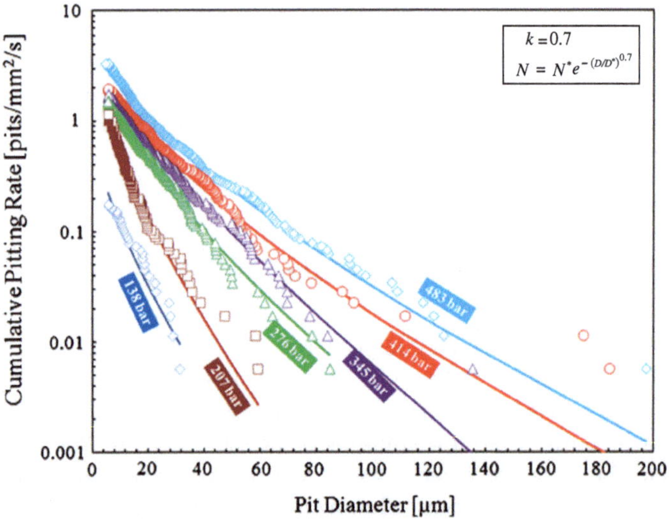

Fig. 3.10 Cumulative pitting rate as a function of pit diameter for different values of the pressure upstream of the DynaJets® orifice on duplex stainless steel A2205 (cut-off depth: 0.3 µm). Line fits correspond to the three-parameter Weibull distribution of Eq. (3.9) with $k = 0.7$ and N^* and D^* fitted to the data. Based on [6], reprinted with permission from Elsevier

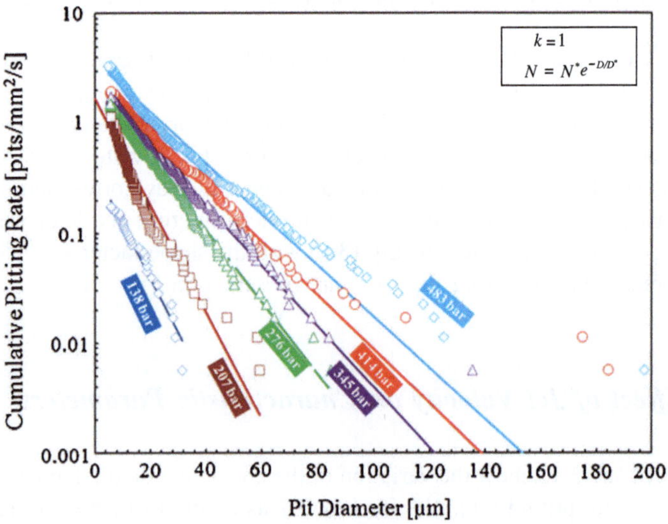

Fig. 3.11 Cumulative pitting rate as a function of pit diameter for different values of the pressure upstream of the DynaJets® orifice on duplex stainless steel A2205 (cut-off depth: 0.3 µm). Line fits correspond to the three-parameter Weibull distribution of Eq. (3.9) with $k = 1$, N^* and D^* fitted to the data. Based on [6], reprinted with permission from Elsevier

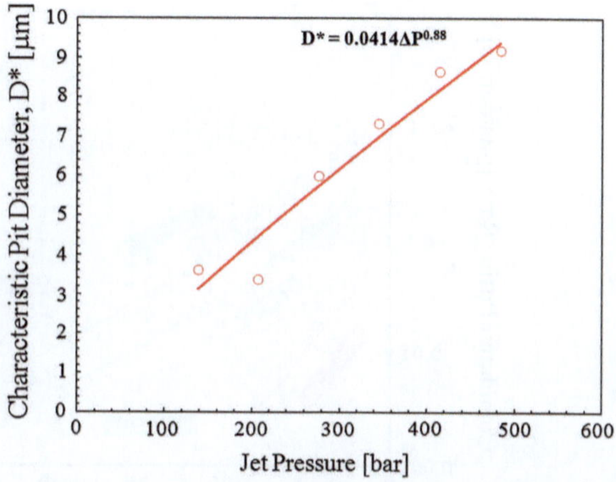

Fig. 3.12 Characteristic pit diameter, D^*, as a function of the pressure drop, ΔP, across the DynaJets® orifice on duplex stainless steel A2205 (cut-off depth: 0.3 μm). $k = 0.7$. Based on [6], reprinted with permission from Elsevier

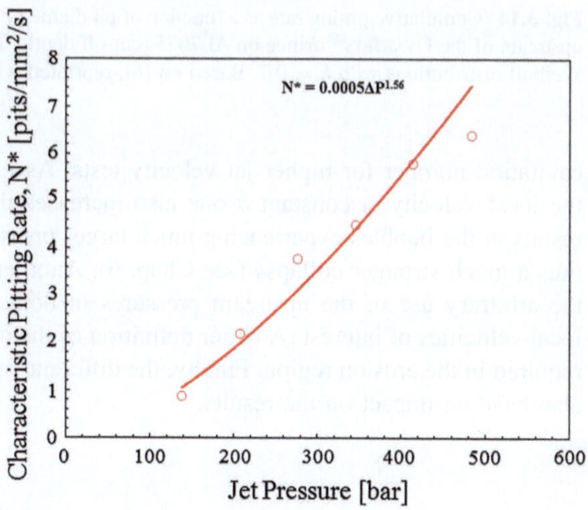

Fig. 3.13 Characteristic pitting rate, N^*, as a function of the pressure drop, ΔP, across the DynaJets® orifice on duplex stainless steel A2205 (cut-off depth: 0.3 μm). $k = 0.7$. Based on [6], reprinted with permission from Elsevier

characteristic pit diameter and the characteristic pitting rate are significantly different between jet cavitation (Figs. 3.12 and 3.13) and tunnel cavitation (Figs. 3.6 and 3.5). Also, cavitation pits appear significantly smaller in the case of jet cavitation (see Fig. 3.12) compared to water tunnel cavitation (see Fig. 3.6). This may be attributed to the smaller length scale of the present jet flow, which produces smaller bubbles and consequently smaller pits (see also Fig. 5.19).

The differences in cavitation generation methods and facilities contribute to this difference in the exponent. As opposed to the constant cavitation number used in LEGI tests, the cavitating jet tests used a constant ambient pressure and a lower

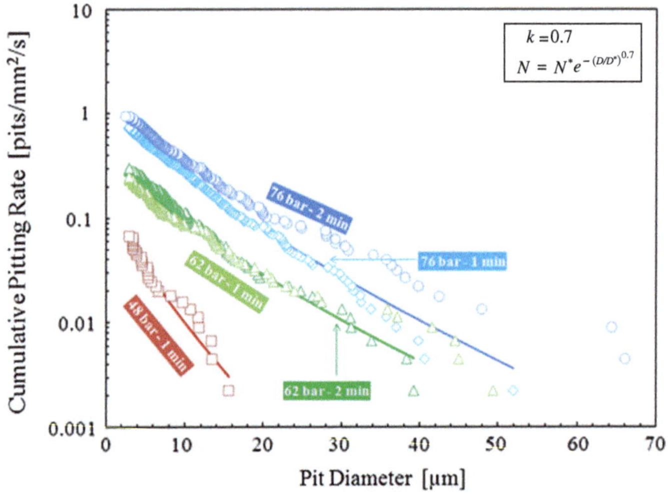

Fig. 3.14 Cumulative pitting rate as a function of pit diameter for different values of the pressure upstream of the DynaJets® orifice on Al 7075 (cut-off depth: 0.3 μm). Lines correspond to fitted Weibull distributions with $k = 0.7$. Based on [6], reprinted with permission from Elsevier

cavitation number for higher jet velocity tests. As mentioned earlier, to increase the local velocity at constant σ one also increases the ambient pressure and this results in the bubbles experiencing much larger pressure during their collapse and thus a much stronger collapse (see Chap. 6). Another potential for discrepancy is the arbitrary use of the upstream pressures in both facilities to characterize the local velocities of interest. A better definition of the characteristic velocity may be required in the erosion region. Finally, the different values of k used for fitting may also have an impact on the results.

3.3.4 Effect of Materials Properties

The same scaling law was applied to the two other tested materials (Al 7075 and NAB). Figure 3.14 shows cumulative pitting rates obtained from pitting tests conducted on Al 7075. Since this aluminum is much softer than the stainless steel A2205, the jet pressures used for the pitting tests were reduced to the range 50 to 80 bars (5 to 8 MPa). Two test durations are shown in Fig. 3.14: 1 and 2 min. As shown in the figure the cumulative pitting rates obtained with the two exposure times are close, indicating that during the incubation period, pitting increases linearly with time. The scaling law (3.9) also fits the experimental data well for Al 7075.

Figure 3.15 shows similar data obtained from tests conducted on NAB samples. The same scaling law (3.9) is shown to provide good experimental data fit.

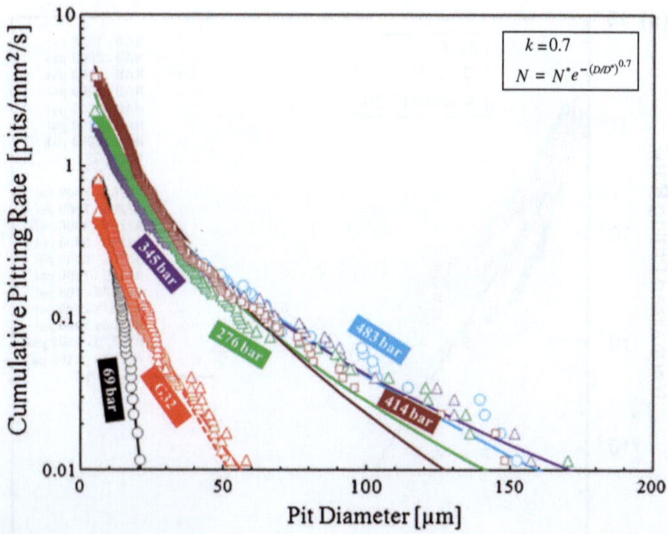

Fig. 3.15 Cumulative pitting rate as a function of pit diameter for different values of the pressure upstream of the DynaJets® orifice on NAB (cut-off depth: 0.3 µm). Lines correspond to fitted Weibull distributions with $k = 0.7$. Comparison of the jet tests with ultrasonic cavitation tests (*red symbols* and *dotted line*) using the ASTM G32 alternative method

Figure 3.15 also compares cavitating jet tests with ultrasonic cavitation G32 test results. The figure illustrates that ultrasonic cavitation erosion data also followed well the Weibull distribution with $k = 0.7$, and that the G32 appears to generate on NAB a cumulative pitting rate, *N*(*D*), close to jet results at about 70 bars (7 MPa). This is consistent with other observations in terms of pressure measurements and material loss curves discussed in Chaps. 4 and 5, which compare the cavitation intensity of the G32 tests to about 60 bars (6 MPa) cavitating jets.

3.3.5 Universal Representation Using Nondimensional Quantities

Since all cumulative pitting rate data follow the same general trends as a function of the pit diameter, independent of the material and the jet pressure, a universal representation is made using non-dimensional quantities. Figure 3.16 shows the non-dimensional pitting rate, $\overline{N} = N/N^*$, as a function of $\overline{D} = D/D^*$ for two different values of the shape factor *k*. $k = 1.0$, which was used in Sect. 3.2.2 for the tunnel data, fits the jet cavitation data well for smaller pit diameters, but not for

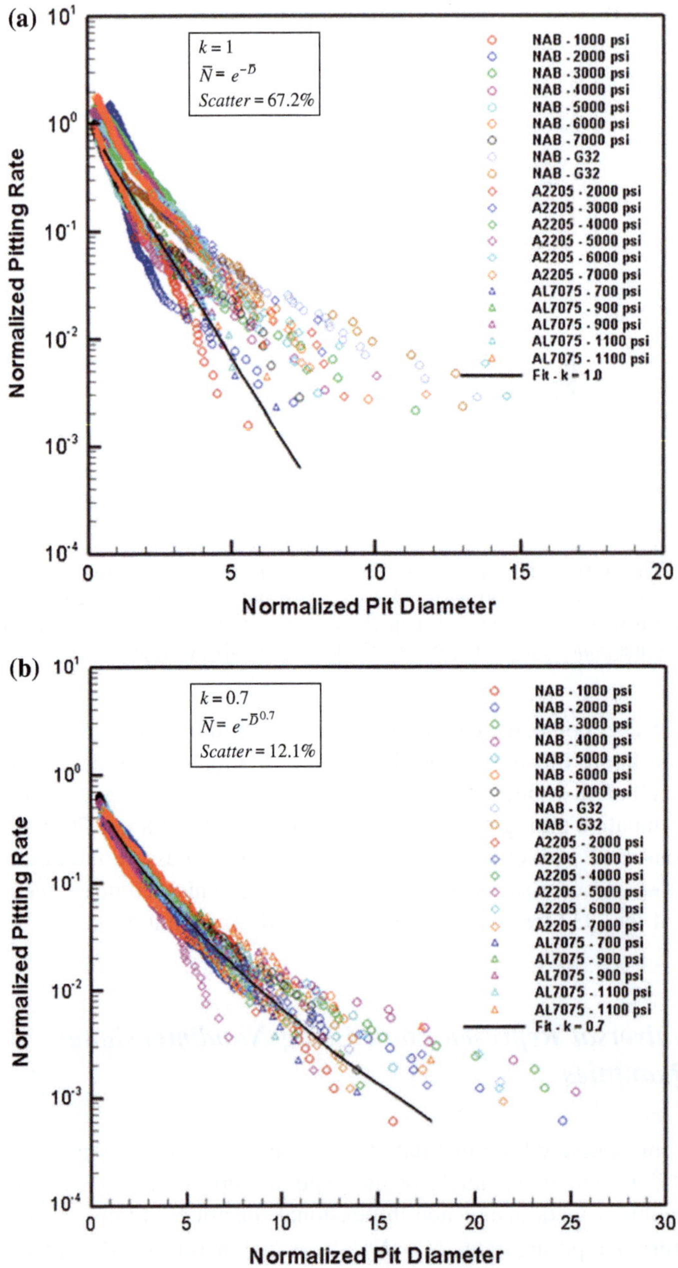

Fig. 3.16 Normalized pitting rate, N/N^*, versus normalized pit diameter, D/D^*, for two values of the fitting shape parameters: $k = 1.0$ (**a**) and $k = 0.7$ (**b**), in the equation $N = N^* e^{-(D/D^*)^k}$. (1000 psi = 6.9 MPa). Based on [6], reprinted with permission from Elsevier

larger diameters. $k = 0.7$, on the other hand, fits the cavitating jet data better than $k = 1.0$ for a wider range of the normalized pit diameter.

The characteristic pit size and the pitting rate can be used as a measure of the intensity level of the cavitating field. For instance, two cavitating fields with the same N^* and D^* obtained with a material should result in the same erosion progression, while a cavitating field with the same D^* but twice N^* should result in twice as fast erosion progression on the same material.

3.4 Estimation of Impact Loads from Pitting Tests

The impact loads generated by the collapsing vapor cavities are basic components of the so-called cavitation intensity [13] that has been used to characterize the hydrodynamic aggressiveness of a cavitating flow. They are key parameters for the prediction of the erosion rate by proper modeling of the material response to cavitation impact loads. It is thus essential to accurately determine impact loads and their statistical distribution in order to correlate them with the flow behavior and material damage.

The most widely used method to detect cavitation impacts and assess their intensity is based on pressure sensitive transducers as discussed in more detail in Chap. 4. Several kinds of such transducers have been used, in particular conventional commercial pressure sensors [14], piezoelectric ceramic discs [15], piezoelectric polymer PVDF (Polyvinylidene fluoride) films [16] as well as fiber optic sensors [17]. In spite of the progress made to improve the accuracy and reliability of pressure transducers by adapting them to impulse loading of cavitation (see Chap. 4), the measurements still suffer from several drawbacks in terms of resonant frequency, rise time, and calibration [14]. In addition, the sensor dimension is often too big compared to the microbubble size and the sensor may be rapidly damaged by cavitation.

Another method to deduce the impulsive pressure is to extract it from the material itself considered as a sensor capable of detecting the spectrum of impulsive loads, which exceed its elastic limit. From this standpoint the material itself is also a filter of the impulsive loads. Every sufficiently strong collapse beyond the elastic limit can induce a pit on the surface of a sample (Fig. 3.17) where its characteristic dimension like depth, diameter, or volume can be associated with the resultant plastic deformation ε. From this deformation the corresponding stress σ responsible for the pit can be deduced if the constitutive equation of material deformation $\sigma = f(\varepsilon)$ is known.

The basic point of this inverse procedure for evaluation of collapse pressure is the need for a proper modeling of material behavior. A complex relation between the loading state and the material response probably occurs because of the high strain rates coupled with triaxial stress fields generated during impulsive loading. Indeed, the material microstructure, grain size, modes of plastic deformation, and

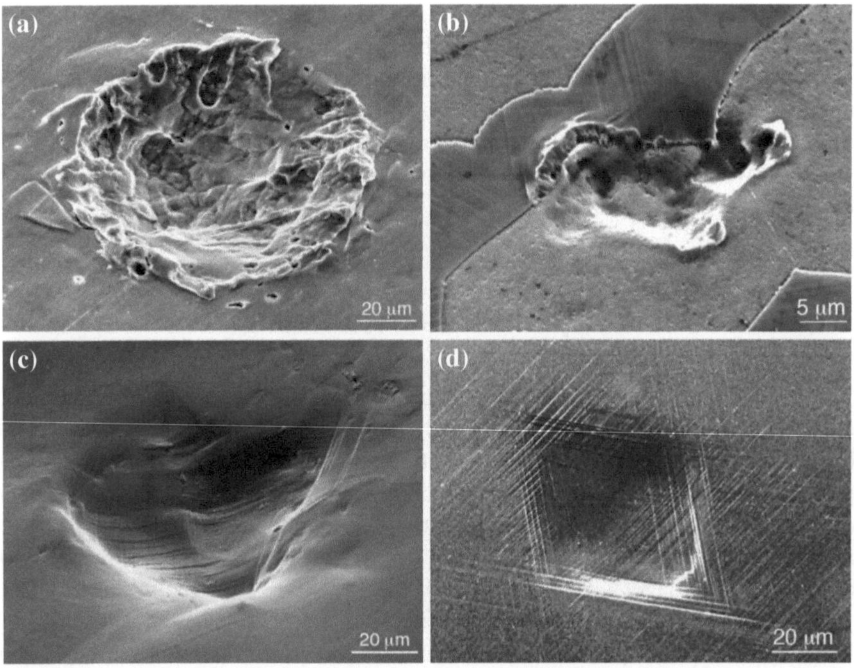

Fig. 3.17 SEM micrographs showing examples of cavitation pit with different geometries on different samples, **a** circular pit surrounded by rims in $\alpha + \beta$ brass (reproduced from Wear 113 (1986) 305, with permission from Elsevier), **b** pit in the region of phase boundary in duplex stainless steel A2205 (reproduced from Mat. Sci. Eng. 86 (1989) 191, with permission from Elsevier), **c** austenitic stainless steel 18–12 (reproduced from Acta Metall. 37(1989) 1079, with permission from Elsevier), **d** Cu-Al sample, 3 slip systems are activated. (EPFL, Lausanne)

several other parameters as described in Sect. 1.4, can affect the material response at various length scales.

From an experimental point of view, the method consists of three steps: (i) perform pitting test to measure pit size, (ii) conduct nanoindentation measurements to establish stress–strain relation, and finally (iii) determine the impulsive loads or pressures by correlating the results of pitting test with nanoindentation test. The principle of the inverse method is explained in detail below.

The inverse method is applied to pitting tests conducted in the cavitation flow loop presented in Sect. 2.4 at different velocities between 45 and 90 m/s, corresponding to upstream pressure of 10–40 bar. For all tests, the ambient pressure was adjusted in order to keep the cavitation number constant ($\sigma = 0.9$), and ensure the same cavity length for all tests and thus maintain a similarity between the different flows at different velocities. Three alloys: aluminum Al 7075-T651, nickel aluminum bronze C95400, and duplex stainless steel A2205, were exposed to the same cavitating flows for comparative analysis. The samples were mechanically polished to obtain a metallographic surface with roughness smaller than 0.1 μm.

3.4.1 Characteristic Deformation of Cavitation Pit

Cavitation pits exhibit a wide variety of forms as can be seen, for example, from the SEM micrographs in Fig. 3.17. The shape of the plastic zone surrounding the pit will not necessarily have a round smooth shape because of the crystallographic orientation dependence of the plastic deformation. However, for simplicity of calculations and modeling, we assume the cavitation pits as spherical caps characterized by their diameter D and depth h as illustrated in Fig. 3.18, analogous to the imprint left by a spherical indenter. With such an assumption, the mean strain, ε, of a cavitation pit can be estimated from Tabor's relation [18] using the following equation:

$$\varepsilon = k \sin\gamma \simeq 0.2 \frac{D/2}{R}, \tag{3.9}$$

where k is a coefficient close to 0.2 for a spherical indent, γ is half the contact angle as shown in Fig. 3.18 and D is the diameter of the cavitation pit. R is the radius of the spherical cap and can be geometrically related to the pit depth, h and pit diameter, D as follows:

$$R = \frac{(D/2)^2 + h^2}{2h}. \tag{3.10}$$

From Eqs. (3.9) and (3.10), the mean strain of a cavitation pit can then be defined by:

$$\varepsilon = 0.2 \frac{D\,h}{(D/2)^2 + h^2} = 0.8 \frac{h/D}{1 + 4(h/D)^2}. \tag{3.11}$$

This equation shows that the mean deformation associated with a cavitation pit is related to the geometric shape factor h/D. Therefore, the measurement of pit diameter D and pit depth h will allow determination of the strain ε of a cavitation pit. Cavitation erosion pits are generally shallow indentations with a depth h much smaller than the diameter D. Consequently, the mean strain of a cavitation pit (3.11) can be approximated by:

$$\varepsilon \cong 0.8 \frac{h}{D}. \tag{3.12}$$

Similar to indentation imprints, the surface of the material may show either sink-in or pile-up around the cavitation pit [19]. The degree of sink-in or pile-up depends upon the strain hardening exponent and the ratio of the Young's modulus E over the yield stress σ_y of the material [20]. For materials with large values of E/σ_y (e.g. metals) the plastic zone has a hemispherical shape and follows the radius of the circle of contact. In this case, plastic deformation occurs mostly near the contact radius and piling up is expected. For materials with low values of E/σ_y (ceramics and glasses) the plastic strains are practically contained within the

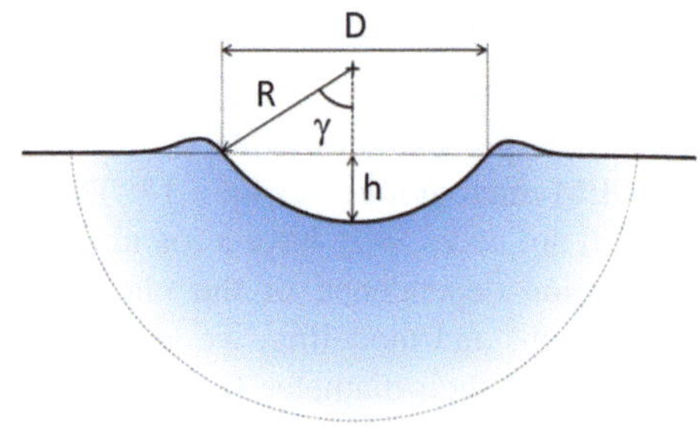

Fig. 3.18 Sketch showing an idealized cavitation pit considered as a spherical cap. The blue region is the plastic zone surrounding the pit

contact circle and the elastic deformation contributes also to the accommodation of impact deformation over a greater distance. In this case, sinking-in is more likely expected. The sketch of Fig. 3.18 represents profile of cavitation pit in ductile materials yielding pile-up.

An example of application of the method is given in Fig. 3.19 which shows the strain as a function of the pit diameter for the three different alloys tested under exactly the same cavitation conditions. Each data point corresponds to a pit. All pits whose depth is smaller than the cutoff depth (0.5 μm here) were ignored by the measuring technique. The corresponding measuring limit is represented by the black dotted line in Fig. 3.19. No measurement is available below this line.

The plots are very similar for the three alloys. This is because the relationship (3.11) between impact strain ε and pit shape factor h/D is material independent and because the values of the shape factor do not depend much on the material.

The plots show that the strongest impulsive loads can induce plastic strains up to $\varepsilon \simeq 20\%$, but the majority of the impulsive loads create deformations less than 5%. Larger strains are apparently associated with impacts of smaller size. This is related partly to the aspect ratio of the small pits being far from spherical shape, and to the measurement errors in defining the profile of submicron pits by the stylus method. However, the volume of deformation remains quite small compared to big impacts. A similar trend was reported by Momma and Lichtarowicz [21] who found that the impingement pressure of submerged cavitating liquid jet significantly increases as the pit size decreases.

3.4.2 Stress–Strain Curve from Nanoindentation Measurements

The estimation of the impulsive stresses from the measured deformations requires the knowledge of the stress–strain relationship $\sigma = f(\varepsilon)$. If this relation is not available for the tested material, it can be derived from nanoindentation measurements as explained below.

The advantage of indentation data relative to conventional uniaxial tensile tests lies in the analogy between material deformation under liquid impingement and

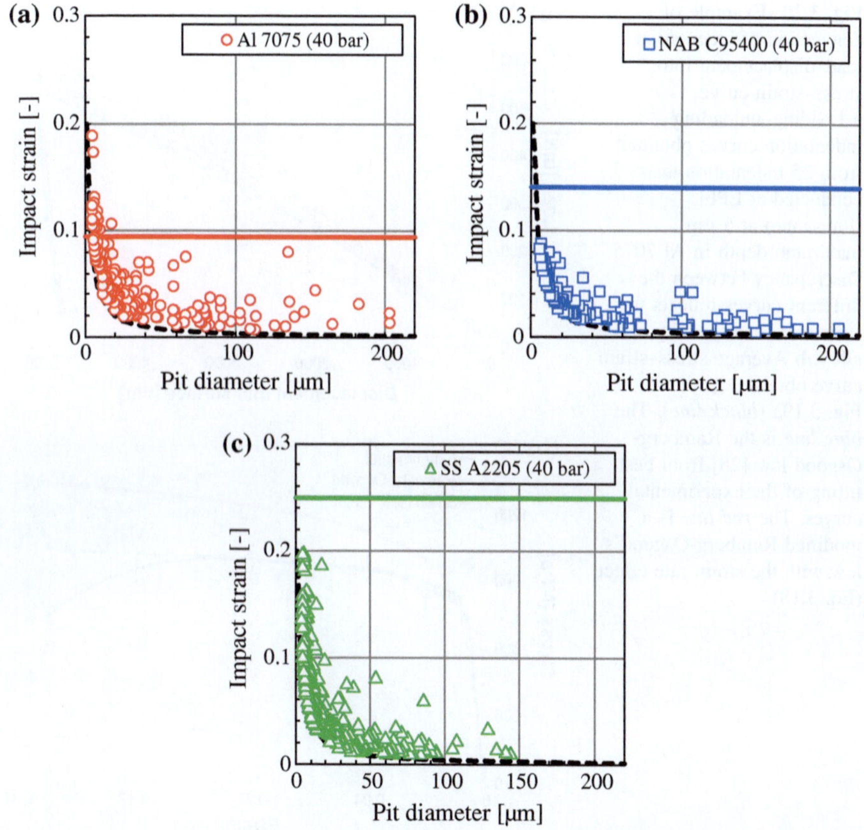

Fig. 3.19 Relation between pit size and strain induced by cavitation impacts for three different alloys. Data are relative to pitting tests conducted in the LEGI cavitation erosion tunnel at the upstream pressure of 40 bar (4 MPa). The *dotted black lines* correspond to the measuring limit associated to the cutoff depth of 0.5 μm chosen here. The *horizontal lines* (*red*, *blue* and *green*) correspond to a typical value of the rupture strain deduced from tensile tests [5]. **a** Aluminum alloy Al 7075. **b** Nickel aluminum bronze alloy C95400 (for this alloy only, an additional filter defined by the condition $h/D < 0.12$ was applied to discriminate between actual cavitation erosion pits and numerous material defects). **c** Stainless steel alloy A2205

under indenter penetration. In both cases, deformation occurs from compressive stress within a confined volume and is in general nonlinear with a complex 3-D state of stress field beneath the contact area.

When a spherical indenter is applied on a surface, the instrumented nano-indenter records load–displacement for a cycle of indentation load-unload, as shown in Fig. 3.20a. The indentation load-depth curve must then be converted to true stress–strain curve and fitted with an appropriate deformation model to obtain a mathematical expression of the stress–strain curve $\sigma = f(\varepsilon)$.

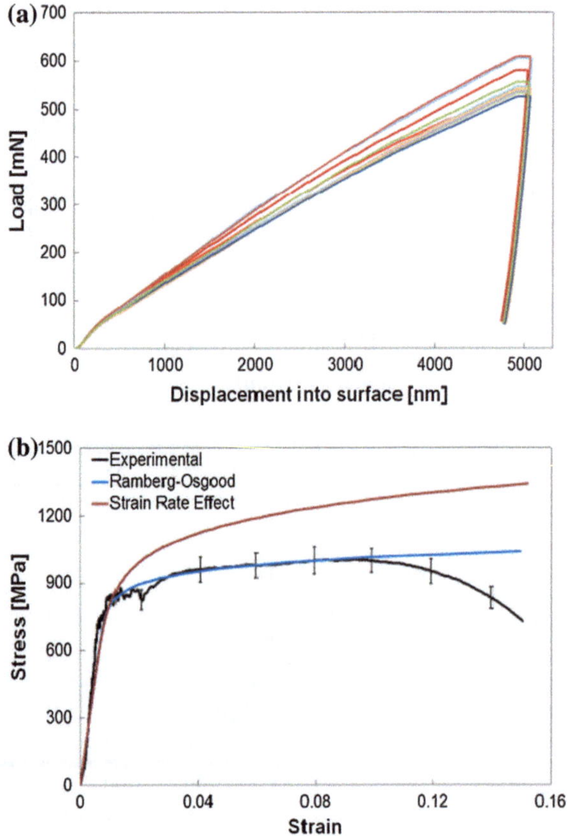

Fig. 3.20 Example of conversion of indentation load–displacement into stress–strain curve. **a** Loading–unloading indentation curves obtained from 25 indentation tests conducted at EPFL (Lausanne) at 5 μm maximum depth in Al 7075. Discrepancy between the different curves reflects the structural heterogeneity of the alloy. **b** Average stress–strain curve obtained from Fig. 3.19a (*black line*). The *blue line* is the Ramberg-Osgood law [28] from best-fitting of the experimental curves. The *red line* is a modified Ramberg-Osgood's law with the strain rate effect (Eq. 3.18)

Definition of true strain is given by Tabor [18] as $\varepsilon = k_1 \sin \gamma$, and Ahn and Kwon [22] as $\varepsilon = k_2 \tan \gamma$. The constants k_1 and k_2 were determined as 0.2 and 0.14 independent of material properties by finite element analysis of various materials [23]. The angle γ is the half-angle between the indenter and the material as shown in Fig. 3.18. Here, we use Tabor's strain definition and apply relations similar to Eqs. (3.9) to (3.11) for strain, ε:

$$\varepsilon = 0.2\,\frac{a_c}{R} = 0.2\,\frac{\sqrt{2Rh_c - h_c^2}}{R}, \tag{3.13}$$

where R is the radius of the spherical indenter, a_c the contact radius, and h_c the contact depth calculated from the penetration depth h_t [24]:

$$h_c = h_t - \beta\frac{P}{S}. \tag{3.14}$$

Here β is a geometric constant that is 0.72, 0.75 and 1.0 for conical, pyramidal and flat punch indenters respectively [25], P is the applied load and S is the harmonic contact stiffness representing the slope of the initial portion of unloading curve in Fig. 3.20a. Therefore, true strain can be obtained by combining Eqs. (3.13) and (3.14).

True stress can be deduced from the following equation [18]:

$$\sigma = \frac{1}{\Phi} P_m = \frac{1}{\Phi} \frac{P}{\pi a_c^2}, \tag{3.15}$$

where Φ is the constraint factor, P_m is the mean pressure of contact, and P is the applied load. The constraint factor is a function of plastic strain and work-hardening exponent and was found to be approximately 1.11 when a material is close to the purely elastic regime and approximately 2.8 when full plasticity has been developed [26]. In the intermediate elastic–plastic regime the constraint factor correlates with h_t/h_c ratio [26]:

$$\Phi = 1.11\left(\frac{h_t}{h_c} - 1\right)^{-1}. \tag{3.16}$$

Therefore, true stress can be easily determined from experimental measurements of the load, P, the displacement into the surface, h_t, and using Eqs. (3.15) and (3.16). The conversion of load–displacement curves (Fig. 3.20a) into stress–strain curve is shown in Fig. 3.20b. Finally, a power law plastic model based on Hollomon equation [27, 28] or Ramberg–Osgood (R-O) relation [29] can be used to obtain a mathematical expression of deformation. Using R-O method, the following equation was obtained for aluminum alloy Al 7075:

$$\sigma = 462\left(\frac{\varepsilon}{0.006} - 1\right)^{0.1}. \tag{3.17}$$

Since the strain rate due to cavitation impulsive loads ($\dot{\varepsilon} \approx 5 \times 10^3 s^{-1}$, see Sect. 1.4) is notably higher than the strain rate of nanoindentation (typically in the range of $\dot{\varepsilon} \approx 5 \times 10^{-2} s^{-1}$), a correction for strain rate effect is recommended in Eq. (3.17). For Al 7075, the strain rate effect on flow stress is reported in [30], from which the following equation can be obtained for an application to cavitation impacts:

$$\sigma = 499.5\left(\frac{\varepsilon}{0.006} - 1\right)^{0.09} \text{ for } \dot{\varepsilon} \approx 5 \times 10^3 s^{-1}. \tag{3.18}$$

The red curve in Fig. 3.20b is obtained from Eq. (3.18) and clearly shows the strain effect.

3.4.3 Distribution of Impact Pressure

Using the stress–strain curve for each material as determined in Sect. 3.4.2, the strain of any pit defined by Eq. (3.11) can be converted into stress, and thereby the distribution of stresses can be obtained for each material. The stress estimates are shown in Fig. 3.21 which was directly deduced from Fig. 3.19 using the stress–strain curve of each material. The measuring limit associated to the cutoff depth (0.5 μm) is plotted as the black dotted curve. It prevents measuring hydrodynamic impacts of small size and small amplitude.

Each material, depending on its strength, detects a limited spectrum of impacts similar to pressure sensitive transducers with different sensitivities. The lower limit is close to the material yield stress (low horizontal dotted line in Fig. 3.21). In principle, there is no upper limit since any impact whose stress exceeds the yield stress is expected to generate a pit. However, it should be observed that the higher the stress, the lower the probability of occurrence. It will be shown later in this section (see Fig. 3.22) that the probability actually decreases exponentially with stress. As a result, impacts of amplitude well beyond the yield stress have a small probability of occurrence. In order to capture a few of them, it would be necessary to increase the exposure time quite significantly. This would result in a dramatic increase of the overall pitting rate and in pit overlapping which would make impossible the identification of the rare pits produced by impacts of high amplitude. Thus, an upper limit exists in practice for the impact stresses that can be measured by a given material. It depends not only on the material itself, but also on the maximum exposure time to avoid overlapping of impacts.

As an example, for the results presented in Fig. 3.21, the exposure time for the pitting test on stainless steel was 60 times larger than that on aluminum. The high-strength stainless steel A2205 obviously filters more impacts than the softer aluminum alloy so that the exposure time could be significantly increased without substantial overlapping. Due to a larger exposure time, pits generated by impact loads of high amplitude are more numerous on stainless steel whereas they are almost absent on aluminum.

Even though the upper limit cannot be precisely defined, it can be concluded from Fig. 3.21 that a pitting test conducted on a given material will make it possible to detect impact stresses whose amplitude roughly lies in the range between its yield stress (low horizontal dotted line) and its ultimate strength (upper horizontal solid line).

Since each material detects a spectrum of impacts in a limited range of stress, it may be useful to assemble the distributions measured by different alloys in order to cover a larger range as shown in Fig. 3.22. This figure gives the distribution of the hydrodynamic impacts as a function of their amplitude. The unit on the vertical axis is that of a conventional impact rate per unit surface area (here impacts/cm^2/s) divided by a stress bandwidth in MPa. In other words, the vertical axis measures the impact rate in a bandwidth of 1 MPa.

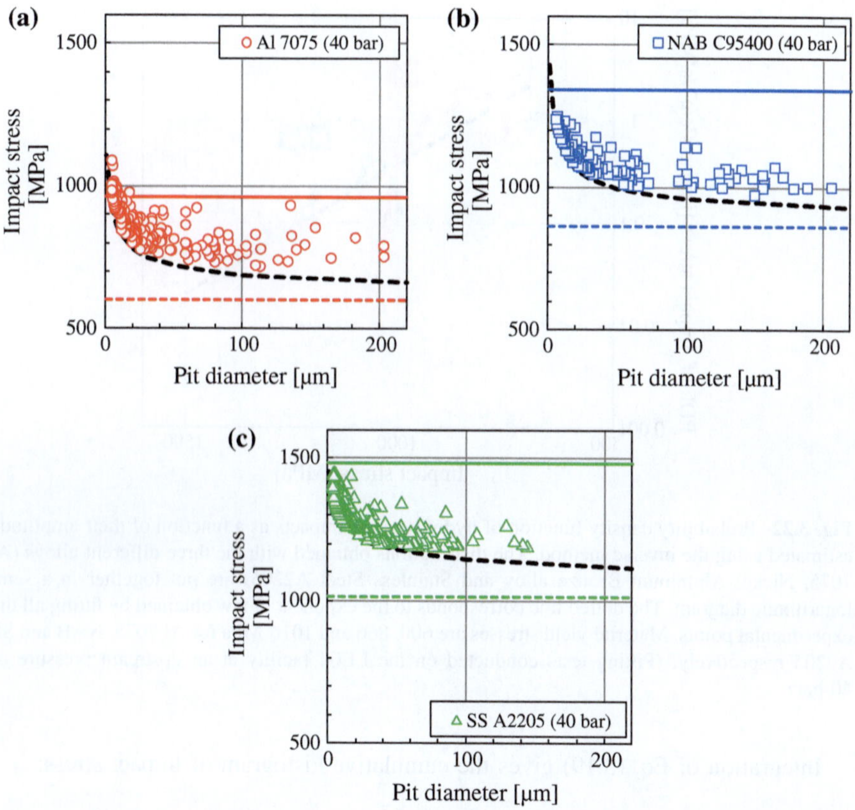

Fig. 3.21 Relation between pit size and stress induced by cavitation impacts for three different alloys. Data are relative to pitting tests conducted in the LEGI cavitation erosion tunnel at the upstream pressure of 40 bar (4 MPa). The dotted black lines correspond to the measuring limit associated to the cutoff depth of 0.5 μm chosen here. The *horizontal solid lines* (*red*, *blue* and *green*) correspond to a typical value of the rupture stress deduced from tensile tests [5] whereas the *horizontal dotted lines* correspond to the yield stress. **a** Aluminum alloy Al 7075. **b** Nickel aluminum bronze alloy C95400. **c** Stainless steel alloy A2205. Corresponding strain was presented in Fig. 3.19

For each material, Fig. 3.22 shows that the distribution n_i of hydrodynamic impacts with respect to the stress amplitude follows roughly an exponential law of the type:

$$n_i = \frac{N^*}{\sigma^*} e^{-\frac{\sigma}{\sigma^*}}, \tag{3.19}$$

where σ^* is a reference stress connected to the slope of the exponential law and N^* a reference rate of hydrodynamic impacts which controls the height of the curve in the diagram.

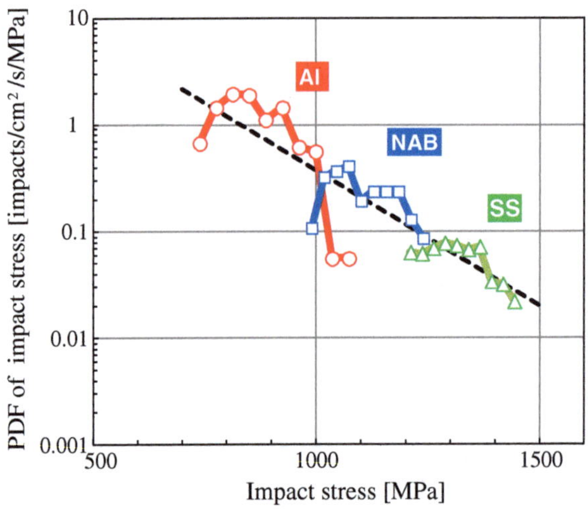

Fig. 3.22 Probability density function of hydrodynamic impacts as a function of their amplitude estimated using the inverse method. The distributions obtained with the three different alloys (Al 7075, Nickel Aluminum Bronze alloy and Stainless Steel A2205) are put together in a semi logarithmic diagram. The dotted line corresponds to the exponential law obtained by fitting all the experimental points. Material yield stresses are 600, 866 and 1016 MPa for Al 7075, NAB and SS A2205 respectively. (Pitting tests conducted on the LEGI facility at an upstream pressure of 40 bar)

Integration of Eq. (3.19) gives the cumulative histogram of impact stress:

$$N_i = \int_{\sigma}^{\infty} n_i d\sigma = N^* e^{-\frac{\sigma}{\sigma^*}}, \tag{3.20}$$

which follows the same type of exponential behavior. As mentioned in Sect. 3.2.4 for pit histograms, the integral may be limited to a finite maximum value of the stress instead of infinity if needed. Equation (3.20) which governs the distribution of impacts with stress is of the same type as Eq. (3.1) which governs the distribution of pits with diameter.

For each material, the distribution tends to drop on the borders of the measurement domain. This might indicate that the pitting test analysis technique underestimates the number of pits at the upper and lower limits of the stress domain.

The three distributions when compared to each other, have a similar slope in a semi logarithmic diagram and join with each other quite satisfactorily to define a common exponential law. This means that the two parameters σ^* and N^* are, to a large extent, material independent.

The three materials considered here give an estimate of the distribution of hydrodynamic impacts in a range of stresses extending from about 0.7 to 1.5 GPa. This range roughly corresponds to the range between the lowest yield stress and the highest ultimate strength of the three materials.

Fig. 3.23 Influence of the hydrodynamic conditions on the distribution of impact pressures for the three alloys. (Tests conducted in the LEGI facility at a constant cavitation number $\sigma = 0.9$)

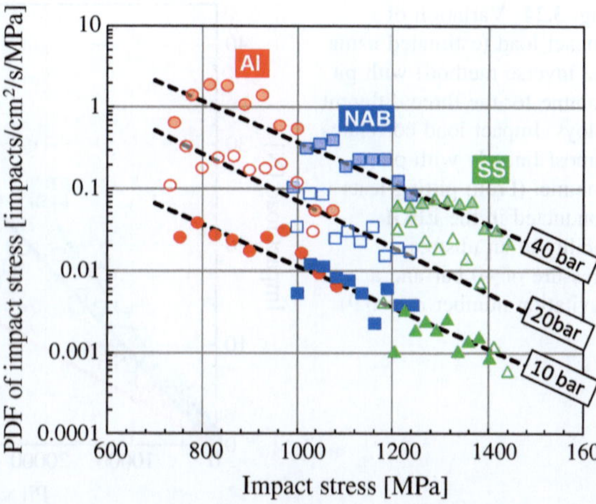

The distribution of impact pressure is a characteristic of the cavitating flow and can be considered, to some extent, as a measure of its aggressiveness. It is then possible to analyze how this distribution changes with the hydrodynamic conditions.

Figure 3.23 presents the distributions obtained in the LEGI facility for three different upstream pressures extending from 10 to 40 bar. The three distributions have very similar slopes in a semi logarithmic diagram so that it can be concluded that the hydrodynamic conditions essentially influence the multiplicative factor of the exponential law N^* but not the reference stress σ^* which governs the slope. In summary, in the present domain of investigation, σ^* appears to be not only independent of the material but also of the hydrodynamic conditions.

3.4.4 Impact Loads

In addition to the impact pressure, the load defined as the product of impact pressure and pit surface can be calculated using this inverse method. The impact loads have a strong and almost linear correlation with pit volumes as can be seen in Fig. 3.24. The correlation coefficient is higher than 0.99 for all three cases. Such a quasi-linear correlation between the impulsive load and the pitting volume was reported for different types of cavitation and material, for example steel [11], pure metals [31] and Al-bronze [32].

The order of magnitude of impulsive loads between 0 and 40 N obtained directly from the material itself by modeling pit induced deformation is comparable to the range of 0–20 N obtained by Hattori et al. [33] in a cavitating liquid jet test chamber (ASTM G134-95 standard) at flow velocities up to 184 m/s, and to the spectrum of 0–50 N obtained by Lee et al. [32] in vibratory cavitation test apparatus (ASTM G32-92) equipped with a piezoelectric sensor.

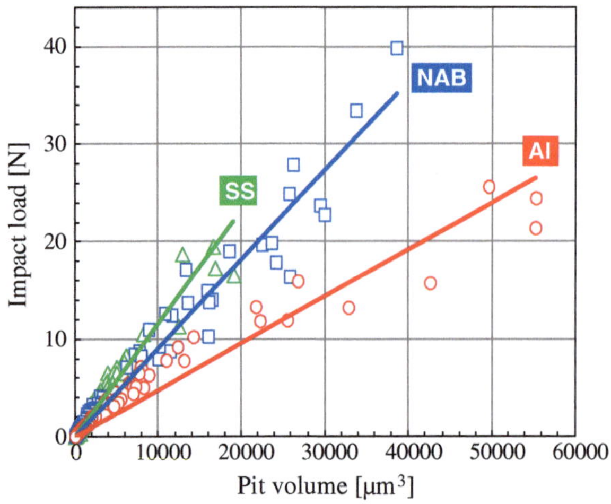

Fig. 3.24 Variation of impact load (estimated using the inverse method) with pit volume for the three different alloys. Impact load correlates almost linearly with pit volume. (From pitting tests conducted in the LEGI facility at an upstream pressure of 40 bar and a cavitation number $\sigma = 0.9$)

Meanwhile, impulsive loads as high as 100–500 N were reported in the literature [21, 34, 35]. Soyama et al. [34] using PVDF transducers measured impulsive loads up to 200 N in a cavitating jet apparatus for a jet velocity at the nozzle outlet in the range 126–155 m/s. Franc and Michel [35] report measurements with various transducers in a cavitating vortex apparatus up to 300 N, which is one order of magnitude higher than the estimates obtained here from deformation of material itself.

This short discussion shows that there is a significant discrepancy between estimates of impulsive loads obtained by different techniques even for the same cavitating flow. Further investigations are needed to reconcile the impulsive load data from different techniques.

3.5 Summary

This chapter was devoted to the initial stage of cavitation erosion known as the incubation period. The incubation period is most often investigated using erosion tests of small duration conducted on metallic samples whose surface has been mirror polished for easy detection of the early damage. Damage during the incubation period is characterized by distributed micro-indentations (or pits) which are shallow plastic depressions. Each pit is believed to be created by a single cavitation event generating an impact load exceeding the material yield stress.

Pitting tests are often used to characterize the cavitation intensity or aggressiveness of a cavitating flow. The specimen surface is carefully analyzed in order to identify each pit and determine its shape including its maximum depth and characteristic diameter. The results are usually presented in the form of histograms or distribution of pitting rates versus pit size.

These histograms are well fitted by a simple exponential or a Weibull exponential distribution. This implies that the density of pits increases very rapidly for small pit sizes and that, as a result, the density of small pits is several orders of magnitude larger than that of the larger pits.

In terms of the surface covered by the pits, the contribution of the smallest pits tends to be negligible because of their very small areas. In the case of an exponential distribution, the contribution of the largest pits to the covered surface also tends to be negligible because of their small number density. As a result, a characteristic pit size can be defined, which corresponds to those pits whose contribution to the eroded surface is the largest. This characteristic pit size is one of the two key parameters of the pit size distribution.

Another key parameter is the coverage time, which is defined as the time required for the material surface to be fully covered once by the pits. The coverage time is an essential characteristic time of the cavitation erosion process. Any measured histogram of pitting rates versus pit diameter is uniquely defined either by these two parameters or by the characteristic pit size and a characteristic pitting rate which is a combination of these two parameters.

For a given erosion device, the histograms obtained for various operating conditions have similar shapes and tend to collapse on a unique curve when the pitting rate and the pit diameter are normalized by the respective characteristic value.

The influence of flow velocity on pitting rate can then be reduced to an analysis of the influence of flow velocity on both the characteristic pit size and the characteristic pitting rate. Even though both increase with flow velocity, the rate of increase depends significantly upon the erosion testing device. This can be attributed to the type of cavitation which is different from one device to another and also to the fact that in some applications, such as the use of cavitating jet, it may not be practical to separate the flow velocity effect from the cavitation number effect, while the two are independently controlled in high-speed cavitation tunnels.

Analysis of a pit dimensions can also provide an estimate of the applied load. By treating a pit in a similar way to a spherical nano-indentation, the mean strain of a pit can be estimated from its shape factor, namely the ratio of its depth to its diameter. Then, using an appropriate stress–strain relationship including strain rate effects, it is possible to estimate the stress associated with the observed strain. The results presented in this chapter show that the distribution of impact loads versus the stress follows an exponential law. As for histograms of pit size, histograms of impact pressures depend upon two parameters, a characteristic impact pressure and a characteristic impact rate. Both parameters appear to be independent of the material used for pitting tests. The interest in using several materials of various yield stresses is to be able to estimate the distribution of impact pressures on a wider range of stresses. By using pitting tests at different velocities and constant cavitation number in a high-speed cavitation tunnel, it was shown that the characteristic impact stress was also independent of the flow velocity in the range of investigation.

References

1. Knapp RT (1955) Recent Investigations of the mechanics of cavitation and cavitation damage. Trans ASME 77:1045–1054
2. Knapp RT (1958) Accelerated field tests of cavitation intensity. Trans ASME 80:91–102
3. Knapp RT, Daily JW, Hammitt FG (1970) Cavitation. McGraw Hill Book Co., New York
4. Hammitt FG (1980) Cavitation and multiphase flow phenomena. McGraw-Hill International Book Co, New York
5. Franc J-P, Riondet M, Karimi A, Chahine GL (2012) Material and velocity effects on cavitation erosion pitting. Wear 274–275:248–259. doi:10.1016/j.wear.2011.09.006
6. Jayaprakash A, Choi J-K, Chahine GL, Martin F, Donnelly M, Franc J-P, Karimi A (2012) Scaling study of cavitation pitting from cavitating jets and ultrasonic horns. Wear 296(1–2):619–629. doi:10.1016/j.wear.2012.07.025
7. Fortes Patella R, Reboud J-L, Archer A (2000) Cavitation damage measurement by 3D laser profilometry. Wear 246(1–2):59–67. doi:10.1016/s0043-1648(00)00446-4
8. Pieralli C, Tribillon G (1987) Traitement d'images 3D appliqué à la profilométrie optique pour l'étude du phénomène d'érosion de cavitation. J Optics (Paris) 18(1):9–18
9. Belahadji B, Franc J-P, Michel J-M (1991) A statistical analysis of cavitation erosion pits. J Fluids Eng 113(4):700–706
10. Ahmed SM, Hokkirigawa K, Ito Y, Oba R (1991) Scanning electron microscopy observation on the incubation period of vibratory cavitation erosion. Wear 142(2):303–314. doi:10.1016/0043-1648(91)90171-p
11. Stinebring DR, Holl JW, Arndt REA (1980) Two aspects of cavitation damage in the incubation zone: scaling by energy considerations and leading edge damage. J Fluids Eng 102(Dec):481–485
12. Weibull W (1951) A statistical distribution function of wide applicability. J Appl Mech-Trans ASME 18(3):293–297
13. Hammitt FG (1979) Cavitation erosion: the state of the art and predicting capability. Appl Mech Rev 32(6):665–675
14. Franc J-P, Riondet M, Karimi A, Chahine GL (2011) Impact load measurements in an erosive cavitating flow. J Fluids Eng 133(12):121301–121308
15. Escaler X, Egusquiza E, Farhat M, Avellan F, Coussirat M (2006) Detection of cavitation in hydraulic turbines. Mech Syst Sig Process 20(4):983–1007
16. Soyama H, Sekine Y, Saito K (2011) Evaluation of the enhanced cavitation impact energy using a PVDF transducer with an acrylic resin backing. Measurements 44:1279–1283
17. Xu R, Zhao R, Cui Y, Lu J, Ni X (2009) The collapse and rebound of gas-vapor cavity on metal surface. Optik 120:115–120
18. Tabor D (1951) The hardness of metals. Clarendon Press, Oxford
19. Fischer-Cripps AC (2004) Nanoindentation. Mechanical Engineering Series. Springer, New-York
20. Elmustafa AA (2007) Pile-up/sink-in of rate-sensitive nanoindentation creeping solids. Modell Simul Mater Sci Eng 15(7):823
21. Momma T, Lichtarowicz A (1995) A study of pressures and erosion produced by collapsing cavitation. Wear 186–187 (Part 2):425–436. doi:10.1016/0043-1648(95)07144-x
22. Ahn J-H, Kwon D (2001) Derivation of plastic stress–strain relationship from ball indentations: examination of strain definition and pileup effect. J Mater Res 16(11):3170–3178. doi:10.1557/JMR 2001.0437
23. Jeon EC, Kim JY, Baik MK, Kim SH, Park JS, Kwon D (2006) Optimum definition of true strain beneath a spherical indenter for deriving indentation flow curves. Mater Sci Eng A 419:196–201
24. Sneddon IN (1965) The relation between load and penetration in the axisymmetric Boussinesq problem for a punch of arbitrary profile. Int J Eng Sci 3(1):47–57. doi:10.1016/0020-7225(65)90019-4

25. Oliver WC, Pharr GM (1992) An improved technique for determining hardness and elastic modulus using load and displacement sensing indentation experiments. J Mater Res 7(6):1564–1583
26. Francis HA (1976) Phenomenological analysis of plastic spherical indentation. J Eng Mater Technol 98(3):272–281
27. Hollomon JH (1945) Tensile deformation. Trans AIME 162:268
28. Hollomon JH, Lubahn JD (1946) Plastic flow of metals. Phys Rev 70:775
29. Ramberg W, Osgood WR (1943) Description of stress-strain curves by three parameters. National Advisory Committee for Aeronautics, Washington
30. Lee WS, Sue WC, Lin CF, Wu CJ (2000) The strain rate and temperature dependence of the dynamic impact properties of 7075 aluminum alloy. J Mater Process Technol 100:116–122
31. Okada T, Iwai Y, Hattori S, Tanimura N (1995) Relation between impact load and the damage produced by cavitation bubble collapse. Wear 184(2):231–239. doi:10.1016/0043-1648(94)06581-0
32. Lee MK, Hong SM, Kim GH, Kim KH, Rhee CK, Kim WW (2006) Numerical correlation of the cavitation bubble collapse load and frequency with the pitting damage of flame quenched Cu-9Al-4.5Ni-4.5Fe alloy. Mater Sci Eng A (Struct Mater Prop Microstruct Process) 425 (1–2):15–21. doi:10.1016/j.msea.2006.03.039
33. Hattori S, Takinami, M., Otani, T. (2009) Comparison of cavitation erosion rate with liquid impingement erosion rate. Paper presented at the 7th international symposium on cavitation, Ann Arbor, Michigan, USA, August 17-22, 2009
34. Soyama H, Lichtarowicz A, Momma T, Williams EJ (1998) A new calibration method for dynamically loaded transducers and its application to cavitation impact measurement. J Fluids Eng 120(4):712–718
35. Franc J-P, Michel J-M (1997) Cavitation erosion research in France: the state of the art. J Mar Sci Technol 2:233–244

Chapter 4
Cavitation Impulsive Pressures

Georges L. Chahine, Jean-Pierre Franc and Ayat Karimi

Abstract This chapter is devoted to the measurement and analysis of impulsive pressures generated by cavitation bubble collapse. The measurement technique makes use of pressure sensors which may be directly exposed to the cavitating flow or protected from it to limit the risk of damage. The unsteady pressure signals detected by the sensors are made of a succession of pulses of various amplitudes and durations. The analysis consists of determining their distribution as a function of amplitude and duration. Typical pressure pulse measurements are presented in the cases of a cavitating jet, an ultrasonic horn and an attached cavity generated in a high-speed cavitation tunnel. Pressure pulse height or duration spectra follow a Weibull (or a simple exponential) distribution that involves a limited number of parameters such as a reference rate and a reference amplitude or duration. The influence of flow velocity on the reference rate and amplitude is investigated and normalized spectra independent of flow velocity are presented. The correlation between peak pressure rate and pitting rate on various materials is analyzed assuming that a pressure pulse generates a pit if its amplitude exceeds a critical value connected to the material yield stress.

G. L. Chahine (✉)
DYNAFLOW, INC., 10621-J Iron Bridge Road, Jessup, MD, USA
e-mail: glchahine@dynaflow-inc.com

J.-P. Franc (✉)
LEGI, Grenoble, France
e-mail: jean-pierre.franc@legi.grenoble-inp.fr

A. Karimi (✉)
EPFL, Lausanne, Switzerland
e-mail: ayat.karimi@epfl.ch

K.-H. Kim et al. (eds.), *Advanced Experimental and Numerical Techniques for Cavitation Erosion Prediction*, Fluid Mechanics and Its Applications 106,
DOI: 10.1007/978-94-017-8539-6_4,

4.1 Introduction

Knowledge of the full three-dimensional pressure field as a function of time is essential to understand and characterize the cavitation field. Stresses on an object exposed to the cavitation field are directly related to the time dependent pressures on the boundary. Acquiring this pressure field is however very difficult in practice, since presently pressure measurements can only be done locally, at limited locations and for limited times. Development of pressure image velocimetry based on Particle Image Velocimetry (PIV) may solve this issue in the future [1]. In addition, transducers or hydrophones are of finite size and bandwidth, which creates limitations on the precision and accuracy of the measurements.

Many researchers sought the correlation between cavitation noise and erosion as an alternative approach to characterize the erosive cavitation field. Summaries of work on such a correlation can be found in [2–7]. Pioneering works have been conducted both in the former Soviet Union and in the US, and the ideas in these pioneering works were further advanced [8–11]. Interest in detecting erosive cavitation fields through acoustic means was high when cavitation in liquid metals became the subject of interest either for fundamental studies or during the development of sodium nuclear reactors [12–16]. These efforts resulted in the development of instrumentation to count pressure peak distributions, such as in [14]. More recently, efforts have been geared toward the development of better measurement techniques and more advanced data analysis [15–18].

4.2 Measurement and Signal Processing Techniques

4.2.1 Pressure Transducers

Different types of pressure transducers can be used to measure pressure pulses in a cavitating flow. Because of the shock-like nature of cavitation induced pressure pulses, it is necessary to use transducers of very short rise time and high natural frequency such that the transducer signal follows as accurately as possible the high frequency loading. For fundamental work, it is also necessary to use miniaturized pressure sensors in order to be able to separate as much as possible pressure pulses caused by different cavitation events. However, considering the very small size of the region significantly affected by a single collapse event (of the order of a pit area, typically measured in micrometers or tens of micrometers in equivalent diameter), there is usually a very large difference between the sensitive area of the transducer and the area of the loaded region (the area ratio can easily exceed 10^4). As a result, the transducer surface is far from being uniformly loaded and pressure transducers cannot give directly a precise estimate of the locally applied stresses.

Pressure pulses can actually be better measured in terms of the load they apply to the transducer. The pressure reading output by the transducer manufacturer

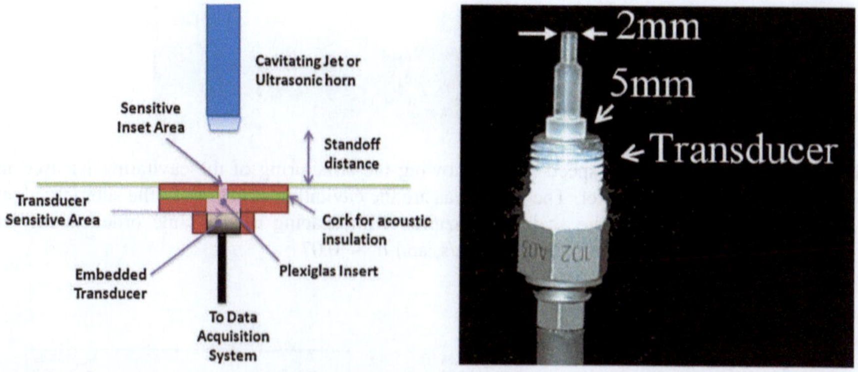

Fig. 4.1 Adaptation of high response pressure transducer with an insert to protect and reduce spatial impulsive pressures overlap

calibration curve is obtained after dividing this load by the transducer sensitive area and assuming a pressure applied uniformly over the whole surface of the transducer sensitive area. The knowledge of the impulsive load spatial extent is needed to extract the actual pressure from the load measurement.

For the studies in this chapter, commercial piezoelectric pressure transducers were used. The one shown in Fig. 4.1 (PCB 102A03) had a rise time of 1 μs, a resonance frequency of 400 kHz and an exposed sensitive area of about 3.14 mm^2 (2 mm diameter). It was positioned in the high erosion region of interest. The transducer sensitivity was 0.5 mV/psi, which is defined assuming a uniform pressure distribution over the transducer surface.

Another important issue with the use of pressure transducers in a cavitating field is the risk of damage to the crystal of the transducer. The transducer may be flush mounted in the cavitation region of interest, but this significantly reduces its lifetime [19]. In order to protect it from the erosive cavitation field, a Plexiglas insert was used as shown in Fig. 4.1. This served multiple purposes. The insert had an acoustic impedance close to that of water, and had a sensitive area one quarter smaller than the transducer sensitive area. This served to minimize overlap in space and time of multiple impulsive pressures on the sensing area (Fig. 4.1). This reduction in size of the impact area is accompanied with losses during the propagation of the acoustic waves through the insert. Using a conventional ball drop technique [19] and comparative submerged spark test calibrations comparing flush mounted transducer and presence of the insert, it was found that the recorded force with this insert should be multiplied by a factor of about six to recover the actual force from the impulse signals at the transducer face. This value was close to, but slightly less than the area contraction ratio due to the insert, acoustic losses, and imperfect transmission.

As illustrated in the high-speed movie picture in Fig. 4.2, submerged cavitating jets exhibit large scale organization shown as regularly separated dark colored structures [20]. This time snapshot shows discrete spacing between moving

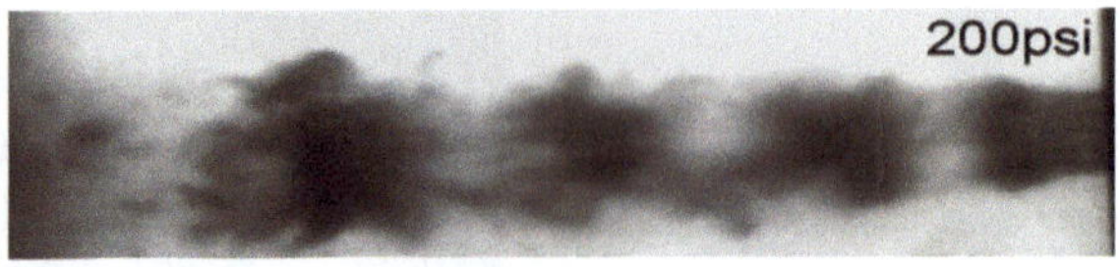

Fig. 4.2 Image from a high-speed movie showing the structuring of the cavitating jet used in studies presented in this chapter. The dark areas are the cavitating structures in the submerged jet shear layer and illustrate large scale organization with spacing of the same order as the jet diameter. CaviJet® nozzle with $V_{jet} \sim 50$ m/s, and $\sigma \sim 0.07$

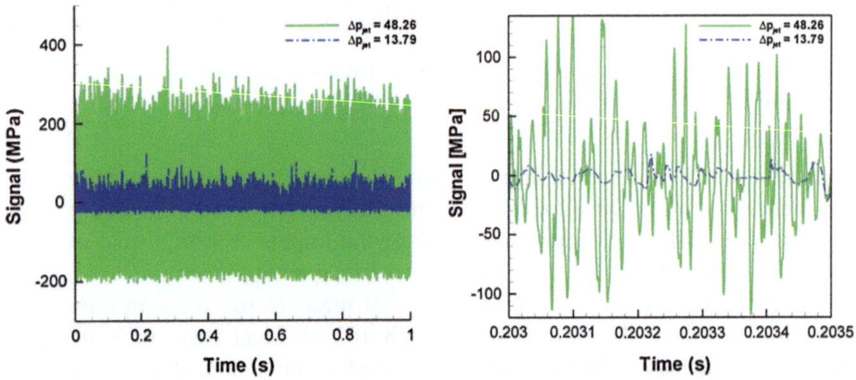

Fig. 4.3 Raw signals for two jet pressures (13.8 MPa, $V_{jet} \sim 160$ m/s and 48.3 MPa, $V_{jet} \sim 310$ m/s). The left hand side shows an example of the raw signal for 1 s. full range used to compute statistics. Superposition of the two signals shows the correlation of the amplitudes with the jet pressure. The right hand figure is a zoom of these signals after the relatively low frequency jet oscillations had been filtered out ($f < 10$ kHz for the 13.8 MPa jet and $f < 40$ kHz for the 48.3 MPa jet)

regions of lower pressures (vortical structures where cavities concentrate) and higher pressures (where cavities are less dense and of smaller size). At a given location, in the time domain, the corresponding pressure versus time signal has a specific discrete frequency (and its harmonics) with high energy content. This corresponds to a normalized frequency or Strouhal number, S_d of 0.3 and its harmonics. The Strouhal number is defined by $S_d = f\ d_{jet}\ /V_{jet}$ where f is the shedding frequency, d_{jet} is the jet orifice diameter, and V_{jet} is the jet velocity. These jet natural pressure fluctuations strongly modulate the measured cavitation field signal and special care has to be taken to analyze the signal.

Figure 4.3 shows two illustrative signals obtained for both a relatively low jet pressure ($\Delta p_{jet} = 13.8$ MPa, jet speed: ~160 m/s) and a relatively high jet pressure ($\Delta p_{jet} = 48.3$ MPa, jet speed: ~310 m/s). The predominant structuring jet frequency was filtered out from each of these signals to isolate the cavitation signature. All frequencies less than 10 kHz were filtered out from the 13.8 MPa jet, while all frequencies less than 40 kHz were filtered out from the 48.3 MPa jet

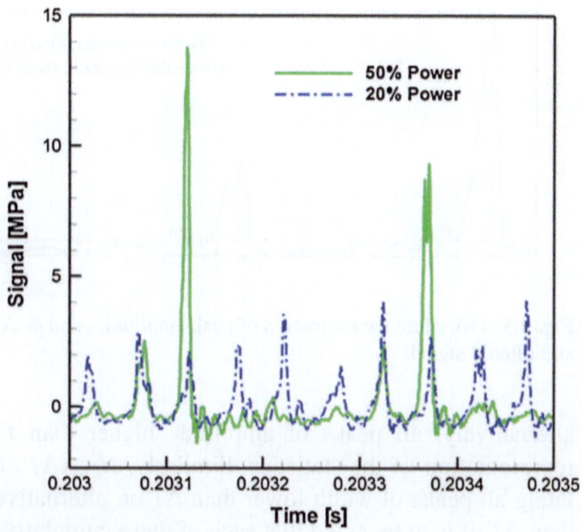

Fig. 4.4 Raw pressure signals for two power settings (20 and 50 % of maximum amplitude) of the ultrasonic horn, illustrating the presence of sharp high frequency impulsive pressures. In this case, the raw signals are predominated by the ultrasonic horn oscillation frequency of 20 kHz on which cavitation events are superposed but seem absent for some periods at the 50 % setting

and only high frequency content due to cavitation was kept. Similarly, Fig. 4.4 shows the pressure signals from the ultrasonic horn generated cavitation for two amplitudes of oscillation. These signals required further analysis as described next.

4.2.2 Signal Processing

In order to extract statistical information from the pressure data and retrieve information about the impulsive pressures characterizing the distribution of cavitation events, the following procedures were used.

To identify the relevant features in a signal, an algorithm was used to filter the spurious background noise from the measured signal and to detect peaks. The heights (amplitudes) and the time widths (durations) of the pressure peaks in the signal were obtained using a threshold amplitude crossing algorithm (Fig. 4.5). The highest pressure and the duration were measured between two crossings (with an upward and a downward slope) of the processed signal above the selected threshold, P_{th}. One disadvantage of this simple approach is the potential of sometimes counting two or more overlapping pulses as a single pulse. Other more complex approaches were also used consisting of cross-correlating with the transducer frequency response in order to eliminate high frequencies resulting from the ringing of the transducer but these did not modify the statistical results significantly.

From the recorded peak height and peak width statistics one can deduce characteristic distributions such as the cumulative number of peak heights, $N_{Cav}(P)$, or cumulative number of peak widths, $N_{Cav}(\Delta T)$. $N_{Cav}(P)$ and $N_{Cav}(\Delta T)$ can be obtained by accumulating all peaks of amplitude lower than P or,

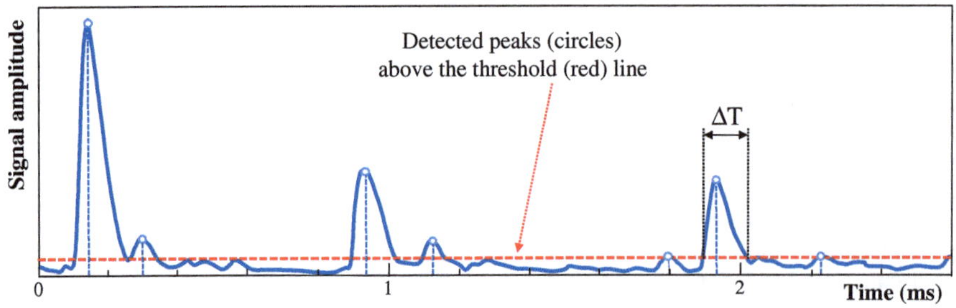

Fig. 4.5 Procedure for estimation of peak amplitudes and peak widths operated on the processed and filtered signal

alternatively, all peaks of amplitude higher than P, depending on the desired representation of the statistics. Similarly, $N_{Cav}(\Delta T)$ can be obtained by accumulating all peaks of width lower than ΔT or, alternatively, all peaks of width larger than ΔT. It is to be noted that each of these cumulative values (using values higher or lower than the current value) can be obtained from the other just by subtracting from N_{Total}, the total number of impulsive pressure peaks detected by unit time. In the mathematical analysis that follows, the peaks are accumulated from the smallest to the current height value, but cumulating from the highest to the current value is as valid and can be sometimes preferred.

A peak height density distribution, $\tilde{N}_{Cav}(P)$, and a period density distribution, $\tilde{N}_{Cav}(\Delta T)$, can be obtained by differentiating the cumulative number distributions, $N_{Cav}(P)$ and $N_{Cav}(\Delta T)$:

$$\tilde{N}_{Cav}(P) = \frac{dN_{Cav}(P)}{dP}, \quad \tilde{N}_{Cav}(\Delta T) = \frac{dN_{Cav}(\Delta T)}{d\Delta T}. \tag{4.1}$$

On the other hand, the number density curves can be integrated between any two values of the pressure, P_1 and P_2, or the time widths, ΔT_1 and ΔT_2 , to obtain the number of pulses, n, of heights between P_1 and P_2 or between ΔT_1 and ΔT_2:

$$n(P_1, P_2) = N_{Cav}(P_2) - N_{Cav}(P_1) = \int_{P_1}^{P_2} \tilde{N}_{Cav}(P)dP, \quad (P_1 \leq P \leq P_2),$$

$$n(\Delta T_1, \Delta T_2) = N_{Cav}(_2) - N_{Cav}(\Delta T_1) = \int_{\Delta T_1}^{\Delta T_2} \tilde{N}_{Cav}(\Delta T)d, \ (\Delta T_1 \leq \Delta T \leq \Delta T_2). \tag{4.2}$$

A two-dimensional number density distribution, $\tilde{N}_{Cav}(P, \Delta T)$, can also be defined encompassing the variations of both the peak height, P, and the peak width, ΔT. The number of pulses occurring within a peak width range, $\Delta T = \Delta T_2 - \Delta T_1$, and a peak height range, $\Delta P = P_2 - P_1$, can then be given as follows:

$$n(\Delta P, \Delta T) = \int_{P_1}^{P_2} \int_{\Delta T_1}^{\Delta T_2} \tilde{N}_{Cav}(P, \Delta T)\, dP\, d\Delta T. \quad (4.3)$$

To obtain the one-dimensional or two-dimensional number density distributions ($\tilde{N}$) numerically, the entire range over which the peaks occur is divided into bins (δP or $\delta(\Delta T)$ in one dimension or $\delta P \times \delta(\Delta T)$ in two dimensions) and the number of pulses occurring within each bin is counted ($n(\delta P)$ or $n(\delta(\Delta T))$ in one dimension or $n(\delta P, \delta(\Delta T))$ in two dimensions). This number is then normalized by the bin size to obtain the number density at discrete locations:

$$\tilde{N}_{Cav}(P,\ \Delta T) = \frac{n(\delta P,\ \delta(\Delta T))}{\delta P \times \delta(\Delta T)}. \quad (4.4)$$

4.3 Peak Number Density Distribution

4.3.1 Effect of Pressure Threshold

Figure 4.6 illustrates the statistical analysis of the cavitation impulsive pressures for a cavitating jet operated at a pressure drop across the nozzle, $\Delta p_{jet} = 13.8$ MPa ($V_{jet} \sim 160$ m/s). The figure shows a plot of the pressure peak number density distributions versus peak height, $\tilde{N}_{Cav}(P)$, for different pressure threshold values. This plot was obtained by counting during 1 s of pressure signal recording, the number of peaks, which are larger than P, in the band $[P - \delta P/2, P + \delta P/2]$, and dividing the result by δP.

As shown in the figure, the number density distribution versus peak height remains almost unaffected by the choice of the threshold, except at very low peak height values, where the smaller peak heights become excluded when a higher value of the threshold is selected. This discrepancy in the curves at low peak height values is actually not an issue, since the lowest amplitude pulses are of the order of the pressure signal noise and these usually are too weak to contribute to cavitation erosion.

Figure 4.7 presents a plot of the peak number density distributions versus peak width, $\tilde{N}_{Cav}(\Delta T)$, for different pressure threshold values. In this case $\tilde{N}_{Cav}(\Delta T)$ is much more sensitive to the selected threshold. Choosing a higher threshold results in smaller time widths of the pulses detected, i.e. the distribution moves slightly to the left with the exception for low peak widths. This is a result of the peak duration becoming shorter as the threshold becomes higher. However, it should be noted that the overall trend is not affected much by the threshold choice.

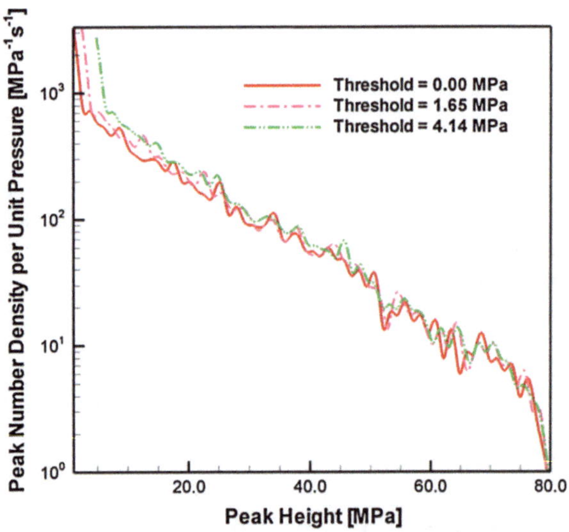

Fig. 4.6 Impulsive pressure peak number density distributions versus peak height, $\tilde{N}(P)$, for different pressure threshold values. Measurements under a CaviJet® cavitating jet at 13.8 MPa ($V_{jet} \sim 160$ m/s). *Note* The peak height pressures shown are "average constant pressures over the transducer sensitive area" and not local concentrated bubble collapse pressures

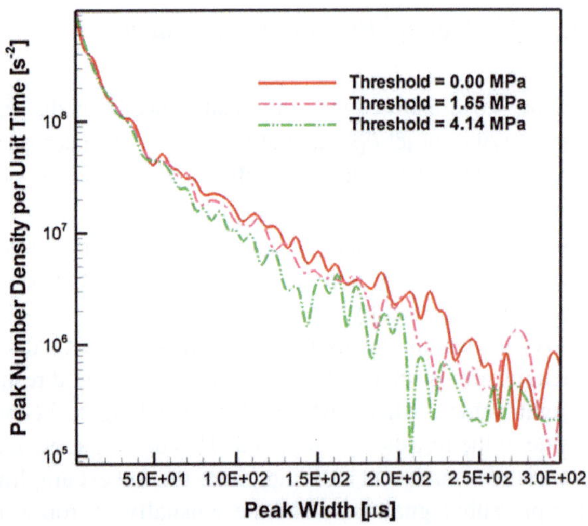

Fig. 4.7 Impulsive pressure peak number density distributions versus peak width, $\tilde{N}(\Delta T)$ for different pressure threshold values. Measurements under a CaviJet® at 13.8 MPa ($V_{jet} \sim 160$ m/s)

4.3.2 Effect of Jet Velocity on Peak Number and Density Distributions

The effect of jet speed (pump pressure) on the cavitation impulsive pressure peak number density distributions versus peak height and versus peak width is presented respectively in Figs. 4.8 and 4.9. Modifying the cavitating jet speed induces a change in cavitation intensity and also a change in the cavitation number and cavitation extent. The figures clearly show that the impulsive pressure peak

Fig. 4.8 Cavitation impulsive pressure peak number density versus peak height for different jet pressures. The flat region at high jet velocities indicates a region where the number of peaks varies little over a range of jet speeds. In that region the cumulative number of peaks increases practically linearly. Measurements under a CaviJet® at different jet velocities or pump pressures, Δp_{jet}. *Note* The pressures shown are "average constant pressures over the transducer sensitive area" and not local concentrated bubble collapse pressures

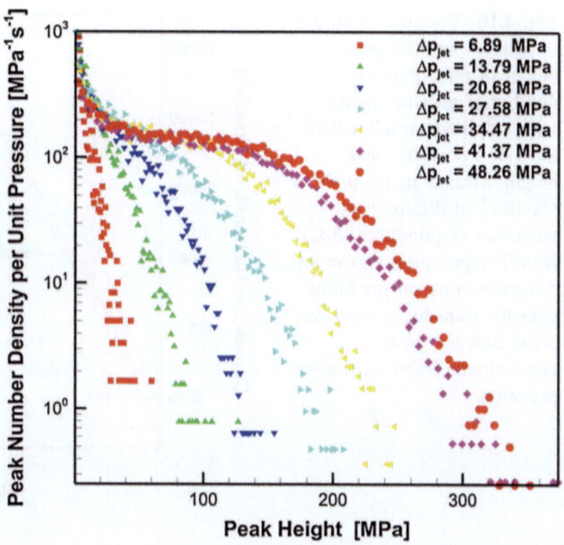

Fig. 4.9 Cavitation impulsive pressure peaks number density versus peak width for different jet pressures. The plot concentrates on the region of smaller peak widths, where the curves are distinguishable. At the larger peak widths, all curves are on top of each other. Measurements under a CaviJet® at different jet velocities or pressures, Δp_{jet}

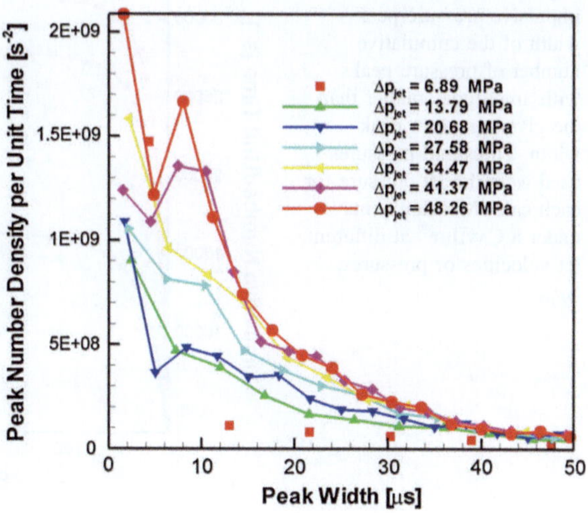

number density increases with the jet pressure. Also, the value of the highest impulsive pressure peak occurring in a given signal increases with the jet pressure. This is expected since the cavitation field becomes more intense as the jet pressure increases. As a result, the peak heights correlate with the jet pressure. Figure 4.9 shows oscillations of the results for the shortest peak widths, $\Delta t < 10$ μs. The oscillations are due to inaccuracies in the measurements in this range due to limitation in the frequency response of the transducer.

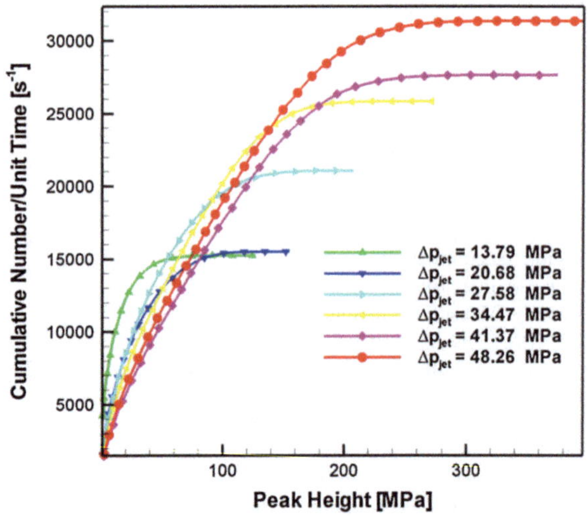

Fig. 4.10 Variation with the impulsive pressure peak height of the cumulative number of pressure peaks with amplitude smaller than the given pressure peak height. Measurements under a CAVIJET® at different jet velocities or pressures, Δp_{jet}. *Note* The pressures shown are "average constant pressures over the transducer sensitive area" and not local concentrated bubble collapse pressures

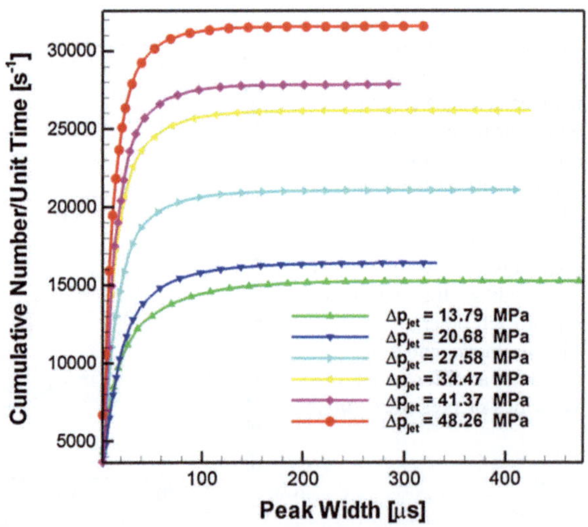

Fig. 4.11 Variation with the impulsive pressure peak width of the cumulative number of pressure peaks with amplitude smaller than the given pressure peak width. Threshold pressures used were the jet pressure for each case. Measurements under a CAVIJET® at different jet velocities or pressures, Δp_{jet}

Figure 4.10 shows plots of cavitation cumulative number of peaks versus peak heights for different jet pressures. In this figure, for a given impulsive pressure peak value, P, the graphs show the total detected number of pressure peaks which are lower than or equal to P. This representation brings out clearly that the cumulative number of high peaks is higher for a signal corresponding to a higher cavitating jet velocity or a higher cavitation intensity. In addition, the total number of peaks (the asymptotic value in each curve for large P) and the highest peak values increase with Δp_{jet}. On the contrary, the slope of the curves for the lower values of P (i.e. the number density) increases with decreasing Δp_{jet}.

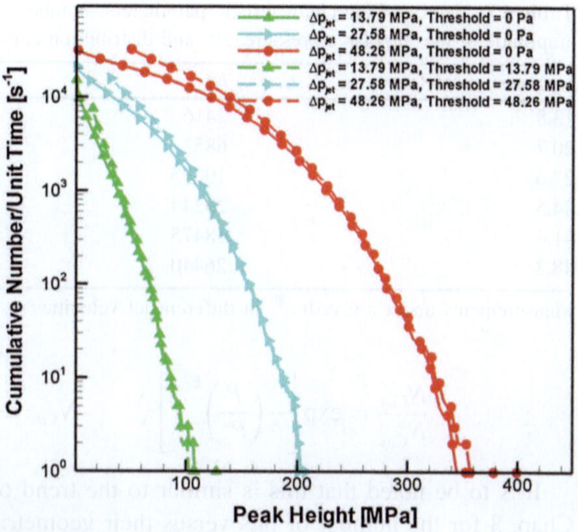

Fig. 4.12 Variation with the pressure peak height of the cumulative number of impulsive pressure peaks with amplitudes larger than the given pressure peak height for various pressure thresholds. Measurements under a CAVIJET® at different jet velocities or pressures, Δp_{jet}. *Note* The pressures shown are "average constant pressures over the transducer sensitive area" and not local concentrated bubble collapse pressures

Figure 4.11 shows plots of the cumulative number of peaks versus peak widths for different jet pressures. Here too the cumulative number of peaks increases with Δp_{jet}, and the peak width becomes smaller for higher Δp_{jet}. However, there is no clear trend of the value of the maximum peak width versus Δp_{jet}.

Figure 4.12 shows plots of cumulative number of cavitation impulsive pressure peaks versus peak height for three different jet pressures when using different pressure thresholds to identify the peaks in the filtered signals. Unlike the plots in Fig. 4.10, in this plot the count for the cumulative number of peaks is started from the highest peak amplitude value. Therefore the zero peak height value includes all peaks. Generating the cumulative curves counting from the largest peak illustrates better the largest peaks that exceed the material yield stress and result in plastic deformation and eventual erosion. It is also less influenced by the smallest peaks which do not contribute to the erosion. This is illustrated in Fig. 4.12, which indicates that there is very little influence of the value of the signal threshold on the overall shape of the curves. Using a threshold of zero for the detection and measurement of the amplitude of the impulsive pressures does not present any particular issues, but is problematic for the pulse width as it can result in counting artificially merged peaks.

4.3.3 Weibull Curve Fit

The shapes of the curves corresponding to different jet pressures also suggest that the following analytical exponential equation [21] for the number of impulsive pressure peaks or cavitation events, N_{cav}, may capture the global behavior of these curves:

Table 4.1 Values of the characteristic parameters: number of impulsive pressure peaks, N^*_{Cav}, amplitude of the impulsive pressure, P^*, and distribution curve fit shape parameter, k_{Cav}

Jet pressure (MPa)	N^*_{Cav}	P^* (MPa)	k_{Cav}
13.8	2416	43	2.54
20.7	6853	61	2.54
27.6	10095	86	2.54
34.5	20344	103	2.54
41.4	18475	141	2.54
48.3	26440	144	2.54

Measurements under a CaviJet® at different jet velocities or pressures, Δp_{jet}

$$\frac{N_{Cav}}{N^*_{Cav}} = \exp\left[-\left(\frac{P}{P^*}\right)^{k_{Cav}}\right], \qquad \overline{N}_{Cav} = \exp\left(-\overline{P}^{\,k_{Cav}}\right). \tag{4.5}$$

It is to be noted that this is similar to the trend observed with pit counting in Chap. 3 for the number of pits versus their geometric size.

In Eq. (4.5), $N_{Cav}(P)$ is the cumulative number of peaks from the highest peak to the current value P. P is the corresponding impulsive pressure peak height. N^*_{Cav} is the characteristic number of impulsive pressure peaks. P^* is the characteristic amplitude of the impulsive pressure, and k_{Cav} is the distribution curve fit shape parameter.

N^*_{Cav} and P^* are two key parameters, which characterize the particular erosive cavitation field and $\overline{N}_{Cav}(\overline{P})$ is the non-dimensional cumulative number of peaks of amplitude larger than or equal to the normalized pressure $\overline{P}$.

While knowledge of the full curve $N_{Cav}(P)$ is necessary to fully characterize the cavitation field, in lack of full experimental data (for example, limitation of the transducer in capturing high amplitudes or frequencies), or for simplified application, N^*_{Cav} and P^*could be used satisfactorily to characterize a cavitation field in conjunction with the general Eq. (4.5).

A value of the shape parameter of $k_{Cav} = 2.54$ provided the best least squares fits for the measured statistical data set of the cavitating jet pressures studied here. Table 4.1 shows these curve-fit parameters for different jet pressures, and Fig. 4.13 shows variations of these parameters with the jet pressure. Figure 4.13 also shows curves of the maximum number of impulsive pressure peaks N^{Total}_{Cav} and the maximum measured amplitudes P_{max}. Interestingly, all four quantities N^{Total}_{Cav}, N^*_{Cav}, P_{max} and P^* vary linearly with the jet pressure, Δp_{jet}. This may be explained as follows.

The effects of changing the jet pressure, ΔP_{jet}, on the number of impulsive pressures include the following: (a) the jet velocity increases as $\Delta P^{1/2}_{jet}$ resulting in an increase of the nuclei supply as $\Delta P^{1/2}_{jet}$, (b) the frequency of the large structures increases to keep the Strouhal number almost constant ($S_d \sim 0.3$), and thus the repetition rate of the impulsive pressure peaks increases also as $\Delta P^{1/2}_{jet}$, (c) the cavitation number varies as the inverse of ΔP_{jet}. The combination of the above appears to result here in an overall trend of N^{Total}_{Cav}, N^*_{Cav} to vary linearly like ΔP_{jet}.

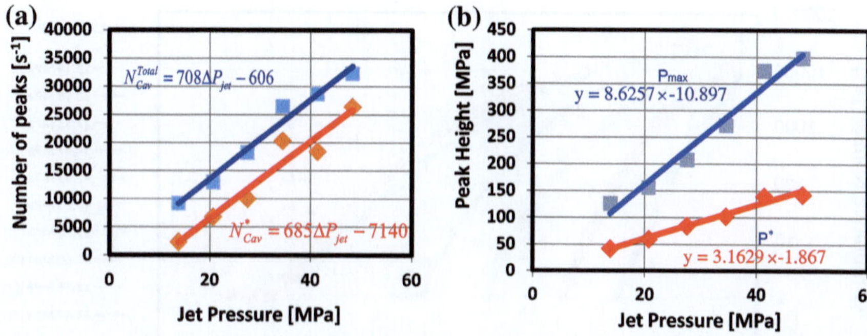

Fig. 4.13 Variation with the cavitating jet pressure, Δp_{jet}, of **a** the characteristic number of the impulsive pressure peaks, N^*_{Cav}, and the maximum number of the impulsive pressure peaks, N^{Total}_{Cav}, and **b** the characteristic impulsive pressure peak height, P^*, and the highest pressure peak, $P_{\max}$. *Note* The pressures shown are "average constant pressures over the transducer sensitive area" and not local concentrated bubble collapse pressures

Concerning the variations with ΔP_{jet} of the amplitude of the impulsive peak pressures, we can reason as follows. The pressures generated by bubble collapse near a boundary are due to either the impact of the re-entrant jet on the structure or to the subsequent collapse of the bubble ring as shown in Chap. 6. The water hammer pressure due to the re-entrant jet is shown to be proportional to $\rho c V_{jet}$, where ρ is the local density of the medium where the bubble is collapsing, c is the sound speed in this medium, and V_{jet} the speed of the re-entrant jet impacting on the boundary. The pressure which drives the collapse of the bubbles in a cavitating jet is the stagnation pressure of the jet, ΔP_{jet}. As a result, $V_{jet} \approx \Delta P_{jet}^{1/2}$. On the other hand, the collapse of the bubble ring generates a shock wave or pressures, which are proportional to the square of the bubble ring collapse velocities, which are proportional to the driving pressure ΔP_{jet}.

The force which is measured by the transducer is an integral of these impacting pressures on the transducer sensitive element and appears to be dominated by the terms proportional to ΔP_{jet}.

The results of fitting the data with Eq. (4.5) are illustrated in Fig. 4.14, which shows the cumulative number of impulsive pressure peaks versus peak amplitude for different jet pressures compared with the corresponding data from the equation. The Weibull curves are shown to represent the measured data very well.

Figure 4.15 shows the normalized cumulative number, $\overline{N}_{Cav} = N_{Cav}/N^*_{Cav}$, versus normalized peak height, P/P^*, for different jet pressures. All curves for the various jet pressures practically superpose. The distribution function (4.5) can be used to represent various cavitation fields once their characteristic values of N^*_{Cav}, P^*, and k_{Cav} are known.

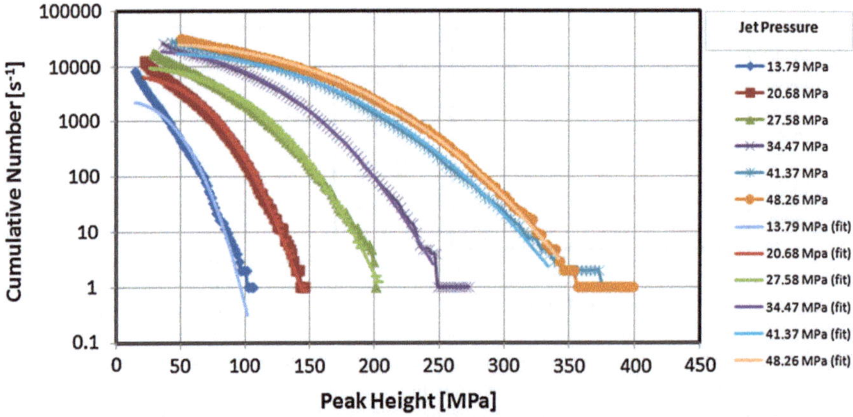

Fig. 4.14 Cumulative number of peaks versus peak amplitude for different jet pressures: curves from data fit of Eq. (4.5) superimposed over actual data. *Note* The pressures shown are "average constant pressures over the transducer sensitive area" and not local concentrated bubble collapse pressures. These would be the values shown here times the ratio of the sensitive area to the impulsive pressure area

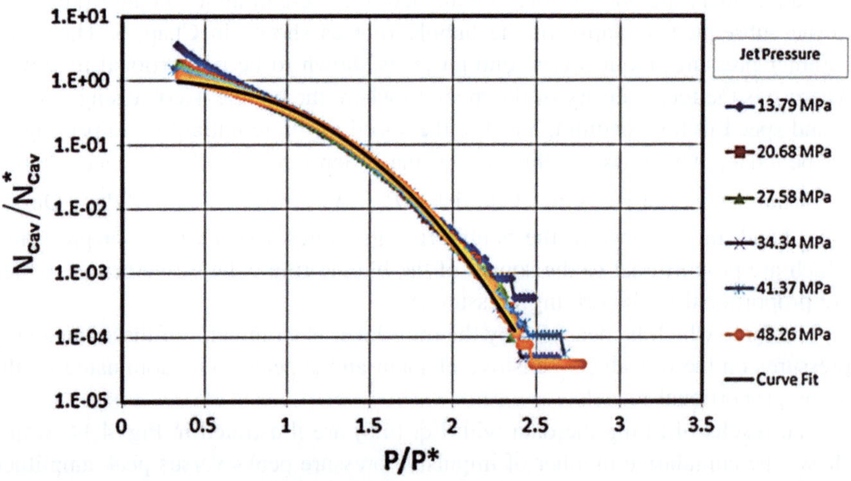

Fig. 4.15 Normalized plot of cumulative number of peaks versus peak amplitude for different jet pressures, compared with the curve of Eq. (4.5)

4.4 Pressure Pulse Measurements at Constant Cavitation Number

The previous sections were devoted to pressure pulse measurements in a cavitating jet discharged in a chamber at atmospheric pressure. As a result, any increase in flow velocity induces a decrease in cavitation number and a subsequent change in

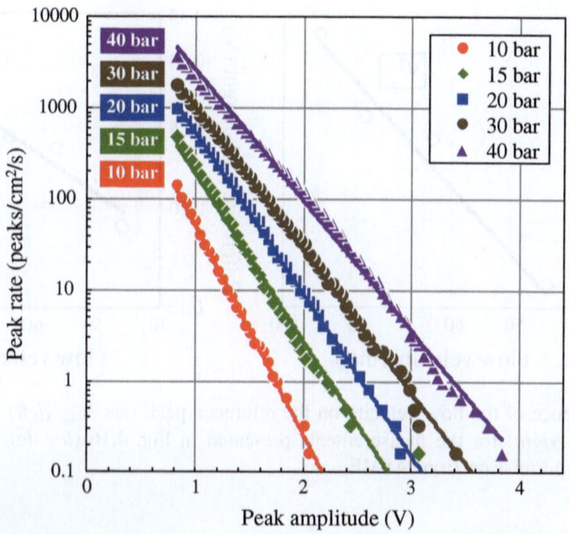

Fig. 4.16 Cumulative peak rate versus peak amplitude. The measurements were conducted in the high-speed cavitation tunnel of the LEGI laboratory (University of Grenoble, France). The sensitivity of the pressure transducer determined by the ball drop test technique is 7.3 mV/N [19]. The various spectra were obtained at different upstream pressures between 10 and 40 bar. The cavitation number was kept constant and equal to 0.9 by adjusting the downstream pressure. The tunnel and the test section were presented in Chap. 2. Adapted from [19], with permission from ASME

the extent of cavitation. Thus, the effect of flow velocity discussed above also includes an effect of cavitation number.

In this section, measurements conducted in a high-speed cavitation loop (see Chap. 2) at a constant cavitation number are presented. Since the cavitation number is kept constant by adjusting the downstream pressure in the facility, the cavitating flows obtained at different velocities are geometrically similar. In this case, the effect of flow velocity does not include any effect of cavitation number or change in cavity extent and is then a pure velocity effect.

The measurements were conducted by means of a piezoelectric pressure sensor (PCB 108A02) of comparable characteristics to the one used for investigating the cavitating jet. The transducer was flush mounted in the region of maximum erosion without any protective insert. The experimental procedure was designed to limit the exposure time and the damage to the transducer. The transducer was dynamically calibrated by the standard ball drop test technique and recovered the manufacturer calibration factor [19].

The cumulative peak rates obtained at different velocities and constant cavitation number are plotted as a function of peak amplitude in Fig. 4.16. The same type of fitting was used as the one presented in Sect. 4.3.3 except that a simple exponential law was quite appropriate to account for the various histograms and the shape parameter k_{Cav} in Eq. (4.5) was then chosen equal to 1:

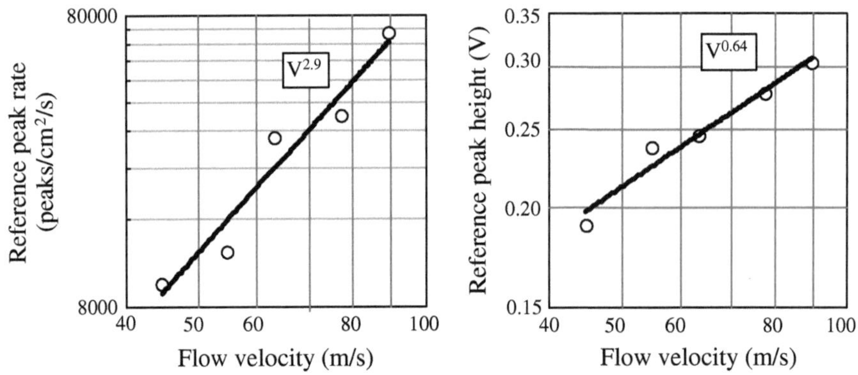

Fig. 4.17 Influence of the flow velocity on the reference peak rate N^*_{cav} (*left*) and the reference peak height P^* (*right*) for the measurements presented in Fig. 4.16 (*log–log scales*). Adapted from [19], with permission from ASME

$$N_{Cav} = N^*_{Cav} e^{-\frac{P}{P*}}. \tag{4.6}$$

As explained in Sect. 4.3.3, this type of fitting provides a means of decomposing the overall effect of flow velocity on the cavitation intensity into two elementary effects, relative first to the rate of cavitation events and second to their amplitude. In the present case, the influence of the flow velocity on the reference peak rate N^*_{Cav} and the reference pressure, P^*, is shown in Fig. 4.17. It appears that both N^*_{Cav} and P^* increase with the flow velocity but in a different way. The amplitude of the pressure pulses increases moderately with the velocity to the power 0.64 whereas the peak rate increases much more rapidly with a power close to three. Thus, the increase in cavitation intensity results from the combination of a marked increase in the rate of impact loads and a moderate increase in their amplitude.

The previous variations with flow velocity are clearly not universal. In the case of cavitating jets, it was shown in Sect. 4.3.3 that the two parameters N^*_{Cav} and P^* follow a different relationship with the flow velocity since they both increase with the velocity squared. Several reasons can be put forward to explain a different behavior.

In addition to a possible difference due to the fit itself and connected to the shape of the histograms ($k_{Cav} = 1$, simple exponential law, against $k_{Cav} = 2.54$, Weibull law), the two cavitating flows are singularly different since one is an attached cavity in a confined flow whereas the other one is a free cavitating jet with cavitation in the shear layer. In addition, the experimental procedures are also different since the cavitating jet is operated at varying cavitation number due to the change of velocity, whereas the cavitation number was kept constant in the case of the tests in the cavitation tunnel despite the increase in velocity. In the case of jet cavitation, a cavitation number effect is then superimposed to the velocity effect as explained above, whereas tunnel cavitation is free from any cavity length effect. These are possible reasons for the differences.

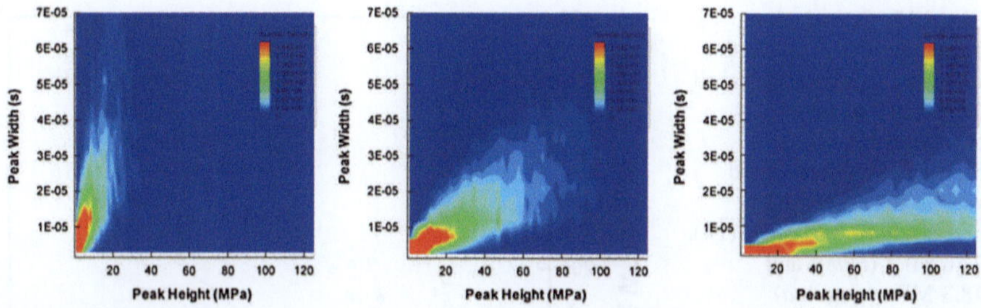

Fig. 4.18 Contour plot of the number density of cavitation impulsive pressure versus pressure peak width and peak amplitude for different jet pressures: 13.8 MPa (*left*), 27.6 MPa (*center*), and 48.3 MPa (*right*)

4.5 Characteristics of Impulsive Pressures

4.5.1 Map of Pulse Width

The pressure peak number density is presented in color contours in Fig. 4.18 as a function of the peak height (amplitude) and width. A scatter plot presenting the peak width versus peak amplitude of all identified individual cavitation impulsive pressures in the recorded signals is also shown in Fig. 4.19. Both figures show three contour plots of the number density distribution for three different jet pressures: 13.8, 27.6, and 48.3 MPa. It is clear from these plots that the slope of the peak widths versus the peak heights decreases significantly as the jet pressure, ΔP_{jet}, or jet speed increases.

Also, the signals from the 13.8 MPa jet have more pulses that are wider (greater peak time width) and lower (smaller peak height) compared to the signal from the 48.3 MPa jet. This reflects the fact that, as the cavitating jet speed increases, i.e. as the cavitation intensity increases, the impulsive pressure peak height increases (as expected) and the width decreases.

4.5.2 Pressure Pulse Distribution Representing Cavitation Field Intensity

A typical cavitation impulsive pressure, P_{rep}, can be represented by a Gaussian function with the following equation:

$$P_{rep} = P_{max} e^{-(t/\Delta t)^2}, \tag{4.7}$$

where P_{max} is the representative pressure peak height, and Δt is the representative pressure peak width. The choice of a Gaussian representation is based on numerical simulations of non-spherical collapse of bubbles near a wall, which

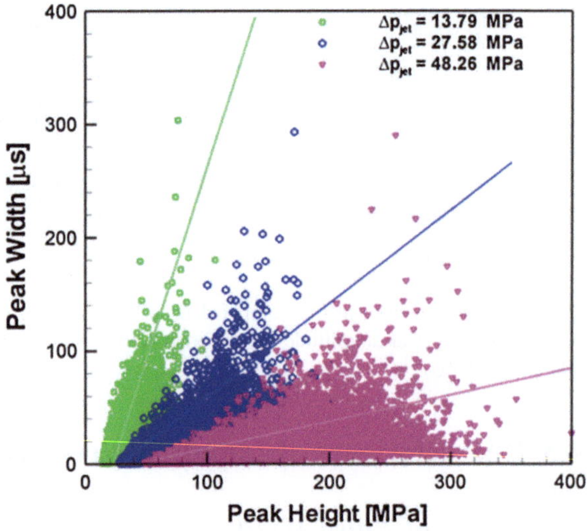

Fig. 4.19 Scatter plot presenting the peak width versus peak amplitude of all identified individual cavitation impulsive pressures in the recorded signals for different jet pressures: 13.8 MPa (*green*), 27.6 MPa (*blue*), and 48.3 MPa (*magenta*)

indicate that the decay over time of the impulsive pressure due to bubble collapse can be fitted well using such a distribution. A similar representation in space can also be made but is not discussed further here.

Since each cavitation field is composed of a rich distribution of impulsive pressures, this distribution can be represented by a characteristic peak pressure height. As in the ocean wave theory, we can use as representative peak height, $P_{1/3}$, which means that 1/3rd of the total number of peaks occurring in the signal have a peak height greater than $P_{1/3}$. From this representative peak height and the scatter plot of Fig. 4.19 of all the pulses measured in the particular cavitation field, a representative peak width can be obtained using the linear curve in the scatter plot. The resulting distributions are shown in Fig. 4.20 for the cavitating jets and reflect the fact that the pressure pulses become higher and narrower (shorter duration) as the jet driving pressure becomes higher.

A similar procedure was followed for the pressure signals measured under the ultrasonic horn described in Chap. 2. Figure 4.21 shows the same trend for three different ultrasonic horn powers in the modification of the illustrative pressures as seen for the cavitating jets. Here, the cavitation intensity is varied by modifying the amplitude of the horn oscillations for the same oscillation frequency. Similar to the jet cavitation, it is shown that the higher the power setting is, the higher and narrower (shorter duration) the dominant pulses are.

4.6 Impact Loads

In the previous sections, cavitation peak pressure measurements were presented as the equivalent pressure that would lead to the same electrical output when uniformly applied to the whole sensitive surface of the transducer. By multiplying this

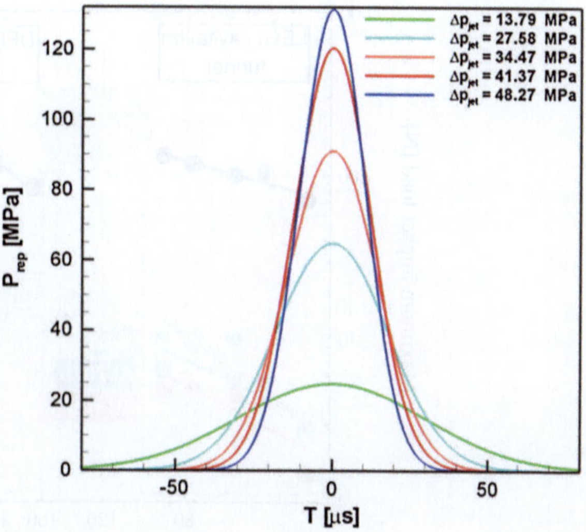

Fig. 4.20 Variation with the cavitation intensity (CAVIJET® at different jet velocities or pressures, Δp_{jet}) of the representative Gaussian pressure pulses capturing the characteristic impulsive pressure peak height and width

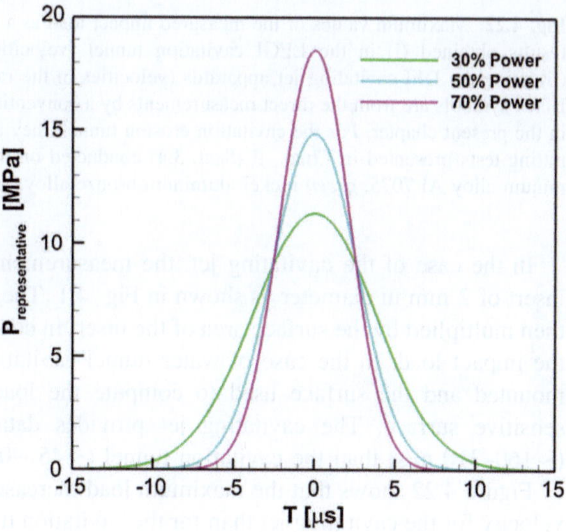

Fig. 4.21 Variation with the cavitation intensity (ultrasonic horn at different oscillation amplitudes for the same frequency) of the representative Gaussian pressure pulses capturing the characteristic impulsive pressure peak height and width

equivalent pressure by the transducer sensitive surface, a load is obtained which is assumed to be the impact load resulting from a bubble collapse and actually applied to a much smaller surface than the transducer sensitive surface typically comparable to the microjet surface area.

The present section is devoted to a discussion of impact load estimates. Emphasis is given to maximum values of impact loads as measured using different techniques and different cavitation devices. Results are summarized in Fig. 4.22 as a function of the flow velocity.

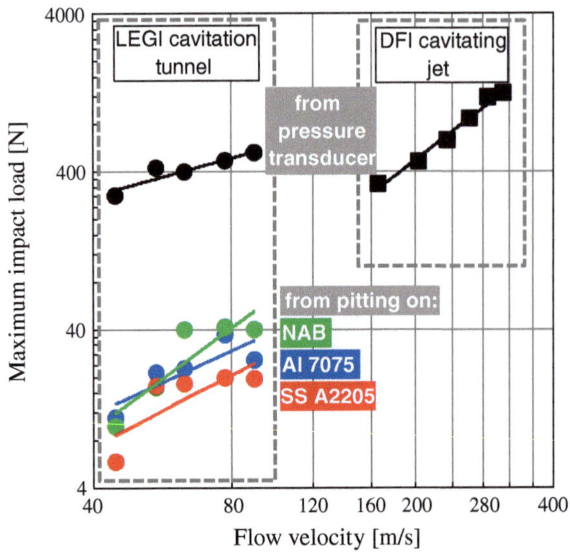

Fig. 4.22 Maximum values of the measured impact load as a function of the flow velocity. The results obtained (i) in the LEGI cavitation tunnel (velocities in the range 45–90 m/s) and (ii) using the DFI cavitating jet apparatus (velocities in the range 160–310 m/s) are compared. Black symbols are from the direct measurements by a conventional pressure transducer discussed in the present chapter. For the cavitation erosion tunnel, they are compared with estimates from pitting tests presented in Chap. 3 (Sect. 3.4) conducted on three different materials (*blue* aluminum alloy Al 7075, *green* nickel aluminum bronze alloy, *red* stainless steel A2205)

In the case of the cavitating jet, the measurement technique used a Plexiglas insert of 2 mm in diameter as shown in Fig. 4.1. The measured peak pressure was then multiplied by the surface area of the insert in contact with the flow to evaluate the impact load. In the case of water tunnel cavitation, the transducer was flush mounted and the surface used to compute the load was the whole transducer sensitive surface. The cavitating jet provides data at a higher flow velocity (~160–310 m/s) than the cavitation tunnel (~45–90 m/s).

Figure 4.22 shows that the maximum load increases more rapidly with the flow velocity for the cavitating jet than for the cavitation tunnel. This is probably due to a different cavitation pattern and also to an additional effect of cavity expansion with the increase of flow velocity which applies to the cavitating jet tests only as discussed in Sect. 4.4. The superposition of such a cavitation number effect to the velocity effect may be partly responsible for a more rapid increase in impact load with flow velocity in the case of the cavitating jet.

In the case of tunnel cavitation erosion, impact loads measured with a conventional pressure transducer are compared with estimates obtained from pitting tests on three different materials in Fig. 4.22. The technique of estimation of impact loads from pitting tests was presented in Chap. 3 (Sect. 3.4). In practice, the impact load was defined as the product of the estimated stress by the pit surface.

The maximum load generally corresponds to the largest pit, as it can reasonably be expected.

In the logarithmic scales of Fig. 4.22, the maximum impact loads obtained from pitting tests on the three materials appear to be in reasonable agreement at all velocities. However, these loads appear to be more than one order of magnitude below direct measurements presented in this chapter using pressure sensors. This highlights that different techniques may lead to significantly different estimates of cavitation impact loads even for the same cavitation device and the same operating conditions. This is a major difficulty when analyzing the response of a material to cavitation impacts since the loading conditions have obviously to be known accurately not only in terms of frequency of occurence, which is expected to be correctly given by most techniques, but also in terms of amplitude which is more difficult to evaluate because of various experimental difficulties. Another explanation, deduced from the Chap. 6 study, is that, due to material response, there is a significant drop between the liquid pressures imposed at the material surface and the maximum effective stresses felt inside the material. Further research is then needed to be able to reliably estimate the amplitude of impact loads.

4.7 Comparison Between Peak Rate and Pitting Rate

Each pressure pulse will generate a pit on the material surface provided it is strong enough with respect to the resistance of the material. A correlation between peak rate and pitting rate is then expected.

Figure 4.23 shows the pressure pulse height spectra already presented in Fig. 4.16 on which the pitting rates measured on three different materials (Aluminum alloy Al 7075, Nickel Aluminum Bronze alloy NAB and Stainless Steel A2205) at various operating pressures have been superimposed.

For each operating condition and each material, the pitting rate, when compared to the pressure pulse spectrum, allows us to define a characteristic peak height for which pressure peak rate and pitting rate coincide. This pressure peak height can be considered as an actual threshold for material pitting.

Figure 4.23 shows that, for a given material, this threshold depends only weakly upon the operating conditions since the pitting curves (black dotted lines) are not far from being vertical.

More precisely, the pulse height thresholds for the three materials at the minimum (10 bar) and maximum (40 bar) values of the operating pressure are plotted in Fig. 4.24 as a function of the material yield stress since yield stress is considered as an appropriate threshold for plastic deformation to occur. On the basis of the three materials considered for pitting tests, Fig. 4.24 shows a strong and almost linear correlation between the pulse height threshold and the yield stress. The slope can be considered as a calibration factor with respect to the material yield stress which makes the pitting rate consistent with the pressure peak rate for all three materials.

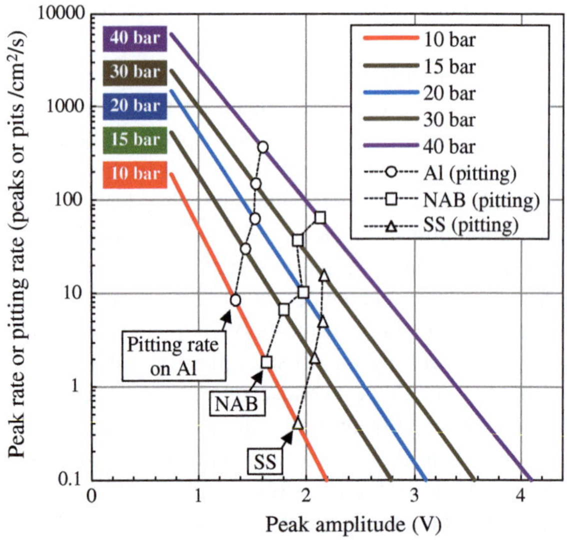

Fig. 4.23 Comparison between peak rate and pitting rate. The experimental conditions are the same as for Fig. 4.16. Pitting tests were made on Aluminum alloy 7075, Nickel Aluminum Bronze alloy NAB and Stainless Steel A2205

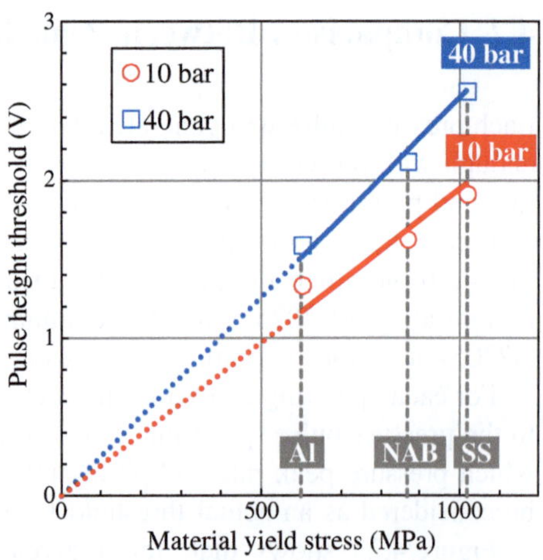

Fig. 4.24 Pulse height threshold as a function of material yield stress. The pulse height threshold is defined as the particular value of the threshold for which peak rate is equal to the measured pitting rate. Data were derived from Fig. 4.23. The values of yield stress considered for the three alloys (Aluminum alloy Al 7075, Nickel Aluminum Bronze alloy and Stainless Steel A2205) in-clude a strain rate effect. More details on the determination of yield stress are available in Chap. 3

According to Fig. 4.24, the slope slightly increases with the operating pressure or flow velocity. This trend might be due to the increase of the average pit size with flow velocity reported in Chap. 3. The increase in pit size should result in an increase of the impact load and a subsequent increase in the signal threshold if it is assumed that the pressure sensor actually measures the impact load rather than the impact stress.

The important conclusion is that a satisfactory correlation is observed between pressure pulse measurements and pitting tests provided an appropriate threshold, linearly correlated with the material yield stress and slightly increasing with the flow velocity, is applied to the pressure signal.

4.8 Summary

In this chapter, pressure signals recorded under various cavitating field conditions were examined. The aim was to identify, count, and characterize the cavitation impulsive pressures potentially responsible for cavitation damage. Operating conditions of a cavitating jet, an ultrasonic horn, and a high speed axisymmetric channel flow were investigated.

The number density and cumulative number distributions of pressure peaks as functions of peak amplitude and peak width were obtained. From the statistics it was found that a cavitation pressure field can be characterized by a cumulative number distribution, $N_{Cav}(P)$, of cavitation impulsive pressure peaks as a function of amplitude, which follows a universal Weibull-like distribution of the type, (see Eq. (4.5)):

$$N_{Cav} = N^*_{Cav} \exp\left[-\left(\frac{P}{P^*}\right)^{k_{Cav}}\right].$$

The same type of law was found in Chap. 3 for the distribution of pitting rates as a function of pit size.

The three parameters, N^*_{Cav}, P^*, and k_{Cav} characterize the cavitating flow field and are respectively a characteristic number of peaks, a characteristic amplitude of the impulsive pressure, and a distribution shape parameter. N^*_{Cav} is interpreted as the total number of pressure peaks corresponding to the limit value $P = 0$. In the particular case of a simple exponential law ($k_{Cav} = 1$), the characteristic amplitude P^* has a simple physical meaning since it is the mean amplitude of all the pressure peaks in the distribution.

While the previous distribution applies to all configurations studied, the variation of N^*_{Cav} and P^* with the flow conditions and particularly with the flow speed appears to be different between various facilities. Moreover, the best fitted value of the shape parameter k_{Cav} was found to be different between jet cavitation ($k \simeq 2.5$) and tunnel cavitation ($k \simeq 1$). This is most probably because the types of cavitation are significantly different between a free cavitating jet which is a shear type cavitation and a confined attached cavity as achieved in the cavitation tunnel. In addition, in the case of a cavitating jet discharged in a container at atmospheric pressure, any increase in flow velocity results in a subsequent decrease in cavitation number so that a cavitation number effect is superimposed to the velocity effect. On the contrary, the control of the operating pressure in the cavitation tunnel made it possible to separate both effects.

The detailed analysis of the cavitation impulsive pressure shapes observed for cavitation in jets and ultrasonic horns resulted in the definition of a typical pressure shape. This can be represented by a Gaussian distribution defined by a significant peak height and a corresponding peak width. These pressure pulses are higher and narrower (shorter duration) as the jet velocity or the ultrasonic horn power increase.

The direct technique of measurement of impact loads presented in this chapter using pressure sensors was compared to the indirect technique presented in Chap. 3 based on combined pitting tests and nano-indentation tests. This showed that the estimates of impact loads obtained from both techniques are significantly different. This observation confirms the real difficulty of measuring the amplitude of impact loads accurately in cavitating flows which is however an essential parameter for the subsequent analysis of the response of the material to the cavitation pressure field.

Finally, it was shown that peak rate data are in good agreement with pitting rate data provided that a suitable threshold is applied to the amplitude of pressure peaks. This threshold increases only slightly with the flow velocity and depends primarily upon the material. It was shown that it is linearly correlated with the material yield stress.

References

1. Liu X, Katz J, Westerweel J, Rockwell D, Tropea C (2006) Instantaneous pressure and material acceleration measurements using a four-exposure PIV system. Exp Fluids 41(2):227–240
2. Knapp RT, Daily JW, Hammitt FG (1970) Cavitation. McGraw Hill Book Co., New York
3. Hammitt FG (1980) Cavitation and multiphase flow phenomena. McGraw-Hill International Book Co., New York
4. Leighton TG (1994) The acoustic bubble. Academic Press, London
5. Blake WK (1986) Mechanics of flow-induced sound and vibration. Academic Press, New York
6. Ross D (1976) Mechanics of underwater noise. Pergamon Press Inc., New York
7. Franc J-P, Michel J-M (2004) Fundamentals of cavitation. Fluid Mechanics and Its Applications, Springer, Boston
8. Illi'ichev VI, Lcsunovskii NP (1963) On the noise spectra associated with hydrodynamic cavitation. Sov Phys Acoust 9:25
9. Morozov VP (1969) Cavitation noise as a train pulse generated at random times. Sov Phys Acoust 14:140
10. De MK, Hammitt FG (1982) New method for monitoring and correlating cavitation noise to erosion capability. Trans ASME. J Fluids Eng 104(4):434–442
11. Hattori S, Sun B-H, Hammitt FG, Okada T (1985) An application of bubble collapse pulse height spectra to venturi cavitation erosion of 1100-0 aluminum. Wear 103(2):119–131. doi:10.1016/0043-1648(85)90128-0

12. Chahine GL, Courbière P, Garnaud P (1979) Correlation between noise and dynamics of cavitation bubbles. In: Kisbosckoi L, Szabo A (eds) Sixth international conference on fluid machinery. Budapest, Hungary, pp 200–210
13. Chahine GL, Courbière P (1986) Noise and erosion of self resonating cavitating jet. In: Saunders D (ed) Eighth jet cutting technology conference, BHRA. Durham, England, pp 167–176
14. Courbière P (1984) An acoustic method for characterizing the onset of cavitation. In: ASME international symposium on cavitation inception in nozzles and pumps, New Orleans, LA, USA, pp 137–145
15. Momma T, Lichtarowicz A (1995) A study of pressures and erosion produced by collapsing cavitation. Wear 186–187(2):425–436. doi:10.1016/0043-1648(95)07144-x
16. Okada T, Iwai Y, Hattori S, Tanimura N (1995) Relation between impact load and the damage produced by cavitation bubble collapse. Wear 184(2):231–239. doi:10.1016/0043-1648(94)06581-0
17. Soyama H, Lichtarowicz A, Momma T, Williams EJ (1998) A new calibration method for dynamically loaded transducers and its application to cavitation impact measurement. J Fluids Eng 120(4):712–718
18. Hattori S, Hirose T, Sugiyama K (2010) Prediction method for cavitation erosion based on measurement of bubble collapse impact loads. Wear 269(7–8):507–514. doi:10.1016/j.wear.2010.05.015
19. Franc J-P, Riondet M, Karimi A, Chahine GL (2011) Impact load measurements in an erosive cavitating flow. J Fluids Eng 133(12):121301–121308
20. Crow SC, Champagne FH (1971) Orderly structure in jet turbulence. J Fluid Mech 48(03):547–591. doi:10.1017/S0022112071001745
21. Weibull W (1951) A statistical distribution function of wide applicability. J Appl Mech-Trans ASME 18(3):293–297

Chapter 5
Mass Loss and Advanced Periods of Erosion

Georges L. Chahine, Jean-Pierre Franc and Ayat Karimi

Abstract This chapter is devoted to the advanced stages of erosion characterized by a mass loss curve of a sample exposed to cavitation as a function of exposure time. Depending upon materials, erosion devices, and operating conditions, different regimes of erosion may be identified on the mass loss curve including incubation, acceleration, deceleration, and steady-state periods. Typical mass (or volume) loss curves obtained for different materials using ultrasonic cavitation, cavitating jets, and a high-speed cavitation tunnel are discussed. They can be normalized by introducing a characteristic volume loss and a characteristic time, which are unique functions of the material and the cavitating field condition. By computing the ratio of characteristic volume loss and characteristic time, the characteristic erosion rate can be deduced for allowing material ranking. The ranking deduced from vibratory cavitation tests and from cavitating jet tests is generally in agreement. Some materials, however, do not rank the same way in a cavitating flow of relatively low aggressiveness (such as vibratory cavitation) as compared to the cavitating flow of higher aggressiveness (high speed cavitating jets).

G. L. Chahine (✉)
DYNAFLOW, INC., 10621-J Iron Bridge Road, Jessup, MD, USA
e-mail: glchahine@dynaflow-inc.com

J.-P. Franc (✉)
LEGI, Grenoble, France
e-mail: jean-pierre.franc@legi.grenoble-inp.fr

A. Karimi (✉)
EPFL, Lausanne, Switzerland
e-mail: ayat.karimi@epfl.ch

K.-H. Kim et al. (eds.), *Advanced Experimental and Numerical Techniques for Cavitation Erosion Prediction*, Fluid Mechanics and Its Applications 106,
DOI: 10.1007/978-94-017-8539-6_5,

5.1 Introduction

The initial stage of cavitation erosion is called the incubation period. During this period the material boundary exposed to the cavitation field undergoes plastic deformation, known as 'pits', without loss of material (see Chap. 3). Under these conditions the cavitation field is characterized by the statistics of its impulsive pressure signature (Chap. 4).

As the exposure time to the cavitation field increases or as cavitation intensity increases, erosion progresses to an advanced stage while the material boundary becomes more profoundly affected through material removal that could in turn influence and modify the pressure field through a modified boundary flow and secondary cavitation.

The advanced stage of cavitation erosion has extensively been studied in the past. Numerous publications and company reports are available on this subject on account of its practical importance to the life of fluid machinery and for design and performance prediction. Summaries of these studies can be found in cavitation review books [1–4], professional symposia and studies such as those by the American Society of Mechanical Engineers (ASME) [5, 6], and the American Society for Testing and Materials (ASTM) [7–9], and references such as [10–15].

5.2 Stages of Erosion Progression

From the extensive studies in the literature, it is commonly accepted that material erosion progresses through the following four successive stages:

- **Incubation period**: Initially the material surface gets deformed and modified microscopically without any loss of material. This is accompanied by work hardening of the surface. Cavitation peening techniques take advantage of this phase to render the material more resistant to stress (e.g. [16]). During this initial phase, permanent deformation may occur, sometimes accompanied by plastic flow and local displacement of the material micro particles, as well as the development of micro-cracks in the later stage. On a typical weight loss versus time curve (Fig. 5.1) this is the very short initial period where material loss cannot be measured. This period can be quite difficult to observe in some accelerated tests, but its duration can be very important to the determination of the life of material subjected to cavitation as some correlations have been presented in the literature between material incubation time and lifetime [17]. Following this period, the erosion process accelerates.
- **Acceleration period**: It is known that the weight or volume loss curve has typically an S shape, as illustrated in Fig. 5.2a, which shows an erosion acceleration phase during which the erosion rate increases until attaining a maximum value. This is illustrated in Fig. 5.2b, which shows the volume loss rate $\dot{V}(t)$. This erosion stage is called the accumulation or acceleration period.

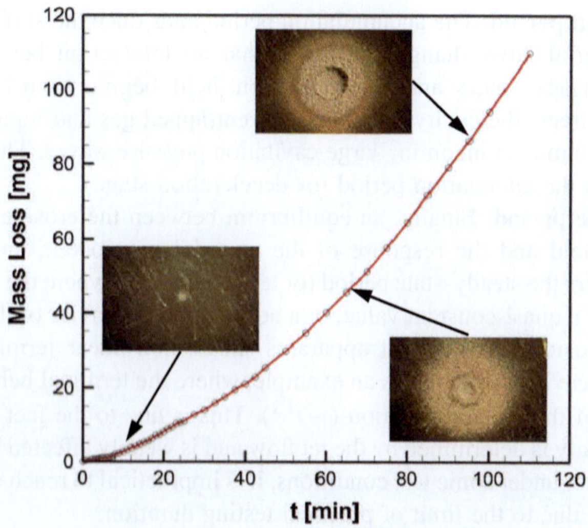

Fig. 5.1 Erosion progression curve for aluminum 1100-0 obtained using the CaviJet® cavitating jet erosion test at Dynaflow (1,100 psi (7.6 MPa), 0.083 in. (2.1 mm) orifice, 1.0 in. (25.4 mm) standoff). Inserts show a picture during the incubation period and two other pictures during the acceleration period. Notice the typical ring-like pattern by the cavitating jet erosion

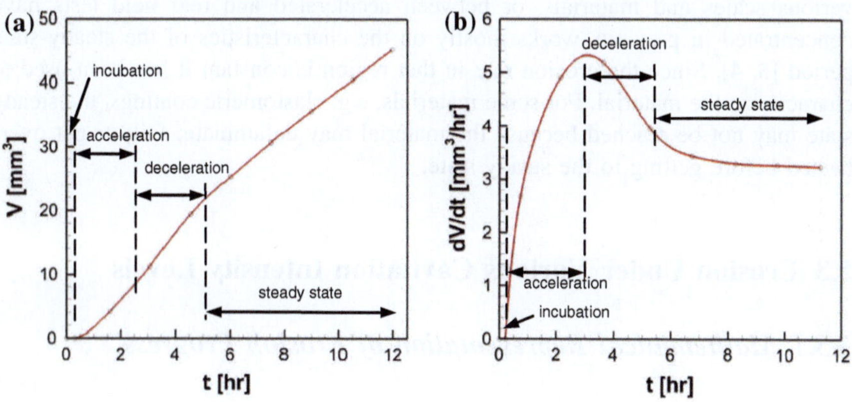

Fig. 5.2 Typical G32 test (20 kHz frequency, 50 μm peak-to-peak amplitude) erosion curves. **a** Volume loss versus time curve (S-curve), **b** rate of volume loss versus time curve. The inflection point in the volume curve and the corresponding maximum point in the volume loss rate characterize the particular combination of erosive field and material

In this phase, the material experiences increased fracture and weight loss following the end of hardening in the incubation period. The extent of this zone depends upon the strain-hardening properties of the material and involves microscopic chunks of material being removed following propagation of large cracks in between the grains of the material.

- **Deceleration period**: The accumulation period ends once the surface properties of the material have changed so much that an interaction between the new material surface shape and the cavitation field begins to occur. The new roughness affects the cavity dynamics and entrapped gas and liquid in the deep craters can damp the incoming large cavitation pressure waves. This time period is known as the attenuation period (or deceleration stage).
- **Steady-state period**: Finally, an equilibrium between the erosive power of the cavitation field and the response of the material may occur, and the erosion process enters the steady-state period (or terminal stage), where the rate of weight loss reaches a quasi-constant value, or a nearly linear behavior of the weight loss S-curve. Erosion in some test apparatus shows non-linear terminal behavior. DynaJets® cavitating jet test is an example, where the terminal behavior follows 1.2 power of the erosion duration ($\sim t^{1.2}$). This is due to the fact that the cavitation intensity is determined by the jet flow and is weakly affected by the surface erosion. Also, under some test conditions, it is impractical to reach or confirm the steady-state due to the limit of practical testing duration.

The clear separation of the four stages described above depends on the type of materials and the erosive cavitation field. For some materials and test conditions, the erosion curve does not always show all four stages. Since the first three stages of the cavitation erosion are quite unsteady and dynamic, correlations between various scales and materials, or between accelerated and real field tests have concentrated in previous works mostly on the characteristics of the steady-state period [3, 4]. Since the erosion rate in that region is constant it has been used to characterize the material. For some materials, e.g. elastomeric coatings, the steady state may not be reached because the material may delaminate, fail, or get overheated before getting to the steady state.

5.3 Erosion Under Various Cavitation Intensity Levels

5.3.1 Mathematical Representation of Erosion Progress

It would be very useful to define the erosion progress using a mathematical formula, especially if the formula includes cavitation intensity parameters such as flow velocity or cavitating jet velocity or pressure. One can then easily transpose experimental data from one operating condition to another. In general, the erosion progress is investigated by measuring the weight loss as a function of time. In this study, however, we use the volume loss in order to avoid the effect of density difference between the different materials tested. The erosion time history is presented here in terms of the volume loss, *V*, versus time, defined as:

$$V(t) = \frac{m(t)}{\rho}, \tag{5.1}$$

where $m(t)$ is the mass loss at time t and ρ is the eroded material density. The progression of the volume loss due to cavitation erosion can be expressed by the following equations based on numerous experimental data obtained at DYNAFLOW with different materials and flow conditions:

$$\begin{cases} V = 0, & for\ t' = t - t_{inc} \leq 0, \\ V = V_1\left(1 - a^{-t'^{n}}\right) + V_2\, t'^{\beta}, & for\ t' \geq 0, \end{cases} \tag{5.2}$$

where t_{inc} is the incubation time, V_1, V_2 are volume loss constants and a, n and β are time parameters which characterize the cavitation erosion progression. The first part of Eq. (5.2) indicates that there is no volume loss when t is less than t_{inc}.

Equation (5.2) can be normalized using a characteristic volume loss, V^*, and a characteristic time, t^* (both to be determined) as follows:

$$\overline{V} = \frac{V}{V^*}, \qquad \overline{t} = \frac{t - t_{inc}}{t^*}. \tag{5.3}$$

This leads to the following normalized volume equation:

$$\overline{V} = 1 - e^{-\overline{t}^{n}} + \alpha\, \overline{t}^{\beta}, \tag{5.4}$$

after setting $V^* = V_1, a = e^{t^{*-n}}$ and $\alpha = V_2 t^{*\beta}/V_1$.

The constants n, α, and β are empirical parameters that characterize the cavitation erosion. From various cavitating jets and ultrasonic cavitation tests database, these values were found to be [15]:

$$n = 2\,, \qquad \alpha = 1/e\,, \qquad \beta = 1.2\,. \tag{5.5}$$

The erosion volume loss (Eq. (5.4)) will then be given by the relatively simple expression:

$$\overline{V} = 1 - e^{-\overline{t}^{2}} + \frac{1}{e}\, \overline{t}^{1.2}. \tag{5.6}$$

The three parameters V^*, t^* and t_{inc} characterize the response of a specific material to a particular cavitation erosion field. Equation (5.6) gives $\overline{V} = 1$ at $\overline{t} = 1$, which means that at $t - t_{inc} = t^*$ we have $V = V^*$.

The first and the second derivatives of the normalized volume loss will then be:

$$\frac{d\overline{V}}{d\overline{t}} = 2\,\overline{t}e^{-\overline{t}^{2}} + \frac{1.2}{e}\,\overline{t}^{0.2}, \tag{5.7}$$

$$\frac{d^2\overline{V}}{d\overline{t}^2} = 2e^{-\overline{t}^{2}}\left(1 - 2\,\overline{t}^{2}\right) + \frac{0.24}{e}\,\overline{t}^{-0.8}. \tag{5.8}$$

These two functions, the velocity and acceleration rates of the volume loss, are shown in Fig. 5.3, which illustrates that the volume loss rate achieves its maximum when $t - t_{inc} \sim 0.75\, t^*$. This relationship provides the definition of the

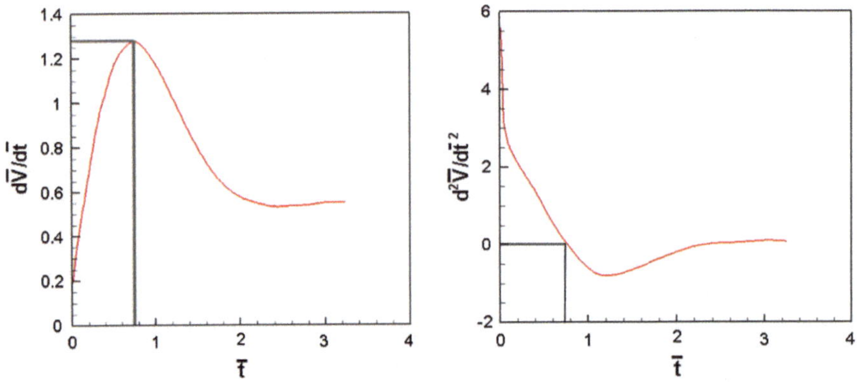

Fig. 5.3 Non-dimensional cavitation erosion volume rate, $d\overline{V}/d\overline{t}$, (*left*) and volume acceleration, $d^2\overline{V}/d\overline{t}^2$, (*right*) versus non-dimensional time, $\overline{t}$ (see Eqs. (5.7) and (5.8)). Maximum volume loss rate, $d\overline{V}/d\overline{t}\big|_{max}$, occurs when $t = t_{inc} + 0.75\,t^*$ providing thus a definition of t^*. Based on [15], reprinted with permission from Elsevier

characteristic time t^*, which is 4/3 the time at which the rate of the material volume loss attains its maximum.

5.3.2 Erosion by Ultrasonic Cavitation and Cavitating Jets

In order to investigate the effects of imparting the cavitation field energy to the material over a long period of time, a set of erosion tests were conducted at DYNAFLOW using cavitating jets at different jet pressures and with an ultrasonic horn. Figure 5.4 shows an example of the volume loss of aluminum alloy Al 7075-T651 subjected to the G32 ultrasonic cavitation field [9] and to a set of cavitating jets [18] with different pressures across the nozzle. The erosion curve marked with G73 is obtained from a liquid impingement erosion test following the ASTM G73 standard [9]. The figure illustrates that both the jets and the ultrasonic device result in volume loss curves with similar shapes. In this figure, G32 results fall a little below the 1,000 psi (6.9 MPa) cavitating jet results, while other tests, not presented here, indicate that G32 results are closer to 700–800 psi (4.8–5.5 MPa) cavitating jet results. The symbols are actual measured data and the solid lines are from Eq. (5.6).

Figure 5.5 shows the normalized volume loss using the same data in Fig. 5.4. The Eq. (5.6) (solid line) is in good agreement with the normalized experimental data. The specific test data cover different ranges of the erosion curve, but all test data fall on one standardized shape of erosion curve.

Figure 5.6 shows the corresponding normalized volume loss rate, $d\overline{V}/d\overline{t}$, curves, which also fall on one curve as expected.

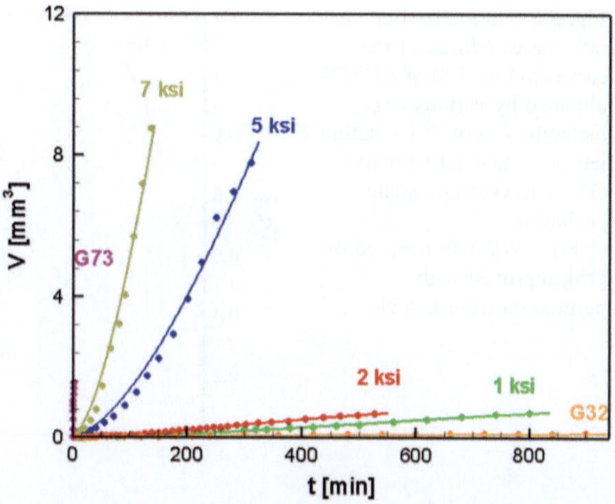

Fig. 5.4 Volume loss versus time of Al 7075 resulting from a CaviJet® nozzle cavitation at different nozzle pressures (different jet velocities), vibrating horn (G32), and liquid impingement (G73). The solid lines are analytical curves obtained from Eq. (5.6). (1 ksi = 6.9 MPa)

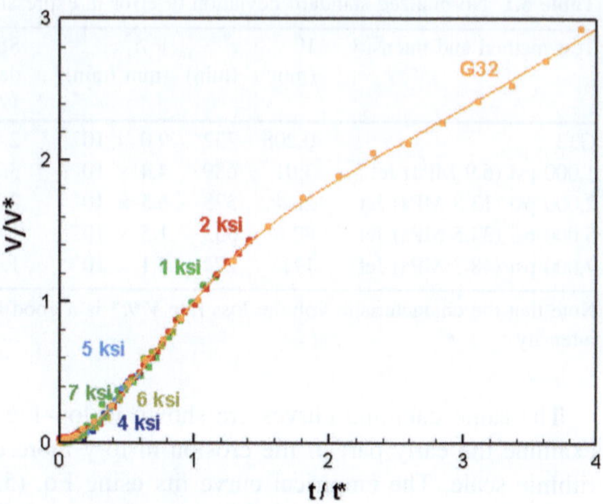

Fig. 5.5 Normalized volume loss versus normalized time on Al 7075 resulting from cavitation at different CaviJet® nozzle pressures (jet velocities) and from G32 ultrasonic cavitation. The solid line was obtained from Eq. (5.6). (1 ksi = 6.9 MPa). Based on [15], reprinted with permission from Elsevier

Table 5.1 shows the standard deviation of the errors when the measured data are expressed by Eq. (5.6). The standard deviation in the normalized volume loss is only 1–3 % for all cases. The table also includes the coefficient of determination which represents how well the prediction, f_i, describes the actual data, y_i:

$$R^2 = 1 - \frac{\sum (y_i - f_i)^2}{\sum (y_i - \bar{y})^2}. \tag{5.9}$$

The coefficients of determination in the table show that Eq. (5.6) is an almost perfect fit to the experimental erosion data with R^2 values larger than 0.988 for different jet pressures and for G32.

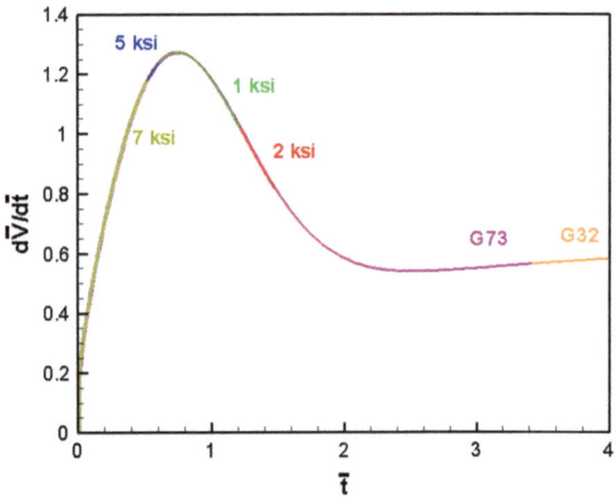

Fig. 5.6 Normalized erosion rate curves (slopes of the curves in Fig. 5.5) of Al 7075 obtained by various test methods: CaviJet® cavitating jets at various jet velocities, G73, and G32 ultrasonic cavitation. (1 ksi = 6.9 MPa). Based on [15], reprinted with permission from Elsevier

Table 5.1 Normalized standard deviation or error in expressing the erosion data with Eq. (5.6)

Test method and intensity	V^* (mm^3)	t^* (min)	V^*/t^* (mm^3/min)	Standard deviation in V/V^* (%)	Coefficient of determination R^2 (Eq. (5.9))
G32	0.208	232	9.0×10^{-4}	2.4	0.9985
1,000 psi (6.9 MPa) Jet	3.01	689	4.4×10^{-4}	3.2	0.9920
2,000 psi (13.8 MPa) Jet	2.44	373	6.5×10^{-3}	2.2	0.9972
5,000 psi (34.5 MPa) Jet	82.6	533	1.5×10^{-2}	1.4	0.9919
7,000 psi (48.3 MPa) Jet	111	274	4.1×10^{-2}	1.4	0.9882

Note that the characteristic volume loss rate V^*/t^* is a good indicator of the relative cavitation intensity

The same data and curves are shown in log–log scale in Fig. 5.7 in order to examine the early part of the erosion history more closely in an expanded logarithmic scale. The empirical curve fits using Eq. (5.6) are good from the acceleration stage to the terminal erosion stage, but do not capture well the incubation period where there is practically no volume loss. This is particularly obvious for the cavitating jet at 2,000 psi (13.8 MPa) and the G32 and G73 tests. In these tests, the time intervals where mass measurements were conducted during the testing were small enough to capture the incubation period.

5.3.3 *Effect of Jet Pressure on Erosion*

As the characteristic volume loss and characteristic time represent the erosion progression of a material, it is interesting to evaluate how these values change when the intensity of the cavitation field is modified, e.g. through jet pressure increase.

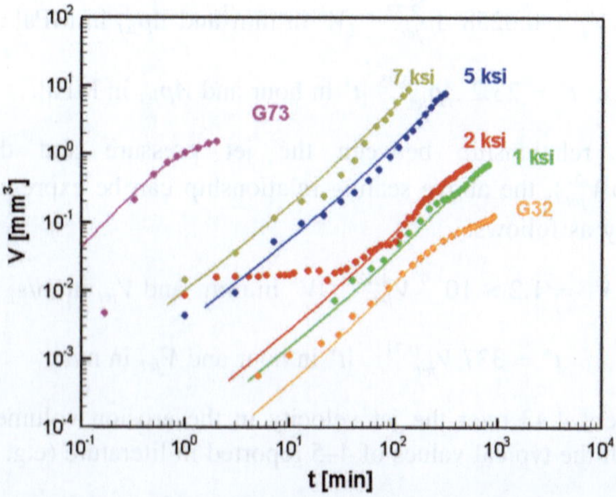

Fig. 5.7 Volume loss versus time on Al 7075 resulting from cavitation erosion with CaviJet® nozzle cavitation at different nozzle pressures (jet velocities) shown in log–log scales. The solid lines are obtained from Eq. (5.6). It is to be noted that the empirical formula, Eq. (5.6), does not fit well the data in the incubation period, particularly for the 2 ksi cavitating jet, G73, and G32. (1 ksi = 6.9 MPa). Based on [15], reprinted with permission from Elsevier

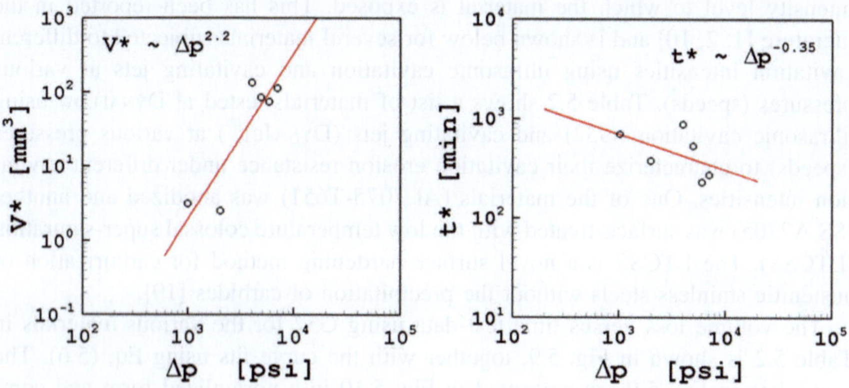

Fig. 5.8 Scaling of the characteristic volume loss, V^* (*left*) and the characteristic time t^* (*right*) with the cavitating jet nozzle pressure. (1000 psi = 6.9 MPa). Based on [15], reprinted with permission from Elsevier

For a given material, the characteristic erosion volume loss, V^*, and the characteristic time, t^*, are directly related to the pressure drop across the nozzle as illustrated in Fig. 5.8. The characteristic erosion volume increases and the characteristic erosion time decreases as the cavitation intensity (i.e. the jet pressure) increases. The following scaling relationship can be derived from the Al 7075 data set:

$$V^* = 0.0258\ \Delta p_{jet}^{2.22} \quad [V^* \text{ in mm}^3 \text{and } \Delta p_{jet} \text{ in MPa}]\ , \tag{5.10}$$

$$t^* = 23.2\ \Delta p_{jet}^{-0.35}\ [t^* \text{ in hour and } \Delta p_{jet} \text{ in MPa}]. \tag{5.11}$$

Using the relationship between the jet pressure and the jet velocity($\Delta p_{jet} = \frac{1}{2}\rho\ V_{jet}^2$), the above scaling relationship can be expressed in terms of the jet velocity as follows:

$$V^* = 1.2 \times 10^{-9}\ V_{jet}^{4.43} \quad [V^* \text{ in mm}^3 \text{ and } V_{jet} \text{ in m/s}]\ , \tag{5.12}$$

$$t^* = 337\ V_{jet}^{-0.71} \quad [t^* \text{ in hour and } V_{jet} \text{ in m/s}]. \tag{5.13}$$

The exponent 4.43 over the jet velocity in the erosion volume expression is consistent with the typical values of 4–5 reported in literature (e.g. [2, 3, 11, 12]).

5.4 Erosion Progression for Different Materials

Comparison of the erosion resistance of different materials should be done carefully. The resistance to cavitation of a material could depend on the cavitation intensity level to which the material is exposed. This has been reported in the literature [1, 2, 10] and is shown below for several materials subjected to different cavitation intensities using ultrasonic cavitation and cavitating jets at various pressures (speeds). Table 5.2 shows a list of materials tested at DynaFlow using ultrasonic cavitation (G32) and cavitating jets (DynaJet®) at various pressures (speeds) to characterize their cavitation erosion resistance under different cavitation intensities. One of the materials (Al 7075-T651) was anodized and another (SS A2205) was surface-treated with the low temperature colossal super-saturation (LTCSS). The LTCSS is a novel surface hardening method for carburization of austenitic stainless steels without the precipitation of carbides [19].

The volume loss versus time test data using G32 for the various materials in Table 5.2 is shown in Fig. 5.9, together with the curve fits using Eq. (5.6). The same data in Fig. 5.9 are presented in Fig. 5.10 in a normalized form and compared with the curve fits using the empirical formula, Eq. (5.6). As for Al 7075 erosion data using cavitating jet shown in Fig. 5.5, the normalized data collapsed nicely on the Eq. (5.6) curves. It is to be noted that various individual erosion data for different materials cover a different portion of a common S-shaped curve depending on the erosion resistance and the test duration. The erosion progression data for all materials fits well with Eq. (5.6), with the characteristic parameters V^* and t^* for each material shown in Table 5.3. These parameters are those obtained from the fits shown in Fig. 5.9.

Table 5.2 Materials tested at DYNAFLOW using ultrasonic cavitation and cavitating jets with varying nozzle pressure

Substrate	Coating
Aluminum 7075	None
Stainless steel 316	None
Stainless steel A2205	None
Nickel aluminum bronze (NAB)	None
High yield stress steel (HY-80)	None
Aluminum 1100-0	None
Aluminum 7075-T651	Anodized
Stainless steel A2205	LTCSS (Low temperature colossal super-saturation)

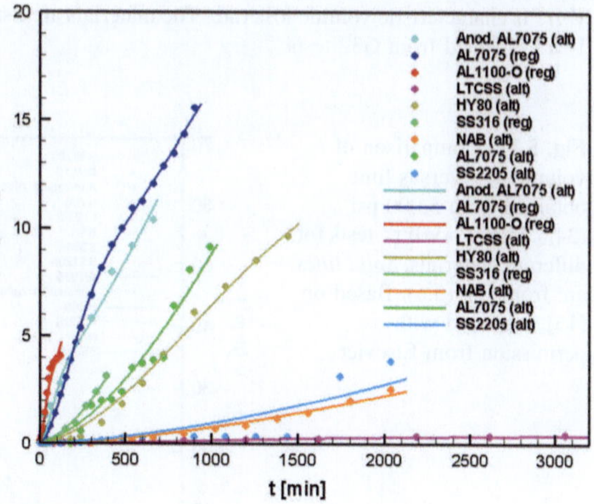

Fig. 5.9 Comparison of volume loss versus time obtained from G32 tests for different materials. *Solid lines* are from Eq. (5.6). In the labels (reg) represents the direct G32 method, (alt) represents the alternative G32 method (see Sect. 2.2). Based on [15], reprinted with permission from Elsevier

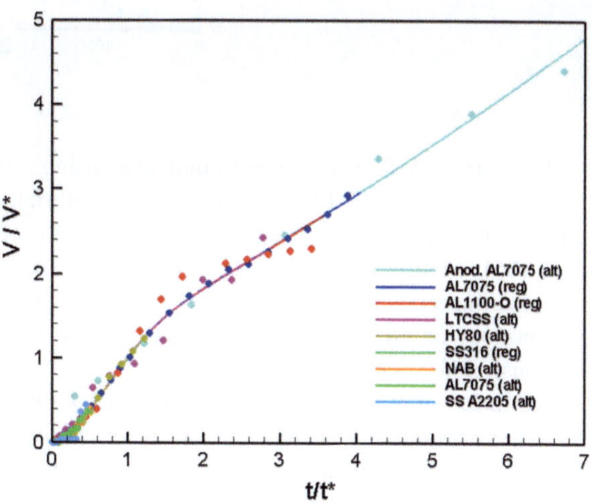

Fig. 5.10 Normalized volume loss as a function of normalized time. Symbols are measurements and *solid curves* are from Eq. (5.6). All curves start from the origin, but for the same physical time, the different materials reached different erosion stages

Table 5.3 Characteristic erosion parameters for the different materials tested with G32 ultrasonic horn and 5,000 psi (34.5 MPa) CaviJet® cavitating jet: V^* is the characteristic volume loss and t^* is the characteristic time

Material	Characteristic volume V^* (mm^3)		Characteristic time t^* (min)		Characteristic erosion rate V^*/t^* (mm^3/min)	
	G32	CaviJet	G32	CaviJet	G32	CaviJet
SS A2205 with LTCSS	0.007	20.3	1100	348	6.5×10^{-6}	0.058
NAB	0.252	21.8	4190	941	6.0×10^{-5}	0.023
SS A2205	0.351	72.0	4520	776	7.8×10^{-5}	0.093
HY-80	0.315	49.1	1180	223	2.7×10^{-4}	0.22
SS 316	0.315	90.7	771	804	4.1×10^{-4}	0.11
Al 7075	0.212	55.7	232	374	9.1×10^{-4}	0.15
Anodized Al 7075	0.0943	76.0	98	533	9.6×10^{-4}	0.14
Al 1100-0	0.0704	–	35	–	2.0×10^{-3}	–

V^*/t^* is characteristic volume loss rate. The materials are listed in the order of increasing value of V^*/t^* obtained from G32 tests

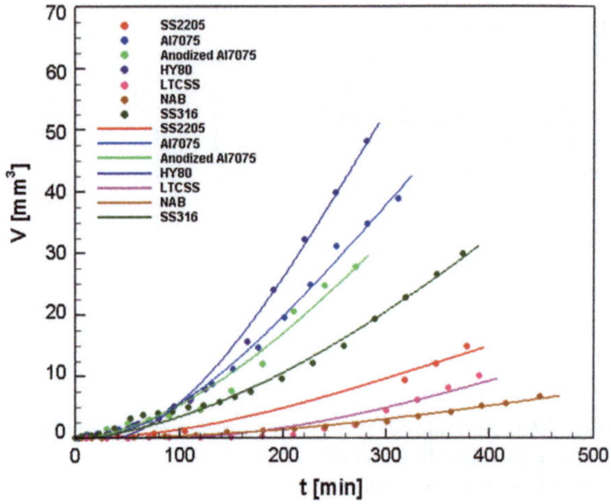

Fig. 5.11 Comparison of volume loss versus time obtained from 5,000 psi (34.5 MPa) CaviJet® tests for different materials. *Solid lines* are from Eq. (5.6). Based on [15], reprinted with permission from Elsevier

Similarly, erosion tests were conducted using cavitating jets with a 5,000 psi (34.5 MPa) CaviJet® for the same set of materials presented in Table 5.2. The cavitating jet results are shown in Fig. 5.11. Here again, all the data fit well with Eq. (5.6). Figure 5.12 shows the normalized erosion time history of the three selected materials (aluminum, nickel aluminum bronze, and stainless steel) under various cavitation intensities. A good collapse onto one normalized S-curve is demonstrated.

Table 5.3 shows the values for the characteristic volume, V^*, and time, t^*, used to fit the erosion data of the different materials tested by the G32 method and the 5,000 psi (34.5 MPa) cavitating jet. These are obtained by minimizing the standard deviation between the measured data and Eq. (5.6). In this table, the materials are

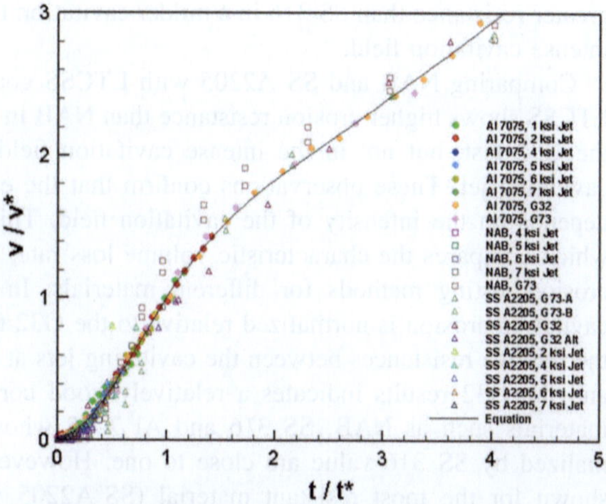

Fig. 5.12 Normalized volume loss versus normalized time for Al 7075, NAB and SS A2205 tested with G32, G73 and CaviJet® at various pressures. Symbols are measured data and *solid line* is from Eq. (5.6). (1 ksi = 6.9 MPa)

Table 5.4 Comparisons of the characteristic volume loss rate, V^*/t^*, for different materials and between the two test methods: G32 ultrasonic cavitation and 5,000 psi (34.5 MPa) CaviJet® cavitating jet tests

Material	Characteristic erosion rate V^*/t^* (mm³/min)		V^*/t^* for G32 tests normalized by the value for SS 316	Ratio of jet to G32 of V^*/t^* normalized by the value for SS 316
	G32	CaviJet®		
SS A2205 with LTCSS	6.5×10^{-6}	0.058	0.02	32.4
NAB	6.0×10^{-5}	0.023	0.1	1.4
SS A2205	7.8×10^{-5}	0.093	0.2	4.3
HY-80	2.7×10^{-4}	0.22	0.7	3.0
SS 316	4.1×10^{-4}	0.11	1.0	1.0
Al 7075	9.1×10^{-4}	0.15	2.2	0.6

The materials are listed in the order of increasing value of V^*/t^* obtained from G32 tests, i.e. by decreasing order of resistance to the G32 erosion field. The last column illustrates deviations between the cavitating jets at 5,000 psi (34.5 MPa) results and the G32 results since good correspondence would have given the value one for all the materials

listed in the order of increasing characteristic erosion rate obtained from G32 testing.

The rankings of erosion resistance in terms of the characteristic erosion rate (V^*/t^*) by the two test methods in Table 5.3 agree in general, but a few materials did not rank the same way. For example, HY-80 was more resistant than SS 316 in G32 tests, but the order is reversed in the 5,000 psi (34.5 MPa) cavitating jet tests. From the observed erosion rates of the two test methods, it is obvious that the cavitation field of a 5,000 psi (34.5 MPa) jet is much more intense and erosive than the cavitation field generated in G32 ultrasonic tests. This implies that HY-80 has greater resistance than SS 316 in a milder cavitation field but is less resistant in an intense cavitation field.

Comparing NAB and SS A2205 with LTCSS coating is also interesting. The LTCSS shows higher erosion resistance than NAB in the milder cavitation field of the G32 test, but not in the intense cavitation field of a 5,000 psi (34.5 MPa) cavitating jet. These observations confirm that the erosion response of materials depends on the intensity of the cavitation field. This is illustrated in Table 5.4, which compares the characteristic volume loss rate, V^*/t^*, for the two cavitation erosion testing methods for different materials. In this table the resistance to cavitation erosion is normalized relative to the G32 tests and SS 316. Comparing the relative resistances between the cavitating jets at 5,000 psi (34.5 MPa) results and the G32 results indicates a relatively good correspondence for the weaker materials such as NAB, SS 316 and Al 7075 whose erosion rates, V^*/t^*, normalized by SS 316 value are close to one. However, significant differences are shown for the most resistant material (SS A2205 with LTCSS) with the G32 results, implying a much stronger resistance to erosion than the 5,000 psi (34.5 MPa) cavitating jets. This is most probably due to the fact that the highest cavitation pressure pulses in the G32 ultrasonic field were not much larger than the material resistance limit, while the 5,000 psi (34.5 MPa) jet was generating much higher pressures than this resistance limit.

5.5 Erosion Profiles

5.5.1 Comparison of Erosion Profiles on Different Materials

Volume loss data presented in Sects. 5.3 and 5.4 were obtained by weighing the samples at different times and converting the mass loss to volume loss by Eq. (5.1). This procedure gives the overall volume loss caused by cavitation to the whole sample. However, the damage is usually far from being uniform on the sample surface. As an example, damage on samples eroded with a cavitating jet has a ring-like distribution centered on the jet axis as shown in Fig. 5.1.

In some cases, it may be useful to characterize the erosion damage more locally. This is the case if the cavitation intensity has also been determined locally from pitting tests or pressure pulse measurements and if a correlation with the local erosion rate is sought. A simple way to assess the local damage consists of using a conventional profilometer and measuring the erosion profile along a line crossing the eroded region or even scanning the whole 3D eroded surface. Such a procedure gives access to the local depth of erosion and then to the local erosion rate. By integration, a total mass loss consistent with the weighing technique can be obtained.

Typical erosion profiles obtained from LEGI cavitation tunnel are shown in Fig. 5.13. The samples were eroded in a cavitation tunnel equipped with a radially

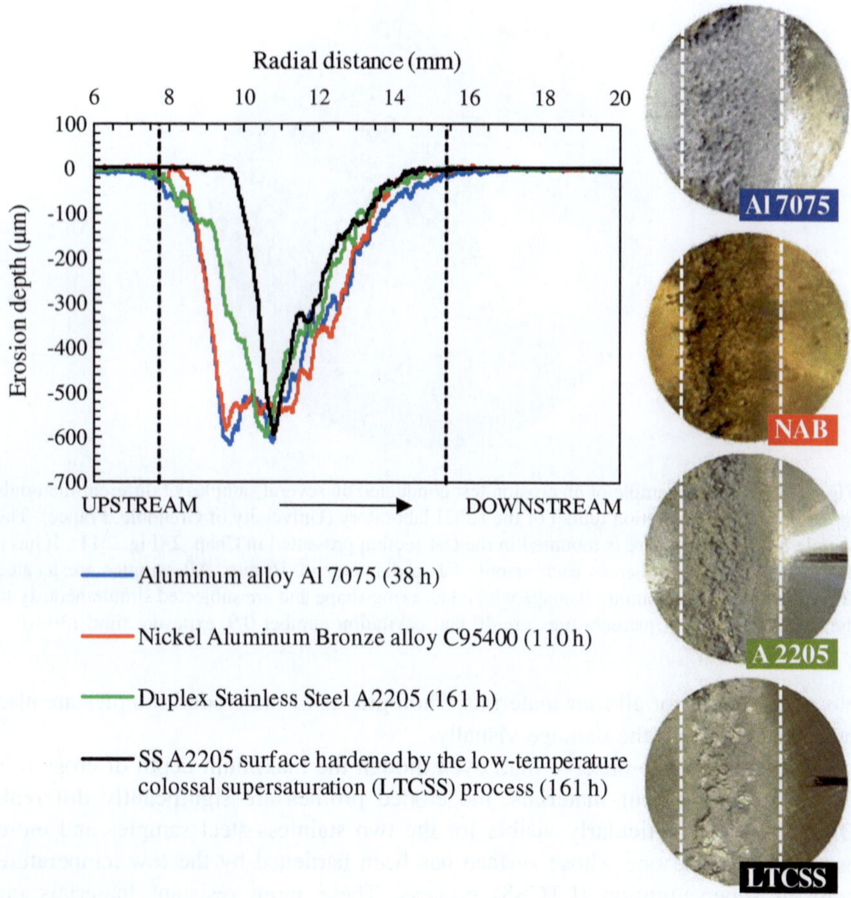

Fig. 5.13 Profiles measured by a contact profilometer on samples of four materials eroded in the high-speed cavitation tunnel of the LEGI laboratory (University of Grenoble, France). The test section is presented in Chap. 2. The upstream pressure was 40 bar (4 MPa) and the cavitation number 0.9. Diameter of samples is 16 mm. All samples were mounted on the same holder as shown in Fig. 5.14 and eroded simultaneously but with different exposure times. The *white dotted lines* on the samples correspond to the *black dotted lines* in the graph and indicate the approximate boundaries of the eroded domain. Erosion is maximum in the region of closure of the cavity and decreases upstream and downstream of the closure line

diverging test section (see Chap.2 for a detailed description of the facility and test section). Several samples of different materials were mounted all together on the same holder as shown in Fig. 5.14 and were eroded simultaneously. Four erosion profiles corresponding to four materials are superimposed in Fig. 5.13. Since the four materials have a different resistance to cavitation erosion, different exposure times are chosen in order to have the maximum erosion depth nearly the same and

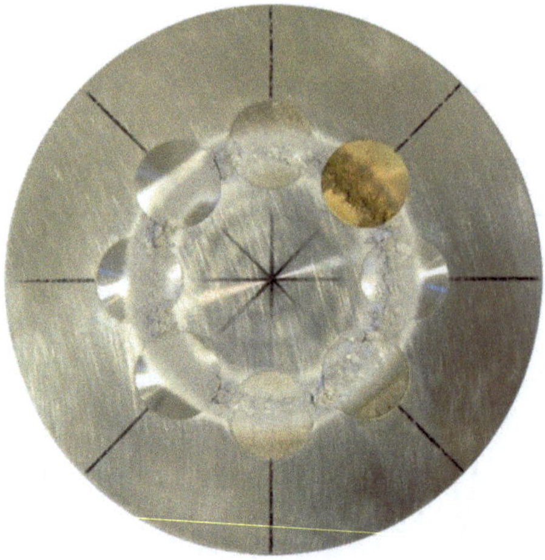

Fig. 5.14 Typical example of an erosion test conducted on several samples of different materials in the high-speed cavitation tunnel of the LEGI laboratory (University of Grenoble, France). The sample holder shown here is mounted in the test section presented in Chap. 2 (Fig. 2.11). It has a diameter of 100 mm whereas each sample has a diameter of 16 mm. All samples are located along the region of maximum damage which has a ring shape and are subjected simultaneously to the cavitation field. (Upstream pressure 40 bar, cavitation number 0.9, exposure time 130 h)

close to 600 μm for all four materials. Photographs of the eroded samples are also presented to assess the damage visually.

It is interesting to observe that, even though the maximum depth of erosion is the same for the four materials, the eroded profiles are significantly different. Differences are particularly visible for the two stainless steel samples and more especially for the one whose surface has been hardened by the low-temperature colossal supersaturation (LTCSS) process. These more resistant materials (as opposed to the Aluminum and NAB samples) have an erosion profile which is narrower. The difference is quite significant upstream of the point of maximum erosion but less significant downstream although noticeable.

Differences in erosion profiles reflect differences in the way the material responds to the local changes in cavitation intensity. Cavitation intensity is maximum in the closure region of the cavity (which corresponds to the region of maximum erosion depth) and decreases upstream and downstream of the cavity closure line. As explained in Sects. 5.3 and 5.4 on the basis of tests conducted with cavitation devices generating a different cavitation intensity (ultrasonic device and jets at various pressures), the resistance to cavitation of a material could depend on the cavitation intensity. In the present case, although the operating conditions are fixed, the cavitation intensity is obviously changing locally around cavity closure. It is thus not surprising that the eroded profiles are different from one material to another.

Moreover, it might be possible that some areas of the most resistant materials which were subjected to a relatively low cavitation intensity are still experiencing

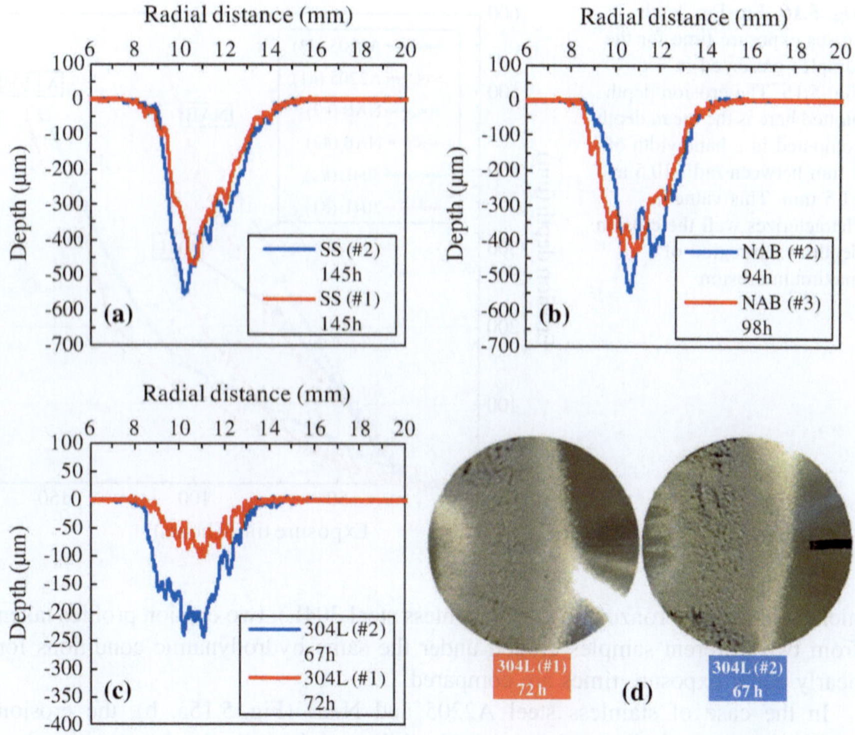

Fig. 5.15 Examples of reproducibility of mass loss tests. Comparison between erosion profiles obtained on two different samples of the same material at equivalent exposure times. **a** Stainless steel A2205. **b** Nickel aluminum bronze alloy. **c** Stainless steel 304L. **d** View of the 304L samples corresponding to the profiles presented in **c**. (Tests conducted in the high-speed cavitation tunnel of the LEGI laboratory, upstream pressure 40 bar, cavitation number 0.9)

the incubation period whereas areas subjected to the highest intensity have already reached the steady-state period. It could even be expected, by a material threshold-type effect, that some regions of the most resistant materials exposed to a cavitating flow of sufficiently small aggressiveness far from the area of maximum erosion are not damaged at all. This could explain qualitatively the reduction in the width of the eroded profiles for more resistant materials. A more quantitative analysis would require characterizing locally the cavitation intensity in terms of amplitude and rate of pressure pulses and analyzing locally the response of the material.

5.5.2 *Reproducibility of Mass Loss Tests*

Figure 5.15 shows three sets of erosion profiles illustrating the degree of reproducibility of mass loss tests. For three different materials (stainless steel A2205,

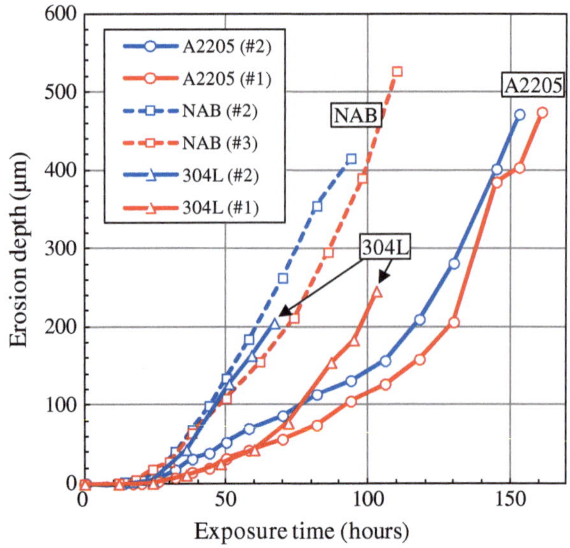

Fig. 5.16 Erosion depth versus exposure time for the samples presented in Fig. 5.15. The erosion depth plotted here is the mean depth computed in a bandwidth of 1 mm between radii 10.5 and 11.5 mm. This value characterizes well the erosion depth in the region of maximum erosion

nickel aluminum bronze alloy, and stainless steel 304L), two erosion profiles taken from two different samples eroded under the same hydrodynamic conditions for nearly equal exposure times are compared.

In the case of stainless steel A2205 and NAB (Fig. 5.15a, b), the erosion profiles are very similar and the reproducibility of the tests can be considered as satisfactory. On the other hand, for stainless steel 304L, Fig. 5.15c clearly shows a lack of reproducibility since the erosion profiles measured on two samples eroded under similar conditions are quite different. The photographs of the two samples presented in Fig. 5.15d confirm visually that the damage is significantly different.

Even though the erosion profiles for SS 304L are different for similar exposure times, the erosion rates during the steady-state period are quite similar. They are respectively 4.8 µm/h (between 51 h and 67 h) and 5.4 µm/h (between 72 and 103 h) for samples #2 and #1 with the relative error on the erosion rate being of the order of 10 %.

Figure 5.16 presents the evolution of the erosion depth with the exposure time for all the samples presented in Fig. 5.15, and confirms good reproducibility of mass loss tests for stainless steel A2205 and NAB since the two series of curves are quite close.

As for the stainless steel 304L, the final slopes (i.e. the terminal erosion rates) are comparable as already mentioned. However, the two curves are clearly shifted in time which suggests that the incubation time for both samples is significantly different. Since there is no difference in the hydrodynamic conditions, this tends to prove that the initial state of the material surface was probably different. It is very likely that the surface of sample #1 of 304L was initially harder than that of sample #2 because of a different degree of superficial hardening resulting from cutting and machining the samples.

In order to avoid the above issue, quality of the erosion samples must be well controlled. In particular, the samples need to be polished prior to exposure to cavitation in order to remove the thin superficial layers of hardened material if any and ensure that the tests are conducted on a virgin surface. It should be noted, however, that any initial material surface hardening has progressively less importance as the exposure time increases and will have no influence on the final erosion rate achieved during the steady-state period.

5.6 Relationship Between Erosion Rate and Incubation Time

A metallic material which offers a high resistance to cavitation erosion will generally show a long incubation time together with a small erosion rate. This trend is visible in Fig. 5.9 for G32 tests and in Fig. 5.11 for the cavitating jet. In the case of the cavitating jet for example, the relatively resistant material SS A2205 with LTCSS (in purple) exhibits a small eventual erosion rate together with a large incubation time whereas a less resistant material such as Al 7075 (in blue) has a large final erosion rate in conjunction with a small incubation time. A correlation between the erosion rate during the steady-state period and the inverse of the incubation time can then be expected [17, 20, 21]. If such a correlation is available, it gives a way to estimate the erosion rate after long exposure times from tests of smaller duration that provide an estimate of the only incubation time.

A relationship between the erosion rate in terms of the mean depth of penetration rate *MDPR* and the incubation time *T* such as:

$$MDPR = k\left(\frac{1}{T}\right)^{\alpha} \tag{5.14}$$

is often assumed with a value of the exponent α of the order of unity [20, 21]. The present section is devoted to a discussion of the relevance and accuracy of this type of relationship as well as of the general trends that can be drawn from such a law with respect to the cavitation intensity.

In the present section, the mass loss curves are fitted using a Weibull type function for the mean depth of penetration rate $\dot{h}$ (μm/h) versus the exposure time t:

$$\dot{h} = \dot{h}_{\max}\left[1 - e^{-\left(\frac{t}{T}\right)^2}\right]. \tag{5.15}$$

Equation (5.15) contains two fitting parameters which are the asymptotic erosion rate $\dot{h}_{\max}$ and a characteristic time T. A non-dimensional plot of this equation is shown in Fig. 5.17 together with the erosion curve obtained by integration of Eq. (5.15) which gives the non-dimensional depth of erosion, $h/(\dot{h}_{\max}T)$, as a function of the non-dimensional exposure time, t/T.

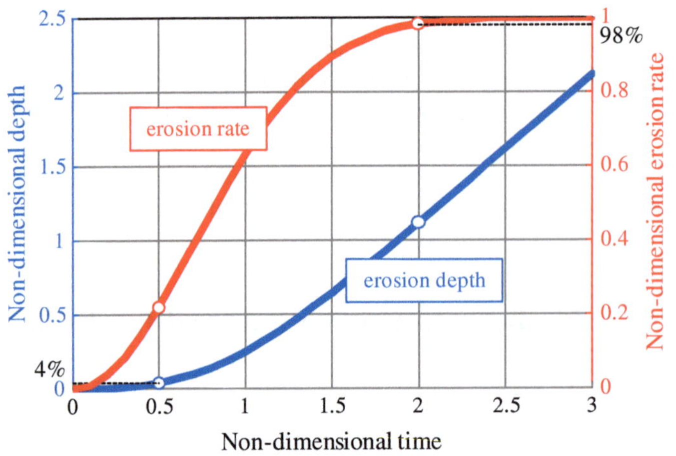

Fig. 5.17 Non-dimensional plot of the erosion rate (Eq. (5.15)) and associated mass loss curve. The *horizontal axis* is the non-dimensional exposure time *t/T*. The *right vertical axis* (in *red*) is the ratio of the instantaneous erosion rate to the maximum erosion rate $\dot{h}_{max}$ during the steady-state regime. The *left vertical axis* (in *blue*) is the instantaneous erosion depth made non-dimensional using the reference depth $\dot{h}_{max}\ T$

Figure 5.17 shows that the unique time scale T considered in this model controls both the incubation time and the acceleration time, the latter being defined as the end of the acceleration period when the erosion rate reaches its maximum value. As an example, after the exposure time 0.5 T (which can be considered as a measure of the incubation time) the non-dimensional erosion depth is only 4 %, and after 2 T (a possible measure of the acceleration time), the erosion rate is already 98 % of its maximum value. Whatever may be the exact threshold values used to define the incubation and the acceleration times, it is clear that both are proportional to the characteristic time T.

According to this model, any mass loss curve is defined by the two parameters $\dot{h}_{max}$ and T. These two parameters were determined for several materials and three different erosion devices, namely the alternative G32 method (or stationary specimen method), a cavitating jet at 5,000 psi (34.5 MPa) and a cavitation tunnel at 40 bar upstream pressure. More details on the cavitation devices are available in Chap. 2. In a diagram $(\dot{h}_{max},\ T)$ as shown in Fig. 5.18, each material is then represented by a point.

It is interesting to observe that all the materials are clustered in different regions of the diagram according to the testing device. Moreover, each cluster associated with a given apparatus is roughly organized around a straight line defined by an equation similar to Eq. (5.14) where we assumed $\alpha = 1$:

$$\dot{h}_{max} \approx \frac{k}{T}. \tag{5.16}$$

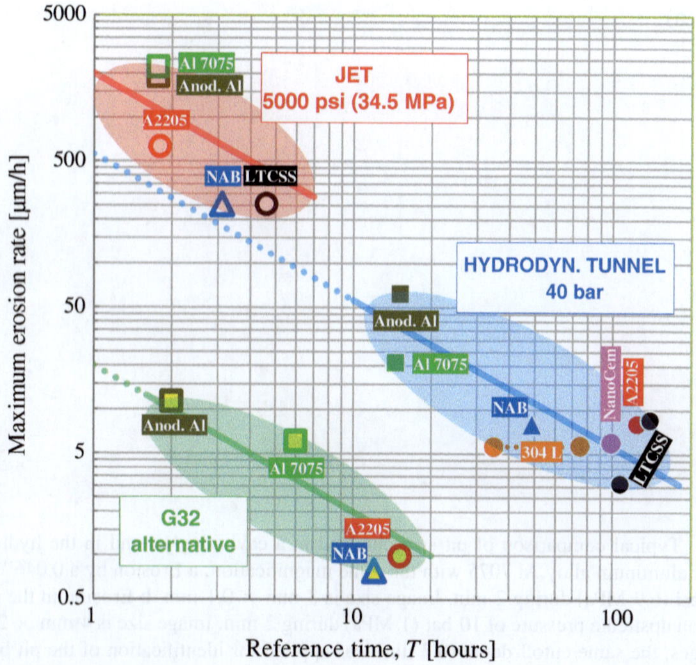

Fig. 5.18 Summarized presentation of mass loss test results in a two-dimensional diagram. The *horizontal axis* is the reference time introduced in Eq. (5.15) which is proportional to the incubation time and the acceleration time. The *vertical axis* is the maximum erosion rate during the steady-state period or mean depth of penetration rate (MDPR). The results obtained using three different erosion devices are compared: G32 alternative method (stationary specimen), high-speed cavitation tunnel at 40 bar (4 MPa) and cavitating jet at 5,000 psi (34.5 MPa). The facilities are presented in detail in Chap. 2. Tested materials are anodized aluminum, aluminum alloy Al 7075, nickel aluminum bronze alloy (NAB), stainless steels 304L and A2205 and stainless steel A2205 surface hardened by the LTCSS process. The *straight lines* are defined by an equation of type 5.16. The value of constant k depends upon the testing device and corresponds to the intercept of the lines with the *vertical axis*

This suggests that the constant k in Eqs. (5.16) and (5.14) is strongly dependent upon the testing facility but less dependent upon the material.

More precisely, the erosion model developed in Chap. 7 shows that the erosion rate during the steady-state period is given by:

$$\dot{h}_{\max} = \frac{L}{\tau}\left\{\left(\frac{\sigma_m - \sigma_0}{\sigma_R - \sigma_0}\right)^{1/n\theta} - 1\right\}. \tag{5.17}$$

In this model, the cavitation intensity is characterized by two parameters which are the mean amplitude of impact loads σ_m (in MPa) and the coverage time τ. The latter accounts for both the rate of impact loads and their characteristic area and measures the time required for the sample surface to be covered by the erosive

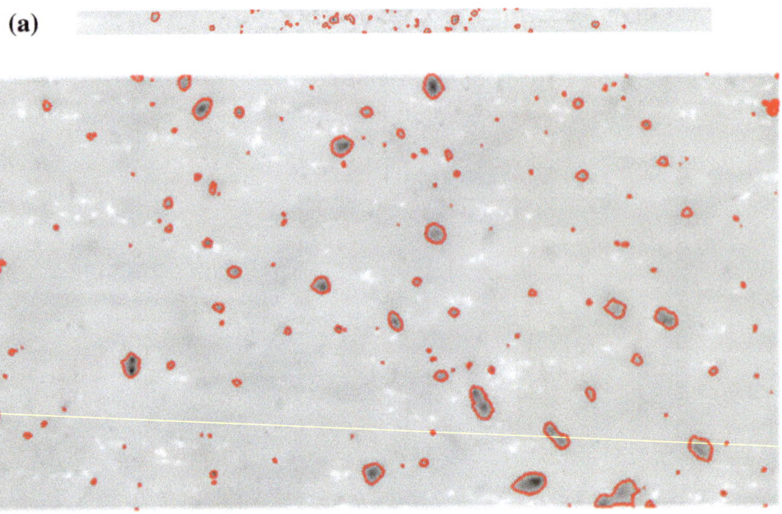

Fig. 5.19 Typical comparison of pit size obtained in a cavitating jet and in the hydrodynamic tunnel for aluminum alloy Al 7075 with the same magnification. **a** Erosion by a 0.046" CaviJet® at 1,000 psi (6.9 MPa) during 2 min. Image size is 3 mm × 0.1 mm. **b** Erosion in the cavitation tunnel at an upstream pressure of 10 bar (1 MPa) during 2 min. Image size is 4 mm × 2 mm. For both images, the same cutoff depth (0.3 μm) was applied for identification of the pit boundaries (plotted in *red*)

impact loads. The other parameters refer to the material and include yield stress σ_0, ultimate strength σ_R, strain hardening exponent n, thickness of hardened layer L and shape factor θ.

Regarding the incubation time, the model predicts that it is proportional to the coverage time τ (see Chap. 7). As a consequence, the coverage time is expected to vanish in the product of the erosion rate by the incubation time, i.e. in constant k. In other words, whereas the erosion rate and the incubation time both depend upon the two parameters σ_m and τ used here to characterize the cavitation intensity, the product would depend upon the only amplitude of impact loads σ_m but not upon the coverage time. This reasoning leads to the conclusion that the product of the erosion rate by the incubation time could be used as an indicator of the cavitation level in terms of the amplitude of impact loads σ_m.

If we refer again to Fig. 5.18, it appears that the value of constant k is actually the intercept of each straight line with the vertical axis. It is then conjectured that the cavitating jet at 5,000 psi (34.5 MPa) is two orders of magnitude more aggressive than the ultrasonic G32 device, whereas the cavitation tunnel at 40 bar is only four times less aggressive than the 5,000 psi (34.5 MPa) cavitating jet. Once more, this applies to the amplitude of impact loads.

As for the coverage time, it can be concluded from Fig. 5.18 that the coverage time for the jet cavitation is much smaller than that for the cavitation tunnel. Since cavitation impacts are significantly smaller in the case of the jet in comparison to the cavitation tunnel because of a smaller overall scale of the flow (Fig. 5.19), it is concluded that the frequency of collapses is much larger. This is consistent with the much larger flow velocity reached in the jet cavitation as compared to the cavitation tunnel.

When comparing the jet cavitation to the vibratory cavitation, it was already noticed that the amplitude of impact loads is several orders of magnitude larger in the cavitating jet. On the other hand, the incubation time for the jet is only slightly smaller. It is then expected that the coverage time in the vibratory device is not so different from that in the cavitating jet.

To conclude, the analysis of mass loss curves in terms of the erosion rate and the incubation time can provide useful information on the hydrodynamic aggressiveness of a cavitating flow in terms of impact load amplitude and coverage time, the latter combining the effects of collapse rate and impact size.

5.7 Summary

The present chapter was devoted to the advanced stages of erosion associated with a measurable loss of material as opposed to the incubation period when the material is subjected only to plastic deformation and pitting. The advanced stages of erosion are commonly characterized by a mass loss (or volume loss) curve that represents the evolution of mass loss of a sample exposed to the cavitating flow as a function of the exposure time. The derivative of this curve is the erosion rate curve.

Several types of erosion rate versus exposure time curves have been reported in the literature depending on tested materials, cavitation erosion devices and operating conditions. The erosion rate curve allows the identification of different regimes of erosion. The incubation period is characterized by a zero erosion rate. It is followed by an acceleration period in which the erosion rate progressively increases. For very large exposure times, a steady state period is generally observed which is characterized by an almost constant erosion rate. The steady-state period may directly follow the acceleration period or a deceleration period may be observed between the acceleration and the steady-state periods.

In the present chapter, we discussed in detail a typical mathematical form of the erosion rate curve based on erosion tests of a set of metallic materials using two cavitation erosion accelerated methods: ultrasonic cavitation following the ASTM G32 method and cavitating jets at various jet speeds. Weight losses versus time were measured at selected time intervals and the results were converted to volume removals for comparison.

All volume loss versus time data was shown to follow a simple law combining a Gaussian and a power law, $\overline{V} = 1 - \mathrm{e}^{-\overline{t}^2} + e^{-1}\overline{t}^{1.2}$. This law includes an acceleration period, a deceleration period and an almost steady state period actually characterized by a very slow increase of the erosion rate with exponent 0.2 of the exposure time which should obviously not be extrapolated beyond the maximum investigated exposure time. In this simple expression, $\overline{V}$ is V normalized by a characteristic volume, V^*, and $\bar{t}$ is time normalized by a characteristic time t^*. The parameters V^* and t^* characterize the response of the specific material to the particular cavitation erosion field.

Values of these parameters for various combinations of materials and cavitation intensities were presented. V^*/t^* characterizes the erosion rate well. For the cavitating jet method and the same material, V^* increases with the jet speed as $V_{jet}^{4.4}$ and t^* decreases like $V_{jet}^{-0.7}$. The overall ranking of erosion resistance of different materials tested in this study showed a general agreement between the two erosion methods; G32 and cavitating jet. The relative ranking for some materials, however, are different between the two methods. This implies that material response to cavitation for some materials depends upon the cavitation method and intensity.

Volume loss is obviously not uniform over the whole eroded area. Experiments conducted in a cavitation tunnel on erosion by an attached cavity have shown that the depth of penetration is maximum in the closure region of the cavity and progressively decreases upstream and downstream of the cavity closure. The erosion profiles in the vicinity of cavity closure do not have, however, a similar shape for different materials tested under the same operating conditions. This is due to a variable cavitation intensity around cavity closure and a differential response of the materials to the varying cavitation intensity.

Volume loss data in the hydrodynamic tunnel may be affected by a non-negligible uncertainty. The reproducibility of erosion tests may be improved by a careful preparation of the samples. If possible, it is recommended to polish the sample surface prior to exposure to cavitation in order to remove the thin superficial hardened layers that might result from cutting and machining the samples.

In this chapter, it was also suggested that the erosion rate during the steady-state period is generally correlated with the duration of the incubation period. The longer the incubation period, the smaller the steady-state erosion rate. In particular, erosion tests conducted on various materials using three different erosion devices (a vibratory device, a cavitating jet and a cavitation tunnel) tend to prove that the product of the steady-state erosion rate by the incubation period does not depend much upon the material but depends primarily upon the device. This kind of correlation makes it possible to predict, to a certain extent, the long term erosion behavior from short term erosion tests.

References

1. Knapp RT, Daily JW, Hammitt FG (1970) Cavitation. McGraw Hill Book Co., New York
2. Hammitt FG (1980) Cavitation and multiphase flow phenomena. McGraw-Hill International Book Co., New York
3. Thiruvengadam A (1974) Handbook of cavitation erosion. Hydronautics, Laurel

4. Eisenberg P, Preiser HS, Thiruvengadam A (1965) On the mechanisms of cavitation damage and methods of protection. Trans Soc Naval Architects Mar Eng 73:241–286
5. Arndt R, Billet M, Blake W (eds) (1993) ASME symposium on bubble noise and cavitation erosion in fluid systems, FED-Vol. 176. New Orleans
6. Rohatgi U, Nagafuji T (eds) (1991) Cavitation erosion power symposium, 1st ASME-JSME fluid engineering conference, Portland, 23–27 June 1991
7. Peterson R (ed) (1962) Symposium on Erosion and Cavitation, ASTM International STP 307. Philadelphia, Pennsylvania (also (1961) 64th annual meeting, ASTM. Atlantic City)
8. Peterson R (ed) (1961) Symposium on Erosion and Cavitation, 64th annual meeting, ASTM. Atlantic City
9. Annual Book of ASTM Standards (2010) Section 3: metals test methods and analytical procedures, vol. 03.02. West Conshohocken
10. Pereira F, Avellan F, Dupont P (1998) Prediction of cavitation erosion: an energy approach. J Fluids Eng 120(4):719–727
11. March PA (1987) Evaluating the relative resistance of materials to cavitation erosion: a comparison of cavitating jet results and vibratory results. Paper presented at the ASME cavitation and multiphase flow forum, Cincinnati, 14–17 June 1987
12. Hattori S, Takinami M, Otani T (2009) Comparison of cavitation erosion rate with liquid impingement erosion rate. Paper presented at the 7th international symposium on cavitation, Ann Arbor, 17–22 Aug 2009
13. Hammitt FG, Chao C, Kling CL, Mitchell TM, Rogers DO (1970) Round-Robin Test with Vibratory Cavitation and Liquid Impact Facilities of 6061–T6511 Aluminum Alloy, 316 Stainless Steel and Commercially Pure Nickel. Materials Research and Standards (ASTM) 10:16–36
14. Chao C, Hammitt FG, Kling CL (1968) ASTM round-robin test with vibratory cavitation and liquid impact facilities of 6061-T6 aluminum alloy, 316 stainless steel, commercially pure nickel, vol 84. The University of Michigan Report MMPP-344-3-T/01357-4-T, Ann Arbor
15. Choi J-K, Jayaprakash A, Chahine GL (2012) Scaling of cavitation erosion progression with cavitation intensity and cavitation source. Wear 278–279:53–61. doi:10.1016/j.wear.2012.01.008
16. Odhiambo D, Soyama H (2003) Cavitation shotless peening for improvement of fatigue strength of carbonized steel. Int J Fatigue 25(9–11):1217–1222. doi:10.1016/s0142-1123(03)00121-x
17. Soyama H, Futakawa M (2004) Estimation of incubation time of cavitation erosion for various cavitating conditions. Tribol Lett 17(1):27–30
18. Chahine GL, Courbière P (1987) Noise and erosion of self-resonating cavitating jets. J Fluids Eng 109(4):429–435
19. Martin F, Lemieux E, Newbauer T, Bayles R, Natishan P, Khan H, Michal G, Ernst F, Heuer A (2007) Localized corrosion resistance of LTCSS-carburized materials to seawater immersion. ECS Trans 3(31):613–621
20. Zhou YK, Hammitt FG (1983) Cavitation erosion incubation period. Wear 86(2):299–313
21. Hammitt FG (1979) Cavitation erosion: the state of the art and predicting capability. Appl Mech Rev 32(6):665–675

Chapter 6
Modeling of Cavitation Dynamics and Interaction with Material

Georges L. Chahine

Abstract This chapter presents a numerical model applied to the collapse of a single cavitation bubble near a rigid or deformable plane material boundary. A hybrid incompressible-compressible method is presented, which accounts for the development of a re-entrant jet as well as the emission of shock waves during the collapse or following liquid–liquid or liquid–solid impact. Wall pressures computed on the bubble axis present a first peak due to the impact of the bubble re-entrant jet followed by a second peak due to the shock wave resulting from the residual toroidal bubble collapse. For small standoff distances, both pressure peaks have comparable amplitudes. The re-entrant jet impact pressure is proportional to the liquid impedance and the jet speed and follows the classical water-hammer equation. The effects of the collapse driving pressure, standoff distance, and bubble size are discussed. The permanent plastic deformation (pit) resulting from the bubble collapse is also computed by using a two-way coupled fluid–structure interaction model. The process of pit formation is described and the influence of various parameters (such as bubble size, collapse driving pressure and load duration) on pit geometry is discussed. The analysis shows that the material experiences stresses that are much lower than the fluid generated impulsive loads and that the resulting pit characteristics depend not only on the impulsive load amplitude but also on its duration and spatial extent.

6.1 Introduction

Previous chapters focused for the most part on presenting and interpreting empirical data obtained from experimental investigations of various aspects of cavitation erosion. The efforts involved characterization of both the cavitation field

G. L. Chahine (✉)
DYNAFLOW, INC., 10621-J Iron Bridge Road, Jessup, MD, USA
e-mail: glchahine@dynaflow-inc.com

K.-H. Kim et al. (eds.), *Advanced Experimental and Numerical Techniques for Cavitation Erosion Prediction*, Fluid Mechanics and Its Applications 106,
DOI: 10.1007/978-94-017-8539-6_6,

and the resulting material permanent deformation and erosion. Interpretation of the results often relied on physical reasoning based on prior knowledge of the cavitating flow configurations studied, the basic dynamics of bubbles, and the response of materials to applied loads. The characterization of cavitation impulsive pressures relied on the response to the impulsive loads of pressure transducers or on the measurement of the permanent deformation of the material itself.

This chapter presents numerical modeling and simulation of the non-spherical dynamics of bubbles near rigid and deformable boundaries with known material elastic–plastic properties and examines the response of the material to the loads generated by bubble collapse. The study aims at isolating the influence of each of the various physical parameters (i.e., bubble size, distance from the wall, collapse driving pressure, and material properties) on the spatial and temporal distribution of impulsive pressures resulting from the bubble dynamics, and also on the material elasto-plastic deformations in response to these loads. Knowledge of the inter-relationships between these parameters allows for the transposition of the results from one condition to another.

The relevant numerical techniques are first described succinctly to provide the assumptions used and their limits of validity for the study of cavitation erosion mechanics, followed by the dynamics of a bubble near a rigid boundary. Scaling rules of the various characteristics of the bubble re-entrant jet parameters and the pressure pulse resulting from the collapse are then analyzed. The unsteady effects of generated impulse loads on the material deformation are subsequently discussed. The time history of material deformation is analyzed and implications for the inverse determination of the loads from permanent deformation in pitting tests are discussed.

6.2 Numerical Models

Cavitating flow field, bubble dynamics, and dynamic material response are simulated numerically in this chapter using several numerical techniques briefly described below. These include:

- a boundary element method potential flow model for 3D bubble dynamics, 3DynaFS-Bem©[1–7],
- a finite difference method compressible Euler flow solver, Gemini [8–10], for bubble dynamics,
- a procedure to link incompressible-compressible flow approaches [11, 12], and
- a finite element structural dynamic code, Dyna3D [13], to model material and structural response.

6.2.1 Flow Field Models

6.2.1.1 Boundary Element Model

3DYNAFS-BEM© is a potential flow solver based on a boundary element method (BEM). The code solves the Laplace equation, $\nabla^2\phi = 0$, for the velocity potential, ϕ, with the velocity vector $\mathbf{u} = \nabla\phi$. A boundary integral method is used to solve the Laplace equation based on Green's theorem:

$$\int_{\Omega} (\phi\, \nabla^2 G - G\, \nabla^2 \phi)\, d\Omega = \int_S n \cdot [\phi\, \nabla G - G\, \nabla \phi]\, dS. \qquad (6.1)$$

In this expression Ω is the liquid domain of integration having elementary volume $d\Omega$. The boundary surface of Ω is S, which includes the surfaces of the bubble and the nearby boundaries with an elementary surface dS. $\mathbf{n}$ is the local surface normal unit vector. $G = -1/|\mathbf{x} - \mathbf{y}|$ is Green's function, where $\mathbf{x}$ is a fixed point in Ω and $\mathbf{y}$ is a point on the boundary surface S. Equation (6.1) reduces to Green's formula with $a\pi$ being the solid angle under which $\mathbf{x}$ sees the domain Ω:

$$a\pi\phi(\mathbf{x}) = \int_S \left[\phi(\mathbf{y}) \frac{\partial G}{\partial n}(\mathbf{x}, \mathbf{y}) - G(\mathbf{x}, \mathbf{y}) \frac{\partial \phi}{\partial n}(\mathbf{y}) \right] dS, \qquad (6.2)$$

with $a = 4$ if $\mathbf{x}$ is a point in the fluid, and $a = 2$ if $\mathbf{x}$ is a point on a smooth surface.

Equation (6.2) provides a relationship between ϕ and $\partial\phi/\partial n$ at the boundary surface S. Thus, if either of these two variables (e.g., ϕ) is known everywhere on the surface the other variable (e.g., $\partial\phi/\partial n$) can be obtained. To solve Eq. (6.2) numerically, the BEM, which discretizes the surface of all objects in the computational domain into panel elements, is applied.

On a bubble surface S_b, ϕ satisfies a kinematic boundary condition, which expresses equality between the normal velocities of the liquid and the free surface:

$$d\mathbf{x}/dt \cdot \mathbf{n} = \nabla\phi \cdot \mathbf{n} \quad \mathbf{x} \in S_b, \qquad (6.3)$$

and a dynamic boundary condition:

$$\frac{\partial \phi}{\partial t} = -\frac{1}{2}|\nabla\phi|^2 + \frac{P_{amb} - P_b}{\rho} - gz|_{x \in S_b}, \qquad (6.4)$$

where P_{amb} is the ambient pressure, $P_b(\mathbf{x},t)$ is the local pressure in the liquid at the bubble interface, and z is the vertical coordinate at point $\mathbf{x}$.

The pressure inside the bubble is assumed to be spatially homogeneous, and the bubble content is assumed to be composed of non-condensable gas, which follows a polytropic expansion/compression law with constant k. This leads to the following form of the normal stress boundary condition at the bubble interface:

$$P_b(\mathbf{x},t) = P_{go}\left[\frac{\mathcal{V}_o}{\mathcal{V}(t)}\right]^k + P_v - \gamma C, \tag{6.5}$$

where P_{go} is the initial pressure of the non-condensable gas and P_v is the water vapor pressure. $\mathcal{V}_o$ is the initial volume of the bubble, γ is the surface tension coefficient, and $C(\mathbf{x},t)$ is twice the local mean curvature at $\mathbf{x}$ given by:

$$C = \nabla \cdot \mathbf{n}, \qquad \mathbf{n} = \pm\frac{\nabla f}{|\nabla f|}, \tag{6.6}$$

where f is the equation of the surface. The appropriate sign is chosen so that the normal always points towards the liquid.

3DynaFS-Bem© was developed specifically to solve three-dimensional potential flow problems including highly non-linear free surface dynamics, such as encountered in bubble dynamics, ocean dynamics, and time dependent interfaces separating two or more fluid phases. It handles the dynamics of cavitation bubbles or explosion bubbles near solid boundaries, deformable cells, or free surfaces [1–7, 11, 12, 14, 15]. 3DynaFS-Bem© can be coupled with a structural code such as Dyna3D (see Sect. 6.2.2) for simulation of fluid–structure interaction problems [10, 13].

6.2.1.2 Compressible Flow Model

Gemini is a compressible Euler equation solver developed by the Naval Surface Warfare Center, Indian Head division, based on a finite difference scheme. The code solves continuity and momentum equations for a compressible inviscid liquid in Cartesian coordinates, which can be written in the following format:

$$\frac{\partial Q}{\partial t} + \frac{\partial E}{\partial x} + \frac{\partial F}{\partial y} + \frac{\partial G}{\partial z} = S, \tag{6.7}$$

$$Q = \begin{bmatrix} \rho \\ \rho u \\ \rho v \\ \rho w \\ \rho e_t \end{bmatrix} \quad E = \begin{bmatrix} \rho u \\ \rho u^2 + p \\ \rho uv \\ \rho uw \\ (\rho e_t + p)u \end{bmatrix} \quad F = \begin{bmatrix} \rho v \\ \rho vu \\ \rho v^2 + p \\ \rho vw \\ (\rho e_t + p)v \end{bmatrix} \quad G = \begin{bmatrix} \rho w \\ \rho wu \\ \rho wv \\ \rho w^2 + p \\ (\rho e_t + p)w \end{bmatrix} \quad S = \begin{bmatrix} 0 \\ 0 \\ 0 \\ \rho g \\ \rho g w \end{bmatrix}, \tag{6.8}$$

where ρ is the liquid density, p is the pressure, u, v, and w are the velocity components in the x, y, z directions respectively, g is the acceleration of gravity, and $e_t = e + 0.5(u^2 + v^2 + w^2)$ is the total energy with e being the internal energy. Equation (6.8) is applied to solve gas, water, or a mixture of both, with an equation of state, which relates pressure, density, and energy for each phase.

The code applies a high order Godunov-type method [16] to accurately capture the location of discontinuities such as shock waves and contact surfaces. A mixed cell algorithm [8] is applied to enable the code to simulate multi-phase and multi-material flows. GEMINI can be coupled with a structural code to predict the response of nearby structures, and can be linked to 3DYNAFS-BEM©, as described in Sect. 6.2.1.3, to model various stages of bubble collapse accurately and efficiently [3–7, 11, 12]. GEMINI has been validated against large scale experiments [8–10, 17, 18].

6.2.1.3 Incompressible/Compressible Link Procedure

Modeling cavitation bubble collapse near boundaries has been extensively done using potential flow boundary methods for axisymmetric cases [19–23] and 3D cases [1–7, 24, 25]. The boundary element method can accurately describe re-entrant jet formation and provide jet characteristics as functions of time since the bubble wall velocities (including the re-entrant jet velocity) are most often small relative to the sound speed in water until the final stage of the bubble collapse. On the other hand, during the bubble explosive growth, rebound, and at jet impact, compressible effects can be non-negligible. These phenomena may lead to shock wave formation and propagation, thereby requiring a compressible approach.

Each of the incompressible and compressible methods has its advantages and disadvantages in addressing cavitation or underwater explosion bubble dynamics. The BEM is efficient since it reduces the dimension of the problem by one: an axisymmetric problem (2D) is reduced to a 1D problem involving line integrals and a 3D problem is reduced to integrals over the surfaces enclosing the 3D domain. This allows very fine gridding for increased accuracy with reasonable computation time. 3DYNAFS-BEM© has been shown to provide re-entrant jet parameters and speed accurately [24, 25]. However, it has difficulty pursuing the computations beyond surface impacts (liquid–liquid when the re-entrant jet hits the opposite side of the bubble and liquid–solid when the re-entrant jet hits the nearby solid wall). On the other hand, compressible finite element models (e.g. [8]) are most adequate to model shock wave emission and propagation, liquid–liquid and liquid–solid impacts. These methods require, however, very fine grids and very small time steps to resolve shock wave fronts. This makes the compressible methods appropriate to model efficiently only the shock wave propagation stage of the bubble dynamics. Concerning the bubble–liquid interface and the re-entrant jet dynamics, the procedure in the finite element models is diffusive since the interface is not directly modeled, and re-entrant jet characteristics are less accurate than obtained with the BEM approach.

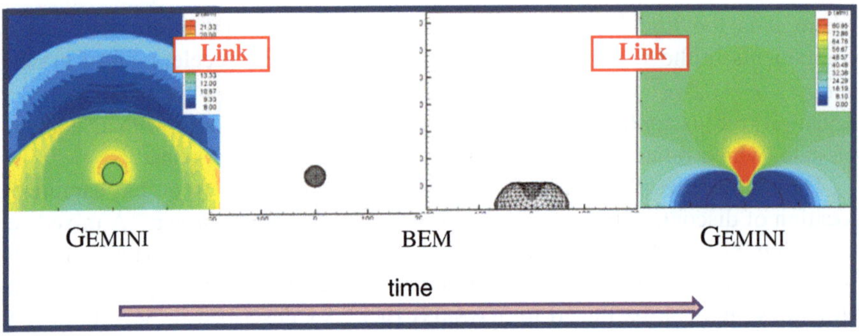

Fig. 6.1 Schematic showing the compressible-incompressible hybrid modeling procedure: link from GEMINI to 3DYNAFS-BEM© during bubble growth and shock propagation and from 3DYNAFS-BEM© to GEMINI following jet impact and shock emission

In order to take advantage of the strengths of both the BEM and compressible Euler methods, a hybrid method was developed at DYNAFLOW to address underwater explosion bubble problems and is used here for cavitation erosion bubble dynamics problems. Underwater explosion bubble studies can provide good visualization and measurement since the maximum bubble sizes are in the range of meters rather than millimeters, and time scales are in seconds rather than microseconds. The large spatial and temporal scales enable one to measure the jet speed and the pressures generated at the boundary with high accuracy both in space and time [26–28].

In the hybrid method, geometry, grids, and flow information are exchanged between the two methods (compressible and incompressible) when needed ("Link" in Fig. 6.1). This link enables one method to pursue the problem resolution starting from the solution given by the other method. The compressible Euler solver GEMINI [8–12] is used during shock formation or during the impact (liquid–liquid or liquid–solid) stage on the materials, while the incompressible potential flow solver, 3DYNAFS-BEM© [1–7], is used during most of the bubble dynamics where the liquid velocities are very small compared to the sound speed. The details of the procedure and validation of this hybrid method can be found in [12].

For a bubble starting with an explosive growth to a maximum bubble size and then collapsing dynamically, the full compressible-incompressible-compressible link procedure in Fig. 6.1 is required, which consists of the following steps as time evolves:

1. The simulation is run with GEMINI until the shock fronts leave the computational domain and the remnant flow field can be assumed to be incompressible.
2. At that instance, the flow field variables (geometry, bubble pressure, boundary velocities) are transferred to 3DYNAFS-BEM© and are used to impose the non-zero normal velocities, $\partial\phi/\partial n$.

3. The bubble growth and collapse are solved using fine grids with 3DYNAFS-BEM© to obtain a good description of the re-entrant jet until this jet moves very close to the opposite side of the bubble.
4. The solution is then transferred back to GEMINI with the required flow variables (velocities, pressures, and densities at all space grid points).
5. The computation is then pursued with GEMINI to obtain the pressures due to re-entrant jet impact and remnant bubble ring collapse.

6.2.2 Structural Model

DYNA3D is a non-linear explicit solid and structural dynamics solver based on the finite element method initially developed by the Lawrence Livermore National Laboratory [13, 29]. Based on a lumped mass formulation it solves the structure momentum equation:

$$\mathbf{M}\ddot{\mathbf{U}} + \mathbf{C}\dot{\mathbf{U}} + \mathbf{K}\mathbf{U} = \mathbf{F}, \tag{6.9}$$

where $\mathbf{U}$ is the displacement vector, $\mathbf{C}$ is the damping matrix, $\mathbf{K}$ is the stiffness matrix, and $\mathbf{F}$ is the matrix of the applied loads on the structure.

DYNA3D is appropriate for problems where high strain rate dynamics or stress wave propagation effects are important. Many material models are available to represent a wide range of material behavior, including elasticity, plasticity, thermal effects, rate dependence, damage, and failure of elements.

6.3 Impulsive Loads Due to Bubble Collapse Near Rigid Boundary

Since the pioneering works of Plesset and Chapman [19], Ellis [20] and Naude and Ellis [21], a large number of numerical and experimental studies have been conducted to investigate the non-spherical collapse of a single bubble near a rigid boundary [22–28]. These studies have established the basis of our understanding of bubble re-entrant jet formation, emission of shock waves, and potential damage on the nearby surface. Simple conceptual extensions of the results from the idealized single bubble case to a multitude of bubbles in a cavitation field have also been done by several researchers [30–39]. More advanced studies involving bubble/bubble interactions and bubble cloud behavior as discussed in [40] remain as the subject of ongoing research.

Despite numerous studies to date, discussions among researchers continue about whether cavitation damage on the material surface results from shock loading or from localized loads of the re-entrant jet impact or from both. Major difficulties stem from the lack of comprehensive and accurate computational methods that account for all the important physics of the liquid, the bubble, the material, and their interactions. In the following sections, detailed analyses of computational results using the above described numerical methods are presented for the dynamics of non-spherical collapse of a single bubble interacting with a nearby solid boundary. The effects of the major parameters governing the bubble collapse dynamics are also presented, including the collapse driving pressure, the standoff distance to the wall, and the bubble size.

6.3.1 Re-entrant Jet Dynamics Using BEM

Bubble collapse close to a rigid boundary can be highly asymmetrical and results in the formation of a fast re-entrant jet towards the boundary. High pressure loading on the boundary will develop from the impact of the jet on the boundary after it crosses the bubble. The intensity of the impact depends mainly on the relative distance between the bubble and the rigid boundary and on the "strength" of the collapse. Figure 6.2 illustrates the bubble dynamics near a rigid boundary using an axisymmetric version of 3DynaFS-Bem©.

Due to the asymmetry of the flow around the bubble created by the presence of the wall, a dipole-like pressure field develops in the bubble region farthest from the rigid wall. A high pressure region in the bulk of the liquid (red region) pushes down the bubble wall where the pressure is low (dark blue region) driving a high velocity flow along the axis of symmetry. This results in a high speed re-entrant jet that pierces the bubble from one side to the other, and then impacts the rigid wall when the bubble standoff to the wall is small. When the re-entrant jet impacts the wall, a water hammer-like pressure is generated and stresses the material. Later on, as the remaining bubble volume continues to shrink, high pressures are generated inside the ring bubble and a pressure wave or shock wave is emitted from the collapsing ring. The relative importance of the water hammer impact and the ring collapse depends on various physical parameters including the standoff distance between the initial bubble center location and the wall, and the bubble collapse driving pressure, $(P_{amb} - P_{bub})_0$.

6.3.2 Pressure Loading from Hybrid Incompressible/ Compressible Model

The hybrid compressible/incompressible approach is used to study the bubble growth and collapse near a rigid boundary in order to investigate the resulting impulsive pressure load. Figure 6.3 through 6.8 illustrate the dynamics and present the pressures generated during the bubble collapse. A bubble with an initial radius

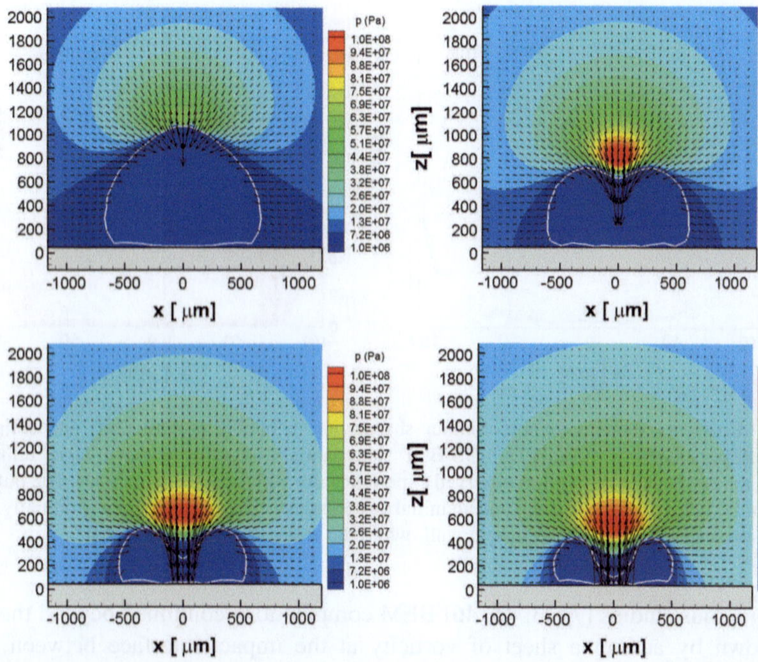

Fig. 6.2 Pressure and velocity field evolution during non-spherical bubble collapse near a wall predicted by 3DynaFS-Bem© for a bubble with an initial radius of 10 µm and a gas pressure of 1 atm subjected to a pressure drop to −1 atm for 1 ms and then a steep pressure rise to 50 atm ($t = 0$). The bubble center is initially at a standoff 0.65 mm away from the wall. The maximum bubble radius is $R_{max} = 1.3$ mm at $t = 0$. The upper left contour is at bubble maximum volume ($t = 20.6$ µs). The other three contours are at $t = 21.1$ µs (*upper right*), $t = 21.6$ µs (*lower left*), and $t = 21.8$ µs (*lower right*)

of $R_0 = 10$ µm located at 75 µm above a rigid boundary is considered. The initial bubble internal pressure is 2.6×10^7 Pa (this could result from a bubble exiting a nozzle with an upstream pressure of 3,800 psi (26.2 MPa)) and the liquid ambient pressure is 10^5 Pa. In the absence of nearby boundaries the spherical bubble would expand to a maximum radius of $R_{max} = 100$ µm, a value easily computed using spherical bubble dynamics equations [41–43]. Figure 6.3 shows the bubble contours at different times during the bubble growth and collapse. During the growth the bubble surface close to the rigid boundary flattens on the wall (actually, a thin water layer remains between the bubble and the wall as observed experimentally [7, 28]) while the surface farther from the wall expands freely until the pressure inside the bubble drops below the liquid dynamic pressure. As illustrated in Fig. 6.2 a localized high pressure region forces the water into the contracting bubble, generating a re-entrant jet which crosses the bubble and impacts the nearby wall. The right hand side of Fig. 6.3 clearly shows the evolution in time of the bubble shape and the re-entrant jet, which advances to strike the wall, while the thin layer between the bubble and the wall becomes thinner.

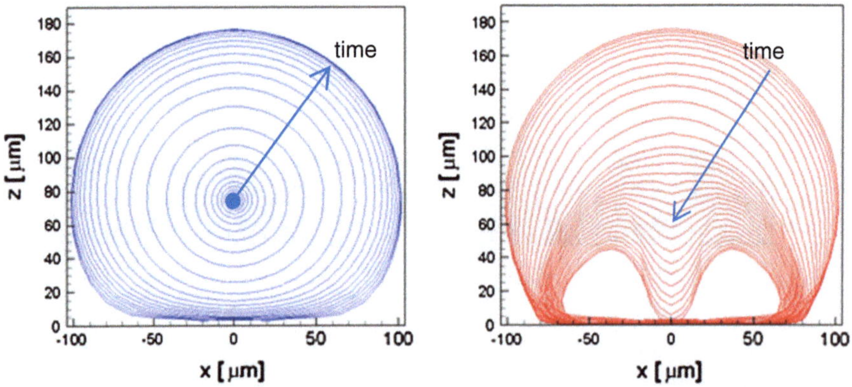

Fig. 6.3 Crosscuts showing bubble contour shapes during bubble growth (*left*) and collapse (*right*). Results obtained from 3DYNAFS-BEM© simulation. Note that a water layer remains between the bubble and the wall as observed experimentally with large scale bubbles. The bubble has an initial radius of $R_0 = 10$ μm and an initial gas pressure of 2.6×10^7 Pa and is initially at a standoff distance of 75 μm from a rigid wall, while the ambient pressure is 1 atm

In previous studies [7, 23, 45, 46] BEM computations continued beyond the jet touchdown by adding a sheet of vorticity at the impact interface between the advancing re-entrant jet and the bulk liquid, or a vortex ring at the edge of this surface. This allowed the computations to proceed beyond the point where the geometry becomes multi-connected and a vortex ring bubble is generated and collapses at the wall. This procedure was able to accurately predict the re-entrant jet impact pressure peak as illustrated in Fig. 6.4. However, the procedure was cumbersome, too sensitive to gridding, and often unstable. The new Link procedure described in the previous section is more robust and easy to use and has also produced validated output [11, 12].

The Link procedure continues the simulations starting from the 3DYNAFS-BEM© solution until a few computation steps before the jet impacts the opposite side of the bubble. The left side of Fig. 6.5 shows the 3DYNAFS-BEM© gridded bubble shape used to proceed with the compressible flow solution. The right side of Fig. 6.5 shows pressure contours from the first time step of GEMINI right after the solution was passed from 3DYNAFS-BEM©.

Figure 6.6 shows the pressure contours and the bubble dynamics from GEMINI at three subsequent times. At $t = 3.63$ μs, the jet impacts the wall generating a localized high pressure impact region on the axis leaving around it a bubble ring (white contours). At $t = 3.85$ μs, the bubble ring has completed its collapse (the white contours reach the minimum) resulting in a ring like high pressure region on the wall around $r = 50$ μm. The high pressure or shock waves travel towards the axis resulting in a high pressure region near the axis, as shown at 3.89 μs in the last contours where the bubble ring is rebounding (white contours increasing area again).

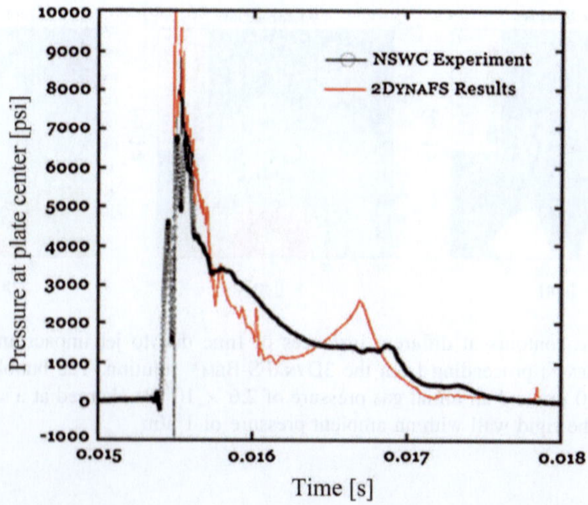

Fig. 6.4 Comparison between experimental measurements and numerical computations of the pressures due to bubble collapse on a rigid wall [7]. The bubble was generated by 10 g of TNT at a depth of 50 m on an instrumented circular plate. The generated bubble had a maximum radius of 0.18 m and the standoff from the plate was 0.14 m. The re-entrant jet speed was as high as 240 m/s. The axisymmetric version of 3DYNAFS-BEM© was used to conduct the numerical simulations and reproduced well the re-entrant jet impact pressure peak at $t \sim 15.6$ ms and to a lesser extent the bubble ring collapse pressure at about 16.7 ms. (1000 psi = 6.9 MPa)

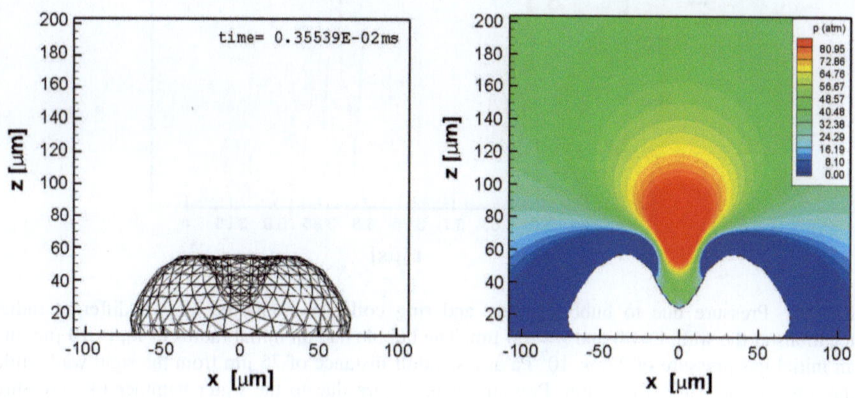

Fig. 6.5 3DYNAFS-BEM© solution (*left*) where the solution is handed back to GEMINI and the first time step from GEMINI (*right*). The bubble has an initial radius of $R_0 = 10$ μm and an initial gas pressure of 2.6×10^7 Pa charged at a standoff distance of 75 μm from the rigid wall with an ambient pressure of 1 atm

Figure 6.7 shows the time history of the pressure computed at two locations on the wall indicated in the insert: the bubble axis of symmetry and a point at a radial distance of 60 μm. There is a marked difference between the signals computed at the two locations.

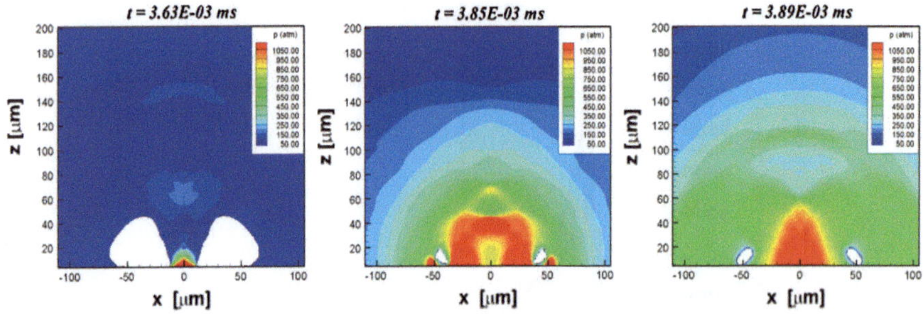

Fig. 6.6 Pressure contours at different instances of time due to jet impact and ring collapse obtained from GEMINI proceeding from the 3DYNAFS-BEM© solution. The bubble had an initial radius of $R_0 = 10$ μm and an initial gas pressure of 2.6×10^7 Pa charged at a standoff distance of 75 μm from the rigid wall with an ambient pressure of 1 atm

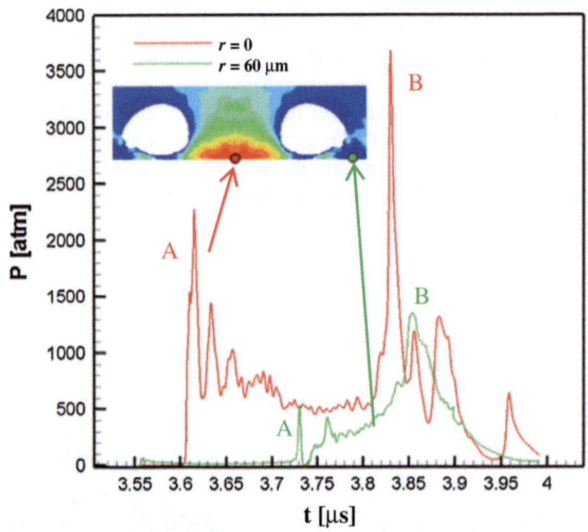

Fig. 6.7 Pressure due to bubble jetting and ring collapse monitored at two different radial locations at the wall, $r = 0$ and $r = 60$ μm. The bubble had an initial radius of $R_0 = 10$ μm and an initial gas pressure of 2.6×10^7 Pa at a standoff distance of 75 μm from the rigid wall while the ambient pressure was 1 atm. Pressure peaks A are due to the water hammer like pressure generated by the re-entrant jet impact, while pressure peaks B are due to ring bubble collapse

The point on the axis sees distinctly two pressure peaks: the first peak at $t = 3.615$ μs (red A) is at the moment of the re-entrant jet impact. The second peak (red B) at 3.83 μs corresponds to the bubble ring collapse. A toroidal pressure wave emanates from the collapsing bubble ring and then focuses and amplifies at the axis of symmetry, $r = 0$. This results in a very sharp high peak (red B) shown in the figure.

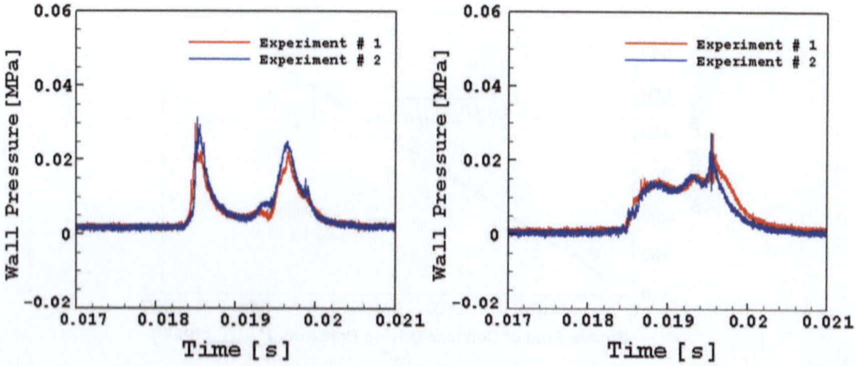

Fig. 6.8 Wall pressures recorded during the collapse and rebound of a spark generated bubble located at 1.8 cm from a vertical wall. The equivalent maximum bubble radius was 2.4 cm and the static pressure at the bubble center was 8,590 Pa for the left figure and 7,900 Pa for the right figure. In the left figure the PCB 101A04 piezoelectric transducer was located at the bubble re-entrant jet impact location. In the right figure the transducer was 0.5 cm off-center. These tests were conducted in the DYNAFLOW spark test facility

The point at $r = 60$ μm does not sense the jet impact at $t = 3.615$ μs as it is outside the jet impact region (~ 10 μm). However, it feels the effect of the jet impact a little later at $t = 3.73$ μs (green A). The second peak (green B) due to the ring collapse is clearly detected because $r = 60$ μm is close to the ring collapse location.

It is to be noted that for such small cavitation bubbles, the distinction between the two events is extremely difficult experimentally using conventional transducers. For spark generated bubbles or underwater explosion bubbles however, these differences in the signal can be clearly measured [25–28]. The difference between the measured signals on-axis and off-axis, further illustrated in Fig. 6.8, could explain why discussions are still on-going about whether cavitation damage is from shock waves or from jets as time histories of the pressure appear different depending on the precise location of the gauge.

6.3.3 Effect of Local Pressure Driving Bubble Collapse

The speed of the bubble collapse depends on the pressure driving the dynamics, e.g. the difference of pressures between the outside and the inside of the bubble. This pressure difference can result from the bubble being driven into a stagnation region where the pressure is very high (e.g. in a cavitating jet), or from very high pressure waves generated by the collapse of bubble clouds. The local pressure driving the individual bubble collapse will vary between scales (see [40]). In order to quantify the effect of the driving pressure on the loads, the hybrid incompressible (3DYNAFS-BEM©) / compressible (GEMINI) approach was used for

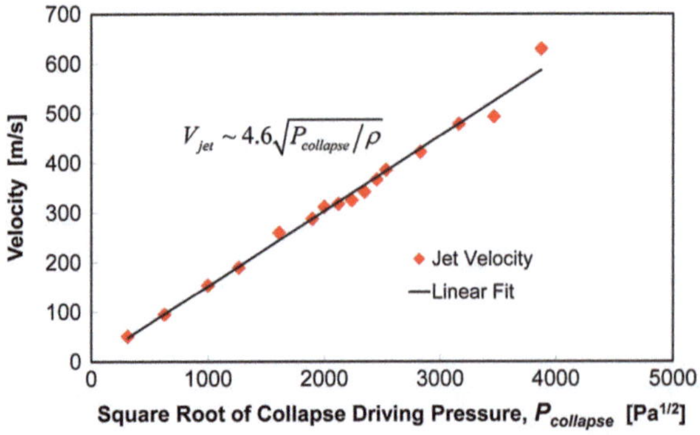

Fig. 6.9 Relationship between the re-entrant jet velocity and the pressure driving the bubble collapse. The *red diamond symbols* are jet impact velocities computed by the 3DynaFS-Bem© for different ambient pressures, and the black curve is a linear fit. Maximum bubble radius 2 mm. Distance of bubble center to wall 1.5 mm (normalized standoff distance of 0.75)

different driving pressures. For illustration, we consider bubbles, which have grown to a maximum radius of 2 mm at which point the internal pressure is close to the liquid vapor pressure, and are then subjected to a collapse driving pressure, $P_{collapse}$, varying from 10^5 Pa (1 atm) up to 1.5×10^7 Pa (150 atm). In all cases the bubble started collapsing when its center was at a distance of 1.5 mm from a rigid wall, that is for a normalized standoff distance of 0.75.

For all values of the driving pressure, a strong re-entrant jet develops and advances to hit the rigid wall with a large impact velocity. As illustrated in Fig. 6.9, the jet velocity increases with the square root of the driving pressure, $P_{collapse}$. This is consistent with previous studies of bubble collapse dynamics [44]. In the present case, the re-entrant jet velocity can be approximated by:

$$V_{jet} \sim 4.6\sqrt{P_{collapse}/\rho}. \tag{6.10}$$

Figure 6.10 shows the peak pressure on the wall from bubble collapse for selected cases in Fig. 6.9. This figure confirms the classical view that the pressures due to bubble collapse are scaled by the water hammer impact pressure [23, 42] which is given by:

$$P_{water\,hammer} = \rho c V_{jet}, \tag{6.11}$$

where ρ is the density of the liquid, c is the speed of sound, and V_{jet} is the jet velocity at impact. The $P_{water\,hammer}$ is the pressure on a flat solid surface exerted by a water column with a flat surface [23, 41–43]. As the flatness of the re-entrant jet tip varies with the intensity of the collapse, so does the efficiency of the impact. As shown in Fig. 6.10, one can estimate the impulsive pressure generated by the jet impact using:

$$P_{impulse} \sim 0.6\rho c V_{jet}. \tag{6.12}$$

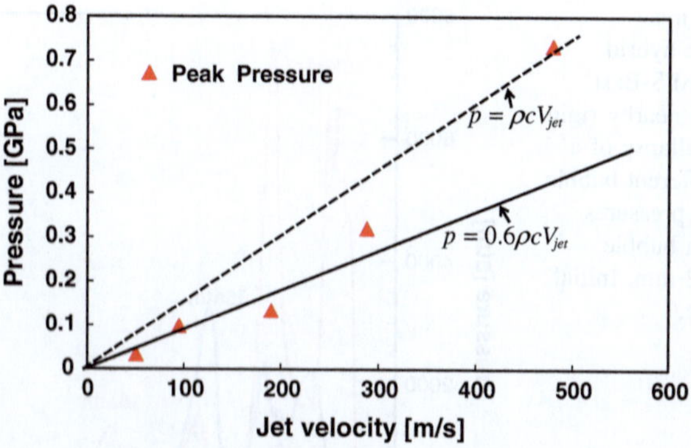

Fig. 6.10 Relationship between the peak impact pressure and the jet velocity. The jet velocities are computed by the 3DynaFS-Bem© at the time of jet impact. The *red symbols* are from computed maximum peak impact pressures by the hybrid approach (3DynaFS-Bem© and Gemini) (such as in Fig. 6.11). Maximum bubble radius 2 mm, distance from bubble center to wall 1.5 mm (normalized standoff distance of 0.75)

The figure shows that the impact efficiency improves for the higher collapse driving pressures and a factor larger than 0.6 applies at these higher pressures.

The details of the impact pressure versus time at the center point on the wall are shown in Fig. 6.11 for a set of collapse driving pressures for the same bubble collapse conditions as in the previous two figures. All pressure histories have been shifted to the same starting point to show times from the moment of jet impact at the wall. As previously discussed in Fig. 6.7, all of the signals but the one corresponding to $P_{collapse} = 4$ atm clearly exhibit at least two pressure peaks following the impact: the first corresponds to re-entrant jet tip liquid impact at the wall, and the second (and sometimes third) is generated by the bubble ring collapse (and sometimes second collapse after rebound) and the focusing of the resulting shock wave on the jet/wall axis.

The characteristic time of the bubble dynamics is the Rayleigh time expressed as:

$$T_{Rayleigh} = R_{max}\sqrt{\rho/P_{collapse}}. \tag{6.13}$$

As the driving pressure increases, $T_{Rayleigh}$ decreases and the spacing between the two peaks also decreases. On the other hand, as the jet speed increases, the resulting jet impact pressure and the bubble ring collapse pressure increase.

The above observations provide a scaling of the level of the impulsive pressure generated on the wall at bubble collapse, $P_{impulse}$, as a function of the ambient pressure driving the bubble collapse, $P_{collapse}$, combining Eqs. (6.10) and (6.12):

$$P_{impulse} \sim 2.76\, c\sqrt{\rho P_{collapse}}. \tag{6.14}$$

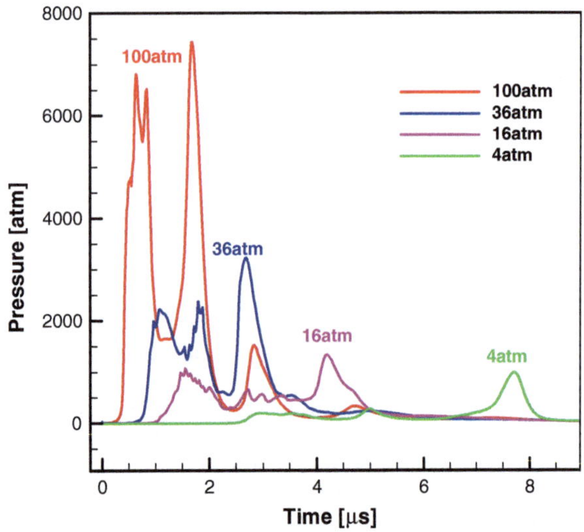

Fig. 6.11 Pressures computed by the hybrid approach (3DynaFS-Bem© and Gemini) at a nearby rigid wall from the collapse of a bubble under different bubble collapse driving pressures. Initial maximum bubble radius, $R_{max} = 2$ mm. Initial standoff distance, $X = 1.5$ mm

The pressure distribution function, $p(\mathbf{M},t)$, at the impact location **M** has a lateral extent, which scales with the bubble maximum radius. This is illustrated in Fig. 6.12, which shows that the maximum pressure drops off rapidly radially away from the jet axis. It decays to practically half its maximum value at a distance of R_{max} from the axis.

6.3.4 Effect of Standoff

The illustrations presented in the previous sections corresponded to a normalized standoff between the bubble center and the wall, $X/R_{max} = 0.75$, which is in the range of maximum pressures generated on a nearby rigid wall [25]. The fastest velocity at the bubble wall increases as the bubbles move away from the rigid boundary. This applies also to the re-entrant jet velocity. However, for a larger standoff, e.g. $X/R_{max} > 1.25$, the re-entrant jet, after penetrating and piercing the opposite bubble side, has to advance in water before reaching the wall. This results in a significant loss of energy that attenuates the jet speed and subsequent impact pressure. On the contrary, for the small standoff distances, $X/R_{max} < 0.75$, the re-entrant jet hits the wall directly but does not have time to fully develop and attain a large speed because of the short distance. Indeed, as we will see later (e.g. Fig. 6.16), the jet speed is a relatively strong function of time.

Figure 6.13 illustrates the effect of standoff by showing bubble contours during collapse for three standoffs: $X = 1.05$, 1.25, and 2.0 R_{max}. The bubble is initially spherical with a radius 100 μm, and is in equilibrium with the surrounding liquid when the ambient pressure is 1 atm. The bubble is then subjected to a sudden

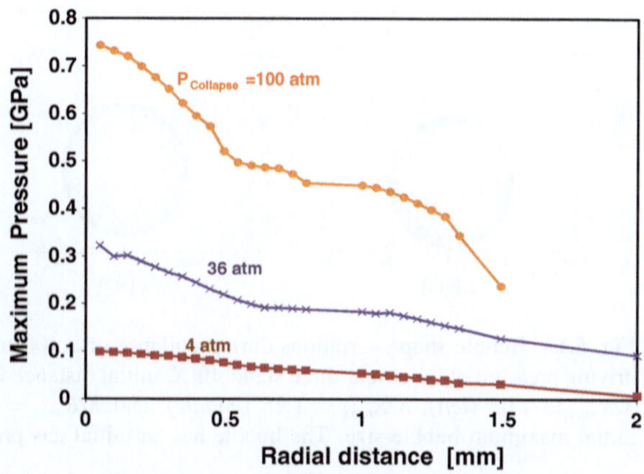

Fig. 6.12 Maximum pressures versus radial distance generated at a nearby rigid wall from the center of a bubble collapsing under different driving pressures. Initial maximum bubble radius, $R_{max} = 2$ mm. Initial standoff distance, $X = 1.5$ mm

ambient pressure rise to 34 atm. For the farthest distance ($X = 2.0\ R_{max}$), the bubble starts a quasi-spherical collapse, then a relatively wide re-entrant jet develops and strikes the opposite side of the bubble far from the wall. As the bubble initial standoff distance is reduced, the bubble elongates perpendicular to the wall during collapse and the re-entrant jet shape becomes more pointed. Moreover, the motion of the bubble surface closer to the rigid boundary is significantly reduced almost to zero for $X/R_{max} = 1.05$, while it is almost equal to R_{max} for $X/R_{max} = 2$.

Figure 6.14 shows the pressures due to bubble collapse at the center of the nearby rigid boundary. The left and right figures present computations for a bubble with initial maximum radii of 10 and 500 μm, respectively. The bubbles are initially spherical and in equilibrium with the surrounding liquid at a pressure of 1 atm. Three standoff distances are considered as in Fig. 6.13: $X/R_{max} = 1.05$, 1.25 and 2.0. Unlike previous cases in Fig. 6.13 where the initial bubble radius was 100 μm, it is here 10 and 500 μm. The reason for the choice of two very different radii is to highlight the fact that the amplitude of the pressure generated by the bubble collapse for the same driving condition is practically independent of the initial bubble radius. This is very well highlighted in Fig. 6.14, where, for the three standoff distances, there is practically no difference in the amplitude of the pressures between the 10 μm (left) and the 500 μm (right) cases. The time scale on the other hand, follows the $T_{Rayleigh}$ scaling (Eq. (6.13)). The first peaks in Fig. 6.14 (~0.18 and ~9 μs on the left and right figure respectively) occur at re-entrant jet impact. The timing of the re-entrant jet impact is seen to scale with R_{max} and is 50 times larger for the 500 μm bubble (~9.5 μs) than for the 10 μm bubble (~0.19 μs). The same applies for the time delay between the jet impact and the bubble ring collapse (~0.2 μs vs. ~0.004 μs).

Figure 6.14 also shows clearly the effect of the standoff distance on the wall pressures. As the standoff decreases from $X/R_{max} = 2.0$ to 1.25 and 1.05, the bubble collapse period (i.e. the time from bubble maximum volume to minimum

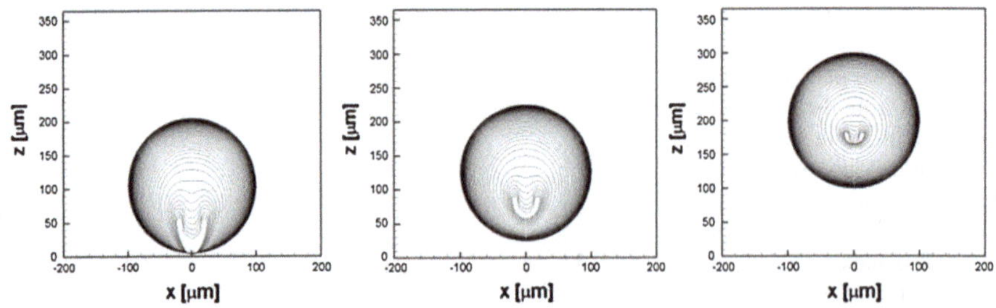

Fig. 6.13 Bubble shape variations during collapse of a 100 μm radius bubble under a collapse driving pressure of 34 atm at three standoffs X, initial distance from the bubble center to the wall: $X/R_{max} = 1.05$ (*left*), $X/R_{max} = 1.25$ (*middle*), and $X/R_{max} = 2$ (*right*). $R_{max} = 100$ μm is the initial maximum bubble size. The bubble has an initial gas pressure of 1 atm

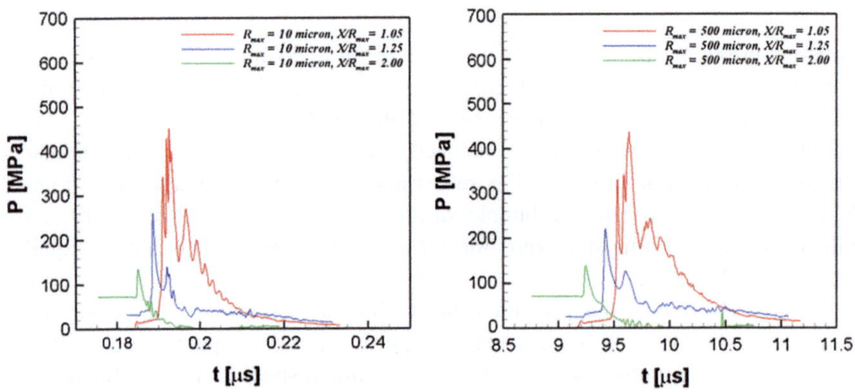

Fig. 6.14 Comparisons for $R_{max} = 10$ μm (*left*) and $R_{max} = 500$ μm (*right*) of the pressure on the wall at the bubble axis due to re-entrant jet impact and bubble ring collapse for different normalized standoffs: $X/R_{max} = 1.05$ (*red*), 1.25 (*blue*), and 2.0 (*green*). The bubbles have an initial gas pressure of 1 atm and are subjected to an ambient pressure of 34 atm. Notice the very close correspondence of the amplitudes between the two cases indicating very small sensitivity of the pressure amplitude to the bubble initial radius, and a ratio of 50 between the times indicating linear relationship between the bubble radius and the time scale

volume) increases, as well documented in the literature [42, 43, 47, 48]. This results in the re-entrant jet impacting the wall later for closer standoffs [47, 48]. This is illustrated in Fig. 6.14 by the first pressure peak occurring at a later time for the smaller standoff. On the other hand, as X/R_{max} increases, the impact pressures decrease, while the pressure peak due to the bubble ring collapse becomes too low to discern in the pressure trace. In the literature [22, 23, 45, 46], $X/R_{max} = 1.5$ is the typical case studied and this may explain why less emphasis has been placed in the past on the presence of the second pressure peak.

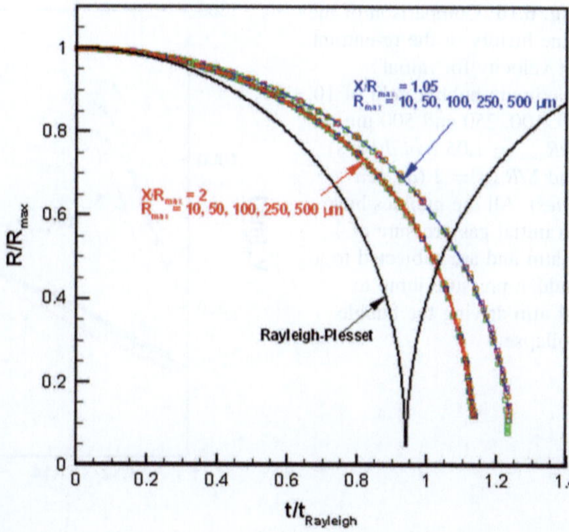

Fig. 6.15 Comparison of the time history of the normalized bubble equivalent radius for initial bubble maximum radii of 10, 50, 100, 250 and 500 μm at $X/R_{max} = 1.05$ (*blue line*) and $X/R_{max} = 2.00$ (*red line*). All the bubbles have an initial gas pressure of 1 atm and are subjected to an ambient pressure of 34.5 atm. The bubble equivalent radius is normalized by R_{max} while the time scale is normalized by the Rayleigh time

6.3.5 Influence of Bubble Size

This section examines the importance of bubble size on the impulsive pressure loads. To do so, bubble radii ranging from 10 to 500 μm are considered. Initial conditions as in the previous section are considered: all the bubbles are assumed to be spherical and in equilibrium with the liquid at an ambient pressure of 1 atm. One should note that this results in different initial internal bubble pressures due to the presence of surface tension. The bubbles are then subjected to a rise in the ambient pressure to 34.5 atm when they are near a rigid boundary located at $X/R_{max} = 1.05$, 1.25 or 2.0.

Figure 6.15 presents the time variations of the bubble equivalent radius (i.e. radius of a sphere with the same volume) for the bubble sizes considered and for two normalized standoff conditions, $X/R_{max} = 1.05$ and 2. Radii and times are normalized respectively by the initial maximum radius, R_{max}, and by the Rayleigh time (Eq. (6.13)). The figure illustrates that for a given standoff, the normalized curves do not distinguish between the different initial bubble sizes. However, the standoff distance shows a distinct effect on the bubble collapse periods, i.e. larger periods for the smaller normalized distances, X/R_{max} and the presence of the wall slows down the collapse. In Fig. 6.15, the solution of the Rayleigh-Plesset equation in an infinite fluid is also presented as a reference.

Figure 6.16 shows the re-entrant jet speed versus time for different bubble sizes and for the two normalized standoff distances, $X/R_{max} = 1.05$ and 2. It is interesting to note that for a given nondimensional standoff, the dimensional jet velocity curves versus normalized times are insensitive to the bubble size especially for the smaller standoff distances, and are mostly controlled by the collapse driving pressure and the normalized distance from the wall. Near the wall,

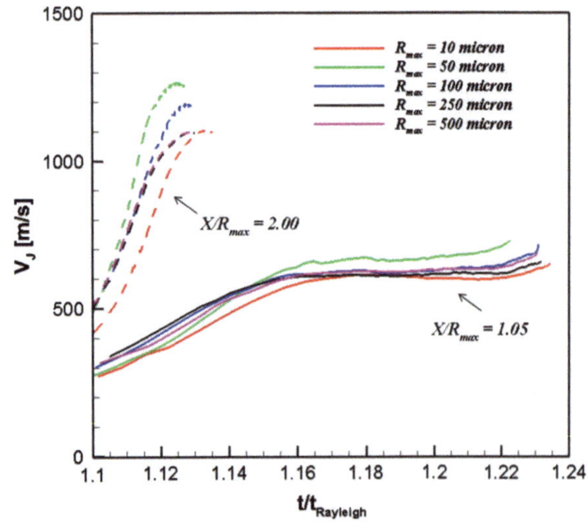

Fig. 6.16 Comparison of the time history of the re-entrant jet velocity for initial maximum bubble radii of 10, 50, 100, 250 and 500 μm at $X/R_{max} = 1.05$ (*solid lines*) and $X/R_{max} = 2$ (*dashed lines*). All the bubbles have an initial gas pressure of 1 atm and are subjected to a sudden pressure jump to 34 atm driving the bubble collapse

asymmetry is predominant and bubble size does not show an important effect. Farther from the wall, surface tension and initial gas pressure, which introduce nonlinear effects not captured by the Rayleigh scaling, have a non-negligible effect on the jet velocity and are responsible for the differences between the curves.

Figure 6.16 also shows that the normalized standoff distance has a very significant effect on the terminal velocities of the re-entrant jet. For instance, for $X/R_{max} = 1.05$ the terminal jet velocity is approximately 650 m/s while for $X/R_{max} = 2.00$ it is in the range of 1,100 to 1,300 m/s.

Figure 6.17 compares the normalized time histories of the wall pressures for $X/R_{max} = 1.05$ and 1.25 for various bubble sizes. The collapse of all pressure loading curves for a given standoff is remarkable. This highlights that, independent of the bubble size, the pressures generated at the wall (see Eq. (6.14)) scale very well with the water hammer pressure due to the re-entrant jet velocity, which depends only on the driving pressure (see Eq. (6.10)). The time also scales very well with the Rayleigh time (see Eq. (6.13)). The normalized peak pressure for $X/R_{max} = 1.05$ is of the order of $0.5\rho c V_{jet}$, while for $X/R_{max} = 1.25$ it decreases to about $0.25\rho c V_{jet}$.

The variations of the pressure on the wall with time and with the radial direction are illustrated in Fig. 6.18. This figure provides a picture of the dynamic evolution of the pressure loading in space and time for two bubble standoff distances. For $X/R_{max} = 1.05$, the radial pressure distribution at different instants shows that as the bubble collapses the pressure on the axis gradually increases until it reaches the first peak (from the re-entrant jet impact) followed by a second peak (from the bubble ring collapse) then it gradually decays and has oscillations from reflected waves through the thickness of the plate (details are discussed in Sect. 6.4). In the radial direction the impact wave propagates away from the axis with the two above described peaks moving outwards while

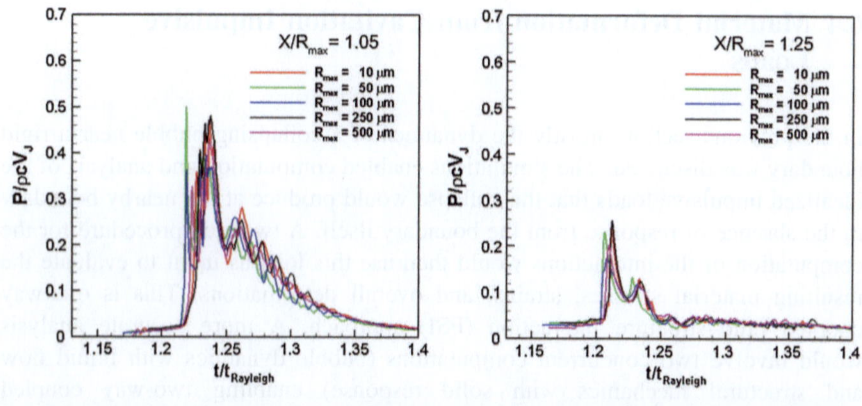

Fig. 6.17 Time history of the normalized pressure on the wall at the axis ($r = 0$) for different bubble sizes at two normalized standoff distances: $X/R_{max} = 1.05$ (*left*) and 1.25 (*right*). All the bubbles have an initial gas pressure of 1 atm and are subjected to a sudden pressure rise of 34.5 atm driving the bubble collapse

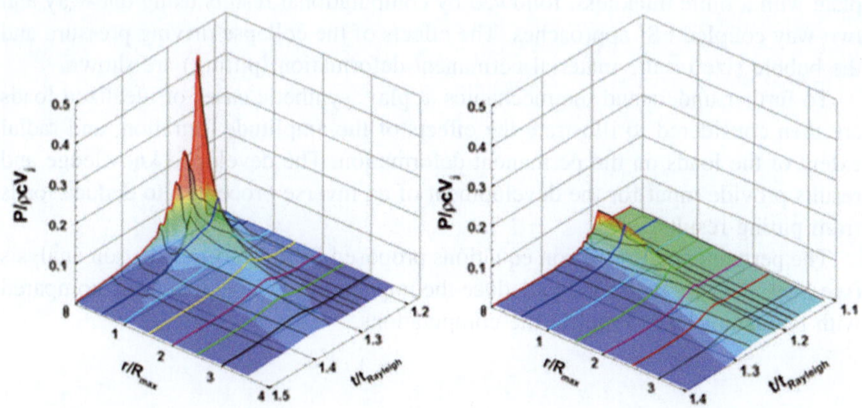

Fig. 6.18 Variations of the pressure at the wall with time and with the radial direction for two standoffs $X/R_{max} = 1.05$ (*left*) and 2.00 (*right*). The bubble has an initial radius $R_{max} = 10$ μm and is in balance with an ambient pressure of 1 atm. It is then subjected to a sudden pressure rise to 34 atm

rapidly decaying. The location of the peak pressure moves outward radially as a ring wave. Due to wave propagation at a finite time after the impact the radial distribution resembles the time trace of the pressure and contains the two peaks described earlier (re-entrant jet and ring collapse). For $X/R_{max} = 2.00$ the peaks are significantly lower, and owing to the larger standoff, only the pressure/shock front generated by the bubble collapse reaches the wall affecting a larger area and then gradually dissipates in time.

6.4 Material Deformation from Cavitation Impulsive Loads

In the previous section, mostly the dynamics of a collapsing bubble near a rigid boundary was discussed. The simulations enabled computation and analysis of the idealized impulsive loads that the collapse would produce at the nearby boundary in the absence of response from the boundary itself. A two-step procedure for the computation of the interactions would then use this load as input to evaluate the resulting material stresses, strains, and overall deformations. This is one-way coupled fluid–structure interaction (FSI) approach. A more accurate analysis would involve two concurrent computations (bubble dynamics with liquid flow and structural mechanics with solid response) enabling two-way coupled dynamics. This two-way coupled approach is necessary when deformations and deformation speeds and accelerations are significant enough to affect the bubble dynamics. This is the case for very large impulsive loads and/or very flexible or deformable materials. In this section, a finite element model is presented first for a canonical problem of a single bubble collapsing in the vicinity of a flat circular plate with a finite thickness, followed by computational results using one-way and two-way coupled FSI approaches. The effects of the collapse driving pressure and the bubble size on the material permanent deformation (pitting) are shown.

To further understand the mechanics at play, synthetic cases of idealized loads are then considered to illustrate the effects of the amplitude, duration, and radial extent of the loads on the permanent deformation. The developed knowledge and results provide input for the development of an inverse procedure to deduce loads from pitting results.

The permanent deformation equations proposed from nano-indentation analysis (see Sect. 3.4) are then used to deduce the imparted impulsive loads and compared with the actual loads input in the computations.

6.4.1 Finite Element Modeling

To enable the fluid–structure interaction simulations, the Link procedure described in Sect. 6.2.1.3 is exercised while the fluid codes are coupled with the structure code briefly introduced in Sect. 6.2.2, DYNA3D [13]. For the present application, the large significant impulsive pressure loads occur only after the re-entrant jet fully penetrates the bubble and strikes the nearby wall, and therefore FSI computations using the compressible code, GEMINI, and DYNA3D are relevant. For the single bubbles considered in this section axisymmetric options of these codes are exercised.

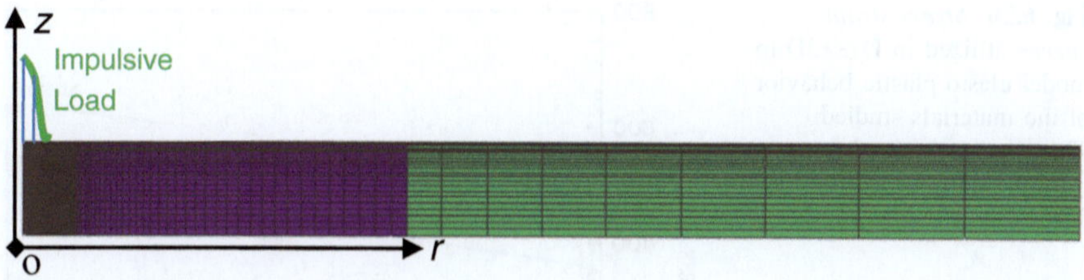

Fig. 6.19 Finite element axisymmetric grid used in DYNA3D to study elasto-plastic material response to loads due to collapsing cavitation bubbles. The cells in the *dark region* near $r = 0$ are 5 μm by 5 μm. The mesh size was stretched by a factor of 10 % in the *magenta region* and 30 % in the *green region*. (*Note r* and *z* do not have the same scale.)

A finite element mesh for a circular plate with 1 m radius and 5 mm thickness is generated using the axisymmetric setup illustrated in Fig. 6.19. The size of the plate (i.e. computational domain) was selected to be sufficiently large relative to the micron size bubbles, in order to minimize numerical reflections from the edges of the plate. The mesh elements near the axis of the plate are 5 μm by 5 μm. This fine element size was maintained up to 1.5 mm radius from the axis and down to 0.6 mm distance from the surface. From there on, the mesh size was gradually increased both in the outward and in the thickness direction. The mesh stretch factor was 10 % up to a radius of 10 cm, and then increased to 30 % beyond 10 cm. This resulted in 377 elements in the radial direction and 258 elements in the thickness direction.

The boundary conditions used for the simulations were as follows:

- An imposed unsteady pressure load at the 'free' material-liquid interface provided by the fluid solution (or a given function in the synthetic runs).
- Rigid boundary condition at the bottom of the plate, i.e. no displacements in the plate thickness direction.
- No reflection boundary conditions at the edge of the plate.

An isotropic elasto-plastic material model was used for the computations. Material properties such as Young's modulus, yield stress, density, and failure strain were specified as input. The materials considered in this study included the materials experimentally studied in previous chapters, i.e. aluminum alloy Al 7075, stainless steel A2205, and NAB. In addition, a softer material, Al 1100-H19, was used in some of the computations. Bilinear representation of the stress–strain curves (see Fig. 6.20) of the four materials were used for DYNA3D. The first line of slope E, the Young's modulus, represented the elastic portion of the behavior. This intersected the linearly modeled plastic behavior beyond the yield point. The slope of the second portion is the tangent modulus. These properties were obtained from data published in the literature [49] and presented in Fig. 6.20 and Table 6.1.

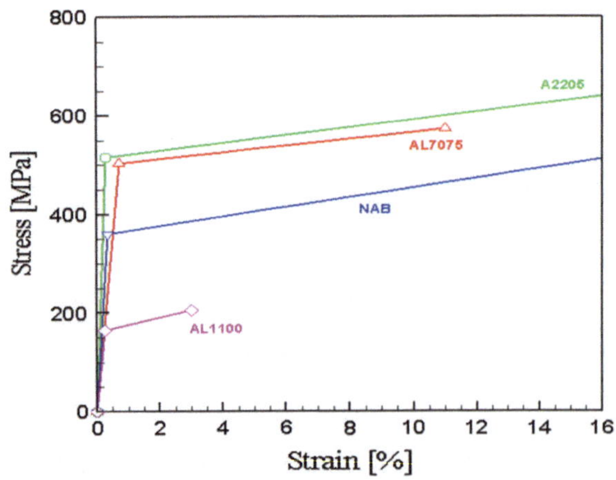

Fig. 6.20 *Stress-strain curves* utilized in DYNA3D to model elasto-plastic behavior of the materials studied. These were obtained from information available online [49]

6.4.2 Process of Pit Formation

In order to study the response of a solid material to impulsive loads from bubble collapse, pressures generated by the bubble collapse were applied to an elasto-plastic material. As long as the local stresses do not exceed the yield stress, the material deforms elastically during the loading and returns to its initial state once the load is relieved. However, once the yield point is exceeded anywhere in the material, permanent deformation occurs. An example is presented in Fig. 6.21 for the collapse of a bubble of maximum radius, $R_{max} = 2.0$ mm, located at an initial normalized standoff distance of 0.75 ($X = 1.5$ mm) from the plate, and subjected to a collapse pressure of 10 MPa. The plate is made of Al 7075 with a thickness of 5 mm. The codes were run in a two-way coupled mode. Figure 6.21 shows the time history of the pressure and the vertical displacement at the plate center point. The interpretation of the pressure versus time profile has been previously discussed in Sect. 6.3.3. As the re-entrant jet impacts the plate the stresses at the surface rise above 600 MPa, largely exceeding the Al 7075 yield stress of 503 MPa used in the computation. Therefore, plastic deformation occurs within the first 0.5 μs and continues to accumulate from then on.

Once the re-entrant jet impact pressures are relieved, the material deformation rate is slowed down until the second large pressures due to the toroidal bubble collapse further increase the deformation. At about 2 μs the bubble ring collapse loading vanishes and the surface starts returning as much as possible towards the initial state. However, since there was significant deviation from the elastic behavior during the loading, a permanent deformation results and the center point in the plate remains depressed by about 10 μm. One can also note successive oscillations at this center point as the stress/strain waves propagate and reflect back and forth at the rigidly held bottom and at the plate-water interface. The above dynamics illustrates the importance of the thickness of a plastic layer on the resulting deformations. No element failure or erosion was observed during this

Table 6.1 Material properties used for DYNA3D computations

Material	Yield stress [MPa]	Young's modulus [GPa]	Tangent modulus [MPa]	Elongation at failure	Density [g/cm^3]
Al 7075	503	71.7	670	0.11	2.81
SS A2205	515	190	705	0.35	7.88
NAB	360	110	1216	0.25	7.64
Al 1100-H19	165	69	1449	0.03	2.71

These were obtained from available information online [49]

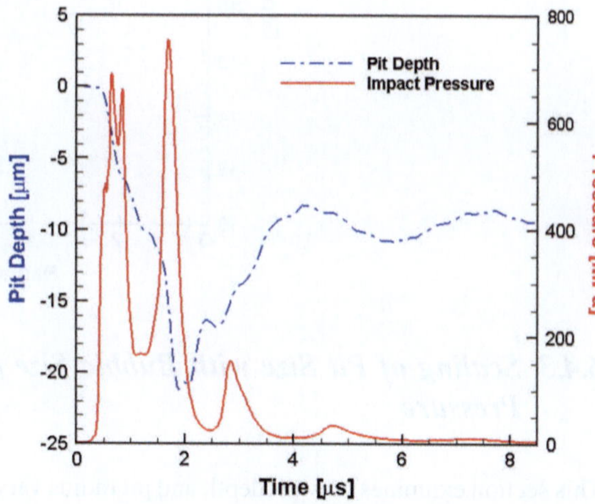

Fig. 6.21 Time history of pressure acting on an Al 7075 plate with 5 mm thickness and resulting pit depth for a collapse driving pressure of 100 atm. Initial maximum bubble radius, $R_{max} = 2.0$ mm. Initial bubble pressure, 1 atm. Initial standoff distance from bubble center, $X = 1.5$ mm

computation as the highest strain observed were 0.025, well below the failure strain limit of 0.11 for Al 7075.

The radial extent of the permanent deformation thus computed is shown in Fig. 6.22. The figure shows the FSI simulation results for three collapse driving pressures: 6, 10 and 15 MPa. In all cases the bubble maximum radius prior to collapse was 2 mm and its standoff from the wall was 1.5 mm. The profiles of the permanent deformation generated on the plate resemble the pit shapes observed experimentally (see Fig. 5.13). This profile reflects the re-entrant jet impact location at $r = 0$, and the location of the bubble ring, which is indicated as the secondary dip at $r \sim 0.5\ R_{max}$ in the Gaussian-like curve. Both the depth and the radius of the pit vary with the driving pressure, i.e. higher collapse pressures result in deeper and wider pits. However, it appears that the pit depth has a much stronger dependence on the collapse driving pressure, while significant permanent deformation appears to mostly occur within a radius not much larger than the maximum bubble radius, R_{max}. For the three different collapse driving pressures, the ratio of depth over radius is on the order of 10^{-2}, indicating very shallow indentations.

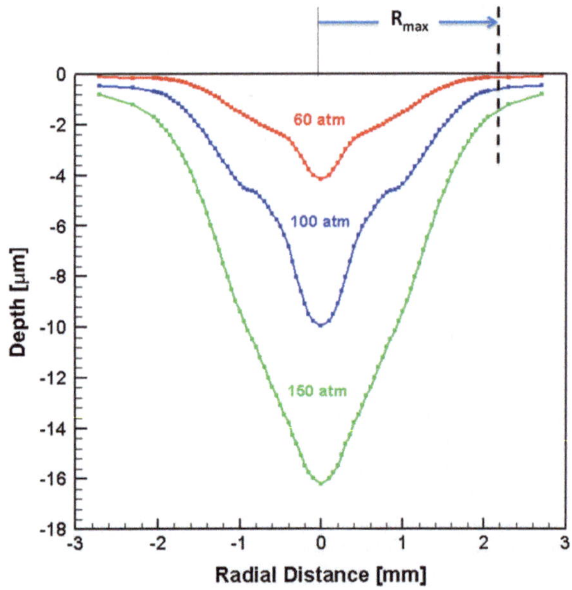

Fig. 6.22 Profile of the permanent deformation (pit depth versus r) on a 5 mm thick Al 7075 plate for different collapse driving pressures. Initial maximum bubble radius, $R_{max} = 2.0$ mm, and initial standoff distance, $X = 1.5$ mm

6.4.3 Scaling of Pit Size with Bubble Size and Driving Pressure

This section examines how pit depth and pit radius vary with the maximum radius of the collapsing bubble and with the amplitude of the collapse driving pressure. Under the same conditions as in Sect. 6.4.2, a bubble with the maximum radius of 2 mm is made to collapse near an Al 7075 plate with a thickness of 5 mm at a standoff distance of 1.5 mm. The bubble is subjected to several collapse driving pressures, $p_{collapse}$, varying from 2 MPa to 15 MPa. As shown in Fig. 6.23 the pit depth appears to vary linearly with $P_{collapse}$ in this range. No permanent deformation occurs if $P_{impulse}$ lies below the value of the yield stress input, 503 MPa, for the Al 7075 in the model, while the permanent depth of the pit attains 24 µm at $P_{impulse} = 1.9$ GPa.

Figure 6.24 shows the pit radius (defined by using a cutoff depth of 1 µm) and the pit depth normalized by the maximum bubble radius. In all these FSI numerical simulations the collapse driving pressure was maintained at 10 MPa. The figure illustrates that the pit radius strongly correlates with the maximum bubble radius. The pit radius remains close to R_{max} in the full range of bubble sizes considered. This is probably a result of the fact that, at a normalized standoff of 0.75, the bubble ring formed following the re-entrant jet impact has a radius close to R_{max} and is responsible for the width (radius) of the pit.

The pit depth on the other hand, appears to be more strongly dependent on R_{max} and increases as R_{max} increases. For $R_{max} < 500$ µm the pit depth is below the computational resolution for a collapse driving pressure of 10 MPa and a normalized standoff of 0.75.

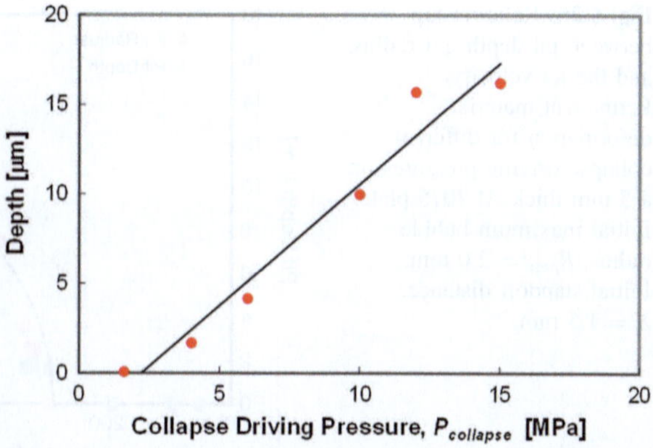

Fig. 6.23 Permanent deformation (maximum depth at the center of a 5 mm thick Al 7075 plate) for different collapse driving pressures. Initial maximum bubble radius, $R_{max} = 2.0$ mm. Initial standoff distance, $X = 1.5$ mm

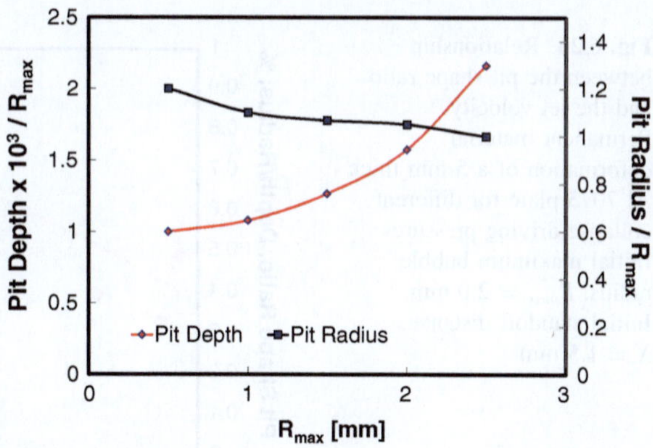

Fig. 6.24 Pit radius and pit depth for different initial maximum bubble radius, R_{max}, for bubble collapse on a 5 mm thick Al 7075 plate with the same collapse driving pressure of 10 MPa. Normalized initial standoff distance for all cases is 0.75

Another interesting information for scaling and/or modeling is the relationship between the pit characteristics and the re-entrant jet velocity at touchdown. Figure 6.25 shows that for jet velocities below 300 m/s there is no permanent deformation. For velocities higher than 300 m/s the pit depth and radius increase almost linearly with the re-entrant jet velocity. As previously discussed, higher jet velocities are associated with higher impact pressures (as was shown in Fig. 6.10) and this results in larger but still very low pit shape ratio (depth/radius) of less than 1 % as shown in Fig. 6.26.

6.4.4 *Effects of Load Duration and Lateral Extent on Permanent Deformation*

The rate of application of the impulsive loads onto the material being eroded has been briefly discussed in Chap. 4, and a nano-indentation experimental technique using quasi-static application of the load to the material was proposed to deduce

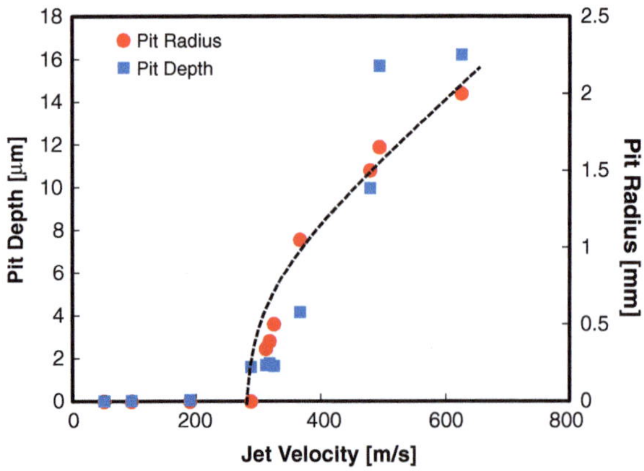

Fig. 6.25 Relationship between pit depth, pit radius, and the jet velocity. Permanent material deformation for different collapse driving pressures on a 5 mm thick Al 7075 plate. Initial maximum bubble radius, $R_{max} = 2.0$ mm. Initial standoff distance, $X = 1.5$ mm

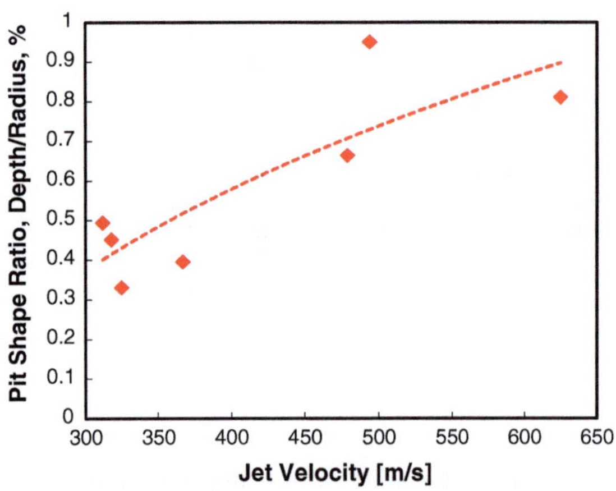

Fig. 6.26 Relationship between the pit shape ratio and the jet velocity. Permanent material deformation of a 5 mm thick Al 7075 plate for different collapse driving pressures. Initial maximum bubble radius, $R_{max} = 2.0$ mm. Initial standoff distance, $X = 1.5$ mm

the impulsive loads in an inverse approach in Chap. 3. From the material response viewpoint the question should be raised as to whether the material would deform in the same way when the load is applied at a rate below or above some material characteristic response time to be sought.

In order to further understand how the impulsive pressure magnitude, duration, and radial extent relate to the material deformation, a simplified synthetic pressure loading was specified without involving the flow field computation. The pressure loading can be described as a function of space and time with the doubly Gaussian profile:

$$P(r,t) = P_0\, e^{-\left(\frac{t}{\Delta t}\right)^2} e^{-\left(\frac{r}{\Delta r}\right)^2}, \tag{6.15}$$

where P_0 is the maximum pressure magnitude, Δt a characteristic duration, and Δr a characteristic radial extent of the pressure loading. Except for the presence of a double peak pressure when bubbles collapse with a small standoff distance, these

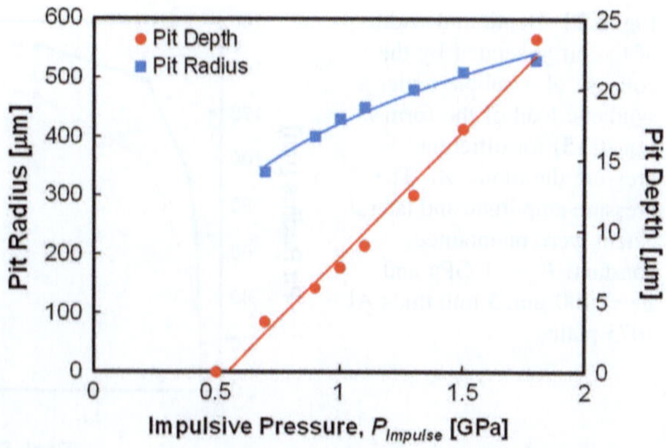

Fig. 6.27 Depth and width of the pit generated by the collapse of a bubble under a synthetic load of the form in Eq. (6.15) for different maximum pressure magnitude, P_0. The duration and radial extent were maintained constant: $\Delta t = 0.1$ µs and $\Delta r = 250$ µm. 5 mm thick Al 7075 plate

idealized profiles are reasonable approximations of those computed in the previous sections of this chapter and of experimental pressure loads recordings.

To extract the importance of P_0, Δt, and Δr, different pressure loads were applied on the Al 7075 plate and the deformations were computed using DYNA3D alone without FSI. Figure 6.27 shows the pit depth and pit radius for different imposed values of P_0 when Δt and Δr are maintained constant: $\Delta t = 0.1$ µs and $\Delta r = 250$ µm.

As predicted earlier with the FSI computations (Fig. 6.23), the pit depth increases linearly with increasing load amplitude. The pit radius is also predicted to increase as the collapse pressure increases but the rate of increase drops beyond an upper limit of the load, probably because the selected Δr is kept the same.

Figure 6.28 shows the effects of the load duration, Δt, on the pit depth and width when P_0 and Δr are kept constant ($P_0 = 1$ GPa and $\Delta r = 100$ µm). The figure clearly shows that both pit depth and pit radius increase nonlinearly with the duration of the applied load, Δt, up to about $\Delta t = 2$ µs in this case. For the considered values of P_0 and Δr, for Al 7075, and $\Delta t > 2$ µs, the permanent deformation becomes independent of the duration of application of the load. This implies that the dynamic effects are reduced with increasing Δt and the results become equivalent to quasi-static loading when the loading duration exceeds a certain period of time.

Figure 6.29 shows the effects of the radial extent of the applied load, Δr, on the pit depth and width when P_0 and Δt are kept constant ($P_0 = 1$ GPa, $\Delta t = 0.1$ µs). Here no saturation effect is shown (as for the peak duration shown in Fig. 6.28) and both pit depth and pit radius vary linearly with Δr. The linear variation of the pit width with Δr reconciles with what we observed in Fig. 6.24, i.e. the pit radius depends mainly on the extent of the area where the pressure load is imposed. The linear relationship between pit depth and Δr implies that the pit depth does not solely depend on the magnitude of the maximum pressure loading; it is also scaled with the radial extent through radial integration of the applied pressure.

Fig. 6.28 Depth and width of the pit generated by the collapse of a bubble under a synthetic load of the form in Eq. (6.15) for different pressure durations, Δt. The pressure amplitude and lateral extent were maintained constant: $P_0 = 1$ GPa and $\Delta r = 100$ μm. 5 mm thick Al 7075 plate

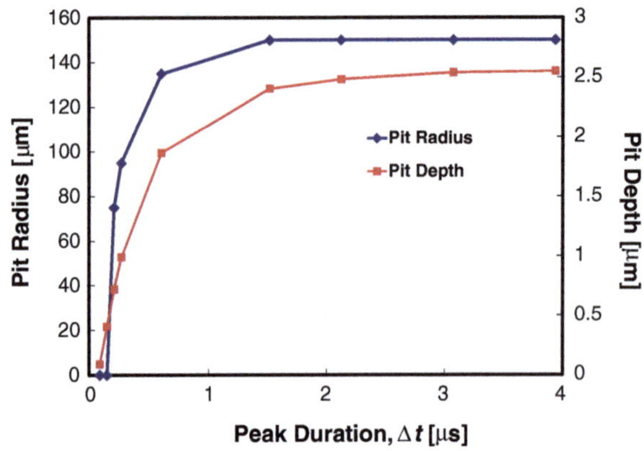

Fig. 6.29 Depth and width of the pit generated by the collapse of a bubble under a synthetic load of the form in Eq. (6.15) for different pressure lateral extent, Δr. The pressure amplitude and duration were maintained constant: $P_0 = 1$ GPa and $\Delta t = 0.1$ μs. Computations on a 5 mm thick Al 7075 plate

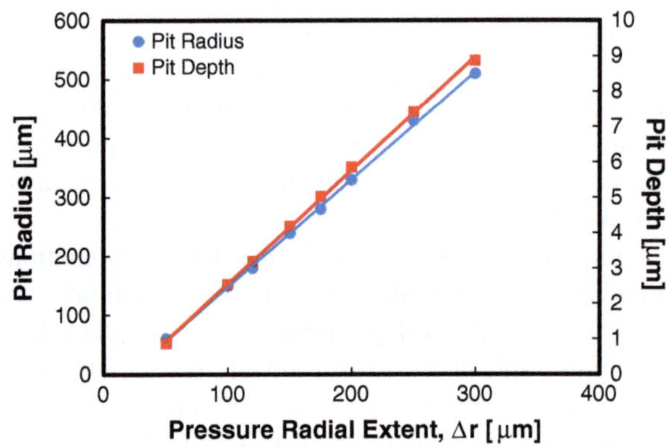

6.4.5 *Recovering Impulsive Load from Material Permanent Deformation*

This section addresses the potential of using pitting tests as a means of characterizing the cavitation flow field intensity. We consider as an illustration a series of computations, where a bubble of initial maximum radius 400 μm at equilibrium in a liquid of ambient pressure 0.1 MPa is subjected to a pressure rise to various maximum pressure levels, $P_{collapse}$, leading to energetic collapse. The bubble center is initially 400 μm away from a plate made of a material with known properties. Two-way fluid–structure interaction computations following the procedures described in the previous sections were conducted using GEMINI and DYNA3D. The computational results include bubble dynamics, pressure loads at the liquid-wall interface, and resulting material stresses, strains, and deformations. Four different values of $P_{collapse}$: 5, 6, 7, and 10 MPa, which resulted in $P_{impulse}$: 0.8, 1.2, 2.15, and 3.4 GPa, were used for all four materials studied earlier: NAB, stainless steel A2205, aluminum Al 7075, and Al 1100-H19.

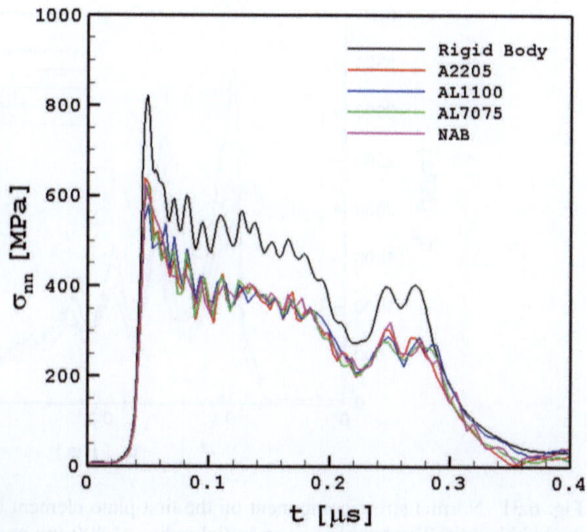

Fig. 6.30 Normal stress component on the first plate element below the interface plate-water at the bubble axis. The stresses are the material response to an impulsive loading due to a bubble collapse over the plate. The bubble had an initial radius of 400 μm and was located above the plate at a distance of 400 μm to the bubble center. The collapsing bubble is driven by pressure rise from 0.1 to 5 MPa

Figures 6.30 and 6.31 present the normal stresses σ_{nn} on the material during the bubble collapse for two values of $P_{collapse}$: 5 MPa and 10 MPa. Both figures show the influence of fluid–structure interaction on the results. When the response of the material, its deformation, and its feedback to the flow field are included, the normal stresses are noticeably reduced as compared to the rigid wall case. For $P_{collapse} = 5$ MPa, all four materials feel practically the same normal stresses, which are about 20 % lower than those felt by the rigid body (Fig. 6.30). For the higher excitation, $P_{collapse} = 10$ MPa, normal stresses show further decrease as the material gets softer, with the peak stresses almost 50 % lower than those felt by the rigid body for Al 1100-H19 (Fig. 6.31). This important effect has been observed experimentally in [50] through water jet impacting on an instrumented deformable plate; the measured pressures were significantly lower than when impacting a rigid plate.

This behavior may originate from the unsteadiness in the case of a deforming target. For a rigid flat wall, there is a direct balance between the normal impact force and the integral of the normal stresses, i.e. balance between the impact pressure, p, and the normal stress:

$$\int p\mathbf{n}ds = \int \sigma_{nn}\mathbf{n}ds, p = \sigma_{nn}. \tag{6.16}$$

For a deforming plate, an additional momentum term exists,

$$\int p\mathbf{n}ds{*}20c{=} \int \sigma_{nn}\mathbf{n}ds + \int \gamma dm, \gamma : \text{normal acceleration.} \tag{6.17}$$

The presence of the acceleration term reduces the value of σ_{nn}.

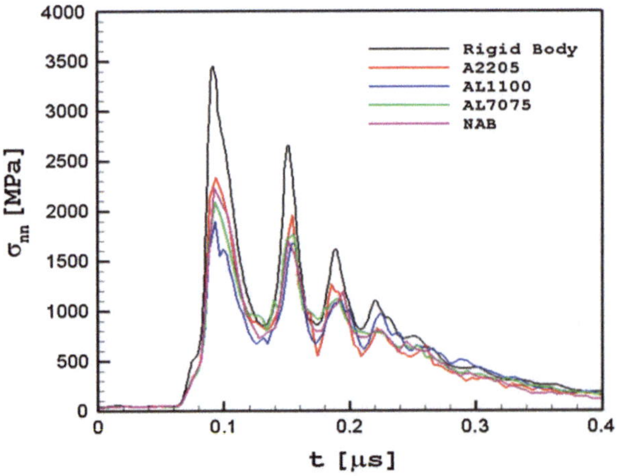

Fig. 6.31 Normal stress component on the first plate element below the interface plate-water at the bubble axis. The bubble had an initial radius of 400 μm and was located above the plate at a distance of 400 μm to the bubble center. The collapsing bubble is driven by a pressure rise from 0.1 to 10 MPa

Another important reason for the reduction in σ_{nn} is geometric. As the boundary deforms it deflects the impacting re-entrant jet and modifies the incoming pressure waves. This results in the otherwise only normal energy on the flat plate to become split between normal and other direction stresses. Figure 6.32 shows the radial stresses σ_{rr} on the first element below the free surface of the plate. The amplitude of σ_{rr} is of the same order of σ_{nn} explaining the noticeable reduction of the normal component. This amplitude is also seen to increase for the softer materials.

For simple uniaxial loading, a material will start yielding if the stress exceeds the material yield stress. For complex loading conditions as is the case for a cavitation impact, a three-dimensional system of stresses including normal and shear stresses develops in the material. To decide whether the combination of stresses will result in yielding, the Von Mises stress or equivalent stress may be computed and compared to the material yield stress. Figure 6.33 shows the equivalent stresses computed in the first element at the surface of each material for the same computations as in Figs. 6.31 and 6.32. One should note the sharp change in slope when the stresses approach the yield stress, partially explaining the stress drop. From that point on, the material follows the plastic portion of the stress–strain curve, which has a much milder slope, thus resulting in a significant change of the slopes of stress versus time and a significant reduction in the value of the maximum equivalent stress relative to the imposed impulsive pressure.

Figure 6.34 shows the vertical displacement versus time for the four materials. It is this quasi-final deformation that remains in the pits after exposure of the materials to the impulsive loads duc to the bubble collapse. The much larger maximum deformation in each of the curves of Fig. 6.34 is not recorded by the material and is unknown to the observer of the pit characteristics. The lengthy

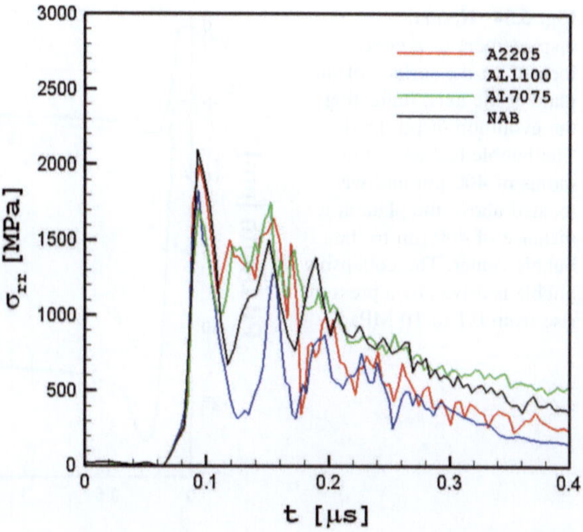

Fig. 6.32 Radial stress component on the first element below the free surface of the plate at the bubble axis. The bubble had an initial radius of 400 μm and was located above the plate at a distance of 400 μm to the bubble center. The collapsing bubble is driven by a pressure rise from 0.1 to 10 MPa

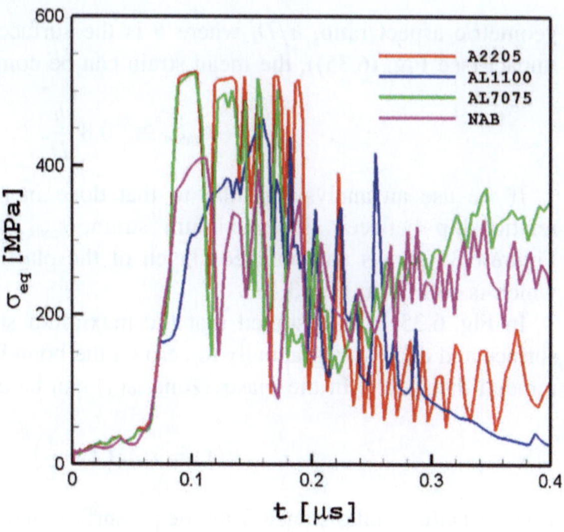

Fig. 6.33 Equivalent stress on the first element below the free surface of the plate at the bubble axis. The bubble had an initial radius of 400 μm and was located above the plate at a distance of 400 μm to the bubble center. The collapsing bubble is driven by a pressure rise from 0.1 to 10 MPa

computations in Fig. 6.34 were stopped at 3 μs, a relatively very long time after the loads due to the bubble collapse have died out (see Figs. 6.31 and 6.32). However, computations for a longer duration are needed until no change in the depth is observed; nevertheless the conclusions drawn here will not change significantly but towards slightly smaller permanent deformations.

An observer will measure the depth of the pit, h, and the pit diameter, D, obtained after using a cutoff depth value (e.g. 1 μm) as described in Chap. 3. An inverse computation of the load could then be made based on the method described in Sect. 3.4. Using the relationship between mean strain, ε_{mean}, and the pit

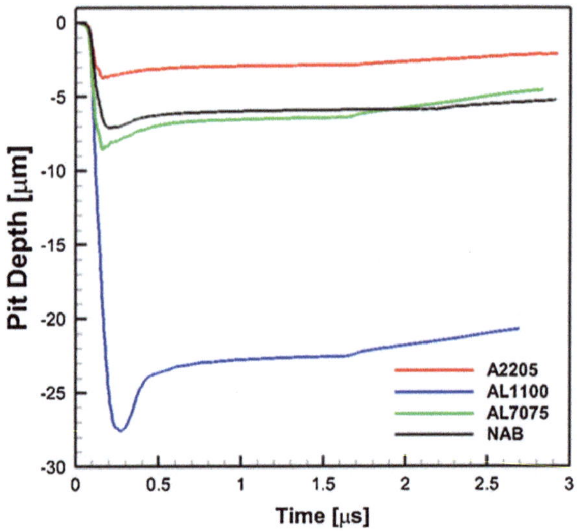

Fig. 6.34 Normal displacement of a node located on the surface of the plate at the axis, indicating the evolution of pit depth. The bubble had an initial radius of 400 μm and was located above the plate at a distance of 400 μm to the bubble center. The collapsing bubble is driven by a pressure rise from 0.1 to 10 MPa

geometric aspect ratio, h/D, where h is the surface deformation induced by an impact (see Fig. (6.35)), the mean strain can be computed as (see Eq. (3.12)):

$$\varepsilon_{mean} \cong 0.8\,\frac{h}{D}. \tag{6.18}$$

If we use an analysis similar to that done in [51, 52], we can deduce the relationship between the maximum strain, $\varepsilon_{\max}$, and the mean strain, ε_{mean}. Figure 6.35 shows an idealized sketch of the plastic zone of the cavitation pit, which is similar to Fig. 3.18.

In Fig. 6.35, it is assumed that the maximum strain, $\varepsilon_{\max}$, occurs on the pit surface and decreases gradually to zero on the boundary of the plastic zone within a radial distance ℓ. In the plastic zone, $\varepsilon(x)$ can be expressed as:

$$\varepsilon\left(x\right) = \varepsilon_{\max}\left(1 - \frac{x}{\ell}\right)^{\theta}, \tag{6.19}$$

where x is the axial distance from the pit surface and θ is a material parameter that defines the sharpness of the strain gradient around the impact.

By integrating $\varepsilon(x)$ from 0 to ℓ using Eq. (6.19) and dividing by ℓ, we obtain the mean strain:

$$\varepsilon_{mean} = \frac{\varepsilon_{\max}}{\theta + 1}. \tag{6.20}$$

Combining Eqs. (6.18) and (6.20), we obtain:

$$\varepsilon_{\max} \cong (\theta + 1)\; 0.8\,\frac{h}{D}. \tag{6.21}$$

Fig. 6.35 Idealized plastic zone diagram due to *spherical* indentation

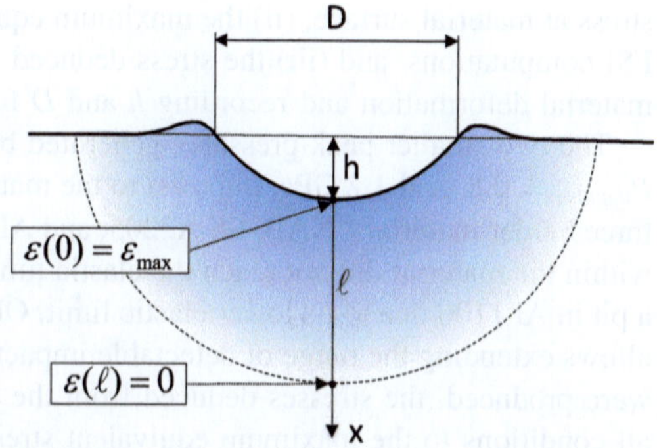

Typically, the maximum strain is 3 to 5 times the mean strain. The actual θ values were obtained experimentally using the nanoindentation method for three different materials: Al 7075, NAB and SS A2205 (see Fig. 7.2). These values are 2.0 for Al 7075, 2.8 for NAB and 2.2 for SS A2205. In the exercise summarized in Table 6.2, we used ε_{max} values obtained from Eq. (6.21) with the experimental θ values for the three materials. We also used, for order of magnitude evaluation, the value $\theta = 2.0$ for Al 1100 (the same value as for Al 7075) since an experimental value is not available.

Using the stress–strain relationship in Fig. 6.20, an effective stress (i.e. surface load) can be deduced from ε_{max}. Table 6.2 shows, for the four values of the maximum impulsive pressure $P_{impulse}$ considered here: (i) the maximum normal

Table 6.2 Impulsive pressure loads, $P_{impulse}$, imposed in the computation, resulting maximum equivalent stresses in the material and stress values deduced from the computed pits resulting from permanent deformation and Eq. (6.21). Pit diameter was computed using a cutoff depth of 1 μm. Bold values correspond to maximum equivalent stresses exceeding yield stress

Material	Yield stress [MPa]	Stress [MPa]	Bubble impulsive pressure peak $P_{impulse}$ [MPa]			
			800	1200	2150	3400
NAB	360	Max normal stress at material surface	606	879	1452	2196
		Max equivalent stress	360	**365**	**396**	**455**
		Stress deduced from pit	No pit	No pit	508	585
SS A2205	515	Max normal stress at material surface	628	888	1506	2310
		Max equivalent stress	511	515	**520**	**536**
		Stress deduced from pit	No pit	No pit	No pit	605
Al 7075	503	Max normal stress at material surface	608	900	1407	2082
		Max equivalent stress	503	503	**510**	**558**
		Stress deduced from pit	No pit	No pit	590	630
Al 1100	165	Max normal stress at material surface	570	829	1247	1878
		Max equivalent stress	**214**	**263**	**390**	**574**
		Stress deduced from pit	280	410	510	555

stress at material surface, (ii) the maximum equivalent stress, both obtained by the FSI computations, and (iii) the stress deduced from Eq. (6.21) after obtaining the material deformation and recording h and D for a cutoff depth of 1 μm.

The two smaller peak pressures generated by the bubble collapse in Table 6.2, $P_{impulse} = 0.8$ and 1.2 GPa, imposed to the materials are completely missed by the three harder materials: NAB, SS A2205, and Al 7075 because the Von Mises stress within the material did not reach the elastic limit. They do, however, both result in a pit in Al 1100 due to its lower elastic limit. Obviously, the use of softer materials allows extending the range of detectable impact loads to smaller values. When pits were produced, the stresses deduced from the pit analysis are close under almost all conditions to the maximum equivalent stresses in the material.

6.5 Summary

This chapter addressed the mechanisms of impulsive load generation on a rigid or deforming solid boundary from the collapse of a single bubble. The effects of the bubble maximum equivalent radius, R_{max}, the collapse driving pressure (i.e. the local pressure driving the bubble collapse), the bubble standoff from the boundary, and the plate material properties were investigated.

By means of a hybrid incompressible-compressible method, the computed pressures on the wall showed two major components: a water hammer pressure from the impact of the re-entrant jet on the solid surface and a shock wave pressure emitted from the bubble at or near its minimum volume (see Fig. 6.6). The re-entrant jet impact pressure is proportional to the product of the liquid impedance (product of sound speed and density of the liquid, ρc) and the re-entrant jet impact speed on the solid surface, V_{jet}. It has in general a lower value than the ideal water hammer pressure resulting from a water column impacting normally on a flat solid surface. It approaches this ideal value, $\rho c V_{jet}$, for very high collapse driving pressures. The shock waves can be spherical for bubbles at large standoff from the wall or toroidal for bubbles collapsing very close to the wall. In the region of normalized standoff, $X/R_{max} < 1$, both pressure types (water hammer and shock wave) appear to generally have the same order of magnitude.

The cavitation impulsive pressures strongly depend on the cavitation collapse driving pressure and to a much lesser extent on the bubble maximum radius, $R_{\max}$. Indeed, both the re-entrant jet velocity and the bubble equivalent radius velocity depend on the square root of the cavitation collapse driving pressure and are independent of $R_{\max}$. However, the loads imparted to the material depend on the area of the impacted or influenced region of the material and depend thus on $R_{\max}^2$. As a result the actual pit diameter depends on $R_{\max}$.

Study of two-way fluid–structure interaction for a bubble collapsing near a plastically deforming material shows the importance of not only the impulsive pressure amplitude but also its duration and radial extent. These finite quantities (duration and radial extent) directly affect the dimensions of the permanently deformed regions of the surface (pits). The unsteady response dependency on the impulsive load duration implies the existence of a material characteristic response time, which should be investigated in future studies.

The idea of deducing cavitation impact loads from basic nano-indentation equations applied to measured or computed pits was investigated. The analysis provided in this chapter shows that such an approach provides a good idea of the generated equivalent stresses in the material. However, in order to obtain the liquid generated pressures and enable accurate quantitative cavitation pressure field characterization, additional considerations and improvements are needed since the deduced pressures are close to the maximum equivalent von Mises stress generated in the material and not close to the imposed peak bubble impulsive pressure.

References

1. Chahine GL, Perdue TO, Tucker CB (1988) Interaction between an underwater explosion bubble and a solid submerged structure. DYNAFLOW INC., Technical Report 89001-1, Sept 1988
2. Chahine GL, Perdue TO (1988) Simulation of the three-dimensional behavior of an unsteady large bubble near a structure. In: Wang TG (ed) Drops and Bubbles, Third International Colloquium, Monterey, pp 188–199
3. Chahine GL, Kalumuck KM (1998) BEM software for free surface flow simulation including fluid-structure interaction effects. Int J Comput Appl Technol 11(3/4/5):177–198
4. Kalumuck KM, Duraiswami R, Chahine GL (1995) Bubble dynamics fluid-structure interaction simulation by coupling fluid BEM and structural FEM codes. J Fluids Struct 9:861–883
5. Kalumuck KM, Duraiswami R, Chahine GL (1995) Analysis of the response of a deformable structure to underwater explosion bubble loading using a fully coupled fluid-structure interaction procedure. SAVIAC Proceedings of the 66th Shock and Vibration Symposium, Vol I. Biloxi, MS
6. Chahine GL, Duraiswami R (1993) Method for calculating 2-D and 3-D underwater explosion bubble behavior in free water and near structures. NSWC Dahlgren Division Report NSWCDD/TR-93/44
7. Chahine GL, Duraiswami R, Kalumuck KM (1996) Boundary element method for calculating 2-d and 3-d underwater explosion bubble loading on nearby structures including fluid-structure interaction effects. NSWC Dahlgren Division Report NSWCDD/TR-93/46
8. Wardlaw AB, Luton JA, Renzi JR, Kiddy KC, McKeown RM (2003) The GEMINI Euler solver for the coupled simulation of underwater explosions. NSWCIHD/IHTR-2500
9. Wardlaw A Jr, Luton JA (2000) Fluid-structure interaction for close-in explosions. Shock Vibr J 7:265–275
10. Wardlaw A Jr, Luton JA, Renzi JJ, Kiddy K (2003) Fluid-structure coupling methodology for undersea weapons. Fluid Structure Interaction II. WIT Press, Boston, pp 251–263
11. Chahine GL, Raju R, Hsiao C-T (2010) Simulation of the impact pressures from non-spherical bubble collapse near a wall: handback from 3DYNAFS© to GEMINI. DYNAFLOW, INC. Technical Report 2M2M9007-NSWCIH-1
12. Hsiao C-T, Chahine GL (2010) Incompressible-compressible link to accurately predict wall pressure. SAVIAC, 81st Shock and Vibration Symposium, Orlando, October 24–28, 2010

13. Whirley RG, Engelmann BE (1993) DYNA3D: A nonlinear, explicit, three dimensional finite element code for solid and structural mechanics—user manual. Lawrence Livermore National Laboratory, UCRL-MA-107254, Revision 1
14. Chahine GL, Hsiao C-T (2012) Modeling microbubble dynamics in biomedical applications. J Hydrodyn 24(2):169–183
15. Hsiao C-T, Chahine GL, Liu H-L (2003) Scaling effects on prediction of cavitation inception in a line vortex flow. J Fluids Eng 125:53–60
16. Jameson A, Schmidt W, Turkel E (1981) Numerical simulation of the Euler equations by finite volume methods using Runge-Kutta time stepping schemes. AIAA paper 81-1259
17. Luton JA, Wardlaw A, Chahine GL, Hsiao C-T (2004) Simulation of surface vessel hull whipping response using fully-coupled hydrocodes. SAVIAC, 75th Shock and Vibration Symposium, October 18–22, 2004
18. Wardlaw AB, Luton JA, Renzi JR, Kiddy KC, McKeown RM (2003) The GEMINI Euler solver for the coupled simulation of underwater explosions. NSWCIHD/IHTR-2500
19. Plesset MS, Chapman RB (1971) Collapse of an initially spherical vapour cavity in the neighborhood of a solid boundary. J Fluid Mech 47(2):283–290
20. Ellis AT (1965) Parameters affecting cavitation and some new methods for their study. California Institute of Technology, Rep E-115.1
21. Naude CF, Ellis AT (1961) On the mechanism of cavitation damage by non-hemispherical cavities collapsing in contact with a solid boundary. Trans ASME, J Basic Eng 83(4):648–656
22. Blake JR, Taib BB, Doherty G (1986) Transient cavities near boundaries. Part I. Rigid boundary. J Fluid Mech 170(September):479–497
23. Zhang S, Duncan H, Chahine GL (1993) The final stage of the collapse of a cavitation bubble near a rigid wall. J Fluid Mech 257(December):147–181
24. Jayaprakash A, Chahine GL, Hsiao C-T (2012) Numerical and experimental study of the interaction of a spark-generated bubble and a vertical wall. J Fluids Eng 134:031301. doi:10.1115/1.4005688
25. Chahine GL, Annassami R, Hsiao C-T, Harris GS (2006) Scaling rules for the prediction of UNDEX bubble re-entering jet parameters. In: Pilkey W (ed) Critical technologies in shock and vibration, 4(1):1–12
26. Harris GS, Ilamni RZ, Lewis W, Rye K, Chahine GL (2009) Underwater explosion bubble phenomena tests near a simulated dam structure. Naval Surface Warfare Center, Indian Head Division IHTR 10-3055, November 2009
27. O'Daniel JL, Harris G, Ilamni R, Chahine GL, Fortune J (2011) Underwater explosion bubble jetting effects on infrastructure. US Army Engineer Research and Development Center report, DTIC, PDF Url: ADA545705, 31 March 2011
28. Harris G (2010) Multiple UNDEX bubble interaction tests. SAVIAC, 81st Shock and Vibration Symposium, Orlando, October 24–28
29. Lin JI (2005) DYNA3D: A nonlinear, explicit, three-dimensional finite element code for solid and structural mechanics. User Manual, Laurence Livermore National Laboratory, UCRL-MA-107254, January
30. van Wijngaarden L (1964) On the collective collapse of a large number of gas bubbles in water. In: Proceedings of 11th International Conference on Applied Mechanics, Springer, Berlin, pp 854–861
31. Mørch KA (1979) Dynamics of cavitation bubbles and cavitating liquids. Treatise Mater Sci Technol 16:309–355
32. d'Agostino L, Brennen CE (1983) On the acoustical dynamics of bubble clouds. ASME Cavitation and Polyphase Flow Forum, Houston, pp 72–76
33. Chahine GL (1983) Cavitation cloud theory. Proceedings, 14th Symposium on Naval Hydrodynamics, Ann Arbor, Michigan, National Academy Press, Washington, D.C, pp 165–194
34. Bremond N, Arora M, Ohl C-D, Lohse D (2006) Controlled multibubble surface cavitation. Phys Rev Lett A 96(22):224501

35. Chahine GL, Liu HL (1985) A singular perturbation theory of the growth of a bubble cluster in a superheated liquid. J Fluid Mech 156(July):257–279
36. Chahine GL, Duraiswami R (1992) Dynamical interaction in a multi-bubble cloud. J Fluids Eng 114(4):680–686
37. Chahine GL (2005) Numerical studies of the interaction of multiple underwater explosion bubbles. In: Pilkey W (ed) Critical technologies in shock and vibrations. 3(1):1–13
38. Harris G (2010) Small scale UNDEX bubble-dam interaction tests. SAVIAC, 81st Shock and Vibration Symposium, Orlando, October 24–28, 2010
39. Chahine GL (1991) Dynamics of the interaction of non-spherical cavities. In: Miloh T (ed) Mathematical approaches in hydrodynamics. SIAM Publications, Philadelphia
40. Chahine GL, Hsiao C-T, Raju R (2014) Scaling of cavitation bubble cloud dynamics on propellers Part II, Chapter 15. In: Kim K-H, Chahine GL, Franc J-P, Karimi A (eds) Advanced experimental and numerical techniques for cavitation erosion prediction, series fluid mechanics and its applications. Springer, Dordrecht
41. Knapp RT, Daily JW, Hammitt FG (1970) Cavitation. McGraw Hill Book Co., New York
42. Hammitt FG (1980) Cavitation and multiphase flow phenomena. McGraw-Hill International Book Co., New York
43. Young FR (1989) Cavitation. McGraw Hill Book Co., New York
44. Blake JR, Gibson DC (1987) Cavitation bubbles near boundaries. Annu Rev Fluid Mech 19:99–123
45. Best JP (1993) The formation of toroidal bubbles upon the collapse of transient cavities. J Fluid Mech 251(June):79–107
46. Brujan EA, Keen GS, Vogel A, Blake JR (2002) The final stage of the collapse of a cavitation bubble close to a rigid boundary. Phys Fluids 14(1):85–92
47. Krieger J, Chahine GL (2005) Acoustic signals of underwater explosions near surfaces. J Acoust Soc Am 118(5):2961–2974
48. Chahine GL (1982) Experimental and asymptotic study of nonspherical bubble collapse. Appl Sci Res 38:187–197
49. www.matweb.com (material property data)
50. Chahine GL, Kalumuck KM (1998) The influence of structural deformation on water jet impact loading. J Fluids Struct 12(1):103–121
51. Karimi A, Leo WR (1987) Phenomenological model for cavitation erosion rate computation. Mater Sci Eng 95:1–14
52. Francis HA (1976) Phenomenological analysis of plastic spherical indentation. J Eng Mater Technol 98(3):272–281

Chapter 7
Modeling of Material Response

Ayat Karimi and Jean-Pierre Franc

Abstract In this chapter, a model is proposed for analyzing the response of ductile materials to repetitive cavitation impacts. This model emphasizes those impacts whose amplitudes exceed the material rupture strength and cause material fracture and mass loss. Impacts whose amplitudes are between the material yield stress and rupture strength are assumed to progressively harden the superficial layers of the material and contribute to the erosion acceleration period. The hardening mechanism is described on the basis of hardness profiles measured on cross sections of eroded samples. In particular, a characteristic thickness of the hardened layers is introduced. The model leads to a simple equation for the prediction of the erosion rate during the steady-state period. The equation shows that it is proportional to the characteristic erosion rate defined as the ratio of the hardened layer thickness to the impact coverage time. This approach tends to prove that the length scale and time scale relevant to the erosion process are respectively the hardened layer thickness (mostly a material property) and the impact coverage time (mostly a flow property).

7.1 Phenomenological Model of Erosion

In spite of substantial progress in understanding mechanisms of cavitation erosion in the last decades, development of a reliable prediction method of material performance under cavitation loading conditions still remains a major challenge.

A. Karimi (✉)
EPFL, Lausanne, Switzerland
e-mail: ayat.karimi@epfl.ch

J.-P. Franc (✉)
LEGI, Grenoble, France
e-mail: jean-pierre.franc@legi.grenoble-inp.fr

K.-H. Kim et al. (eds.), *Advanced Experimental and Numerical Techniques for Cavitation Erosion Prediction*, Fluid Mechanics and Its Applications 106,
DOI: 10.1007/978-94-017-8539-6_7, © Springer Science+Business Media Dordrecht 2014

Most of the works dealing with the prediction of cavitation erosion are limited to establishing correlation functions between erosion rate and material properties obtained under conventional mechanical testing [1, 2], as well as between erosion rate and cavitation intensity index [3, 4] or cavitation acoustic power [5, 6]. Models based on curve-fitting methods are also available that attempt to predict the long-term erosion behavior from short-term erosion results [7]. A review of such models with the parameters necessary for the computation of erosion rate has been given in [8]. None of these methods gave satisfactory results over a wide range of applications, probably because of difficulties in accurately defining the erosive capability of a cavitating flow, and inability in precise identification of material response and relevant properties to cavitation loading.

In contrast to the above-mentioned attempts, we propose here an analytical erosion model which describes the erosion rate as an explicit function of cavitation loading and material response. The model takes into account essentially deformation processes and failure mechanisms that take place under cavitation loading which could be significantly different from bulk deformation under conventional mechanical testing such as tensile or fatigue tests.

We assume that during cavitation erosion the material is exposed to repeated impulsive loading whose amplitude, frequency, and distribution over the exposed area are determined by the flow conditions as described in the previous chapters. The impulsive loads can induce local deformation in the subsurface layers at high strain rates of about 10^3 to 10^4 s^{-1} and create pits of different shapes as shown in the examples of Fig. 3.17. The accumulation of local strains and the process of mechanical hammering of the workpiece gradually harden the near surface layers and, when the overall plastic strain of surface layers reaches the rupture limit of the workpiece material, failure occurs and cracks propagate through the subsurface layers leading to material removal and erosion.

As described in Chap. 5, the erosion rate versus time curve consists of different stages. The early stages comprising the incubation period and the acceleration period correspond to the accumulation of surface deformation and development of surface topography. During the acceleration period, the erosion rate continuously increases with exposure time (Fig. 7.1a) until it reaches the maximum. During the steady state period, the work hardening profile does not change much and the erosion rate remains almost constant (see Chap. 5). The steady-state period may be longer or shorter depending on the ductility of the material and the liquid flow pattern in the cavitation zone. Although not shown in Fig. 7.1a, an attenuation period can be observed in long term erosion experiments where the accommodation of surface profile and cushioning effect of liquid or diffused air trapped in the pockets and crevices of the eroded surface result in the reduction of erosion rate. The attenuation period is more dependent on the cavitation facility than on the intrinsic properties of material and is not considered in the present modeling.

Topography of the eroded surface at the early stage of erosion consists mainly of small dimples and tearing types rupture with microvoids as reported in Fig. 7.1c. As erosion progresses, surface roughness becomes more pronounced leading to the formation of typical hill and valley morphology with large scale

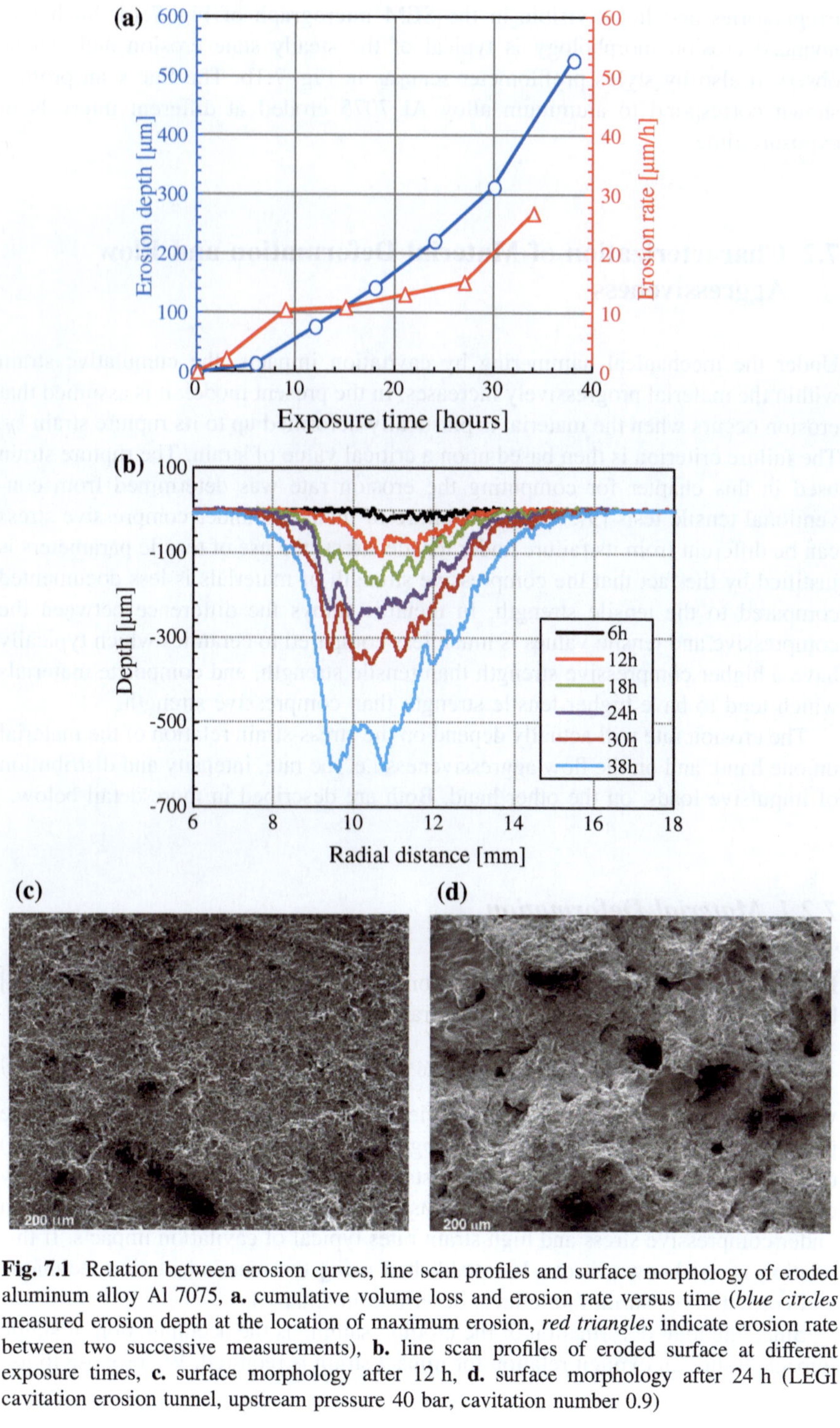

Fig. 7.1 Relation between erosion curves, line scan profiles and surface morphology of eroded aluminum alloy Al 7075, **a.** cumulative volume loss and erosion rate versus time (*blue circles* measured erosion depth at the location of maximum erosion, *red triangles* indicate erosion rate between two successive measurements), **b.** line scan profiles of eroded surface at different exposure times, **c.** surface morphology after 12 h, **d.** surface morphology after 24 h (LEGI cavitation erosion tunnel, upstream pressure 40 bar, cavitation number 0.9)

irregularities and holes visible in the SEM micrograph of Fig. 7.1d. Such pronounced erosion morphology is typical of the steady state erosion and can be observed also by stylus profilometer scan as in Fig. 7.1b. The line scan profiles shown correspond to aluminum alloy Al 7075 eroded at different intervals of exposure time.

7.2 Characterization of Material Deformation and Flow Aggressiveness

Under the mechanical hammering by cavitation impacts, the cumulative strain within the material progressively increases. In the present model, it is assumed that erosion occurs when the material is plastically deformed up to its rupture strain ε_R. The failure criterion is then based upon a critical value of strain. The rupture strain used in this chapter for computing the erosion rate was determined from conventional tensile tests [9]. Although failure of materials under compressive stress can be different from its failure under tensile stress, the use of tensile parameters is justified by the fact that the compressive strength of materials is less documented compared to the tensile strength. In metallic alloys the difference between the compressive and tensile values is much less compared to ceramics which typically have a higher compressive strength than tensile strength, and composite materials which tend to have higher tensile strength than compressive strength.

The erosion rate will actually depend on the stress-strain relation of the material on one hand, and on the flow aggressiveness i.e. the rate, intensity and distribution of impulsive loads, on the other hand. Both are described in more detail below.

7.2.1 Material Deformation

For material deformation, the Hollomon-type power law relationship [10, 11] between stress and amount of plastic strain will be used:

$$\sigma = \sigma_0 + k\varepsilon^n, \tag{7.1}$$

where σ is the applied stress, σ_0 is the yield stress, k is the strength index, ε is the plastic strain, and n is the strain hardening exponent. It is to be noted that Eq. (7.1) does not necessarily refer to tensile tests neither to compression tests, but may correspond more appropriately to the constitutive equation of material deformation under compressive stress and high strain rates typical of cavitation impacts. If this relation is not known, it can be established using the nanoindentation technique and correction of strain rate effects described in Chap. 3.

Since the total deformation of the eroding sample is the accumulation of single impact strains, an explicit relation for impact strain is required. We propose to use a power law for description of the radial distribution of strain in the plastic zone around impact [12, 13]:

$$\Delta\varepsilon_r = \Delta\varepsilon_0\left(1 - \frac{r}{l}\right)^{\theta}, \qquad (7.2)$$

where, $\Delta\varepsilon_0$ is the increment of strain at the impacted point, $\Delta\varepsilon_r$ is the strain induced by the impact at a radius r from the impact point, l is the depth of the hardened layer, which measures the radius of the plastic zone around impact, and θ is a material parameter that defines the sharpness of strain gradient around the impact. The parameter θ is dependent on stacking fault energy denoting microscopic aspect of deformation, e.g. planar dislocations or dislocation tangles and cells.

A relation similar to Eq. (7.2) is used to describe the deformation profile in an eroded sample:

$$\varepsilon_x = \varepsilon_R\left(1 - \frac{x}{L}\right)^{\theta}, \qquad (7.3)$$

where ε_R is the rupture strain, ε_x is the strain at depth x from surface, and L is the thickness of the hardened layer.

Figure 7.2 presents typical examples of hardness profiles measured in the cross-sections of eroded samples using the nanoindentation technique. For this measurement, the samples were cut using a wire saw in order to minimize the surface hardening. Then, they were mirror polished to remove the possible hardened layers. The experimental points are well fitted by the power law in Eq. (7.3). The thickness of the hardened layer L is a key parameter in the erosion rate computation since the erosion rate is actually proportional to L as shown in Sect. 7.3.2.

7.2.2 Cavitation Aggressiveness

Cavitation aggressiveness is characterized by a statistical distribution of impacts in terms of frequency, intensity, and size. For a steady state cavitation, the number of impacts per unit time and unit surface area can be considered as constant on the average. Regarding the radial extent of the load, even though the loads are distributed over a certain range of diameters, it will be assumed here that the surface area of the loads can be correctly characterized by a mean surface area denoted S_m.

The cavitation aggressiveness can be expressed by a probability density function of impacts. In a previous study [12], using a curve-fitting method it was shown that the Gamma-function type distribution fits well with the relationships between impact pressure and impact number:

$$\mathcal{P}(\sigma) = A\,\sigma^{\alpha-1}e^{-\frac{\sigma}{\beta}}, \qquad (7.4)$$

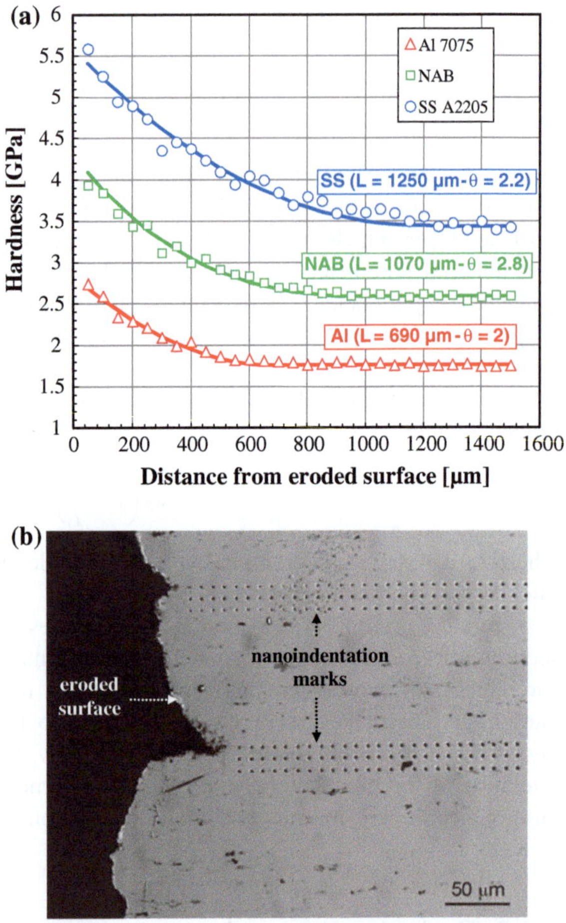

Fig. 7.2 Micro-hardness profile in eroded samples. **a.** Hardness versus depth for three materials (aluminum alloy Al 7075, nickel aluminum bronze alloy NAB and stainless steel A2205). *Symbols* are measured data on the cross-section of eroded samples using the nanoindentation technique. *Lines* are obtained from Eq. (7.3). The profiles clearly show the surface hardening effect due to cavitation impacts. Parameters L and θ were obtained by best fit. **b.** Optical micrograph showing location of indentation rows with respect to the eroded surface in Al 7075 sample (Measurements made at EPFL, Lausanne, Switzerland)

where σ is the impact pressure, $\mathcal{P}(\sigma)$ is the probability density of impacts whose amplitude is σ, α is a shape parameter and β is a scale parameter. The parameters A, α and β depend on the cavitation facility and flow conditions and should be determined experimentally using pressure transducers or pitting tests.

By changing α and β, a wide variety of distribution shapes can be obtained. In particular, by choosing $\alpha = 1$and $\beta = \sigma^*$, a simple exponential law is found which can be written, according to Eq. (3.19), as follows:

$$\mathcal{P}(\sigma) = \frac{N^*}{\sigma^*}\, e^{-\frac{\sigma}{\sigma^*}}, \tag{7.5}$$

As discussed in Chap. 3, σ^* is a reference stress connected to the slope of the exponential law and N^* is a reference rate of hydrodynamic impacts which controls the height of the curve. $\mathcal{P}(\sigma)$is interpreted as the rate of impacts of amplitude

σ within a bandwidth of 1 MPa and per unit surface area. This law proved to correctly account for the distribution of pressure pulses measured in a hydrodynamic tunnel (see Chaps. 3 and 4).

Two categories of impacts are distinguished with respect to the fracture strength of the eroded material, σ_R. The impacts with amplitude smaller than σ_R contribute only to work hardening and fatigue. The impacts with amplitude above the rupture strength can create local fracture and thereby generate erosion in addition to plastic deformation.

According to this assumption, the number of impacts per unit time directly involved in erosion, $\dot{n}_R$, is defined as the total number of impacts whose amplitude exceeds the material rupture strength σ_R. It is given by:

$$\dot{n}_R = \int_{\sigma_R}^{\infty} \mathcal{P}(\sigma) d\sigma. \tag{7.6}$$

Although the erosive impacts have various amplitudes, their amplitude will be characterized in the model by a mean value σ_m which can be determined from the probability density function (7.4):

$$\sigma_m = \frac{1}{\dot{n}_R} \int_{\sigma_R}^{\infty} \sigma \mathcal{P}(\sigma) d\sigma. \tag{7.7}$$

In the particular case of an exponential distribution (7.5), one has a simple equation for the number of erosive impacts per unit time and unit surface area:

$$\dot{n}_R = N^* e^{-\frac{\sigma_R}{\sigma^*}}, \tag{7.8}$$

and for their mean amplitude:

$$\sigma_m = \sigma_R + \sigma^*. \tag{7.9}$$

7.3 Erosion Rate Computation

7.3.1 Principle of the Model

The principle of the erosion rate computation is graphically presented in Fig. 7.3. The model proceeds by time steps corresponding to the repeated coverage of the material surface by cavitation impacts.

Initially, when the material is virgin, the damaging impacts are all the impacts whose amplitudes exceed the yield stress σ_0. Their rate is defined by:

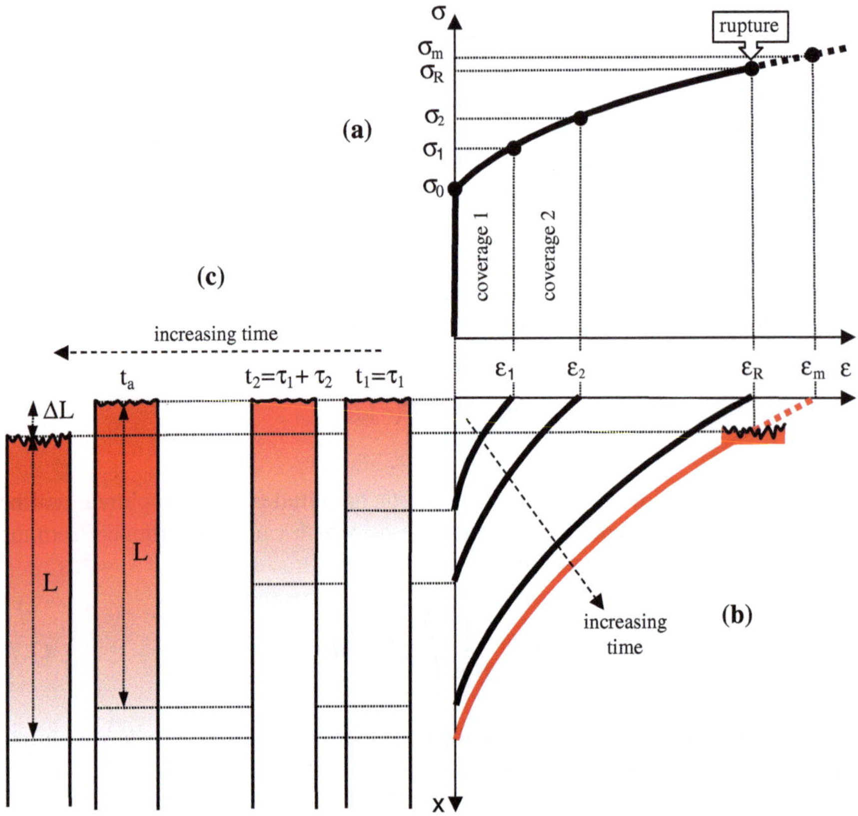

Fig. 7.3 Diagrams showing the principles of the erosion model, **a.** stress-strain variation at the end of each coverage time, **b.** impact induced strain profile at the end of each coverage, **c.** change in geometry of specimen as a function of time. t_a is the duration of the acceleration period, i.e. the time required to reach the steady state, as explained in Sect. 7.3.3

$$\dot{n}_1 = \int_{\sigma_0}^{\infty} \mathcal{P}(\sigma)d\sigma. \tag{7.10}$$

The time required for the surface to be covered by these impacts is the first coverage time given by:

$$\tau_1 = \frac{1}{\dot{n}_1\, S_m}, \tag{7.11}$$

where S_m is the average surface area of cavitation impacts. The idea of a coverage time was already introduced in Sect. 3.2.2.

The mean stress amplitude for the impacts whose amplitudes exceed the yield stress σ_0 is defined by:

$$\sigma_1 = \frac{1}{\dot{n}_1} \int_{\sigma_0}^{\infty} \sigma \mathcal{P}(\sigma) d\sigma. \tag{7.12}$$

Using the stress-strain curve (Fig. 7.3a), a strain ε_1 can be defined which corresponds to this σ_1 value. At the end of the first coverage, the strain will be ε_1 on the whole material surface and a strain gradient develops within the material as shown in Fig. 7.3b. The material flow stress increases from σ_0 to σ_1 during the first coverage. Here, the flow stress refers to the instantaneous value of yield stress that sustains plastic deformation according to stress-strain curve of Fig. 7.3a.

The second coverage involves only cavitation impacts whose amplitudes exceed the new flow stress σ_1. Their rate is defined by an equation similar to Eq. (7.10):

$$\dot{n}_2 = \int_{\sigma_1}^{\infty} \mathcal{P}(\sigma) d\sigma, \tag{7.13}$$

and their mean amplitude by an equation similar to Eq. (7.12):

$$\sigma_2 = \frac{1}{\dot{n}_2} \int_{\sigma_1}^{\infty} \sigma \mathcal{P}(\sigma) d\sigma, \tag{7.14}$$

At the end of the second coverage time defined by:

$$\tau_2 = \frac{1}{\dot{n}_2 \, S_m}, \tag{7.15}$$

the new flow stress for the material is σ_2 (Fig. 7.3a) and the surface strain is ε_2 corresponding to σ_2 (Fig. 7.3b).

The work hardening process is continued until the rupture strain ε_R is reached on the material surface. A steady-state regime of mass loss is then achieved.

7.3.2 Erosion Rate During the Steady-State Period

During the steady-state period, only impacts whose amplitudes exceed the material rupture strength σ_R are expected to damage the material. Their rate is $\dot{n}_R$ and their mean amplitude is σ_m as defined by Eqs. (7.6) and (7.7).

A key parameter in the model (in addition to the thickness L of the hardened layer mentioned above) is the coverage time i.e. the time required for the material surface to be covered by the erosive impacts. It is given by the following equation:

$$\tau_R = \frac{1}{\dot{n}_R \, S_{mR}}, \tag{7.16}$$

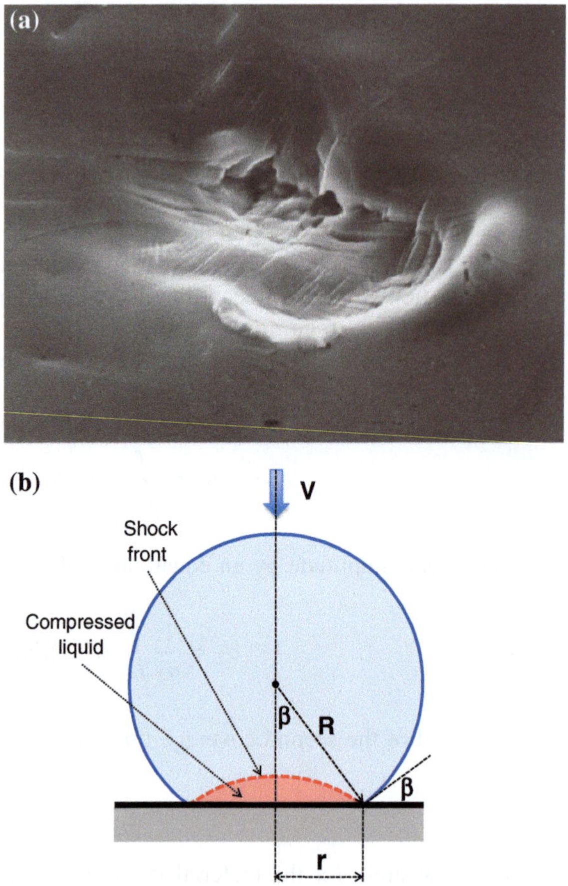

Fig. 7.4 Illustration of single impact damage by liquid impingement. **a.** Cavitation impact on stainless steel surface shows the ruptured area is only a fraction of impact area. **b.** Impact of spherical liquid drop on a rigid surface and zone of highly compressed liquid

which is very similar to Eqs. (7.11) and (7.15) except that it refers to the only impacts whose amplitudes exceed the rupture strength. In addition, S_{mR} is not a simple average surface area of impacts, but an average of ruptured areas. This differentiation arises from the fact that in an impacted area, only a fraction of the surface is ruptured and not the whole impacted surface as shown in Fig. 7.4a. So, S_{mR} in Eq. (7.16) is much smaller than S_m in Eqs. (7.11) and (7.15).

To estimate S_{mR}, a method based on the principles of plasticity is given in [12]. Here, an alternative solution based on high speed liquid impact theories [14–16] is proposed. Such an assumption is justified by the fact that the impingement of a cavitation re-entrant jet is comparable to the impingement of a high speed liquid drop with respect to the development of impact pressure, and deformation of the material.

When a liquid drop impacts a rigid surface, the initial phase of impact involves compression of the liquid and generation of shockwaves [14, 16] as illustrated in Fig. 7.4b. The liquid confined between the shock front and the solid surface near the contact zone is compressed, but the rest of the liquid drop is unaware of impact

[17]. Both the contact line and shock front expand over the impacted surface, but at some point the shock front detaches from the contact edge because the velocity of the contact line decreases in time and falls below the shock speed.

It is assumed that the ratio of the compressed area πr^2 to the impact area πR^2 is proportional to the ratio of the ruptured area to the whole contact area. The condition for the shock front to overtake the contact edge is [17]:

$$\sin \beta = \frac{V}{C}, \tag{7.17}$$

where, β is the contact angle, V is the impact velocity, and C is the shockwave velocity. The compressed area radius, r, at this stage is given by:

$$r = R \sin \beta = R \frac{V}{C}. \tag{7.18}$$

From Eq. (7.18), the ratio of ruptured surface, S_{mR}, to impact surface, S_m, can be estimated as:

$$\frac{S_{mR}}{S_m} = \left(\frac{V}{C}\right)^2. \tag{7.19}$$

In the acoustic limit, the shock velocity is assumed to be equal to the ambient speed of sound ($\sim$1,500 m/s), but during the first stage of impact the shock velocity can exceed the sound speed by several orders of magnitude as described in [15]. Assuming for example the mean velocity of shock waves is 2,000 m/s and the velocity of cavitation jets impacts is $V \approx 500$ m/s, then $S_{mR}/S_m \approx 0.062$. This typical order of magnitude shows that the ruptured area is significantly smaller than the impacted area. In terms of radius, the radius, r_{mR}, of the ruptured area is about 25 % of the radius r_m of the impacted area, $r_{mR}/r_m \approx 0.25$.

The steady-state regime of erosion occurs as soon as the strain on the material surface reaches the rupture strain ε_R. From this time, the strain profile inside the material does not change further and steady-state conditions are achieved. However, after each new coverage, the strain profile sinks into the eroded sample to a depth ΔL (see Fig. 7.3c) which corresponds to the thickness of the most superficial layer where the strain exceeds the rupture strain. This depth is supposed to be removed by cavitation erosion according to the failure criterion adopted here. This is the principle of computation of the erosion rate.

More precisely, let us assume that the material surface has been exposed to cavitation for a sufficient time so that the steady-state regime of erosion has been reached and the surface strain is ε_R. During further exposure to cavitation, new impacts of mean amplitude σ_m hit the material surface. After complete coverage of the surface, i.e. after one more time step τ_R, the surface strain is increased and its value reaches the mean value ε_m ($\varepsilon_m > \varepsilon_R$) associated with the mean stress σ_m (see Fig. 7.3a) and defined according to Eq. (7.1) by:

$$\frac{\varepsilon_m}{\varepsilon_R} = \left(\frac{\sigma_m - \sigma_0}{\sigma_R - \sigma_0}\right)^{1/n}. \tag{7.20}$$

The new strain profile inside the material is given by an equation similar to Eq. (7.3) where the thickness of the hardened layer is now $L + \Delta L$:

$$\varepsilon = \varepsilon_m \left(1 - \frac{x}{L + \Delta L}\right)^{\theta}. \tag{7.21}$$

To estimate ΔL, one observes that $\varepsilon = \varepsilon_R$ at depth $x = \Delta L$. This condition when introduced in Eq. (7.21) leads to the following equation:

$$\varepsilon_R = \varepsilon_m \left(1 - \frac{\Delta L}{L + \Delta L}\right)^{\theta}, \tag{7.22}$$

which allows the computation of the eroded depth ΔL:

$$\Delta L = \left\{ \left(\frac{\varepsilon_m}{\varepsilon_R}\right)^{1/\theta} - 1 \right\} L. \tag{7.23}$$

Using Eq. (7.20), this equation can be written in terms of stress rather than strain in the form:

$$\Delta L = \left\{ \left(\frac{\sigma_m - \sigma_0}{\sigma_R - \sigma_0}\right)^{1/n\theta} - 1 \right\} L. \tag{7.24}$$

The steady state erosion rate $\dot{E}$ is then given by:

$$\dot{E} = \frac{\Delta L}{\tau_R} = \frac{L}{\tau_R} \left\{ \left(\frac{\sigma_m - \sigma_0}{\sigma_R - \sigma_0}\right)^{1/n\theta} - 1 \right\}. \tag{7.25}$$

This equation shows that the erosion rate is strongly correlated with the ratio L/τ_R which is a reference erosion rate computed on the basis of the maximum thickness of hardened layer, L, and the coverage time, τ_R.

7.3.3 Duration of the Acceleration Period

As mentioned in Sect. 7.3.1, work-hardening develops during the acceleration period and the flow stress progressively increases. As an example, in the particular case of an exponential distribution for the impact loads such as Eq. (7.5), the successive flow stresses during the acceleration period are $\sigma_1 = \sigma_0 + \sigma^*$, $\sigma_2 = \sigma_1 + \sigma^* = \sigma_0 + 2\sigma^*$, and more generally $\sigma_i = \sigma_0 + i\,\sigma^*$ for coverage i. The acceleration time ended when σ_i exceeds the rupture strength σ_R. The number m of coverages required to reach the steady-state regime is then such that $\sigma_m > \sigma_R$ whereas $\sigma_{m-1} < \sigma_R$. Still in the particular case of an exponential distribution, the number of times full coverage is achieved, m, is then the lowest integer exceeding the ratio $(\sigma_R - \sigma_0)/\sigma^*$.

The duration of the acceleration period t_a is the time required to reach the steady-state regime. It is the sum of the coverage times until the end of the acceleration period, i.e.:

$$t_a = \sum_{i=1}^{m} \tau_i, \tag{7.26}$$

where the ith coverage time τ_i is defined by an equation similar to Eqs. (7.11) or (7.15):

$$\tau_i = \frac{1}{\dot{n}_i \; S_m}. \tag{7.27}$$

When introducing the coverage time τ_R during the steady-state period defined by Eqs. (7.16), (7.27) can be written in the form:

$$\tau_i = \frac{\dot{n}_R}{\dot{n}_i} \frac{S_{mR}}{S_m} \tau_R. \tag{7.28}$$

This equation shows that each coverage time during the acceleration period is proportional to the coverage time, τ_R, used for computing the erosion rate during the steady-state period. As a result, the acceleration time t_a is also proportional to τ_R. Thus, the coverage time, τ_R, is a relevant characteristic time not only for the computation of the steady-state erosion rate but also for the computation of the acceleration period.

It is to be noted that a coverage time can be defined for any stress value σ. It gives the time required for the material surface to be covered by impacts with an amplitude larger than σ. When pitting tests are considered, the characteristic stress to be chosen is the yield stress σ_0 since any impact whose amplitude is larger than the yield stress is supposed to make a permanent deformation. When mass loss is concerned, the present model suggests considering the rupture strength σ_R for defining the coverage time since it assumes that only the impacts whose amplitudes exceed the rupture strength will result in damage. In addition, it suggests using the ruptured surface area as opposed to the hydrodynamic impact area as the reference surface.

7.3.4 Erosion Rate During the Acceleration Period

There are two main reasons for lower erosion rate during the early stages of erosion, and its gradual increase with the exposure time to a maximum at the steady state period: (i) work hardening of near surface layers, and (ii) roughening of surface and propagation of micro-cracks.

At the very beginning of exposure to cavitation, the material is in general more ductile and dissipates a greater part of the impact energy by plastic deformation.

As exposure time increases, the eroded surface becomes work hardened and more brittle, and this ductile damping effect decreases while the part of impact energy dissipated by elastic deformation and rupture becomes more important.

In parallel to work hardening, surface roughness increases and micro-cracks develop, which can enhance the probability of micro-fractures and subsequent material loss.

Then, the erosion rate, $\dot{e}$, during the acceleration period after coverage i can be expressed as:

$$\dot{e} = \delta_i \chi_i \dot{E} \quad for \quad t < t_a, \tag{7.29}$$

where $\dot{E}$ is the final erosion rate during the steady-state period and δ_i and χ_i are the coefficients of work hardening and surface morphology at step i that both approach to 1 at the end of the acceleration period. A brief description of estimating both factors is given below, but more detailed procedure can be found in [12].

To estimate the coefficient of work hardening, the damage due to an impact striking a virgin surface is compared to that of an impact striking an already strained surface. The total work done by the impact in both virgin and strained samples is the same, and follows the relation below:

$$W_{total} = W_{elastic} + W_{plastic} + W_{rupture}. \tag{7.30}$$

At the beginning of a test, ductile displacement in the virgin material reduces the probability of local fracture and crack formation. With work hardening, material displacement becomes more difficult whereas local fracture and crack formation is favored. This is equivalent to an enhancement of the apparent deformation for the same impacts on the hardened sample. With this argument, one can obtain the following relation for the coefficient of work hardening [12]:

$$\delta_i = \frac{(n+1)\sigma_0 + \sigma_i}{(n+1)\sigma_0 + \sigma_R} \left(\frac{\sigma_i - \sigma_0}{\sigma_R - \sigma_0} \right)^{1/n}. \tag{7.31}$$

The coefficient of surface morphology, χ_i, is taken to be the fraction of the exposed surface which is ruptured or, in an equivalent manner, which has been subjected to impacts with amplitudes larger than the rupture strength. This reflects that, in any cavitation situation, only some impacts are able to produce a rupture or to propagate a crack in the impacted area. The already ruptured and cracked zones are regions more likely to break away and to originate erosion because of stress concentration effects.

7.4 Application of the Model

To test the validity of the erosion model, it is applied to three different alloys and the prediction is compared with the experimental measurements. The test materials are aluminum alloy Al 7075, nickel aluminum bronze C95400, and stainless steel

Table 7.1 Mechanical properties of test materials as obtained by nanoindentation and corrected for strain rate effects

	Aluminum alloy Al 7075	Nickel aluminum bronze C95400	Stainless steel A2205
E (GPa)	73	104	181
σ_0 (MPa)	600	866	1016
k (MPa)	1041	1180	724
n (-)	0.45	0.46	0.31
ε_R (MPa)	0.095	0.14	0.25
σ_R (MPa)	961	1344	1487
θ (-)	2.0	2.8	2.2
L (μm)	690	1070	1250

Rupture strain $\varepsilon_R$was deduced from conventional tensile tests and used to estimate the rupture strength on the basis of the stress—strain curve

A2205. The mechanical properties of the alloys are listed in Table 7.1. The parameters of the stress—strain curve (E, σ_0, σ_R, k and n) were determined using nanoindentation techniques as explained in [18] and corrected for strain rate of $\dot{\varepsilon} = 5 \times 10^3\ \mathrm{s}^{-1}$. Plastic zone parameters (L and θ) were obtained by performing nanoindentation measurements on cross-sections of highly eroded samples (see Fig. 7.2).

The erosion tests were conducted in the cavitation flow loop presented in Sect. 2.4 at a flow velocity of 90 m/s equivalent to an upstream pressure of 40 bar. The cavitation intensity was determined from combined pitting tests and nanoindentation tests. As shown in Sect. 3.4, the impact loads follow an exponential law given by Eq. (7.5) with the following parameters:

$$\begin{cases} \sigma^* = 171\ \mathrm{MPa}, \\ N^* = 22,200\ \mathrm{impacts/cm^2/s}. \end{cases} \tag{7.32}$$

The corresponding cumulative impact rate is shown in Fig. 7.5. For each material, the yield stress, σ_0, and rupture strength, σ_R, are indicated. The impact rate corresponding to the yield stress is actually the pitting rate whereas the impact rate corresponding to the rupture strength is the number of impacts directly involved in the erosion process previously denoted $\dot{n}_R$.

As for the mean diameter of impacted area, it was taken as the mean diameter of all pits measured on all three materials and was determined to be equal to 40 μm. This parameter is a purely hydrodynamic parameter which should be material independent. The mean surface area of impacts is then $S_m = 1,257\ \mu\mathrm{m}^2$ and the mean ruptured area is $S_{mR} = 78.5\ \mu\mathrm{m}^2$ assuming that the ratio S_{mR}/S_m is 0.062 as given by Eq. (7.19).

Table 7.2 gives the values of the different parameters which are involved in the computation of the erosion rate as well as a comparison between the computed and the measured erosion rate. The computed erosion rate is in good agreement with the measured one in the case of nickel aluminum bronze and stainless steel.

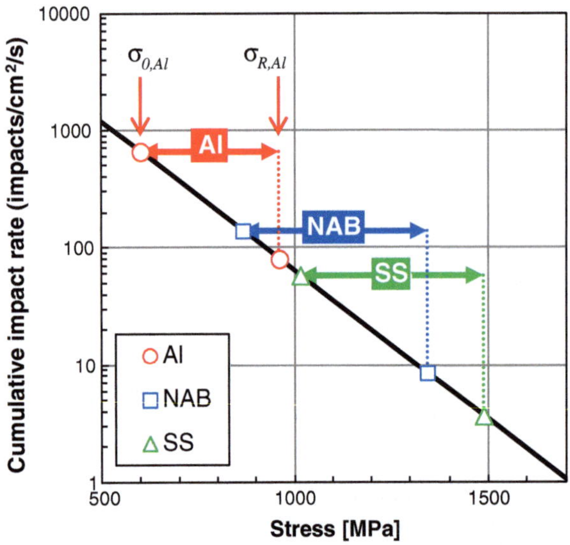

Fig. 7.5 Distribution of hydrodynamic impacts as a function of amplitude. The distribution is approximated by an exponential law as demonstrated in Chap. 3. The distribution is a purely hydrodynamic characteristic which measures the intensity of the cavitating flow. It is material independent and was determined from joint pitting tests and nanoindentation tests conducted on three different materials (see Chap. 3) (LEGI cavitation erosion facility, upstream pressure 40 bar, cavitation number 0.9)

Table 7.2 Comparison of the predicted and measured erosion rate for three different materials

	Aluminum alloy Al 7075	Nickel aluminum bronze C95400	Stainless steel A2205
Pitting rate based on yield stress σ_0 (pits/cm^2/s)	664	140	58
$\dot{n}_R$ from Eq. (7.8) (impacts/cm^2/s)	80	8.6	3.7
S_m (μm^2)	1257	1257	1257
S_{mR} (μm^2)	78.5	78.5	78.5
S_{mR} (cm^2)	78.5×10^{-8}	78.5×10^{-8}	78.5×10^{-8}
τ_R from Eq. (7.16) (h)	4.4	41	95
L/τ_R (μm/h)	157	26	13
σ_m from Eq. (7.9) (MPa)	1132	1515	1658
ε_m [-]	0.225	0.272	0.679
$K = \left(\frac{\sigma_m-\sigma_0}{\sigma_R-\sigma_0}\right)^{1/n\theta} - 1 = \left(\frac{\varepsilon_m}{\varepsilon_R}\right)^{1/\theta} - 1$	0.538	0.268	0.575
Predicted erosion rate $\dot{E} = KL/\tau_R$ from Eq. (7.25) (μm/h)	**84**	**7.0**	**7.5**
Measured erosion rate $\dot{E}$ (μm/h)	**21**	**8**	**8**

The table gives the values of the various parameters used in the model (LEGI cavitation erosion facility, upstream pressure 40 bar, cavitation number 0.9)

The prediction however is less accurate for the aluminum alloy, due probably to its relative high ductility that may lead to overestimated pitting rates. Even though the model still needs to be improved, it appears that Eq. (7.25) gives a reasonable order of magnitude of the erosion rate.

This tends to confirm that the thickness of the hardened layer and the coverage time are actually two key parameters in cavitation erosion prediction. The thickness of the hardened layer is a metallurgical parameter characterizing the work-hardening process and the deformation within the material whereas the coverage time which is a combination of the rate and size of impacts is a hydrodynamic parameter characterizing the flow aggressiveness. Flow aggressiveness is characterized additionally by the mean amplitude of impact loads exceeding the material rupture strength which appears in the multiplicative factor K.

In its current version, the model is still crude and based on a series of simplifying assumptions in order to keep it analytical. In particular, the whole distribution in size and amplitude of the impact loads is ignored and replaced by average values. The amplitude of impact loads itself was determined using a simple analogy between a cavitation pit and spherical indentation. The three-dimensional nature of the strain field below the impact is ignored. Failure is accounted for by a simple criterion based upon a threshold value of the strain determined from conventional tensile tests. The fraction of the impacted surface which is actually ruptured is still largely unknown. In spite of all these simplifications or unknowns, the model gives a general scheme for predicting cavitation erosion damage which drastically differs from conventional correlative techniques. It is based upon a quantification of the aggressiveness of the cavitating flow in term of impact loads and a physical analysis of the response of the material to the repetitive and overlapping impact loads.

7.5 Summary

In this chapter, a model was proposed to analyze the response of a ductile material to cavitation impacts. According to this model, the cavitation erosion rate $\dot{E}$ during the steady state period is given by the following equation (see Eq. (7.25)) in terms of both fluid and material properties:

$$\dot{E} = \frac{L}{\tau_R}\left\{\left(\frac{\sigma_m - \sigma_0}{\sigma_R - \sigma_0}\right)^{1/n\theta} - 1\right\}.$$

On the fluid side, the model emphasizes those impacts whose amplitudes exceed the material rupture strength σ_R, and which are supposed to fracture the material, thus responsible for mass loss. Although those impacts are distributed over a range of amplitudes, the model assumes for simplicity that their amplitudes can be characterized by a mean value, σ_m, that can easily be deduced from their statistical distribution. In addition to the amplitude parameter σ_m, the damaging impacts are characterized by their coverage time, τ_R, defined by Eq. (7.16). The coverage time is the time required for the material surface to be covered once by the cavitation impacts. Equation (7.16) combines the rate of impacts, $\dot{n}_R$, that hit a unit surface area of the material and the mean surface area, S_{mR}, which is ruptured under each impact.

On the material side, in addition to the fracture strength, σ_R, and the yield stress, σ_0, the ductile behavior of the material is accounted for by a surface hardening effect due to cavitation impacts. A power law was chosen to fit the variation of micro-hardness with the depth from the eroded material surface. It is characterized by two parameters, a typical thickness of the hardened layer, L, and a shape factor θ, both appearing in the above equation. The parameter n is the strain hardening exponent of the Hollomon-type stress-strain relationship chosen here.

An essential conclusion drawn from the above equation is that the erosion rate is proportional to the ratio, L/τ_R, which has the dimension of an erosion rate. In other words, this model indicates that the characteristic length of the erosion phenomenon is the thickness of the hardened layer, which is mainly a material parameter, whereas the characteristic time of the phenomenon is the coverage time, which is for the most part a hydrodynamic parameter. The model also leads to the conclusion that the coverage time is a relevant time scale to compute the duration of the incubation period.

References

1. Heymann FJ (1970) Towards quantitative prediction of liquid impact erosion. In: Characterization and development of erosion resistance, ASTM Specical Technical Publication vol 474. pp 212–248
2. Hammitt FG, Huang YC, Kling CL, Mitchell TM, Salomon LP (1970) A statistically verified model for correlating volume loss due to cavitation or liquid impingement. In: Characterization and determination of erosion resistance, ASTM Specical Technical Publication vol. 474. pp 288–322
3. Steller J, Krella A, Koronowicz J, Janicki W (2005) Towards quantitative assessment of material resistance to cavitation erosion. Wear 258(1–4):604–613. doi:10.1016/j.wear.2004.02.015
4. Kato H, Konno A, Maeda M, Yamaguchi H (1996) Possibility of quantitative prediction of cavitation erosion without model test. J Fluids Eng 118(3):582–588
5. Lush PA, Selim SMA, Studd LW, Angell B (1979) The relation between cavitation noise and erosion. In: Field J (ed) Proceedings 5th international conference on erosion by liquid and solid impact, Cambridge, UK, Paper 62–1
6. Hattori S, Sun B-H, Hammitt FG, Okada T (1985) An application of bubble collapse pulse height spectra to venturi cavitation erosion of 1100–0 aluminum. Wear 103(2):119–131. doi:10.1016/0043-1648(85)90128-0
7. Noskievic J (1983) The extended mathematical model of cavitation and erosion wear. In: Field JE, Corney NC (eds) Proceedings 6th international conference on erosion by liquid and solid impact, Cambridge, UK, paper no 7, pp 1–9
8. Rao PV, Buckley DH (1984) Predictive capability of long-term cavitation and liquid impingement erosion models. Wear 94:259–274

9. Franc J-P, Riondet M, Karimi A, Chahine GL (2012) Material and velocity effects on cavitation erosion pitting. Wear 274–275:248–259. doi:10.1016/j.wear.2011.09.006
10. Hollomon JH (1945) Tensile deformation. Trans AIME 162:268
11. Hollomon JH, Lubahn JD (1946) Plastic flow of metals. Phys Rev 70:775
12. Karimi A, Leo WR (1987) Phenomenological model for cavitation rate computation. Mater Sci Eng 95:1–14
13. Francis HA (1976) Phenomenological analysis of plastic spherical indentation. J Eng Mater Technol 98(3):272–281
14. Heymann FJ (1969) High-speed impact between a liquid drop and a solid surface. J Appl Phys 40(13):5113–5122. doi:10.1063/1.1657361
15. Lesser MB (1981) Analytic solutions of liquid-drop impact problems. Proc R Soc Lond A: Math Phys Sci 377(1770):289–308. doi:10.1098/rspa 1981.0125
16. Lesser MB, Field JE (1983) The impact of compressible liquids. Annu Rev Fluid Mech 15(1):97–122. doi:10.1146/annurev.fl.15.010183.000525
17. Dear JP, Field JE (1987) Paper 4. In: Field JE, Dear JP (eds) 7th international conference on erosion by liquid and solid impact, Robinson College, Cambridge, UK, 7–10 Sept 1987
18. Carnelli D, Karimi A, Franc J-P (2012) Application of spherical nanoindentation to determine the pressure of cavitation impacts from pitting tests. J Mater Res 27(1):91–99. doi:10.1557/jmr.2011.259

Part II
Selected Papers from the International Workshop on Advanced Experimental and Numerical Techniques for Cavitation Erosion Prediction (Grenoble, March 1–2, 2011)

Chapter 8
Hydrodynamic Processes Controlling Cavitation Erosion

Göran Bark and Rickard E. Bensow

Abstract In this paper we consider development of cavitation erosion having its origin in the sheet cavitation on the hydrofoils and propeller blades. The discussion includes generation of cloud cavitation by processes related to sheet cavity dynamics and how a cloud collapse can be enhanced by energy cascading from the collapse of a sheet cavity into the collapse of a cloud. A decomposition of the cavitation process leads to the basis for a conceptual model for description and analysis of the generation of erosion by mixed glassy and cloud cavitation. This model is made up of ten Analysis Models identifying the large to moderately small-scale events to be traced for unfolding behaviors that can control erosion.

8.1 Background and Present Approach

We start with a brief review of the generic example of erosive cavitation shown in Fig. 8.1. This cavitation in the propeller root region develops from a large sheet cavity generating three erosion regions, of which two are shown in the figure. The root cavity is beneficial for isolating relevant events and this selection is assumed not to limit the conclusions.

The propeller is mounted in the upstream end of the shaft having an 8-degree inclination relative the incoming homogenous flow. This means that the propeller blades experience a periodic unsteady inflow. The narrow sheet cavity in frame 1 is initially attached to the leading edge but is moving slowly downstream in frame 2. Due to a jet flow filling of the sheet cavity, its closure region is moving upstream. This jet starts as a re-entrant jet, but is now enhanced by flows induced by the shed vortices and by the increasing collapse forcing pressure on the blade.

G. Bark (✉) · R. E. Bensow
Department of Shipping and Marine Technology, Chalmers University of Technology, Gothenburg, Sweden
e-mail: bark@chalmers.se

K.-H. Kim et al. (eds.), *Advanced Experimental and Numerical Techniques for Cavitation Erosion Prediction*, Fluid Mechanics and Its Applications 106,
DOI: 10.1007/978-94-017-8539-6_8, © Springer Science+Business Media Dordrecht 2014

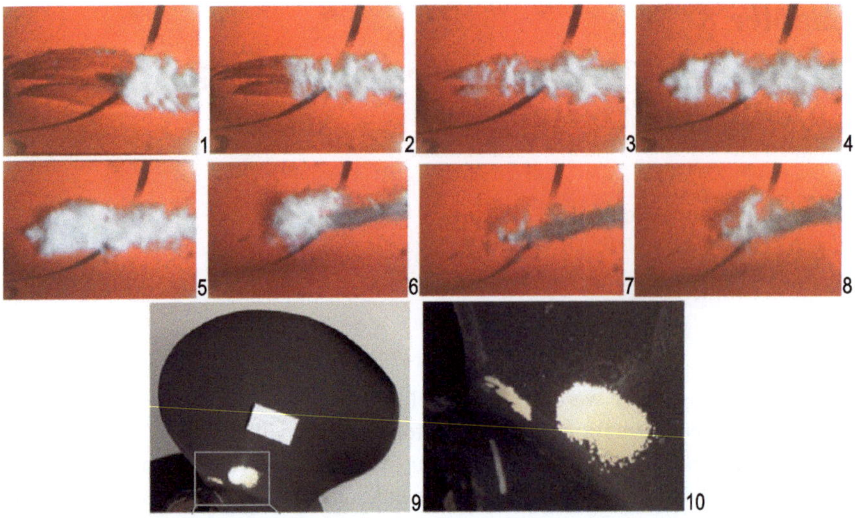

Fig. 8.1 Frames 1–8 are samples from a high-speed film of a sheet cavity in the root region of a propeller in hė SSPA cavitation tunnel. The smaller upstream white patch in the root region, photos 9 and 10, shows the erosion of soft paint due to collapses of the glassy sheet cavity with the attached bubble cloud, frames 2–3. The larger white patch is due to the collapses of the clouds in frames 5 and 8. Frame 8 shows the rebound of the cloud in frame 5. From [2]

Shear interaction between the filling flow and the flow outside the sheet cavity generates shedding of vortex cloud cavities in the closure region. There are two erosive collapses shown in the frames. The first is the collapse of the two glassy parts finished in frame 3, and the second is the cloud collapse occurring around frame 7. The collapse of the glassy sheet cavity contributes to the erosion (paint wear) in the small upstream patch in the photographs 9 and 10 and the pure cloud collapses generate the larger downstream patch. The area of the downstream patch is enlarged by scattering of the cloud collapses, and by a partial overlap with the collapse area of the later collapsing rebounded cloud that is visible in frame 8.

Presently, assessments of propeller erosion are usually based on the analysis of model scale experiments by:

(a) Visual assessment of the wear/erosion of a soft paint being exposed to cavitation for a certain time, Fig. 8.1, and
(b) Visual assessment of the cavitation aggressiveness based on high-speed video recording of cavity collapses.

None of the methods are strictly quantitative, but some assessments of the cavitation aggressiveness are often made in both methods, under the assumption that the soft paint is a rough but useful scaling to the full-scale propeller material. Without supplementary observations of the cavitation a bare paint test does not reveal much information about the hydrodynamics behind the erosion. Assessment of propeller erosion by paint tests, empirically calibrated by model to full-scale correlation of erosion data, can be quite reliable however.

The high-speed video analysis of the cavitation development leads to useful information about the involved hydrodynamics, but in this method the erosion sensor is replaced by an approximate and partly subjective analysis of the collapse kinematics. The video and paint methods do, however, supplement each other, partly because they usually fail for different reasons. A third method, based on a recording of the structure noise created by cavity collapses on a propeller blade, has so far been applied at full-scale propellers [9]. An emerging technique is advanced numerical simulations of cavitation, permitting analysis of kinematics and collapse pulses; [6, 11, 23, 24, 26, 28, 30]. It is to be noted in particular that Li presented a method to use the collapse pulse steepness as an erosion indicator.

Principles for erosion assessment based on cavitation kinematics and statistics are described in the EROCAV observation handbook [2], supplemented by [4, 13, 14]. The guidelines were primarily written for the analysis of experimental model or full-scale data, while keeping numerical simulations also under consideration. The present paper updates and extends the analysis by [1] and covers processes that can be investigated by numerical simulation.

We further develop the decomposition of the cavitation process as was previously suggested by [4]. The study is primarily based on experimental recordings, but numerical simulations are also used for supplementary visualization of events that are extremely difficult to record experimentally. Implicit Large Eddy Simulations (ILES) for cavitating flow are used for these supplementary studies [7].

From high-speed video recordings of cavitation on propellers and foils, certain processes have been identified as crucial in the development of erosive cavity collapses. To support the analysis and description of cavitation developments as in Fig. 8.1, idealized definitions of main and subprocesses are provided in ten "Analysis Models". The Analysis Models are presented as propositions based on experimental observations, some numerical results and imbedded hypotheses about the processes. The main and subprocesses constitute a particular decomposition of the cavitation process into processes that can be observed to a large extent.

The Analysis Models cover large to moderately small scales, thus excluding extreme details of the final collapse and rebound of a cavity, such as individual bubbles in clouds and very late and small-scale cavity disintegrations. This limitation of scales resolved in the analysis matches the resolution of numerical and experimental methods needed to capture the large scales related to design.

8.2 Erosion Due to Sheet and Cloud Cavitation

The classic assumption is that cavitation erosion primarily results from a collective collapse of the bubbles in a cloud and is influenced then by the small-scale bubble–bubble interactions, e.g. [16, 17]. An issue in experimental recording and

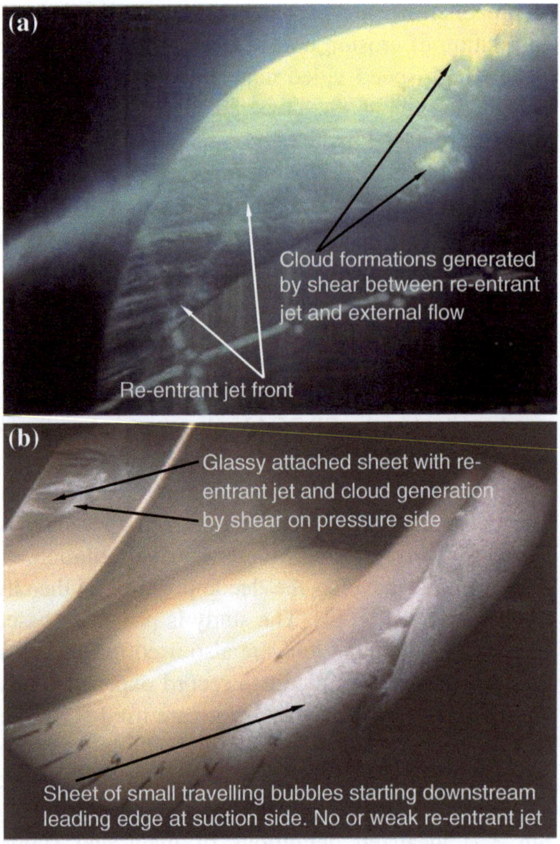

Fig. 8.2 Cavitation on full scale propellers. **a** A mainly glassy sheet cavity with a re-entrant jet on the suction side. Photo by SSPA. **b** A glassy sheet cavity with a re-entrant jet on pressure side (*left*) and a sheet of travelling bubbles, a cloud sheet, on the suction side (*right*). The bubbly sheet far to the right starts close to leading edge. Photo by DTMB. From [20, 21]

numerical simulation of erosive cavitation lies in determining the most large and small-scale events that need to be captured with appropriate resolution in time and space in order to understand, control and predict the hydrodynamics of erosion. This problem is present in all cases illustrated in Figs. 8.1–8.12.

The main ideas applied in the EROCAV handbook to determine the required range of scales and resolution at tracing cavity collapses are summarized as follows. To trace the energy cascading to the last collapsing bubble in a collective cloud collapse requires capturing the initial conditions, i.e. large scale quantities, and hence we understand that erosion is not only a small-scale event. Actually it was assumed that the initial conditions and the cascading process significantly control the amount of energy that finally is exposed to the solid body. For the collapse of a glassy sheet cavity, the experimentally based hypothesis introduced was that collapse energy could be cascaded by a sheet collapse as well, to the

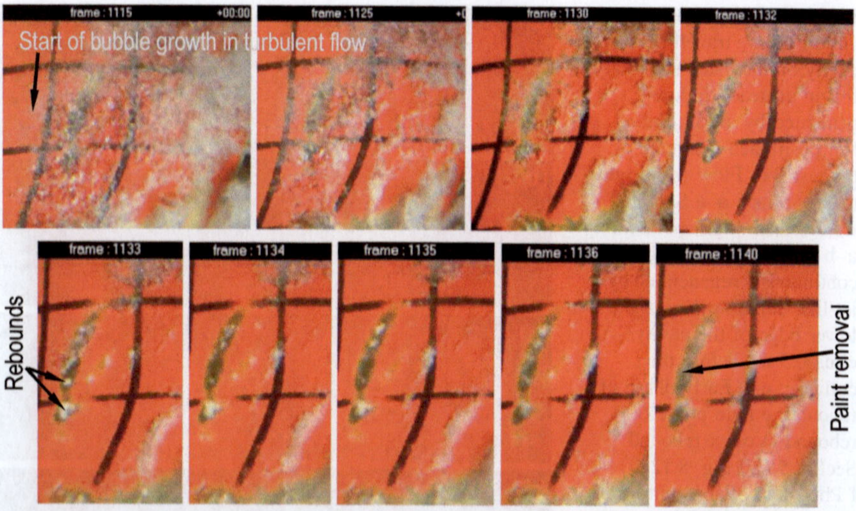

Fig. 8.3 Sheet of travelling bubbles performing erosive scattered collapses on a model propeller blade. Flow from left to right. Notice dispersion into several collapse points, indicated by the intensely white rebound spots, frames 1133–1135. Unit step in frame number equals 1/30,000 s. SSPA cavitation tunnel. From [4], with permission from AMT11

collapse of a cloud attached to the sheet, thereby enhancing the cloud collapse. The approach is to trace the energy cascading as close as possible, to the last collapsing bubbles, looking for a fast and possibly accelerating collapse into a small area, a collapse "point". In this process losses of collapse energy, due to cavity disintegration and scatter of the collapse point over cavitation cycles, are traced. The collapse pulse should preferably be recorded as well. In experiments, confirmation of a fast collapse into the small area is made by observation of the rebound. If the rebound cannot be checked, erosion assessment gets increasingly uncertain at an earlier completion of the collapse tracing.

A hydrodynamically generated cavity attached to or travelling on a body surface usually contains cloud formations. Depending on the amount of cloud formations in a cavity, it is here called a mixed glassy and cloudy, partially, mainly or almost glassy cavity. Sheet cavities at full scale often start as a bubble cloud growing in a turbulent boundary layer but in other cases sheet cavities can be mainly glassy also at full scale, Fig. 8.2. A sheet of travelling bubbles can also develop at model scale, Fig. 8.3.

Assuming thus that cloud cavitation is necessary to obtain sufficient energy focusing and that a cloud collapse can be enhanced by a sheet collapse, we list basic ways by which cloud cavitation can be created:

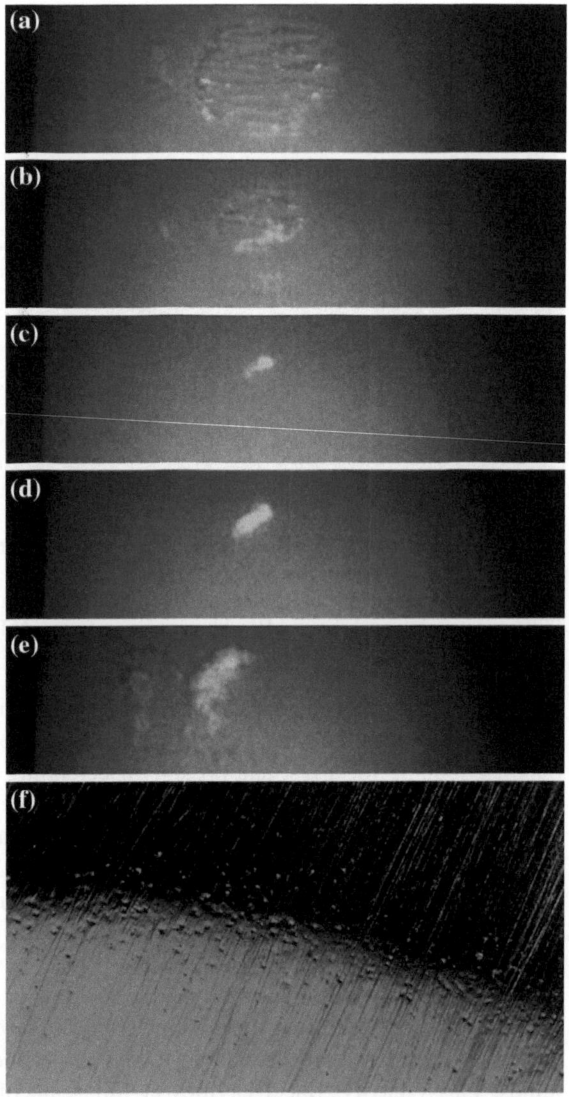

Fig. 8.4 Collapse of an almost glassy travelling sheet cavity on a twisted stationary foil in unsteady inflow generated by an oscillating foil positioned upstream of the test foil. Flow from right to left, 5 m/s. **a–e** Sampled frames from high-speed film. **a–b** A tiny bubble cloud is continuously regenerated by a jet-like collapse flow in the lower edge of the cavity, circa 0.5 ms before the collapse. **c–d** Early phase of rebound. **e** Late phase of rebound (Vortex rebound, Sect. 8.5 and Fig. 8.12). **f** Photo of collapse pitting of the bronze foil. SSPA cavitation tunnel. From [31], with permission

(a) By direct excitation of cavitation nuclei, outside a vibrating body surface, or in a turbulent boundary layer, as the bubble sheets in Figs. 8.2b and 8.3.
(b) By the shear and mixing between external and internal flow that can occur when a sheet cavity is filled by re-entrant jets or reversed flows of other origins, Figs. 8.6 and 8.8.
(c) In the rebound of a glassy cavity, by compression and rebound of nuclei, enhanced by growing interface instability during the collapse and by the rarefaction phase of the collapse and rebound pulse, Fig. 8.6, frame 7064.

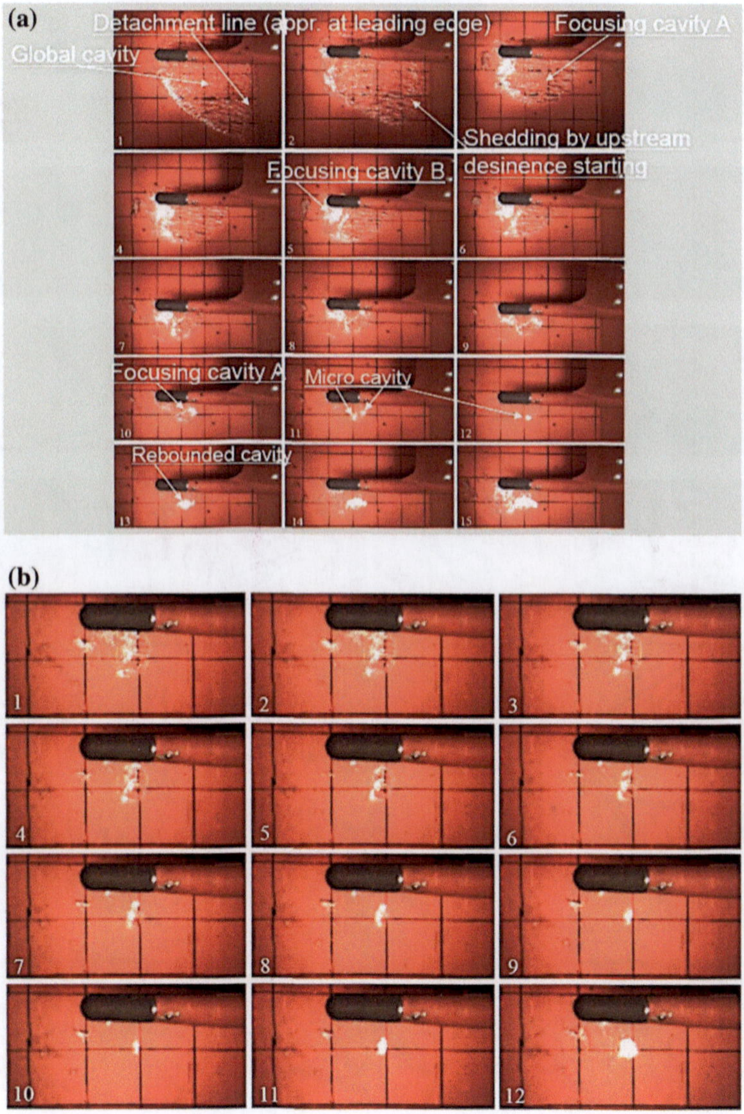

Fig. 8.5 Collapse of an almost glassy travelling sheet cavity on a twisted stationary foil in unsteady inflow generated by an oscillating foil upstream the test foil. Flow from right to left, 5 m/s. Samples from high-speed video. **a** Illustration of the concepts of "global", "main focusing" and "micro focusing" cavities introduced in Analysis Model 3. **b** Increased time and spatial resolutions of the final stage of the collapse and the early rebound of the focusing cavity A. Time interval between frames 8, 9, 10 and 11 is 1/30,000 s. SSPA cavitation tunnel. From [2]

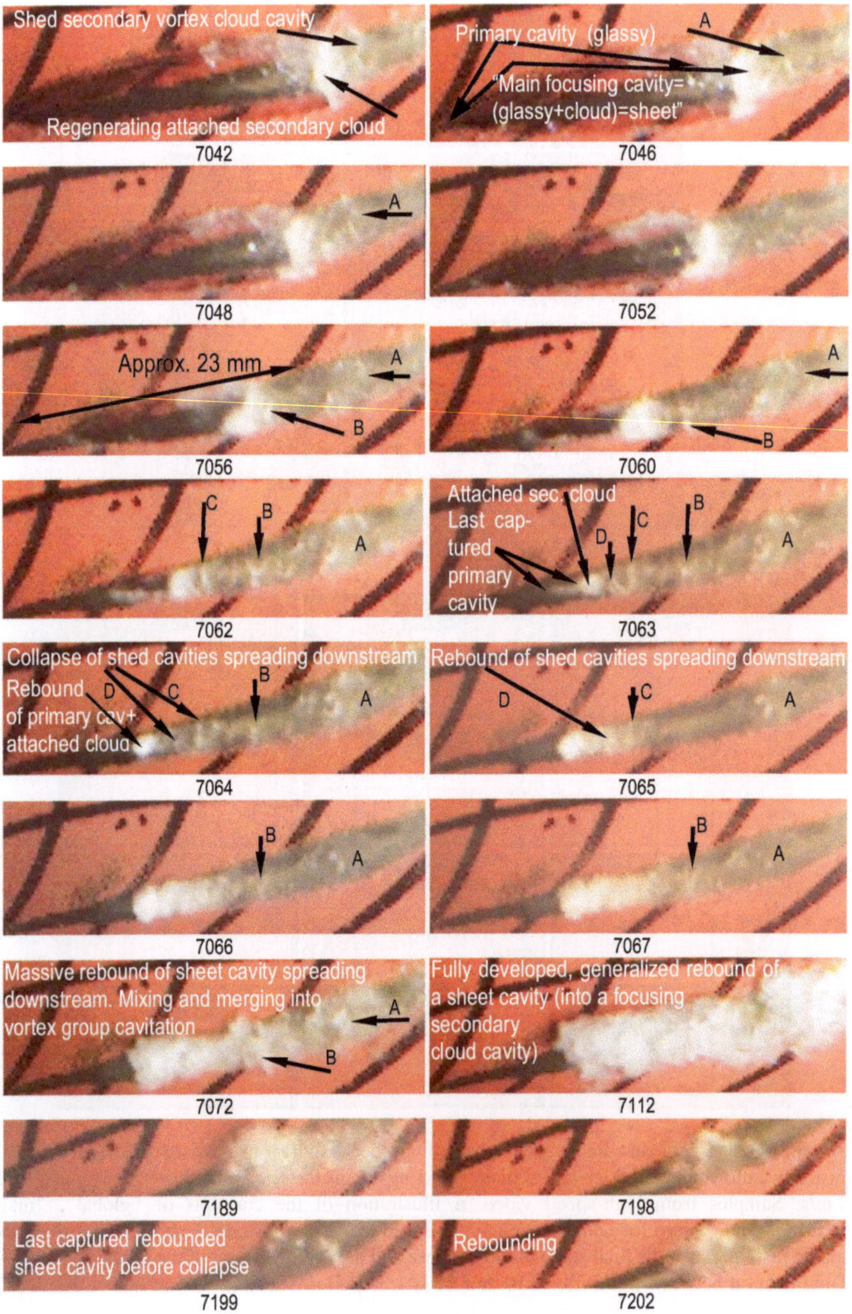

Fig. 8.6 Collapse of a root sheet cavity and its generalized rebound on a propeller in inclined flow. Flow from left to right, 9 m/s. High-speed video. Unit step in frame number equals 1/75,000 s. Length scale in frame 7056. The development of the shed vortex cloud cavities A–D is discussed in the introduction to Analysis Model 9 in Sect. 8.6. Erosion in Fig. 8.7. SSPA cavitation tunnel

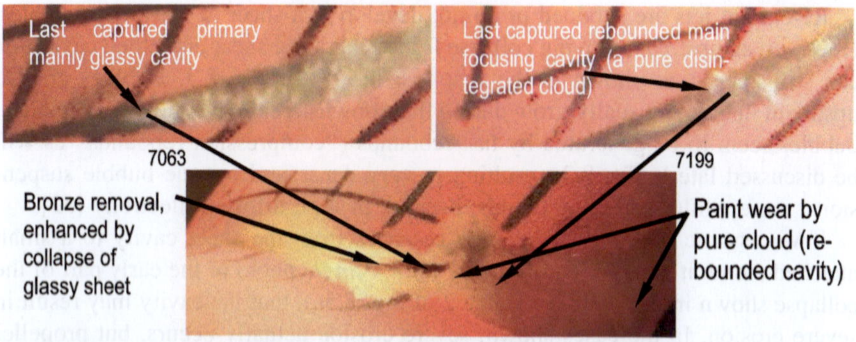

Fig. 8.7 Erosion created by the cavitation shown in Fig. 8.6

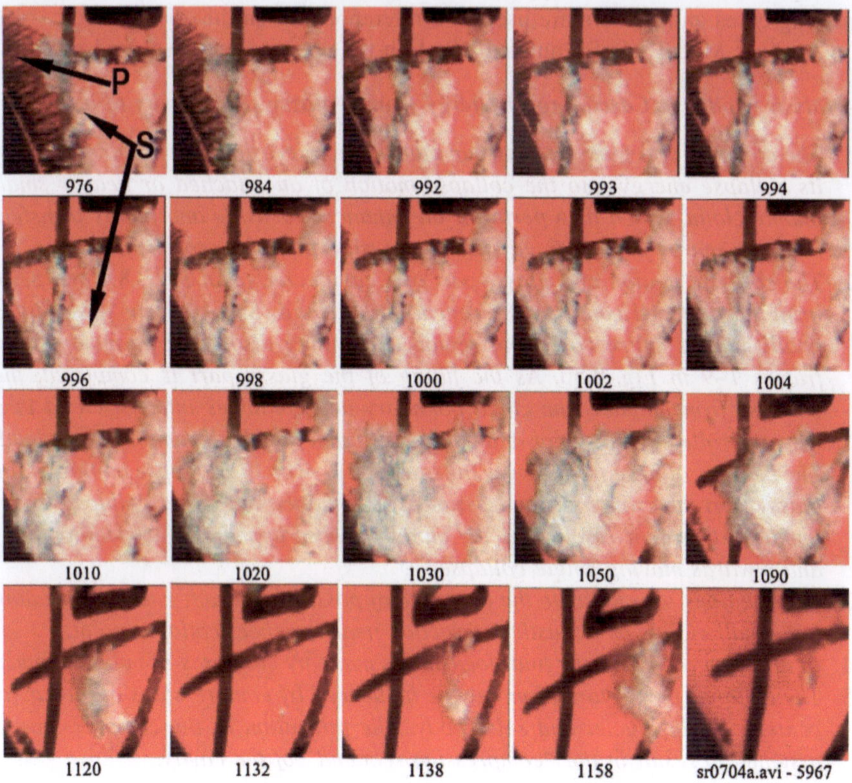

Fig. 8.8 Propeller operating in inclined flow. Flow from left to right, 8 m/s. Samples from high-speed video recording. Unit step interval of frame number equals 1/90,000 s. Upstream moving collapse of primary sheet cavity (P) and growth and merging of secondary cloud cavities (S) into "vortex group cavitation", as described by [10]. The vortex group cavitation rebounds in frames 1132–1158. The last frame is taken from a different recording of the same test. The different collapse positions in frames 1132 and 5967 show a typical collapse position scatter. SSPA cavitation tunnel. From [3], with permission from CAV 2009

Re-entrant jets are subdued or do not develop in a sheet of travelling bubbles. However, an upstream moving collapse front generating shear and cloud cavitation can still appear, as shown in the first two frames of Fig. 8.3. Bubbles created upstream in a sheet usually grow during their downstream advection. The smallest bubbles seem to be generated by fast rebounds ("compression rebounds" as will be discussed later), Fig. 8.3, resulting in very dense and opaque bubble suspensions, more white than other clouds, because of high diffuse reflectivity [32].

If we assume that collapse energy cascades from the sheet cavity to a small attached cloud, it is possible to state by even a *single* photo of the early part of the collapse shown in Fig. 8.4a (or frame 2 of Fig. 8.5a), that the cavity may result in severe erosion. In the cases shown, severe erosion actually occurs, but propeller studies also demonstrate that a proper design can sometimes force this type of cavity into a slow and non-erosive collapse on the blade surface. Although the two examples are extreme, this type of development can be found in ordinary engineering cases and is therefore used as a reference case in this discussion.

Analysis Model 1 is formulated based on the observations discussed above:

Analysis Model 1: Cascading of energy from a glassy sheet cavity to a cloud

1. *The collapse of a glassy cavity can contribute to erosion by cascading part of its collapse energy into the collapse motion of an attached or nearby small bubble cloud, which then performs an enhanced focusing and finally transfers collapse energy from both cavities to the solid body for erosion work.*
2. *The collapse motion of a glassy part of a cavity can, as in Figs.* 8.4 *and* 8.5b, *bring an attached continuously regenerating cloud (See Analysis Model 6) to finally fill the entire cavitating region, containing also some permanent gas, frames 1–9 in Fig.* 8.5b. *As the filling of the glassy part is completed, the collapse front impacts the cloud boundary and continues into the cloud in a way controlled by the vapor and gas and internal bubble dynamics in the cloud. The bubble collapse front moves inwards,* [27]*—with the collapse motion of the glassy part superimposed and assumed to enhance the cloud collapse by an impact. At present resolutions the cloud collapse appears as a monotonously and inwards moving single collapse.*
3. *Observations also indicate the existence of possibly relevant energy cascading into small clouds by acoustic interaction near the sheet collapse point.*
4. *In all observed cases the final focusing is made by collapses of bubble clouds.*
5. *The collapse flow retardation may be influenced by compressions of the cavity content, liquid, nuclei and solid body. The relaxation of the compressions is assumed to result in the "compression rebound" of the cavity.*

The resolution addressed in point 2 shows the cloud as a continuum in shades of grey and does not resolve individual bubbles, corresponding to [16, 17]. The smallest clouds forced to erosive collapses by sheet collapses are assumed not necessarily to be erosive without the energy gained from the sheet.

Analysis Model 2: Cloud cavitation characteristics influencing erosion

1. ***Inhomogeneity of bubble distribution in the cloud****, caused by inhomogeneous cavitation nuclei distribution and the cloud generation and mixing processes. This results in multiple collapse points that are also scattered over cavitation cycles. The collapse energy is dispersed in space and time, and erosion is reduced, Fig.* 8.3*, frame 1132 and Fig.* 8.8*, frames 1132 and 5967.*
2. ***Synchronization of different parts of cloud cavities****, by flow and pressure fields. This synchronization controls the extent to which different parts of a cloud contribute to erosive collapses. The tiny clouds attached to the glassy cavity in Fig.* 8.5b *are effectively synchronized by the converging flow of the glassy cavity collapse.*
3. ***Lifting of cloud cavities from the body surface*** *can reduce erosion. This can occur particularly if the cloud develops in a vortex shedding, as in Figs.* 8.8, 8.13a, 8.14e *and* 8.14g. *This can also happen in the final closure of a sheet cavity in an upstream moving collapse, Fig.* 8.13g.
4. *Transformation of a shed vortex cloud cavity into a horseshoe vortex cavity can enhance the erosion by synchronization and focusing of the cloud collapse.*

8.3 Focusing of Collapse Energy

The flow converging towards a point in a spherically symmetric cavity collapse results in a kinematic focusing into a local and temporal maximum of kinetic energy on the cavity interface. By the stopping of the collapse motion a part of the kinetic energy released by the collapse is transferred to potential energy in the compressed cavity content and surroundings. Even a non-accelerating and non-converging motion of a planar liquid interface hitting a solid body means a focusing of collapse energy at the impact. If bubbles are imbedded at the interface inside the liquid the focusing can be further enhanced.

Energy focusing has been discussed by a number of authors, for example by [16, 17] in relation to erosion and by [12, 34] in relation to sonoluminescence, the latter effect being a striking visualization of energy focusing. Based on the paper by [8] the concept of energy focusing suggested in the EROCAV handbook was extended to cover also collapse sequences. For a single collapse of a cavity, which may pass through different stages, the concept of energy focusing created by converging collapse flows and cascading of energy to small scale collapses are summarized in Analysis Model 3 and illustrated in Fig. 8.5.

Analysis Model 3: Focusing cavities and collapses—related energy cascading

1. *A **global cavity** is a cavity of any type and configuration from which the main focusing cavity defined in next point is formed. An example is the global sheet cavity, initially covering a large part of the propeller blade and later disintegrating into three erosive focusing cavities, one of which is the root cavity shown in frame 7042 of Fig.* 8.6.
2. *A **main focusing cavity** is a part of the global cavity that performs a main focusing collapse to the size below which the collapse cannot anymore be described by the resolution that is used in the recording or simulation. Examples of main focusing cavities are the cavity in Fig.* 8.4a, *the cavities A and B in frames 3–5 of Fig.* 8.5a, *the sheet in frame 7046 and the cloud in frame 7112 of Fig.* 8.6, *the cloud in frame 1050 of Fig.* 8.8, *the bubble in Fig.* 8.10, *the travelling sheet in Fig.* 8.12 *and the cloud in frame 522 of Fig.* 8.15.
3. *A **micro focusing cavity** is a cavity continuing the energy focusing started by the main focusing cavity, into time and spatial scales that are not adequately resolved but still may be sparsely sampled in the recording or simulation. The micro focusing cavity may be a regenerating cloud continuing the energy focusing into the final collapse, Fig.* 8.5b, *frames 9–10 (See Analysis Models 1 and 6). (If erosive the micro focusing is followed by a fast rebound to a dense cloud of very small bubbles, Fig.* 8.5b, *frames 11–12).*
4. *The focusing may proceed by cascading of kinetic energy from the collapse motion of the glassy sheet cavity to an attached cloud (Analysis Model 1, Fig.* 8.5b, *frames 1–9), or by acoustic interaction to nearby clouds. Acoustic interaction forces the collapse of shed clouds downstream of the sheet collapse point in frames 7064–7066 of Fig.* 8.6. *In the limit of a homogeneous cloud the focusing may proceed continuously from the main into the micro focusing stage, but often the cloud will split into more than one collapse point, as in Fig.* 8.3.
5. *Due to the limitation of the selected resolution, there may be cascading stages in the micro focusing that are not captured. Although the assessment quality typically increases with recording resolution, it is assumed that approximate and relative assessments of erosion risks of different designs sometimes can be made from recordings that do not show or resolve the later part of the collapse.*

Although discrimination between the “main” and “micro” focusing processes that depends on the recording or simulation resolution may seem to be inappropriate, it is necessary, however, to allow for the analysis of experimental or simulation data with different resolutions For typical propeller erosion, focusing cavities are usually large enough to be detected and analyzed by a model test. At full scale, however, erosion can also be generated by smaller cavities that are not properly captured or are even missing from the model test, due to scale effects or incorrectly given conditions, etc.

Although some details are left for later discussions, the review of collapse energy focusing is completed by listing processes in Analysis Model 4 that can generate focusing cavities.

Analysis Model 4: Formation of focusing cavities
To become a focusing cavity, this cavity or a part of it has to be transformed into a configuration from which a focusing collapse motion will start, or be initially created into such a configuration. This creation or transformation can be made by:

1. ***Direct creation of a travelling cavity*** *in a flow by the:*

(a) *Excitation of* ***advected nuclei*** *in a pressure field, as the sheet of travelling bubbles in Figs.* 8.2b *and* 8.3, *and the travelling bubble in Fig.* 8.10.
(b) *Excitation occurring in a* ***compression rebound*** *and the continued vortex rebound, Fig.* 8.6 *frames 7063–7112, as explained in Analysis Model 10.*

2. ***Shedding/detachment*** *of a part from a cavity due to:*

(a) ***Re-entrant jets*** *or similar flows locally filling the cavity and resulting in shear and* ***shedding of secondary cavitation*** *(Analysis Model 6). Examples are the shed clouds in Fig.* 8.1 *and the cavity B in Fig.* 8.5a.
(b) ***Local cavitation desinence and shedding*** *at the cavity detachment point or elsewhere, due to changing pressure distributions. An example is the sheet cavity A in frames 2–3 of Fig.* 8.5a *shed from the leading edge due to a decreased angle of attack. The root sheets in Figs.* 8.1, 8.6 *and* 8.8 *were split into two parts collapsing towards the blade root and the blade tip, respectively.*

3. ***Formation of an upstream moving collapse*** *of an attached cavity, as the sheet cavity P in Fig.* 8.8 *and the sheet cavity in Fig.* 8.13.

Some of these listed ways towards collapse energy focusing are related to the development of already started collapses and thus also to concepts introduced in the sections below

8.4 Primary and Secondary Cavitation

The term secondary cavity/cavitation has appeared in the literature to address subprocesses related to a main cavity. The definition of secondary cavitation introduced in this paper highlights a class of processes that are critical for generation of cloud cavitation created by the dynamics of a sheet cavity, for example.

Analysis Model 5: Primary cavities

A ***primary cavity*** *is the glassy, cloud or mixed cloud and glassy cavity that remains when secondary cavitation volumes have been subtracted from the total cavity volume, according to Analysis Model 6 below.*

Examples of primary cavities are the sheet of travelling bubbles in Fig. 8.2b, *most of the bubble sheet in Fig.* 8.3, *the glassy part of the mixed glassy and cloud sheet cavity in frames 7042 and 7046 in Fig.* 8.6 *and the glassy part P of the sheet cavity in Fig.* 8.8. *Some streaks of primary cloud cavitation are created in the far upstream part of the sheet cavity in Fig.* 8.2a.

Analysis Model 6: Secondary cavities

A ***secondary cavity*** *is created by the flow and pressure created by motions of the primary cavity interface, particularly the collapse/filling motions, as re-entrant jets or similar flows. Secondary cavitation processes resulting from collapse motion are here alternatively interpreted as rebounds of the cavity, appearing as a "compression rebound", a "vortex rebound" or a mixture of both types. These two rebound types constitute together a* ***generalized rebound*** *as will be described in the Analysis Models 7 and 10. Secondary cavities are generated by the:*

1. ***Flow kinematics and shear*** *developing in a jet-like filling of a cavity. This generates a secondary vortex cavity, usually of cloud type, here defined also as a* ***vortex rebound of at least a part of the primary cavity*** *(e.g. a sheet). Three typical stages/types of the cloud development are:*

(a) *A small cloud attached to a glassy cavity and continuously regenerating due to shear created by a jet-like collapse flow penetrating into the glassy cavity, frames 1–8 in Fig.* 8.5b, *and frame 247 in Fig.* 8.10.

(b) *A cloud of moderate extent, attached to the downstream region of a glassy sheet cavity, where the cloud continuously regenerates as in point 1(a), frames 1–2 in Fig.* 8.1, *frame 7042 in Fig.* 8.6 *and frame a in Fig.* 8.13. *Some shedding usually occurs.*

(c) *A cloud occurring as shed vortex cavitation, sometimes merging into vortex group cavitation, forming usually a new focusing cavity downstream of the sheet cavity, frames 2–6 in Fig.* 8.1 *and frames 976–1132 in Fig.* 8.8.

2. ***Compression rebound*** *of a cavity, frame 12 in Fig.* 8.5b *and frame e in Fig.* 8.12.

Secondary cavitation includes new cavitation from new nuclei as well as re-opening of cavity residues from a collapsed cavity. The new or recreated vapor volumes can be mixed into the already existing vapor volume, as seem to happen in the shear and mixing zone in frame 7042 of Fig. 8.6. Although secondary cavitation is usually of cloud type, it can be glassy in limiting cases, as in Fig. 8.9 and in frames 260–303 of Fig. 8.10. A reason for introducing the concept of a generalized rebound is the difficulty to discriminate vortex rebounds from compression rebounds in observations. A special type of secondary cavities is a part shed from a sheet due to vanishing of the local sheet thickness.

A glassy primary cavity alone can be a focusing cavity and contribute to erosion. If the primary cavity is of cloud type, it alone can generate erosion. A secondary cloud can become a focusing cavity and generate erosion, either alone or when attached to a primary cavity. Primary and secondary cloud cavitation can coexist in the same cavity, as in the sheet cavity in Fig. 8.2a. Primary cloud cavitation is in Fig. 8.2 generated in the upstream region by the process (a) described in Sect. 8.2 and secondary cloud cavitation is generated by process (b) in the downstream part of the sheet. Discrimination of primary and secondary cloud cavitation may not always be possible, and is usually not necessary, but still the creation of both types should be captured in experiments and advanced simulations.

8.5 Asymmetry and Vortex Formation

Collapse asymmetry and the related vortex formation can influence erosion. [5] demonstrated theoretically the jet and vortex formation due to an asymmetric bubble collapse close to a wall in a liquid at rest and later bubble collapses in the flow close to a solid boundary were experimentally investigated by [33]. Considering two cases of travelling bubbles on an oscillating foil, Figs. 8.9 and 8.10, and the attached sheet cavity in Fig. 8.11, we will make some introductory observations of collapse asymmetry and related vortex formation.

Figure 8.9 shows a collapse and compression rebound of a travelling bubble on an oscillating foil in a region with negligible streamwise pressure gradient. The bubble, having some similarity with a traveling sheet cavity, is flattened from above and makes the first collapse and rebound as a glassy torus, with a minimum size just after frame 11644. The almost conserved shape, glassiness and lack of shedding indicate that no significant vortex motion is created. Nor is the collapse violent enough to trigger the cavity to rebound as a cloud. The rebound is completed in frame 11654. In the next collapse and rebound, started in frame 11668, the bubble becomes increasingly asymmetric, due to the streamwise increasing pressure.

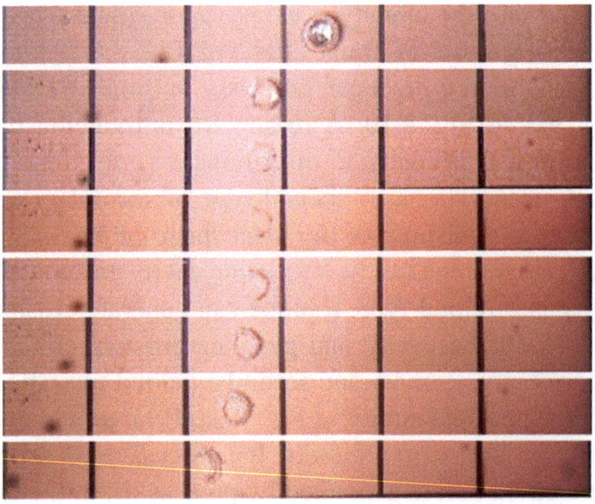

Fig. 8.9 Almost symmetric collapse and compression rebound of a travelling bubble to a glassy secondary cavity on an oscillating foil. The first collapse and rebound, completed in frame 11654, are almost symmetric, while an asymmetry appears in the next collapse started in frame 11668. Flow from right to left, 5 m/s. Chord length of foil is 120 mm and the frame rate was 12,000 frames/s. SSPA cavitation tunnel. From [4], with permission from AMT11

The bubble in Fig. 8.10 collapses in a region of streamwise increasing pressure. The downstream edge of the bubble has almost stopped moving in frame 235, and the jet-like flow filling the bubble results in shear generated cloud cavitation, and shedding of a small vortex cavity in frames 235–247. A glassy, spanwise extending part of the bubble with the cloud in the center still exists in frame 250.

The collapse symmetry may increase in the end but some asymmetry may remain also in the central part of the cavity after frame 251 in Fig. 8.10, resulting in some vortex motion. This vortex motion is assumed to survive the minimum size of the cavity and result in a possibly cavitating *"closure vortex"* superimposed on the compression rebound seen as an intensely white central cloud spot in frame 252. According to point 1 of Analysis Model 6, a cavitating "closure vortex" can alternatively be interpreted as a *"closure vortex rebound"*. This is visible in frame 252 of Fig. 8.10, as the greyish cavity extending spanwise from the white compression rebound spot. In frame 254 the shed vortex cavity has rebounded as a partly glassy cavity, indicating a slow collapse. Continued interaction between rebounding cavities and the recovering flow results in a *"continued vortex rebound"*, shown in frames 254–285. Other examples of closure vortex rebounds are shown in Figs. 8.11 and 8.12. Notice particularly the closure vortex rebound due to the spanwise moving collapse front in Figs. 8.11a and 8.12c–d. Based on these observations we formulate Analysis Model 7.

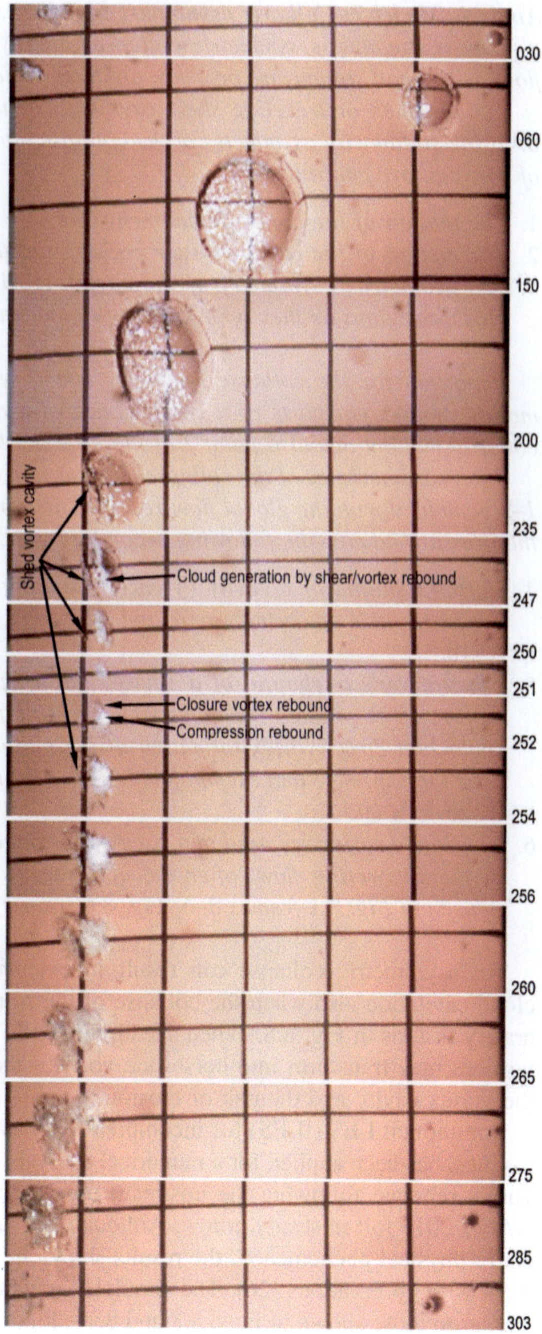

Fig. 8.10 Asymmetric collapse of a travelling bubble and generation of vortex and compression rebounds. Flow from right to left, 5 m/s, oscillating foil, 12,000 frames/s. Same condition as in Fig. 8.9. SSPA cavitation tunnel. From [4], with permission from AMT11

Analysis Model 7: Collapse asymmetry and vortex formation

Almost streamwise symmetric collapses of travelling cavities can develop in flows with small streamwise pressure gradient, as for the sheet cavity in Fig. 8.4a.

For attached or traveling sheet cavities, as shown in Figs. 8.5, 8.6, 8.8 *and* 8.12, *the streamwise asymmetry of the collapse motion, and the related formation of vortices, are controlled by:*

1. *The* ***motion of the cavity detachment line***,
2. *The* ***motion of the cavity closure region, and the re-entrant jet flow***,
3. *The* ***geometrical symmetry of the cavity***, *particularly in the detachment and closure regions as they approach each other in the final collapse motion.*

Important for the collapse symmetry is the ***synchronization*** *of the collapse motion of different parts of a cavity. Lack of synchronization can disperse the collapse motion. See also point 2 of the Analysis Models 2 and 9.*

The characteristics of the collapse motions and cavity geometry listed in points 1–3 combined with the global flow result in the formation of spanwise vortices. If the pressure admits, the following secondary vortex cavitation/rebounds occur:

4. ***Vortex rebound of a cavity by early shedding*** *and formation of independent vortex cavities or vortex group cavitation, Fig.* 8.6 *frame 7042, Figs.* 8.8 *and* 8.11b.
5. ***Closure vortex rebound of a cavity***, *due to spanwise or chordwise collapse motions. The process emanates from the collapse asymmetry and the related shear and direct vortex formation at the closure of the primary cavity, visualized in Fig.* 8.13 *and exemplified in Fig.* 8.10 *frames 251–252, Figs.* 8.11a–b, 8.12b, c, e *and* f.
6. ***Continued vortex rebound*** *of a cavity, due to the shear and pressure conditions in the recovering flow, after the collapse of the primary cavity. Examples appear in Fig.* 8.1 *frames 3–6, Fig.* 8.6 *frames 7072–7112 and Fig.* 8.12h.

An asymmetric collapse can result in extensive development of secondary cloud cavitation also when the collapse of the primary sheet cavity is only moderately fast, as in Fig. 8.8. When shedding occurs from a convex sheet cavity the vortices may transform into horseshoe vortices that can enhance the focusing of the vortex cavity and the risk of erosion.

An implicit LES (ILES) for incompressible flow, without permanent gas in the cavities, has been applied for a numerical analysis of the development of a closure vortex rebound following the upstream moving collapse of a sheet cavity on a NACA 0015 foil in steady inflow. Although the simulation does not represent the entire physical phenomenon, the results shown in Fig. 8.13a–f are in good agreement with observations [26]. The simulated condition generates a collapse that is similar to those shown in Figs. 8.8 and 8.11. Figure 8.13a-f shows the finishing of the collapse of the primary sheet cavity and the formation of vortex group cavitation. The development shown in Fig. 8.13 is a continuation from Fig. 8.14g.

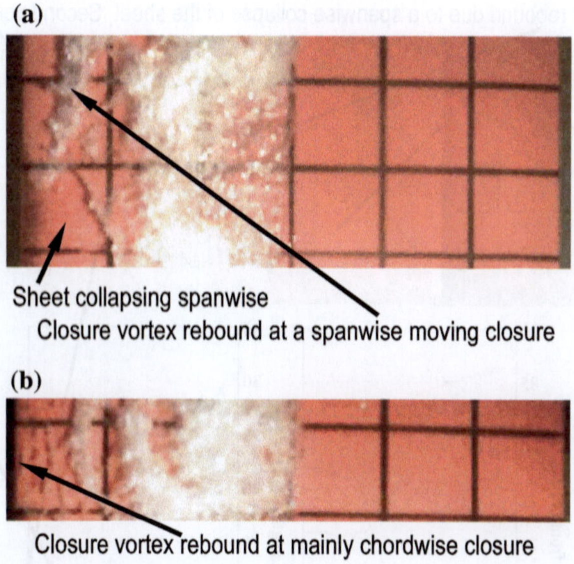

Fig. 8.11 Two types of closure vortex rebounds. **a** A sheet cavity attached to the leading edge of an oscillating foil generates a "closure vortex rebound" due to a spanwise moving collapse. Flow from left to right, 5 m/s. **b** The far left vortex cavity is a "closure vortex rebound", created by the final asymmetric closure of the sheet cavity in an upstream moving collapse, a development explained in Fig. 8.13. The interacting and cavitating vortices (vortex group cavitation) will later merge into a single cloud. This process occurs also on the propeller shown in Fig. 8.8. SSPA cavitation tunnel. From [4], with permission from AMT11

Figure 8.13a–b shows the penetration into the sheet cavity of a reversed flow induced mainly by earlier shed vortices still existing downstream of the sheet shown in Fig. 8.14g. The interaction between the penetrating jet and the flow outside the sheet cavity as described by [10] creates a vortex at the upper corner of the "open" closure region of the sheet. Some vapor from the sheet cavity is captured by the vortex, but condensates partly and temporarily, as can be observed also in experiments, e.g. in Fig. 8.6. In Fig. 8.13c three such vortices are created, and a new vortex appears upstream. In Fig. 8.13d the last two cavitating vortices are created far upstream during the final filling of the sheet. The two vortices merge into a single cavitating vortex, Fig. 8.13e, interpreted to be the cavity that can be observed also in experiments and is identified as a closure vortex rebound, Fig. 8.11b.

This vortex moves downstream and after Fig. 8.13f it typically joins the next downstream vortex, as can be observed also in Figs. 8.1 and 8.8. Figures 8.13e and 8.13g, show how the closure vortex is lifted to the outer side of the still upstream moving liquid jet. A similar development can be imagined at the leading edge in frames 994–1010 in Fig. 8.8. By lifting the collapse point away from the body and dispersing the collapse in time and over a larger area, the closure vortex rebound has a potential to reduce erosion. The simulation indicates that a closure vortex rebound can develop very fast, a result indicating that discrimination between a

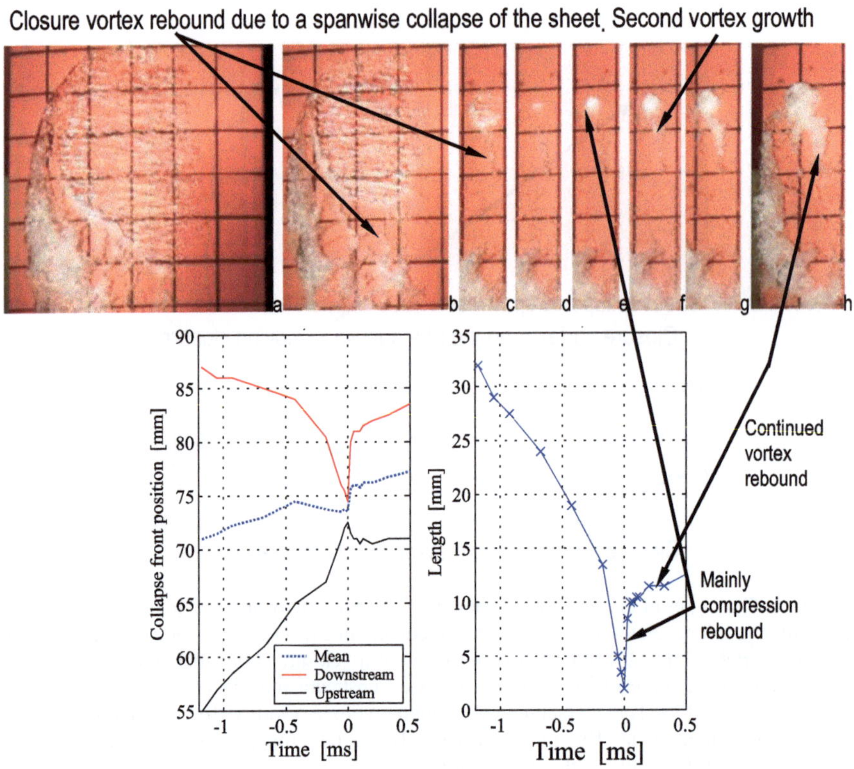

Fig. 8. 12 Travelling sheet cavity performing a collapse that finally becomes almost symmetric. Oscillating foil, flow from right to left, 5 m/s. The early part of the rebound is assumed to be a mainly compression rebound. A slow continued vortex rebound occurs at times >0.1–0.2 ms. The plots of streamwise collapse fronts and streamwise cavity length are made from a video with 40,000 frames/s. SSPA cavitation tunnel. [4], with permission from AMT11

closure vortex rebound and a compression rebound can be uncertain, particularly at moderate resolution in experimental recordings. Experiences from the EROCAV studies indicate that a fast, intensely white and *repeatable* compression rebound *on the body surface* is a reliable erosion indicator. This may still be a useful criterion but the discrimination from a vortex rebound is necessary then.

Based on the collapse asymmetry visible in frames 200–252 of Fig. 8.10 a superposition of compression and closure vortex rebounds is assumed to exist in the central part of the rebounded cavity. The whitest central part of the cloud in frame 252 is interpreted as a compression rebound and the spanwise extending greyish part as a closure vortex rebound. Developing of a more white and fast rebound of the central part is also clear in Fig. 8.12, as is the continued vortex rebound of long duration shown in the cavity length plot in Fig. 8.12 and in the frames g–h. An enhancement of the compression rebound by a fast closure vortex rebound superimposed on the compression rebound as in Fig. 8.10 cannot, however, be excluded when an asymmetry of the collapse is present, as often can be the case.

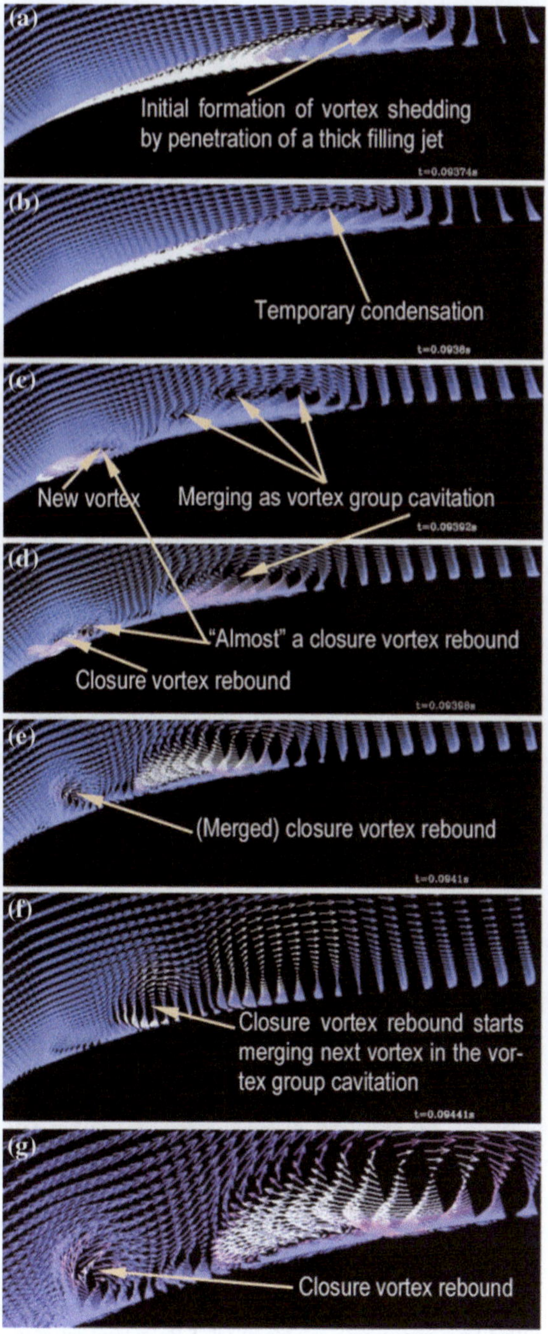

Fig. 8.13 Implicit 2-D, LES, [25, 26]. White velocity vectors represent pure vapor and blue pure water. Simulation of the later part of the upstream moving collapse of an attached sheet cavity on a 2-D NACA 0015 foil in stationary inflow from left, at 8° angle of attack and $\sigma = 1.2$. **a–c** Formation of vortex shedding and vortex rebounds in the downstream region. **d–f** Formation of a closure vortex rebound close to the sheet cavity detachment point. **g** Enlarged frame e showing the closure vortex rebound. From [4], with permission from AMT11

(a)

(b)

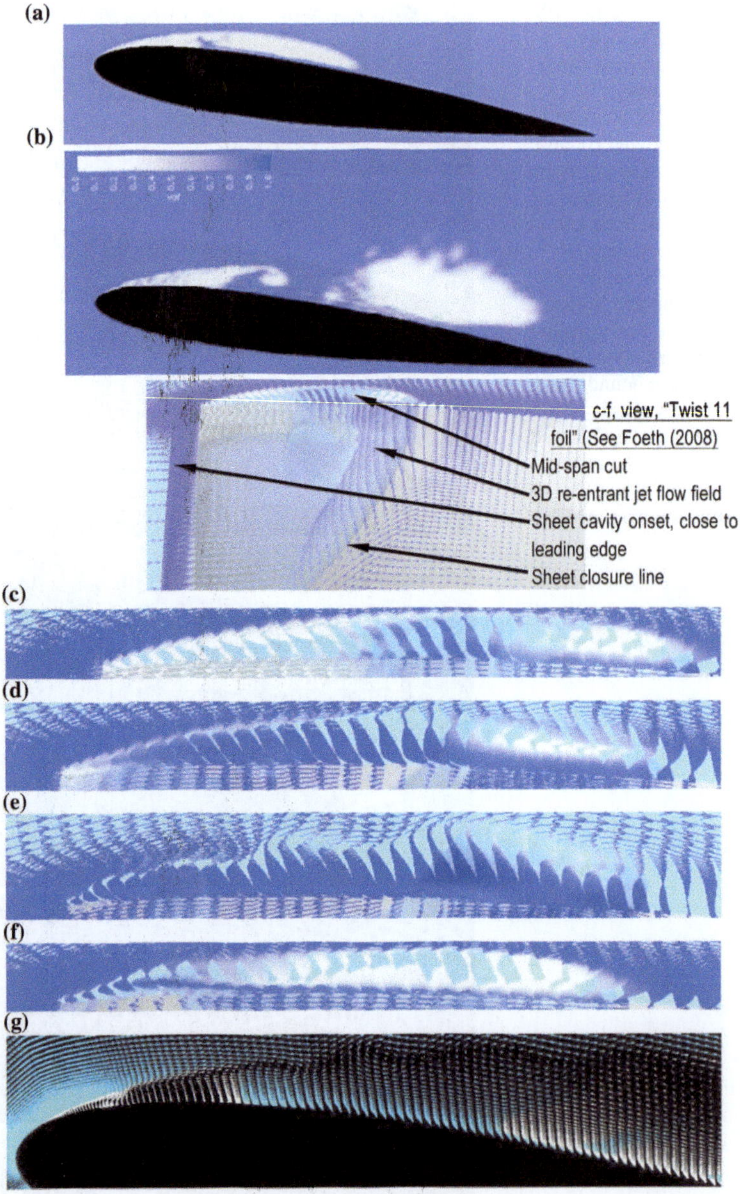

(c)

(d)

(e)

(f)

(g)

Fig. 8.14 Simulations of filling flows of jet-type using ILES. White is vapor and blue is water. **a** 2-D NACA 0012 foil and 2-D simulation, re-entrant jet is accumulating upstream. **b** As in frame a, but with a thick jet, induced mainly by the shed vortex cloud. **c–f** Twisted 3-D foil and 3-D simulations. **c–e** Velocity profiles showing accumulation of liquid inside the sheet cavity due to filling by a converging re-entrant jet. **f** Re-entrant jet is reaching the leading edge. **g** Filling by thick jet, induced by shed vortices on 2-D, NACA15 foil, 2-D simulation (The continuation is shown in Fig. 8.13). Frames **a** and **b** show simulation by [35], **c–f** by [19] and **g** by [25, 26]. Frames from [4], with permission from AMT11

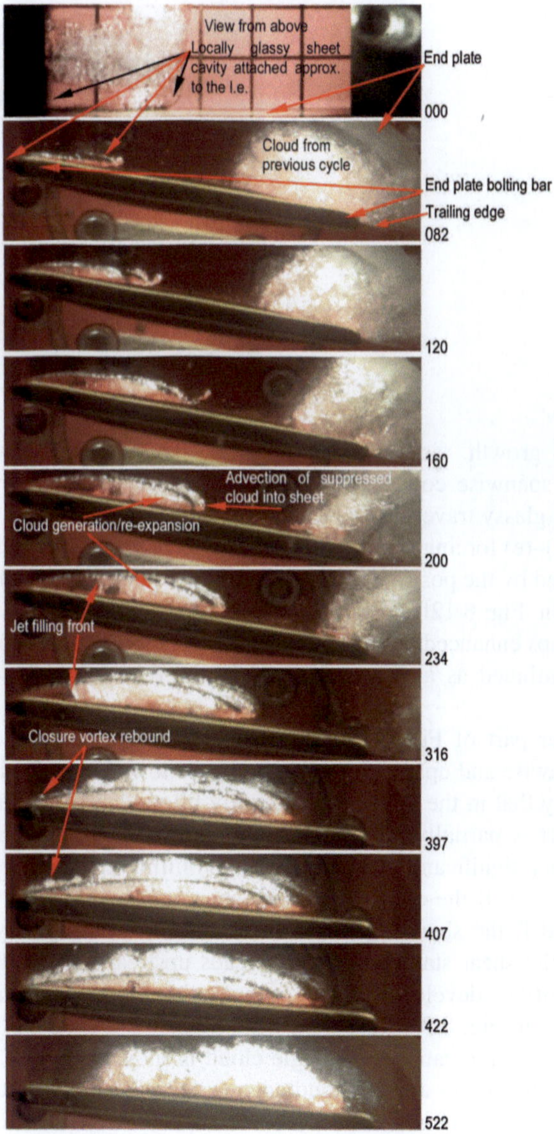

Fig. 8.15 Development of cloud cavitation inside a glassy sheet cavity, 10,000 frames/s. Flow from left to right. First frame shows view from above with the transparent end plate at the free end of the foil. Oscillating 2-D foil of a propeller section in the region with sheet cavitation shown in Fig. 8.8, but here with pressure side cavitation, chord 120 mm, 5 m/s, cav. numb. = 1.0, mean angle of attack 7°, oscillation amplitude 3° and oscillation frequency 15.9 Hz. SSPA cavitation tunnel

Fig. 8.15 continued

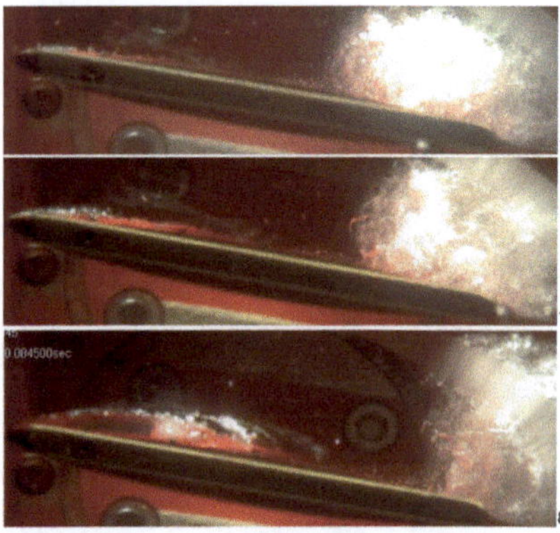

The second growth, visible in Fig. 8.12f, of the closure vortex rebound generated by the spanwise collapse, Fig. 8.12c, may be influenced by the collapse pulse from the glassy travelling cavity (See the discussion in Sect. 8.6 of pressure components (a)–(e) forcing shed cavities). Briefly the vortex cavity is first assumed to be suppressed by the positive part of the collapse pulse between Figs. 8.12d and e, and then, in Fig. 8.12f, assumed to possibly perform a weak compression rebound, perhaps enhanced by the rarefaction part of the collapse pulse. Finally the rebound is continued as a vortex rebound in the recovering flow and pressure, Fig. 8.12g–h.

In the lower part of Fig. 8.11a, downstream of the glassy sheet cavity performing a spanwise and upstream moving collapse towards the leading edge, there is a shed cavity that in the upper part is totally filled with cloud cavitation and in the middle part is partially glassy. At least initially the cloud fractions in shed cavities can vary significantly, depending on the filling kinematics and the related shear development. If the cloud generating shear appears only far upstream as in Figs. 8.14a and f, the shed cavity can initially be partially glassy, as shown in Fig. 8.11a. If the shear starts early and spreads upstream, Fig. 8.14c–e, the shed cavities may initially develop as vortex group cavitation that later transforms into a single cloud, as in Fig. 8.8.

A direct and local creation of a single cloud is shown in Fig. 8.15. Frame 000 (of different scale) shows a local cloud advancing upstream close to the endplate. The locally 3-D converging re-entrant jets will, as on the Twist 11 foil in Fig. 8.14, accumulate liquid inside the sheet cavity and then cloud generation by shear starts close to the closure line, frame 120 in Fig. 8.15. In frame 200 it is noted that, after some suppression in the sheet closure region, cloud formations are also advected back into the glassy sheet cavity, thereby increasing the cloud fraction inside the sheet. In frame 397 the re-entrant jet has almost reached the detachment point of the sheet cavity, which is now almost filled by cloud cavitation. After this filling there remains no glassy sheet cavity, the collapse of which can synchronize and accelerate the cloud into a single collapse of the type that develops after frame 9 in Fig. 8.5b. Instead the large cloud in frame 522 of Fig. 8.15 that looks fairly homogenous, is lifted during the formation of the cloud vortex from the foil surface and finally becomes less homogenous with imbedded vortex cavities, frames 522–845. The erosion in this case was not systematically investigated but there is a tendency of development of horseshoe vortex cavities with the vortex ends pointing normal to the foil, a potentially erosive configuration. Video recordings reveal also the existence of scattered, possibly erosive collapses.

8.6 Generalized Collapses and Rebounds

We introduce the concepts of the generalized collapse and rebound to highlight mechanisms that can reduce the erosive energy. We also sum up the growth of secondary potentially erosive cavitation and extend the view on the rebound of a sheet cavity, briefly introduced in Analysis Model 6. As an idealized reference in the definition of the generalized collapse, we first define the simple collapse.

Analysis Model 8: Simple collapses

A collapse of a focusing cavity containing a small amount of non-condensable gas is called a simple collapse if:

1. *The decrease of the cavity volume is forced only by a pressure jump* $\Delta P \geq 0$, *over the cavity interface, with* ΔP *controlled by the global conditions,*
2. *The rebound that follows is mainly a compression rebound,*
3. *The loss or dispersion of focusing volume and collapse energy focusing due to the following processes is negligible:*

(a) *Vortex formation and vortex rebound due to collapse asymmetry.*
(b) *Filling of the focusing cavity by re-entrant jets or similar flows, which are not synchronized with the collapse flow that is finally filling the cavity.*
(c) *Disintegration of the focusing cavity (Analysis Model 4.2).*
(d) *A temporarily reduced collapse velocity of the cavity interface due to an inflow change resulting in a reduced collapse forcing pressure jump* ΔP.

The simple collapse is the worst-case reference collapse with negligible loss of erosion potential and its definition addresses particular processes controlling erosion. The definition is in this sense a basis for collapse analysis. A basic example and limiting case of a simple collapse is the spherically symmetric collapse of a cavity containing some gas, the compression of which contributes to the rebound motion. A symmetric collapse of a sheet cavity can approach this behavior and the high focusing efficiency of a spherical collapse. Cavities approaching a simple collapse are the cavity in Fig. 8.4, cavity A in Fig. 8.5 and the root cavity in Fig. 8.6.

The vortex formation addressed in point 3(a) leads to a dispersion of energy focusing that is exemplified by the difference between the collapse of a spherical cavity in an unbounded environment and the collapse of an initially spherical cavity close to a wall. Because of the asymmetry developing in the latter type of collapse, the energy is dispersed by formation of a jet filling and impact, and a later collapse of remaining parts. In the case of an attached sheet cavity the jet impact may be directed into the incoming liquid in the sheet detachment region and formation of a vortex cavity occurs, Fig. 8.13b–d. Although different, the jet impact in the sheet cavity shares some features with the shock forced jet impact into the liquid on the opposite side of a bubble simulated by [22].

Jet fillings, point 3(b), of different types are shown in Figs. 8.13, 8.14 and 8.15. In Figs. 8.13 and 8.14 the jet filling is the only flow creating a volume decrease/collapse, because the angle of attack and global pressure are constants. Although not shown by the single frame in Fig. 8.14g, most of the sheet cavity is in this case filled by a mainly non-accelerating jet flow, implying that most of the sheet volume is lost as a focusing volume. The only remaining focusing made by the sheet cavity occurs later during the jet filling, when some acceleration is noticed. The jet impact occurs in the sheet detachment point, as can be imagined in Fig. 8.14g. In this case, a very small part, if any, of the large sheet cavity can contribute to erosion by focusing of collapse energy that is cascaded to an attached cloud. The only contribution by this sheet cavity to erosion is most likely the secondary cloud cavitation generated by the sheet cavity filling and the related cloud shedding.

A similar development is shown in Fig. 8.8 with generation of closure vortex rebounds, although with some tiny scattered compression rebounds close to the leading edge. The collapse front, a thick jet flow, moves upstream with a constant velocity that slowly increases in the later part of the collapse (Fig. 5 in [14]). This is a weak focusing, but if enhanced by a higher spanwise focusing stabilized by larger spanwise pressure gradients, some erosion may occur far upstream due to sheet collapses similar to those in Figs. 8.1 and 8.11a. This mechanism is assumed to result in the scattered rebounds close to the leading edge in Fig. 8.8.

Significant loss of focusing cavity volume can occur by disintegration of the focusing cavity into parts, point 3(c). Shed parts can form new erosive focusing cavities, as the cavity B shed by re-entrant jet action in frame 5 of Fig. 8.5a. Shed cavities can also collapse without noticeable energy focusing or erosion risk, as the large cloud in frame 1090 in Fig. 8.8. Most of these clouds make scattered collapses at some distance from the blade surface. Frame 1 in Fig. 8.1 shows the disintegration of the primary sheet cavity into two glassy focusing parts. Disintegration by local vanishing of a sheet (locally increased pressure, Analysis Model 4.2(b)), is the mechanism by which the focusing root cavity in frame 7046 in Fig. 8.6 is created from a large sheet cavity disintegrating into two major glassy sheet cavities.

Point 3(d) refers to the situation where the collapse is slowed down for some time, or even interrupted or reversed into cavity growth due to variations of the inflow. If the collapse motion is stopped and then restarts, the kinetic energy gained in the first part of the collapse is missing in the restarted collapse. For different reasons usually only a small part of the initially available potential energy is finally focused into an erosive collapse, and only sometimes is there a cascading of energy from a glassy sheet to an attached cloud. In assessing the energy focusing, it is thus important to capture how and to which extent the initially available potential energy is redistributed into kinetic energy during the collapse.

If forced by similar pressure fields, we assume that some characteristics of a sheet cavity collapse, such as the collapse pulse development, are qualitatively similar to the corresponding data in the collapse of a spherically symmetric cavity. This assumption is supported by the observation that a significant part of the non-dimensional collapse history of the travelling sheet cavity A in Fig. 8.5 and a spherically symmetric collapse are fairly similar [14]. For analysis of shed cavities and the development of secondary cloud cavitation, we recall some basics of the spherically symmetric collapse.

According to Lord Rayleigh [29] an empty spherical cavity generates as it vanishes an infinitely high pressure peak at 1.59 $R(t)$ from the cavity center, $R(t)$ being the cavity radius. Because there is no viscosity, compressibility or internal pressure retarding the collapse, this pressure pulse is interpreted as an inertial effect of the converging flow. This pulse, referred to as the "inertial collapse pulse" in the following, grows from zero amplitude early in the collapse at distances $>1.59R$ and reaches high amplitudes only during the later part of the collapse.

The cavity interface motion and the pressure in the surrounding compressible liquid during the collapse and rebound of a spherical cavity containing some permanent gas are shown by [18] (their Figs. 8.5 and 8.7). As the cavity approaches the minimum radius the internal gas pressure controls the maximum pressure in the liquid at the bubble interface and thus also defines the initial maximum of the stopping pulse. Earlier during the collapse the inertial pulse with its maximum at a small distance outside the bubble interface defines the maximum pressure in the liquid. If the collapse includes an impact normal to a body surface, an impact pulse controlled by the impact geometry and kinematics will develop and be a part of the stopping pulse in the liquid. A moving stagnation point and a jet may develop, similar to the formation of a re-entrant jet.

In the first frame of Fig. 8.6, the collapse of the focusing cavity, i.e. the glassy primary sheet cavity with its attached secondary cloud, is approaching its end as the blade leaves the wake peak. The pressure forcing the development of the cavities shed from this sheet cavity is assumed to be the superposition of:

(a) The *globally controlled pressure* on the blade, varying with blade position.
(b) The *stagnation pressure*, occurring downstream of the sheet cavity at re-entrant jet flow formation and shedding of cavities.
(c) The *inertial collapse pulse* due to the sheet cavity collapse.
(d) The *collapse stopping pulse* due to the stopping of the cavity interface, including a possibly occurring impact pulse.
(e) The *rarefaction phase* (occurring during the retardation of the rebound motion after the compression phase) of the stopping pulse, sometimes also enhanced by the pressure in a recovering flow.

The erosion under the generic condition in Fig. 8.6 is related to the collapse of the sheet cavity after frame 7063, including the attached cloud. The aim is to understand the development resulting in the large secondary cloud cavity shown in frame 7112. This cloud, forming a separate focusing cavity, will later create erosion by its collapse after frame 7199 in Fig. 8.6.

The flow filling the sheet cavity from the closure end is a combination of an early re-entrant jet and a collapse motion forced by the global pressure. A number of small vortex cloud cavities, A, B, C and D in Fig. 8.6, are shed from the attached cloud. Cavity A is actually a group of cavities while the others are mainly single vortex cavities. Identification and ranking of the mechanisms (a)–(e) in video recordings is an uncertain analysis but still we make the following observations and interpretations:

1. Close to the attached cloud, the shed cavities are suppressed, pulsate slightly, or collapse completely and rebound. Occasionally some weak pulsations are seen later. The suppression and motion of the shed cavities may be forced by a combination of the inertial collapse pulse close to the attached cloud, the unsteady stagnation pressure and the global pressure. Frame 7063 is the last frame showing a glassy part of the sheet cavity. (The arrowheads point at the upstream and downstream edges of the glassy part).
2. The early phase of the compression rebound of the sheet cavity is visible in frame 7064, as an intensely white cloud. In this frame the latest shed cavity, D, has almost collapsed and cavity C is suppressed. The suppression of the shed cavities spreads downstream from the collapse point, a process assumed to be forced by the collapse stopping pulse from the sheet cavity.
3. In frame 7065 the collapse pulse suppression has reached the cavity group A. Cavity B and the diluted bubble cloud in the collapse wake have collapsed. The outward spreading rebound is assumed to be partly a compression rebound and partly a growth forced by the rarefaction phase of the stopping pulse from the sheet collapse. In this way cavity D and the diluted cloud up to cavity C are expanded. The small bubbles now being excited to form the cloud were initially shed from the sheet and have locally weakened the liquid in the collapse wake.
4. In frame 7066 the rebound has reached cavity B and the upstream part of cavity A. The downstream part of cavity A is further suppressed and the diluted cloud downstream of cavity A has condensed, forced by the outwards spreading collapse pulse.
5. In frame 7067, the rebound has propagated further downstream. The rebound in the sheet collapse point is now very strong, almost explosive. This continues at least to frame 7072 where some of the originally shed cavities can still be identified. During a strong interaction with the incoming flow, the cloud is spreading outwards from the blade surface.
6. A striking feature up to frame 7072 is the slow downstream motion of the shed cavities. During a short time after shedding, shed cavities are sometimes observed to move upstream. Cavity B is, for example, observable from frame 7056 (scale in this frame) to 7067 virtually at the same position on the blade. During this time a small bubble assumed to be advected with the flow has moved circa 1.5 mm on the propeller blade. After approximately frame 7072

the transport of the cloud adapts to the flow, and a rotational motion of the cloud groups is noticed in the video. This behavior is assumed to result in further cloud generation in the recovering flow and pressure after the sheet cavity collapse.
7. The cloud finally undergoes an erosive collapse in frame 7199, dispersed in this blade passage into four collapse points.
8. Acoustic interaction forced by the collapse pulse is not limited to cavities in the collapse wake but because of the short duration of this pulse a significant influence will only be visible on small cavities. Mainly small nearby clouds can be expected to collapse by this interaction. The influence of the rarefaction phase is more extended in time, as indicated by simulation of two acoustically interacting bubbles [15].

A significant part of cloud growth occurs in the recovering flow and pressure, in Fig. 8.6 mainly after the frame 7067. The recovering flow can also generate a new growth of the glassy sheet during a short time before the propeller blade has left the wake peak. This "false rebound" of the sheet cavity, visible close to the leading edge in frames 1090–1120 of Fig. 8.8, is simply a limited realization of the cycling of the sheet cavity by the re-entrant jet action that can occur in steady inflow.

Observed with lower time resolution the development between frames 7063 and 7112 in Fig. 8.6 would be interpreted as a massive compression rebound. The presently applied recording resolution reveals, however, the existence of a number of partly overlapping sub-events, resulting in a rebound-like development and generation of potentially erosive cloud cavitation. The entire process can actually be described as a "generalized rebound", briefly introduced in Analysis Models 6 and 7 and fully defined below in Analysis Model 10.

The mechanisms identified in Fig. 8.6 can also be found in Fig. 8.8 having different collapse symmetry. A short time after shedding there is also a suppression of the shed cavities by the inertial and stopping pulses from the sheet collapse, and some tiny and spatially scattered collapses and rebounds occur. The early shed vortex cavities grow to a moderate size and collapse individually, frames 984–994 of Fig. 8.8. Due to the sheet collapse asymmetry the rebound at the leading edge is mainly a vortex rebound, followed by a strong cloud generation in the recovering flow. Some acoustic interaction effects may be weaker in this case. The result is the large but weakly erosive vortex group cloud shown in frame 1050 of Fig. 8.8.

Compared to the cases in Figs. 8.1, 8.6 and 8.7, however, the cloud collapses in the case of Fig. 8.8 are observed to occur at a slightly larger distance above the blade surface. There is also an observed larger cloud disintegration and collapse point scatter in Fig. 8.8. Although not stated to be significant in this case, it is to be noted that the cloud in Fig. 8.8 does not contain the small bubbles generated in a compression rebound. In comparison, significant cloud collapse pulses generated in the Fig. 8.8 case are also lower and less frequent ([13], paper VI).

As a continuation of Analysis Model 8, we first define the generalized collapse of a focusing cavity in Analysis Model 9, addressing dissipative events to look for in assessing the sheet cavitation.

Analysis Model 9: Generalized collapses

The focusing cavity under consideration is a part of an entirely glassy cavity, a glassy cavity with an attached cloud, or a pure cloud cavity. The generalized collapse of these types of cavity is composed of some of the following processes:

1. ***The simple collapse****, according to Analysis Model 8, with added complexity due to non-negligible effects of the mechanisms (a)–(d) in point 2 below.*
2. ***Dissipative or dispersive collapse mechanisms****, reducing the collapse energy or focusing by reduction of the cavity volume or changing the kinematics:*

(a) ***Vortex formation****, due to a collapse asymmetry that can reduce the energy focusing. This can occur particularly in the final collapse phase by creating a closure vortex, as in the upstream moving collapse of an attached sheet cavity in Figs.* 8.8 *and* 8.11b, *with the kinematics shown in Fig.* 8.13.

(b) ***Filling of the cavity by re-entrant jet flow****, or similar flows, which are not synchronized with the collapse flow that is finally filling the cavity. Examples include the filling of the sheet cavity by the re-entrant jet flow in Fig.* 8.2a, *most of the filling of the sheet cavity by the thick jet flow in Fig.* 8.8, *most of the filling made by the re-entrant jets shown in Figs.* 8.13 *and* 8.14.

(c) ***Disintegration*** *of the focusing cavity (Analysis Model 4.2).*
Examples include frames 1 and 7 in Fig. 8.1 *and* 8.3, *cavity B in Fig.* 8.5a, *the sheet cavity in frames 7042–7063, the secondary cloud in frame 7199 of Fig.* 8.6 *and the late disintegration of the large cloud into a thin cloud unsynchronized with the imbedded vortices in frames 755–845 of Fig.* 8.15.

(d) ***Collapse velocity reduction****, resulting e.g. in a temporal interruption of the collapse (See point 3(d) in Analysis Model 8 and comments below).*

In a generalized collapse, effects 2(a)-2(d) may dominate over the simple collapse. Erosive cavities often rebound into new erosive secondary cavities.

For the rebound, we recapitulate that both the almost non- or weakly accelerating sheet collapses in Figs. 8.8 and 8.15 and the moderately accelerating sheet collapse in Fig. 8.1 generate clouds that are approximately as massive as the cloud in frame 7112 in Fig. 8.6. Recorded at low resolutions all these clouds can be misinterpreted as resulting from strong compression rebounds of the sheet cavities. Only the case in Fig. 8.6 does actually include a strong compression rebound, frame 7064. In the cases of Figs. 8.1 and 8.8 the compression rebounds are moderate and negligible, respectively. To get a unified view on the rebound and creation of secondary cavitation, we define a generalized rebound, addressing the following issues:

(a) The difficulty to discriminate a closure vortex rebound from a compression rebound when the latter is used to detect erosive collapses.
(b) The closure vortex rebound as a process that can reduce the erosion risk.
(c) The generalized rebound as a source of possibly erosive cloud cavitation.

The full description of the generalized rebound, part of which was briefly introduced in Analysis Models 6 and 7, is presented below in Analysis Model 10.

Analysis Model 10: Generalized rebounds

A generalized rebound is here defined as the superimposed compression and vortex rebounds of a sheet cavity and its shed cloud cavities, including a possible merging of rebounded cavities. An example is the rebound of the sheet in frame 7042 in Fig. 8.6, *starting in 7042, continuing in 7063–7064, and mainly finished in frame 7112. In this development the main processes 1 and 2 are identified:*

1. ***Generalized rebound of the sheet cavity, including attached clouds****. The sheet cavity can be of any configuration in which re-entrant jets or similar flow may develop. The process is decomposed into the following sub processes:*

(a) ***Vortex rebound of the sheet cavity by shear generation during jet filling of the sheet****. The single vortex cavities thereby generated and shed may merge into a single vortex group cloud cavity. Examples are shown in Fig.* 8.1 *frames 2–5, Fig.* 8.6 *frames 7042–7063, Fig.* 8.8 *frames 976–1090, Fig.* 8.10 *frame 247 and the limiting case shown in frame 522 of Fig.* 8.15, *where the entire sheet cavity is transformed and shed as a cloud due to the jet filling. Vortex rebound of a sheet of travelling bubbles is weak or non-existing.*

(b) ***Closure vortex rebound of the sheet cavity****, due to a chordwise asymmetric closure of the sheet. This can occur in an upstream collapse point, Figs.* 8.11b *and* 8.13g *or in a downstream collapse point, Fig.* 8.10 *frames 251–252. This type of rebound can also occur in a spanwise moving collapse, Fig.* 8.8 *frames 994–1002, Fig.* 8.10 *frame 252, Fig.* 8.11a *and* 8.12.

(c) ***Compression rebound of the sheet cavity****, Fig.* 8.3 *frames 1133–1136 (cloud sheet), Fig.* 8.4d, 8.5b *frames 11–12, Fig.* 8.6 *frame 7064, Fig.* 8.10 *frame 252, superimposed on the closure vortex rebound, and Fig.* 8.12 *frame* e.

(d) ***Continued vortex rebound of the sheet cavity****, forced by the shear and pressure conditions in the recovering flow after the processes 1(a)-(c), Fig.* 8.1 *frames 4–5, Fig.* 8.6 *frames 7066–7112, Fig.* 8.10 *frames 254–265 and Fig.* 8.12 *frames* g–h.

2. ***Generalized rebounds of clouds shed from the sheet cavity****, as the clouds A-D in Fig.* 8.6. *This development starts by process 2(a), partly while the collapse of the sheet cavity and cloud shedding from the sheet are still going on according to process 1(a):*

(a) ***Compression rebounds of shed secondary clouds****, following possible collapses after shedding, Fig.* 8.1 *frames 2–4 and Fig.* 8.6 *frames 7042–7067.*

(b) ***Continued vortex rebound of shed secondary clouds****, controlled by shear and pressure conditions in the recovering flow. This development starts after processes 1(a)–(c) for the sheet cavity and is typically joined with the development of the sheet described in process 1(d). Examples are found in Fig.* 8.1 *frames 4–5, Fig.* 8.6 *frames 7065–7112 and Fig.* 8.8 *frames 994–1090.*

Often cavities generated by the processes 1–2 will merge. Finally there is the mainly compression rebound, following a possibly joined collapse of shed, merged clouds and the rebounded sheet cavity, as shown in frame 8 of Fig. 8.1, *frame 7202 of Fig.* 8.6 *and frames 1132–1158 of Fig.* 8.8. *In the latter case only the shed clouds are involved. Further collapses and rebounds of decaying intensities may follow.*

8.7 Summary and Conclusions

It is certainly possible to describe cavitation processes, as those in Figs. 8.1 and 8.6, in the classic terms of bubble, sheet, vortex and cloud cavitation, etc. By adding the present concepts, however, it is also possible to identify more specific processes controlling the development towards erosive cavity collapses.

The concepts introduced are condensed into ten conceptual Analysis Models that serve as informal propositions to be applied to experimental or numerical data for identification and analysis of large to moderately small-scale events involved in erosive processes. The ten specific processes to trace are related to:

1. Cascading of energy from a glassy sheet cavity to a cloud.
2. Cloud cavitation characteristics influencing erosion.
3. Focusing cavities and collapses—related energy cascading.
4. Formation of focusing cavities.
5. Primary cavities.
6. Secondary cavities.
7. Collapse asymmetry and vortex formation.
8. Simple collapses.
9. Generalized collapses.
10. Generalized rebounds.

The essential process is No. 3, related to collapse energy focusing, while the other processes provide supplementary understanding about why and how erosive cavity configurations are created and developing, and how the cascading and focusing of collapse energy proceed. The definitions introduced by the Analysis Models constitute a nomenclature for analysis and communication.

The possibility of cascading of collapse energy from the main focusing cavity of glassy, cloud or mixed type to a smaller cloud creating the erosive impact is the core of the present approach. The energy cascading mechanism making this possible includes particularly a regenerating cloud that survives the collapse of the glassy part, as is described in the Analysis Models 1, 3 and 6 and illustrated in Fig. 8.5b. This crucial regenerating cloud is created by a filling flow, often of jet type, as shown by the example in frames 7042–7063 of Fig. 8.6. The cloud formation has been related to collapse asymmetry, Figs. 8.1 and 8.6, but slightly different kinematic conditions to be further investigated are noted as well, Figs. 8.5b and 8.12.

Although erosion is considered to be primarily the result of cloud collapses, the present study demonstrates that it can be significantly enhanced by the collapse of a glassy sheet cavity. In the present study cloud collapses enhanced by sheet collapses were by far the most erosive cases. It is emphasized that erosion created by cloud cavitation alone can be significant, dominating or the only erosion that occurs, as shown in Figs. 8.1, 8.3, 8.7 and 8.8. Figure 8.8 shows an example where the collapse of the secondary cloud is the only erosive collapse. In this case the erosion was weak, depending on specific properties of the condition.

Although the resolutions applied in Figs. 8.5b and 8.6 are higher than what can be reached sometimes in engineering applications, they mainly reveal only the initial conditions for the final energy focusing. It is remarked in point 2 of Analysis Model 1 that there may exist further cascading and focusing stages, hidden inside the seemingly homogenous unresolved cloud in frame 9 of Fig. 8.5b. The missing resolution of events between the frames 9 and 11 in Fig. 8.5b, the interval in which the rebound starts, leads to an uncertainty in the assessment. If, however, no major disintegration of the cloud cavity is indicated and a fast compression rebound is confirmed, it would be expected that repetition of such events is most probably erosive.

The fact that collapses of this type can generate not only paint wear but also real erosion, photo f in Figs. 8.4 and 8.7, at typical model testing velocities means the generation of bubbles that are small enough to create significant energy focusing. The smallest bubbles seem to be created in a compression rebound. Very small bubbles are created in many vortex rebounds/shear flows as well, but in some cases the bubbles created in the later part of a continued vortex rebound can be larger, at least at low Reynolds numbers, as in frames 265–285 in Fig. 8.10. This indicates a possible existence of scale effects related to the Reynolds number and the nuclei distribution. This type of scale effect may reduce the micro focusing and erosion/paint wear produced by the bubbles, although an adequate large scale focusing may exist and can be evaluated. This may apply to model tests of rudders.

Based on the assumption that tracing the collapse, and the loss of collapse energy according to point 2 in Analysis Model 9, it would be possible to understand, control and predict the development towards erosive collapses. The listed loss mechanisms are active at scales that are related to design but extend also into processes of smaller scales, as some types of cavity disintegration. For detecting erosive cavitation development, the analysis of cavitation means also tracing of collapse symmetry and cloud synchronization, ideally at temporal and spatial resolutions as shown in Figs. 8.5b and 8.6. Together with an observation of the rebound, this allows an estimate and some confirmation of the unresolved continuation, the small-scale and fast "micro focusing" that creates the pulse finally resulting in erosion on the solid body. Detection of a compression rebound on a body surface is a preliminary alarm of an erosion risk, provided it is clear that it is not a pure vortex rebound of a cavity, a process that can indicate a reduced erosion risk. Both types of rebounds are usually present, partly overlapping, and are major and coupled sources of typically erosive secondary cloud cavitation. Finally, mention is made of the strong influence of random scatter of collapse points, and scale effects on this scatter and cavitation intermittency that can appear at model testing.

The present study indicates that in a typical model test much of the critical mechanisms may be captured by a video recording of 10,000 frames/s or sometimes less, but in some cases recordings with higher frame rates are useful. Similar resolutions are obtained by numerical simulations by, for example, [30] for foils. [26] captured the design influence by investigating the influence of focusing cavities by local vortex formations on the blade and by blade camber on the development of potentially erosive cavities on propellers.

Acknowledgments The presented work was prepared under the European Union project Hydro Testing Alliance (HTA), JRP6, with input from the earlier European projects EROCAV and VIRTUE. Results from these projects, reproduced with permission, are disseminated by the proceedings of CAV 2009 and AMT11, referred to in the related figures. Part of the simulations was made with support from the Rolls-Royce University Technology Centre at Chalmers, The Swedish Armed Forces (FM), and The Swedish Defense Research Agency (FOI).

References

1. Bark G, Bensow RE (2012) Hydrodynamic mechanisms controlling cavitation erosion. In: Proceeding of 29th symposium on naval hydrodynamics. Also published in the special issue, International Shipbuilding Progress, vol 60, 2013, Gothenburg, 26–31 Aug 2012. doi:10.3233/ISP-130097
2. Bark G, Berchiche N, Grekula M (2004) Application of principles for observation and analysis of eroding cavitation. The EROCAV observation handbook. Department of Shipping and Marine Technology, Chalmers University of Technology, Sweden, www.chalmers.se/smt
3. Bark G, Grekula M, Bensow RE, Berchiche N (2009) On some physics to consider in numerical simulation of erosive cavitation. In: Proceeding of 7th international symposium on cavitation (CAV2009), Ann Arbor, 17–22 Aug 2009
4. Bark G, Grekula M, Lu N-X (2011) Analysis of erosive cavitation by high speed video records. In: Proceeding of 2nd international conference on advanced model measurement technology for the EU maritime industry, (AMT11), Newcastle University, Newcastle upon Tyne, 4–5 April 2011
5. Benjamin TB, Ellis AT (1966) The collapse of cavitation bubbles and the pressure thereby produced against solid boundaries. Royal Soc of London Philos trans A 260:221–240
6. Bensow RE, Bark G (2010) Implicit LES predictions of the cavitating flow on a propeller. J Fluids Eng 132:041302-1
7. Bensow RE, Huuva T, Bark G, Liefvendahl M (2008) Large eddy simulation of cavitating propeller flows. In: Proceeding of 27th symposium on naval hydrodynamics, Seoul, 5–10 Oct 2008
8. Berchiche N, Grekula M, Bark G (2003) Concept of focusing of the collapse energy—application in cavitation observations. In: Proceeding of 5th international symposium on cavitation (CAV 2003), Osaka, 1–4 Nov 2003
9. Boorsma A, Fitzsimmons P (2009) Quantification of cavitation impacts with acoustic emissions techniques. In: Proceeding of 7th international symposium on cavitation (CAV 2009), Ann Arbor, 17–22 Aug 2009
10. Foeth E-J (2008) The structure of three-dimensional sheet cavitation. Ph.D. Thesis, Delft University of Technology, Delft

11. Fukaya M, Tamura Y, Matsumoto Y (2010) Prediction of cavitation intensity and erosion area in centrifugal pump by using cavitating flow simulation with bubble flow model. J Fluid Sci Technol 5(2):2010
12. Gong C, Hart DP (1998) Ultrasound induced cavitation and sonochemical yields. J Acoust Soc Am, Vol 104
13. Grekula M (2010) Cavitation mechanisms related to erosion; studies on kaplan turbines, foils and propellers. Ph.D. Thesis, Department of Shipping and Marine Technology, Chalmers University of Technology, Gothenburg
14. Grekula M, Bark G (2009) Analysis of high-speed video data for assessment of the risk of cavitation erosion. In: Proceeding of 1st international conference on advanced model measurement technology for EU maritime industry (AMT09), Nantes, 1–2 Sept 2009
15. Hallander J, Bark G (2002) Influence of acoustic interaction in noise generating cavitation. In: Proceeding of 24th Symposium on Naval Hydrodynamics, Fukuoka, 8–13 July 2002
16. Hansson I, Mørch KA (1980) The dynamics of cavity clusters in ultrasonic (vibratory) cavitation erosion. J Appl Phys 51(9):4651–4658
17. Hansson I, Mørch KA (1981) Erratum: The dynamics of cavity clusters in ultrasonic (vibratory) cavitation erosion [J Appl Phys 51:4651 (1980)]. J Appl Phys 52:1136
18. Hickling R, Plesset MS (1964) Collapse and rebound of a spherical bubble in water. Phys Fluids 7(1):7–14
19. Huuva T (2008) Large eddy simulation of cavitating and non-cavitating flow. Ph.D. Thesis, Department of Shipping and Marine Technology, Chalmers University of Technology, Gothenburg
20. ITTC (1993) The cavitation committee report. In: Proceeding of 20th international towing tank conference, vol 1, The University of California, Berkeley, pp 191–255
21. ITTC (1996) The cavitation committee report. In: Proceeding of 21st international towing tank conference, vol 1, The Norwegian University of Science and Technology, Trondheim, pp 63–126
22. Johnsen E, Colonius T (2009) Numerical simulations of non-spherical bubble collapse. J Fluid Mech 629:231–262
23. Li ZR (2012) Assessment of cavitation erosion with a multiphase reynolds-averaged navier-stokes method. Ph.D. Thesis, Delft University of Technology, Delft
24. Li ZR, Van Terwisga T (2012) On the capability of a RANS method to assess the cavitation erosion risk on a hydrofoil. In: Proceeding of 8th International Symposium on Cavitation (CAV 2012), Singapore
25. Lu N-X (2010) Large eddy simulation of cavitating flow on hydrofoils. Licentiate Thesis, Department of Shipping and Marine Technology, Chalmers University of Technology, Gothenburg
26. Lu N-X (2013) Modelling cavitation mechanisms using large eddy simulation. Ph.D. Thesis, Department of Shipping and Marine Technology, Chalmers University of Technology, Gothenburg
27. Mørch KA (1982) Energy considerations on the collapse of cavity clusters. Applied Scientific Research, No. 38, pp 313–321
28. Ochiai N, Iga Y, Nohmi M, Ikohagi T (2010) Numerrical prediction of cavitation erosion intensity in cavitating flows around a Clark Y 11.7% hydrofoil. J. Fluid Sci Technol 5(3):2010
29. Lord Rayleigh (1917) On the pressure developed in a liquid during the collapse of a spherical cavity. Phil Mag 34:94–98
30. Schmidt SJ, Thalhamer M, Schnerr GH (2009) Inertia controlled instability and small scale structures of sheet and cloud cavitation. In: Proceeding of 7th international symposium on cavitation (CAV2009), Ann Arbor, 17–22 Aug 2009
31. Schöön J (2000) A method for the study of unsteady cavitation—observations on collapsing sheet cavities. Ph.D. Thesis, Department of Shipping and Marine Technology, Chalmers University of Technology, Gothenburg

32. Schöön J, Bark G (1998) Some observations of violent collapses of sheet cavities and subsequent cloud cavitation on a foil in unsteady flow. In: Proceeding of ASME fluids engineering summer meeting, Washington, DC, 22–25 June 1998
33. Van der Meulen JHJ, Van Renesse RL (1988) The collapse of bubbles in a flow near a boundary. In: Proceeding of 17th symposium on naval hydrodynamics, The Hague, 28 Aug to 2 Sept 1988
34. Weninger KR, Camara CG, Putterman, SJ (1999) Energy focusing in a converging fluid flow: implications for sonoluminiscence. Phys Rev Lett 83(10):2081–2084
35. Wikström N (2006) Approaching large eddy simulation of cavitating flows for marine applications. Ph.D. Thesis, Department of Shipping and Marine Technology, Chalmers University of Technology, Gothenburg

Chapter 9
On the Kinematics of Sheet and Cloud Cavitation and Related Erosion

Peter F. Pelz, Thomas Keil and Gerhard Ludwig

Abstract The influence of flow parameters such as cavitation number and Reynolds number on the cavitating cloud behavior and aggressiveness is analysed in an experimental work. The focused geometry is a convergent-divergent nozzle with a given radius of curvature at the minimum cross section. By means of a high-speed camera the kinematics of cloud cavitation is visualized. The shape of the cloud is a horse shoe (U-shaped) with two legs ending at the material surface which is in agreement with the Helmholtz vortex theorem. Indeed it is worthwhile to look at the cavitation cloud as a ring vortex whose second half is a mirror vortex within the material. Due to the convection flow, the legs of the vortex are elongated and hence the rotational speed of the vortex core will increase. Thus cavitation bubbles will concentrate within the legs of the vortex and that behavior is observed in the cavitation experiments. The aggressiveness of the cloud is quantified by using soft metal inserts adapted on the nozzle geometry. The interpretation of the plastic deformation, called pits, is done with a 2-dimensional optical measurement system, which is developed to scan large and curved surfaces. In this way damage maps are obtained. Consequently dimensional analysis is used to analyse and generalize the experimental results. Thus a critical Reynolds number is found for the transition from sheet to cloud cavitation. Further an upper limit for the Strouhal number exists for the given geometry. A physical model for the critical Reynolds number is given elsewhere [1]. Also a model for the dynamics of the observed stretched cloud with circulation is published by Buttenbender and Pelz [2].

P. F. Pelz (✉) · T. Keil · G. Ludwig
Technische Universität, Darmstadt, Germany
e-mail: peter.pelz@fst.tu-darmstadt.de

K.-H. Kim et al. (eds.), *Advanced Experimental and Numerical Techniques for Cavitation Erosion Prediction*, Fluid Mechanics and Its Applications 106,
DOI: 10.1007/978-94-017-8539-6_9,

9.1 Introduction

One of the more aggressive cavitation forms is cloud cavitation which is observed above a critical Reynolds number as will be shown in this paper. The high aggressiveness of cloud cavitation was reported by several researchers (e.g. [3–6]). It is now understood that the cloud plays a key role in the temporal and spatial energy focusing cascade which is essential for the cavitation erosion process. The stationary energy flux of the incoming flow is concentrated in time and space in the form of clouds. The state of the collapse and the time delay depends on the growth rate, the initial size and the initial bubble population of the cloud [2]. When the cloud collapses the second energy focusing process starts: Like a single bubble when collapsing the cloud imposes a sink flow and focuses the kinetic energy of the surrounding fluid in its center.

The intensity of the collapse and thus the aggressiveness is determined by the dynamic behavior and by the type of the structures occurring. Kawanami et al. [7] employed a model to describe the shape of a cloud by a U-shaped vortex. Their investigations base on holographic experiments in a cavitation tunnel. Similar observations were done by Joussellin et al. [8]. The circulation of the cloud can be observed by high-speed visualization (Kawanami et al. [9]).

The aggressiveness of cavitation erosion is assessed by using soft metal layers, as material sensors. This approach enables the evaluation of damage distributions and damage intensities. To quantify the effective damage energy, a pit-count measurement system is used, which detects the size and the position of each pit on the damaged surface. The effective damage energy and thereby the cavitation aggressiveness is quantified by a dimensionless damage function.

9.2 Experimental Set Up and Dimensionless Products

The needs for the test rig are (i) an excellent optical accessibility, (ii) the possibility to vary the geometry curvature and (iii) to provide an installation-opportunity for soft metal layers which serve as material sensors.

The optical access (i) is realized by a channel having rectangular cross section and walls completely made of acrylic glass. The high-speed observations are illuminated with a stroboscope for the top view perspective. For the side view perspective a laser light sheet is used. The copper layer (iii) is placed directly downstream of the narrowest cross-sectional area. The analyzed area is close to the cloud collapse location downstream of the sheet cavity.

To initiate evenly the cavitation over the whole nozzle width, an obstacle with rectangular cross section seen in Fig. 9.1 is placed in the smallest cross section area. First experiments show that the side wall friction influence due to surface roughness is small in comparison to the obstacle influence and is not visible in the experiments. Flows without this obstacle would generate single bubbles over the

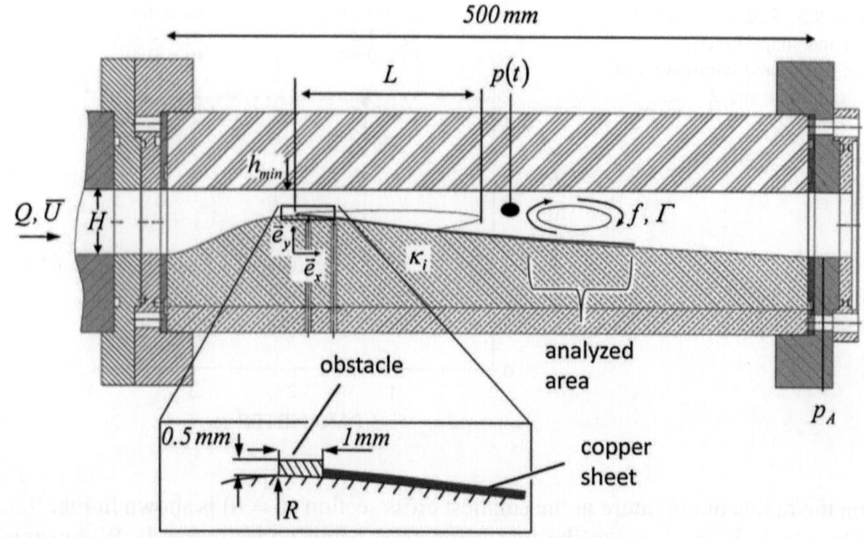

Fig. 9.1 Nozzle test rig at the Chair of Fluid Systems Technology

nozzle width and two cavitation zones in the corners between contour and vertical side walls of the tunnel instead a close cavity zone as shown in Fig. 9.2.

The nozzle shape is given by the following function where the height h and the axial length x are made dimensionless by the channel height $H = 50$ mm:

$$h^+(x^+) = \frac{const.}{\exp\left(\frac{x^+}{\kappa_1}\right) + \exp\left(-\frac{x^+}{\kappa_2}\right)}, \tag{9.1}$$

where superscript + indicates a non-dimensional quantity.

The non-dimension radius of curvature in the smallest cross section area is determined by the two free parameters κ_1 and κ_2. The influence of both parameters

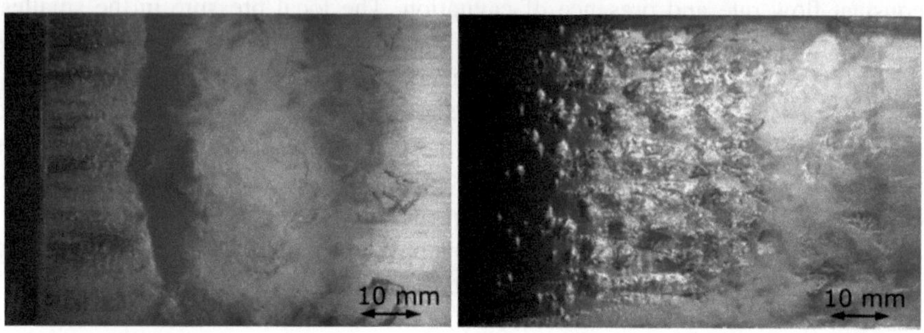

Fig. 9.2 Influence of the obstacle in the smallest cross section area for $R^+ = 2.5$, $Re = 3.1 \times 10^5$ and $\sigma = 6.05$. *Left* with obstacle, *right* without obstacle

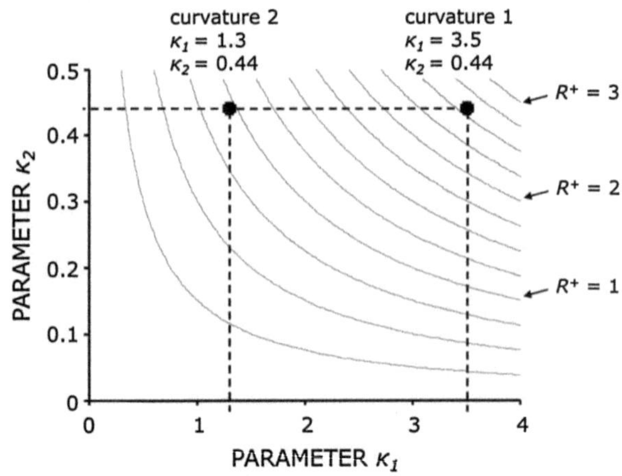

Fig. 9.3 Radius of curvature for the shape function, Eq. (9.1), and parameters of curvature 1 and 2

on the radius of curvature at the smallest cross section ($x = 0$) is shown in Fig. 9.3. The filled circles indicate the two geometries being under research. Within this contribution two cavitating flows with a radius of curvature of $R^+ = 2.5$ and $R^+ = 0.95$ are presented and discussed. The damage analysis was only done for the radius of curvature $R^+ = 2.5$.

The cavitation behavior is influenced by the Reynolds number and the cavitation number, based on the outlet pressure p_A and the cross section averaged flow velocity $\overline{U}$, which is varied between 3.9 and 7 m/s (Fig. 9.1):

$$\sigma = \frac{p_A - p_V}{\frac{1}{2}\rho\,\overline{U}^2}, \tag{9.2}$$

$$Re = \frac{\overline{U}H}{\nu}, \tag{9.3}$$

where ν denotes the kinematic viscosity, ρ the density and p_V the vapor pressure of water. The channel outlet is used as the reference location because only the pressure downstream of the cavitation zone influences the cavity in the case of constant flow rate and presence of cavitation. The local pressure in the smallest cross section is approximately the vapor pressure while the pressure difference is the same for constant velocity in the convergent part.

A single pressure sensor with an eigenfrequency of 125 kHz is flush-mounted in a fixed position in the side wall of the channel 65 mm downstream of the smallest cross section. The sensor records pressure fluctuation caused by the cloud collapses and helps to detect typical frequencies of the flow.

For cloud cavitation the most relevant physical parameters are the shedding frequency f of the cavitation cloud, the typical size L of the sheet and the circulation Γ of the cloud which is discussed in the next section. The measured associated dimensionless relations are:

$$\frac{fH}{\overline{U}} = St(Re, \sigma, \kappa_i), \tag{9.4}$$

$$\frac{L}{H} = \lambda(Re, \sigma, \kappa_i), \tag{9.5}$$

$$\frac{\Gamma}{\overline{U}H} = \gamma(Re, \sigma, \kappa_i). \tag{9.6}$$

With Eqs. (9.4–9.6) the largest scales of the periodic cavitation process are given.

9.3 Experimental Results

9.3.1 Observations

The high-speed images from the top view of the channel (Fig. 9.4) show the sheet and the cloud cavitation for a radius of curvature $R^+ = 2.5$. The operating conditions given by $\sigma = 5.7$ and $Re = 3.13 \times 10^5$ are such that cloud cavitation is dominant. This corresponds to a flow velocity of $\overline{U} = 6.3$ m/s in the outlet section area. The process is evaluated by a software tool detecting the front edge of the sheet and of the cloud.

The cycle starts by the growth of the cavitation sheet to its maximum sheet length L (Fig. 9.4a–e). The cavity starts growing with its origin fixed at the inserted wire along the rectangular cross section. While the sheet cavity reaches its maximum length, the re-entrant jet flows from a stagnation line on the surface. The front of the re-entrant jet is clearly visible in Fig. 9.4e–f. In the case of periodic cloud cavitation ($Re > Re_{crit}$) there is enough time for the upstream spread of the re-entrant jet to reach the leading edge of the sheet (Fig. 9.4a next period).

During the growth process, the cloud is observed in Fig. 9.4a–c. The white dotted line indicates the most downstream edge of the cloud. The cloud remains in the wake of the growing sheet cavitation, which can be seen by the stagnation of the rolled-up cloud. During this period, the cloud is imposed a circulation and forms a vortex with contact to the side walls. When this happen a cloud separates and the sheet starts again with the growth as described above.

Due to the Helmholtz vortex theorem the vortex has either to be closed at infinity or being a ring at every moment. This is the physical reason for the often seen U-shaped cloud (Fig. 9.4c–d).

The lower pressure in the vortex core leads to a decrease of the static pressure, which is measured by the pressure sensor installed at the channel wall downstream of the smallest cross section area. Actual pressure value for each picture is shown in the middle of the cross of the diagram. Past pressure values are demonstrated on the left hand side of the cross.

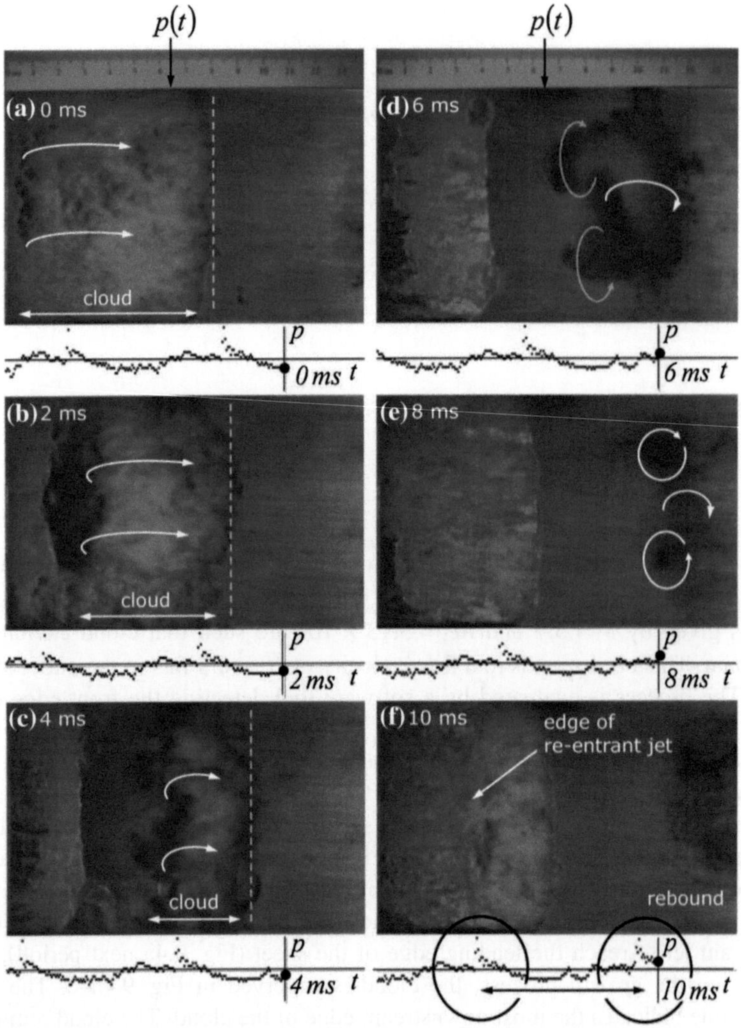

Fig. 9.4 High speed visualization for $Re = 3.13 \times 10^5$, $\sigma = 5.7$ and $R^+ = 2.5$. The actual pressure value is shown in the cross of the diagram *below* the pictures. The past pressure signal corresponds to the *left hand side* of the cross. The shedding frequency is about 49 Hz

The cloud collapse produces a pressure peak, which is also recorded and qualitatively shown in Fig. 9.4f. The collapse of the cloud seen in the pressure peak of Fig. 9.4f is between 8 and 10 ms pictures. The periodic behavior of the flow is visible from the pressure timeline shown in Fig. 9.4 as well.

The Fourier transform of time histories of pressure signal and of position of typical edges (detect by image processing) in high-speed videos are shown in Fig. 9.5 for one typical operating condition in which the characteristic frequency is 60 Hz. The cloud shedding frequency depends on the time delay until the sheet

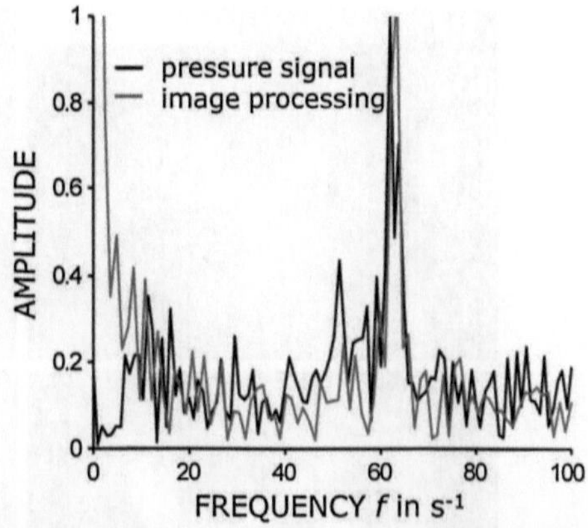

Fig. 9.5 Fourier transform of the pressure history and the position of the sheet cavity front for $R^{+} = 2.5$, $Re = 3.13 \times 10^{5}$ and $\sigma = 5.7$ for the first curvature

reaches its maximum length plus the time delay until the re-entrant jet reaches the leading edge. Both times are determined by the velocity field of the sheet cavity and the sheet cavity length. The analysis of both signals show a clear typical frequency associated with the cloud shedding.

The cavitation behavior changes by a change of the geometry. In the following, the influence of the curvature is discussed on the base of high-speed videos. Compared to the geometry above, the second one has a different parameter κ_1 whereas κ_2 is constant and thus changes the dimensionless radius of curvature in the smallest cross section area (Fig. 9.3). In this case the divergent part of the nozzle is changed dramatically which leads to more detachment flow downstream of the smallest cross section area.

The sheet cavitation could not be detected, which makes the analysis of the cavitation behavior of curvature 2 ($R^{+} = 0.95$) more difficult. Instead of a sheet cavity a formation of cavitating vortices close to the obstacle could be seen. The vortices merge together and form a bigger one, which also collapses in the downstream flow. This vortex merging process is marked in Fig. 9.6. Maybe this is the reason that the frequency is irregular so that no typical periodic cloud shedding can be detected. In Fig. 9.7 the Fourier transformation of the pressure signal shows a frequency band from 0 to 100 kHz and its harmonic frequencies.

In the following the influence of Reynolds number and cavitation number on the dimensionless products for the first curvature ($R^{+} = 2.5$) are discussed.

9.3.2 Strouhal Number Versus Reynolds and Cavitation Number

Figure 9.8 shows the Strouhal number (see Eq. (9.4)) as a function of the cavitation and Reynolds numbers. There are two characteristic flow regimes (shown in Figs. 9.8, 9.9). For a given cavitation number there is a critical Reynolds number.

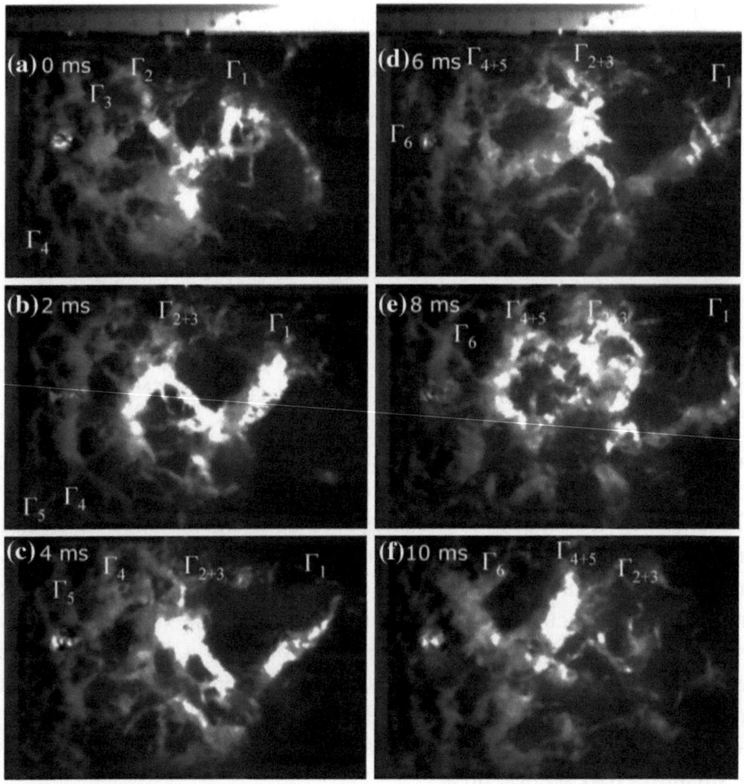

Fig. 9.6 High-speed video of the cavitating flow for a smaller radius of the curvature $R^+ = 0.95$ and for $Re = 3.13 \times 10^5$ and $\sigma = 5.7$

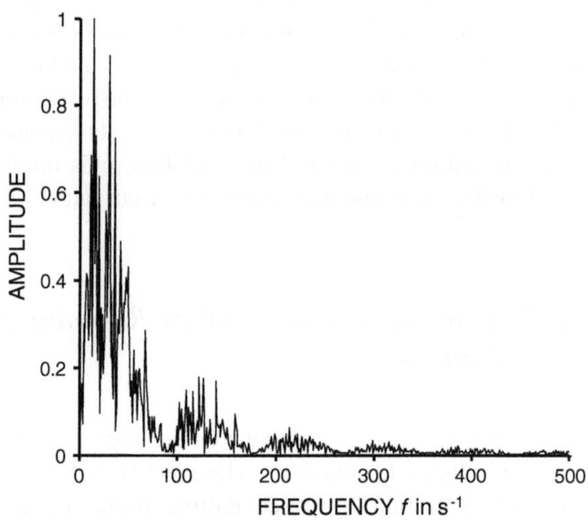

Fig. 9.7 Fourier transform of the pressure history for $Re = 3.13 \times 10^5$ and $\sigma = 5.7$ for $R^+ = 0.95$

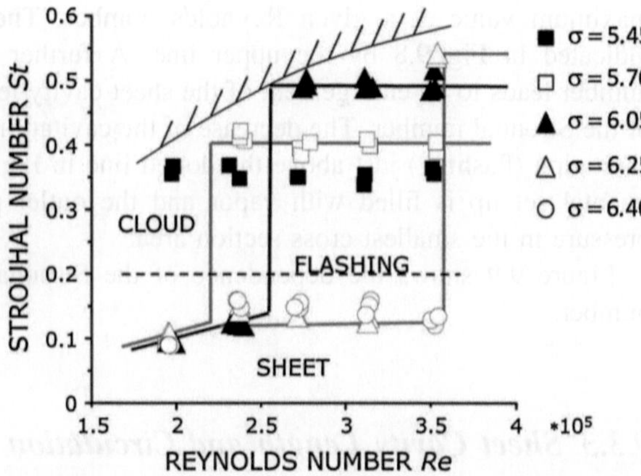

Fig. 9.8 Transition from sheet to cloud cavitation for $R^{+} = 2.5$

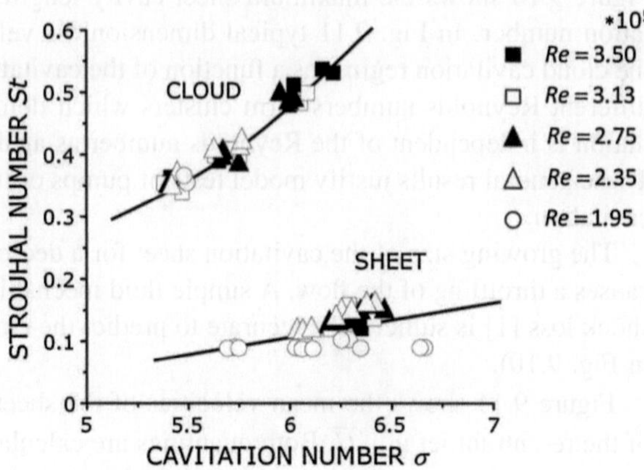

Fig. 9.9 Strouhal number versus cavitation number for $R^{+} = 2.5$

(i) Below the critical Reynolds number sheet cavitation without cloud shedding could be observed. (ii) Above that critical Reynolds number cloud cavitation is observed. The Strouhal number is in that case independent of the Reynolds number. Hence there should be no dependence on the size of the device in that flow regime. The research about the critical Reynolds-effect is not finished yet. Further experiments with glycol–water mixture to analyse a viscosity effect are in the planning stage.

A critical velocity was first described by Arndt [10]. In this paper, the velocity is made dimensionless by the kinematic viscosity and the channel height (i.e. Reynolds number).

For the highest cavitation number of 6.4, a critical Reynolds number could not be observed. (But there might be such a number outside the operating range of our test rig.)

By reducing the cavitation number at a given Reynolds number the sheet cavity length increases until the stability limit is reached again. The transition is associated with a sudden increase of the Strouhal number from its minimum to its maximum value at a given Reynolds number. The limiting curve is roughly indicated in Fig. 9.8 by the upper line. A further reduction of the cavitation number leads to an enlargement of the sheet cavity length and thus to a reduction of the Strouhal number. The decrease of the cavitation number finally leads to full cavitation (flashing) just above the dotted line in Fig. 9.8. In this case the experimental set up is filled with vapor and the outlet pressure corresponds to the pressure in the smallest cross section area.

Figure 9.9 shows the dependence of the Strouhal number on the cavitation number.

9.3.3 Sheet Cavity Length and Circulation of the Cloud

Figure 9.10 shows the maximum sheet cavity length L as a function of the cavitation number. In Fig. 9.11 typical dimensionless velocities are characterized for the cloud cavitation regime as a function of the cavitation number. The markers for different Reynolds numbers form clusters which demonstrate that the cloud cavitation is independent of the Reynolds number as again Figs. 9.8 and 9.9 indicate. These general results justify model tests of pumps or turbines with respect to cloud cavitation.

The growing size of the cavitation sheet for a decrease of the cavitation number causes a throttling of the flow. A simple fluid mechanical model based on Carnot's shock loss [1] is sufficiently accurate to predict the experimental results (solid line in Fig. 9.10).

Figure 9.11 shows the mean velocities of the sheet cavity growth $u_{sheet}/\overline{U}$ and of the re-entrant jet $u_{jet}/\overline{U}$. Both quantities are calculated with the time needed for the sheet cavity to reach its maximum length and the time needed for the re-entrant jet to reach the leading edge. The sheet and jet velocities are determined by means of visualizations and are time averaged velocities. As Fig. 9.11 shows, the velocities do not depend on the Reynolds number. Another point to mention is the difference between the velocities which cannot be explained by classical free-

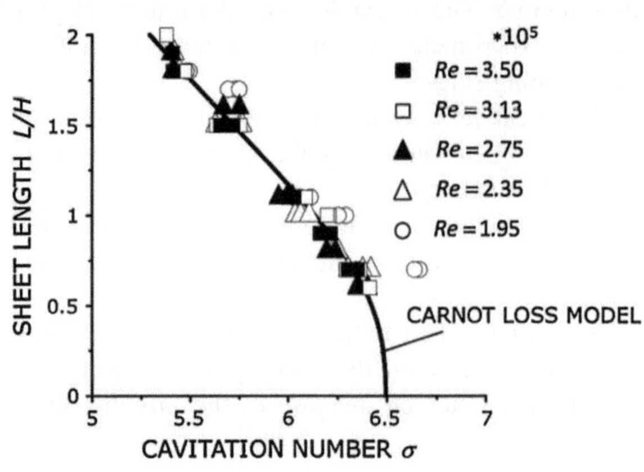

Fig. 9.10 Sheet length versus cavitation number for $R^{\prime} = 2.5$. The Reynolds number has no influence on the sheet length. The *solid line* indicates a Carnot loss model

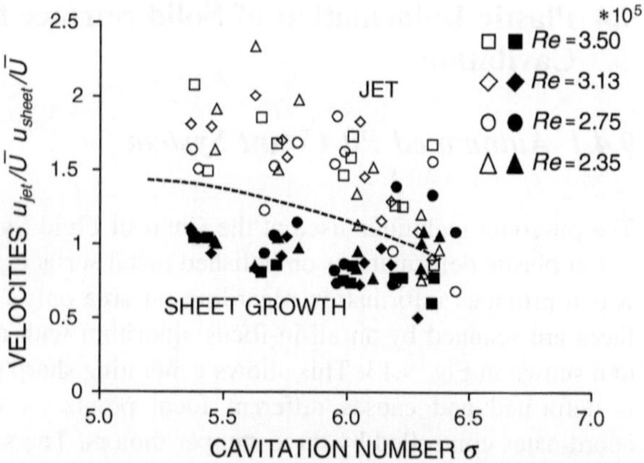

Fig. 9.11 Mean velocity of the growing sheet cavitation and the velocity of the re-entrant jet for $R^+ = 2.5$

streamline theory and depends on other physical effects which are beyond the scope of this work. The mean sheet cavity growth velocity is close to the velocity of the incoming flow. The mean flow velocity at the minimal cross section is by a factor of 2.5 higher than the sheet cavity growth velocity.

To quantify the circulation, the closed integral over the sheet is evaluated by using the mean values of the velocity in the smallest cross section area which is nearly the velocity above the sheet cavity and of the determined velocity of the re-entrant jet u_{jet},

$$\gamma = \frac{\Gamma}{\overline{U}H} = \frac{L}{H}\left(\frac{H}{H_{\min}} + \frac{u_{jet}}{\overline{U}}\right) \tag{9.7}$$

An increased cavitation length, i.e. decreased cavitation number, results in an increased circulation number as shown in Fig. 9.12.

To conclude, as soon as a critical Reynolds number is reached, there is a periodic cloud shedding and the Reynolds number vanishes on the right hand side of Eqs. (9.4–9.6) which then become $f\overline{U}/H = St(\sigma, \kappa_i), L/H = \lambda(\sigma, \kappa_i)$, $\Gamma/\overline{U}H = \gamma(\sigma, \kappa_i)$.

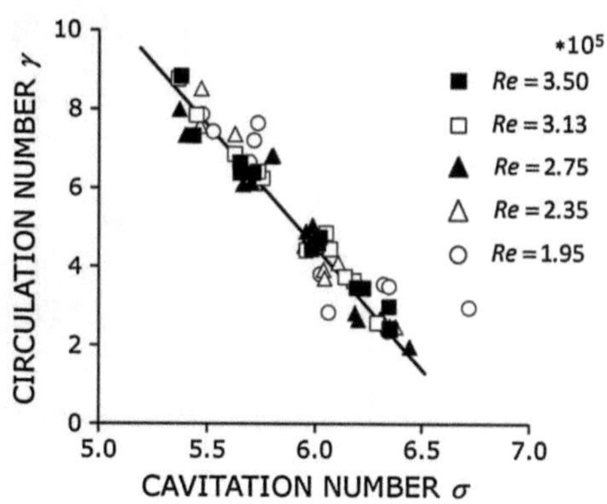

Fig. 9.12 Dimensionless circulation number versus cavitation number for $R^+ = 2.5$

9.4 Plastic Deformation of Solid Surface Due to Cloud Cavitation

9.4.1 Automated Pit-Count System

The pit-count technique used at the Chair of Fluid Systems Technology is able to detect plastic deformations on polished metal surfaces. It is a 2-D optical technique which provides information regarding pit area only (Lohrberg [11]). Curved surfaces are scanned by an all-in-focus algorithm with a completely automated system shown in Fig. 9.13. This allows generating sharp pictures although the surface is deformed and causes different focal points. A microscope moves in three coordinates controlled by three stepper motors. The surface is scanned before and after a cavitation experiment. By comparing the two scans, a filter algorithm is implemented to distinguish between material defects and pits, which are the result of collapsing cavitation bubbles. Such a pit is shown in Fig. 9.14.

The pit-count software gives finally a damage map of the curved surface. The creation of damage maps with a local damage distribution with known radius and position of all plastic deformation are possible. The damage energy can be calculated by using an empirical relationship developed and used by Lohrberg [11],

$$E = \vartheta L_m^3 \sum \left(1 + \frac{r_i}{l_m}\right)^3, \tag{9.8}$$

where ϑ is the yield stress of the material, L_m and l_m are two material typical lengths. The product ϑL_m^3 is the minimum energy to generate a pit, which is equal

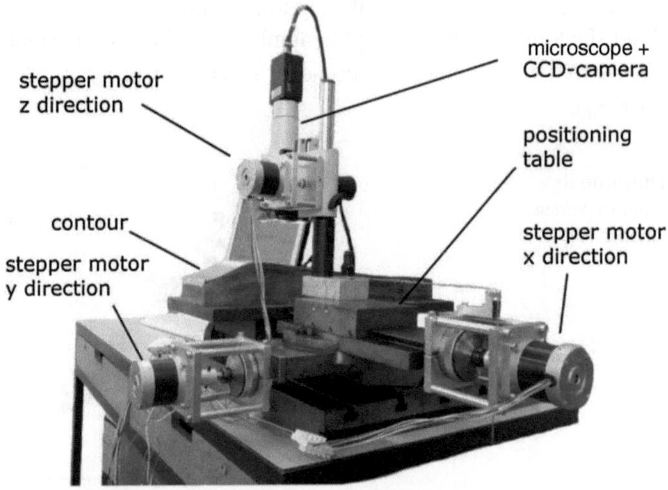

Fig. 9.13 Fully automated pit-count system at the Chair of Fluid Systems Technology

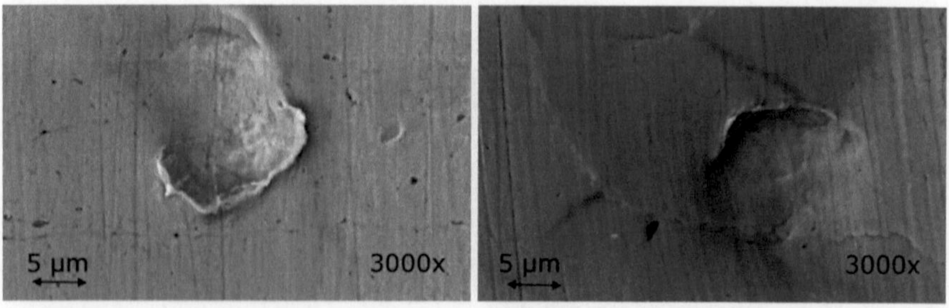

Fig. 9.14 Plastic deformations of a copper surface after 30 min (*left*) and dislocation movements of grains after 300 min (*right*), $\sigma = 0.112$

to the energy to get over the elastic limit. r_i is the pit radius which is made dimensionless with a material dependent constant l_m which gives an empirical relation between the pit radius and the plastic deformed volume. All empirical relations are found out by Lohrberg [11].

9.4.2 Damage Maps

Pit count measurements are carried out for Reynolds numbers 2.75×10^5 and 3.5×10^5, three cavitation numbers 5.45, 5.7, 6.05 and a dimensionless radius $R^+ = 2.5$. Figure 9.15 (left) shows those operation points to lay completely within the cloud regime.

Figure 9.15 (right) shows the damaged surface in percent for a Reynolds number of 2.75×10^5 and the three cavitation numbers. The exposure time T= 90 min is made dimensionless by the shedding frequency. Thus fT gives the number of clouds shed by the sheet cavity during the exposure time. The number of load cycles is given for each experiment in Figs. 9.15 and 9.16.

As expected, a smaller cavitation number results in an increased cavitation zone and thus an increased damage during the exposure time. A smaller sheet cavity with a higher frequency and thus more damage cycles leads to less damage than a large sheet with a lower frequency as can be seen qualitatively in Fig. 9.15.

Figure 9.16 shows the experimental results for the larger Reynolds number of 3.5×10^5. The load cycles are given within the pictures. In all damage pictures a typical erosion of a collapsed U-shaped cavitation vortex can be observed.

To compare the experimental results the deformation energy is accumulated for the whole polished surface using the empirical relation (9.8).

To make different flow conditions comparable, it is useful to normalize the deformation as:

$$\frac{E}{\rho \overline{U}^2 H^3 f T} = \pi(Re,\ \sigma,\ \kappa_i)\,. \tag{9.9}$$

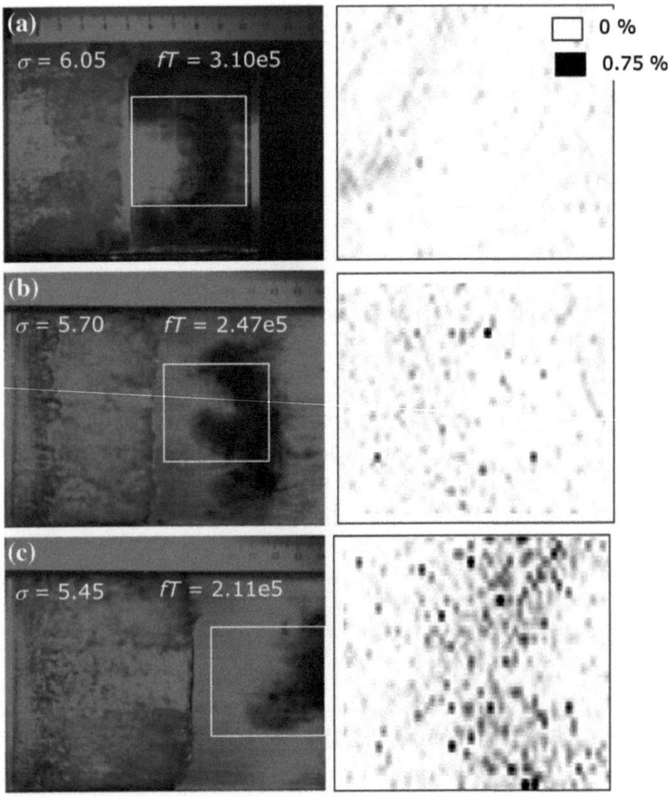

Fig. 9.15 Damage maps for the cavitation numbers 5.45, 5.7, 6.05 at a small Reynolds number of 2.75×10^5 and $R^+ = 2.5$. Each pixel represents the information of several single pictures. The *colour* indicates the damage evaluated by an analysis of the single pictures

In order to compare the aggressiveness of operating points, the non-dimensional energy is based on the number of damage cycles fTin Eq. (9.9). The result is shown in Fig. 9.17 where the accumulated dimensionless damage energy per cycle is given as a function of the cavitation number.

Up to $Re = 3.5 \times 10^5$ at the lowest cavitation number (see damage map shown in Fig. 9.15c), all results are on one line. To show the influence of the Reynolds number for small cavitation numbers an operating point for a Reynolds number at 2.35×10^5 is added in the measurement program.

Figure 9.18 shows the position of center of the damage given in Figs. 9.15 and 9.16 and the maximum sheet cavity length normalized with the channel height. The damage position is primarily influenced by the cavitation number and thus by the sheet cavity length.

The position of the maximum cavity length corresponds to the location of the cloud collapse.

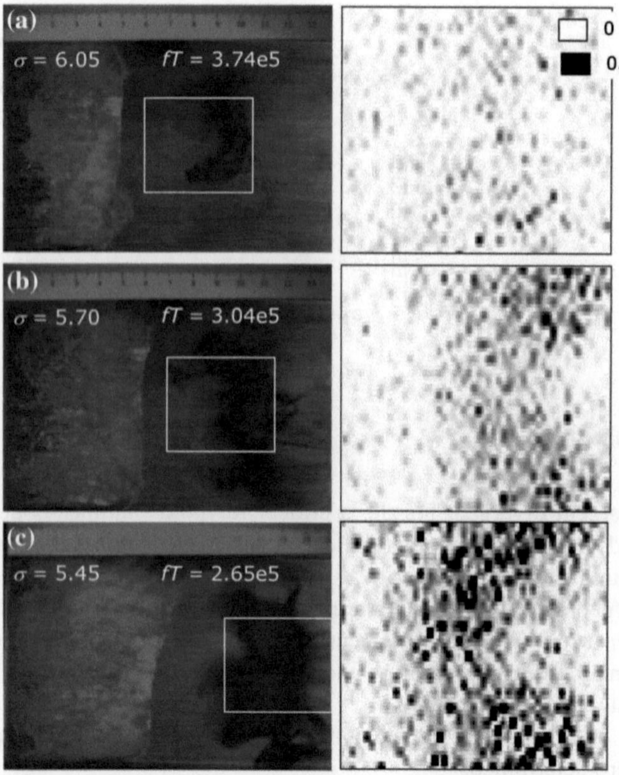

Fig. 9.16 Damage maps for the cavitation numbers 5.45, 5.7, 6.05 at a small Reynolds number of 3.5×10^5 and $R^+ = 2.5$. Each pixel represents the information of several single pictures. The *colour* indicates the damage evaluated by an analysis of the single pictures

Fig. 9.17 Dimensionless damage energy per damage cycle for $R^+ = 2.5$

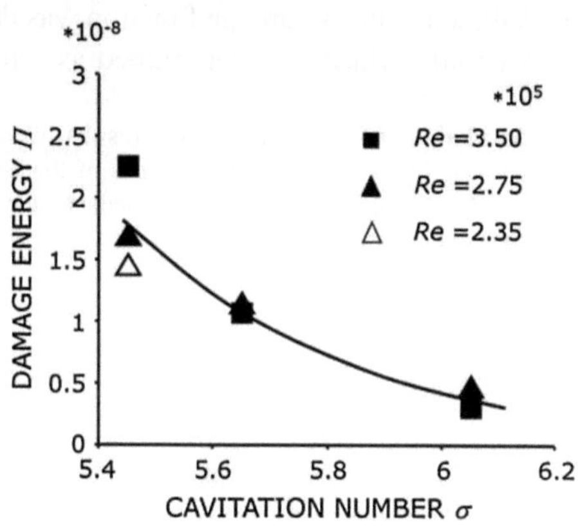

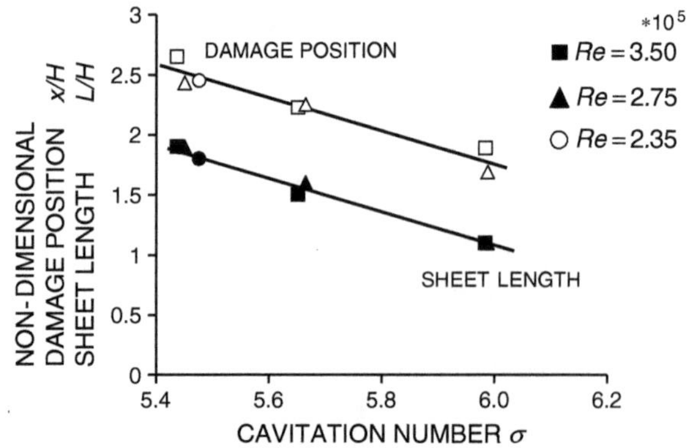

Fig. 9.18 Dimensionless damage position and non-dimensional sheet cavity length for $R^+ = 2.5$

9.5 Conclusions

- For a convergent divergent nozzle, a critical Reynolds number is determined. For Reynolds numbers below that critical value sheet cavitation is observed. For Reynolds numbers above that critical value cloud cavitation is observed.
- The nozzle geometry is such that the minimal radius of curvature can easily be changed. Up to now the radius of curvature was set to 2.5 times the channel height. A small radius of curvature and thus a more divergent nozzle part leads to a more stochastic cavitation.
- By the aid of optical measurements the largest scales of the cavitation phenomena were determined.
- An upper limit for the Strouhal number was observed. That upper limit is dependent on the Reynolds number.
- The circulation of clouds was determined.
- Within the cloud regime there is no influence on the Reynolds number, which justifies model tests of pumps and turbines.
- A dimensionless damage function was defined and the plastic deformation of the material surface was determined as a function of cavitation number.

Acknowledgments The presented results were obtained within the research project No. 16054 N/1, funded by budget resources of the Bundesministerium für Wirtschaft und Technologie (BMWi) approved by the Arbeitsgemeinschaft industrieller Forschungs-vereinigungen "Otto von Guericke" e.V. (AiF).

References

1. Keil T, Pelz PF (2012) On the transition from sheet to cloud cavitation. In: Proceedings of the 8th international symposium on cavitation, CAV2012, Singapore, 13–16 August 2012
2. Buttenbender J, Pelz PF (2012) On the bubble dynamics in cavitation clouds under circulation and strain. In: Proceedings of the 8th international symposium on cavitation, CAV2012, Singapore, 13–16 August 2012

3. Ida M (2004) Investigation of transition frequencies of two acoustically coupled bubbles using a direct numerical simulation technique. J Phys Soc Jpn 73(11):3026–3033
4. Ida M (2009) Multibubble cavitation inception. J Phys Soc Jpn 21(11):113302
5. Konno A, Kato H, Yamaguchi H, Maeda M (2002) On the collapsing behavior of cavitation bubble clusters. JSME Int J Ser B 45(3):631–637
6. Wang YC (1996) Shock waves in bubbly cavitating flows (Part I) Shock waves in cloud cavitation (Part II) Bubbly cavitating flows through a converging-diverging nozzle. PhD thesis, California Institute of Technology, Pasadena
7. Kawanami Y, Kato H, Yamaguchi H, Maeda M, Nakasumi S (2002) Inner structure of cloud cavity on a foil section. JSME Int J Ser B 45(3):638–645
8. Joussellin F, Delannoy Y, Sauvage-Boutar E, Goirand B (1991) Experimental investigations on unsteady attached cavities. ASME Fluids Eng Div 116:61–66
9. Kawanami Y, Kato H, Yamaguchi H, Tagaya Y, Tanimura M (1997) Mechanism and control of cloud cavitation. J Fluids Eng 119:788–794
10. Arndt REA, Hambleton WT, Kawakami E (2009) Creation and maintenance of cavities under horizontal surfaces in steady and gust flows. J Fluids Eng 31:111301-1
11. Lohrberg H (2001) Messung und aktive Kontrolle der erosiven Aggressivität der Kavitation in Turbomaschinen. Fortschritts-Bericht VDI Reihe 8 Nr. 942 ISBN 3-18-394208-5

Chapter 10
Investigations into Dependence Between Cavitation Structures and Cavitation Erosion

Matevž Dular

Abstract Experiments within the cavitation erosion incubation period were performed on simple and modified two-dimensional hydrofoils with circular leading edges. A pit-counting method, based on computer-aided image processing, was used for direct measurement of the cavitation erosion by evaluating the damage of the surface. Cavitation phenomenon above hydrofoils at different flow conditions (pressure, water gas content, flow velocity) was observed. A clear relation between characteristics of cavitation structures and cavitation damage was established. A study of influence of gas content in water and flow velocity on the cavitation erosion aggressiveness was performed. There we found a clear influence which shows a drop in aggressiveness of cavitation erosion as the gas content of water is increased. Also a power law was confirmed for velocity influence on cavitation erosive aggressiveness. Due to the extreme length of experiments, many studies tend to perform tests only within the incubation period and the mass loss rate is then predicted by extrapolation. A rotating disc test rig that generates a very aggressive cavitation and pure copper specimens, as erosion sensors, were used to investigate the correlation between the damage within the incubation period and mass loss rate. Like in the case of a single hydrofoil we also observed dependency of the cavitation erosive aggressiveness on the size and dynamics of cavitation structures. Results presented in these studies will serve as a basis for achieving the final goal of the ongoing work—to develop a method that will enable accurate prediction of cavitation erosion with minimal experimental effort or even solely by using computational fluid dynamics.

M. Dular (✉)
Laboratory for Water and Turbine Machines, University of Ljubljana, Askerceva 6, 1000 Ljubljana, Slovenia
e-mail: Matevz.Dular@fs.uni-lj.si

K.-H. Kim et al. (eds.), *Advanced Experimental and Numerical Techniques for Cavitation Erosion Prediction*, Fluid Mechanics and Its Applications 106,
DOI: 10.1007/978-94-017-8539-6_10,

10.1 Introduction

Cavitation phenomenon, characterized by vapor generation and condensation, occurs frequently in hydraulic machines. It causes vibration, increase of hydrodynamic drag, changes the flow hydrodynamics, generates noise, thermal and optical effects such as luminescence, and produces cavitation erosion. It was Rayleigh [1] who first introduced the problem of cavitation erosion of the ship propellers. Since then a wide range of studies that deal with problems from bubble dynamics to material testing, have been made.

It is known from previous experimental and theoretical studies [2] that shock waves emitted during bubble collapse have the potential to cause erosion. But it is possible that a number of other events takes place in the process of pit creation and later mass loss.

Figure 10.1 shows the process of cavitation cloud separation from the attached part of cavitation. The images and velocity fields were obtained by a combination of visualization and PIV measurements [3]. The cloud collapse in the higher pressure region downstream, causes the formation of the back flow, which causes a new separation of the cavitation cloud. Also a pressure wave of magnitude of several 100 kPa is emitted [4] at cloud collapse which can trigger a sequence of events that consequently lead to erosion.

There are several still unconfirmed theories as to what is the primary cause for the erosion. One is the shock wave that is emitted at the collapse of the bubble [2]. The second theory suggests that the shape of the bubble near the surface becomes unstable as the pressure wave from the cloud collapse reaches it. The fluid that surrounds the bubble takes a shape of jet through the bubble in the direction towards the solid surface [5]. Finally Tong et al. [6] reported that the collapse of a toroidal cavity, which sometimes forms after the liquid jet impact, also generates high pressures which can be far higher than those associated with the initial jet impact.

10.2 Experimental Set-Ups

Experiments were conducted in two set-ups—on single copper coated hydrofoils in a cavitation tunnel and on copper specimens inserted into a rotating disc test rig. For evaluation of the damage, pit counting and measurements of mass loss were used.

10.2.1 Single Hydrofoils and Incubation Period

The basic geometry used is a 50 mm wide, 107.9 mm long and 16 mm thick symmetric hydrofoil with circular leading edge and parallel walls (CLE—Circular

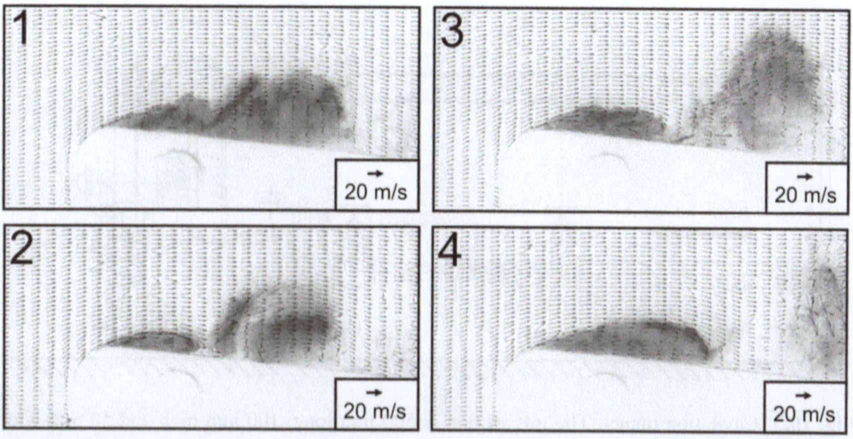

Fig. 10.1 Cavitation cloud separation

Fig. 10.2 Copper coated Circular Leading Edge (CLE) and Asymmetrical Leading Edge (ALE) hydrofoils used for experiments. Based on [7], reprinted with permission from Elsevier

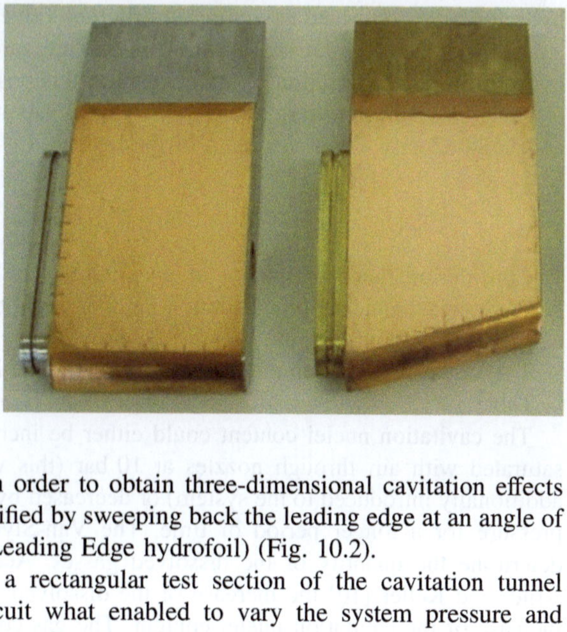

Leading Edge hydrofoil). In order to obtain three-dimensional cavitation effects the basic geometry was modified by sweeping back the leading edge at an angle of 15° (ALE—Asymmetrical Leading Edge hydrofoil) (Fig. 10.2).

Hydrofoil was put into a rectangular test section of the cavitation tunnel (Fig. 10.3) with closed circuit what enabled to vary the system pressure and consequently the cavitation number. The test section of the cavitation tunnel is 500 mm long, 100 mm high and 50 mm wide. Two observation windows are mounted for top and side view observation.

The velocity in the reference plane upstream of the hydrofoil was held constant during the tests at 10, 13 or 16 m/s. An incidence angle of 5°, where cavitation was the most aggressive on the sample, was chosen for all the tests (at smaller incidence angle the cavitation remained attached, on the other hand—at a higher angle, the position of cloud collapse moved too far away from the surface of the

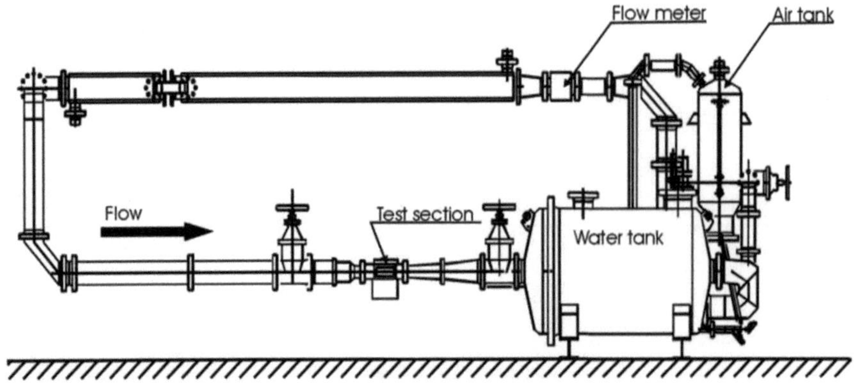

Fig. 10.3 Cavitation tunnel. The test section is 500 mm long, 100 mm high and 50 mm wide. The maximal flow velocity is 20 m/s. The pressure at the inlet to the test section can be adjusted between 0.2 and 5 bar. Based on [7], reprinted with permission from Elsevier

hydrofoil). We observed cavitation at different values of cavitation number (2.5, 2.3, 2.0), which was defined by the pressure measured 400 mm upstream of the hydrofoil (the cross-section of the channel at this position was the same as in the test section—100 mm high and 50 mm wide), vapor pressure p_v, fluid density ρ and flow velocity V just upstream of the hydrofoil:

$$\sigma = \frac{p - p_v}{0.5\rho V^2}. \tag{10.1}$$

Considering the combination of uncertainties of pressure, velocity and temperature measurements, the cavitation number could be determined within ± 0.02 of global uncertainty.

The various experimental conditions investigated in this work are summarized in Table 10.1.

The cavitation nuclei content could either be increased by injection of water, saturated with air, through nozzles at 10 bar (this way no larger bubbles were additionally introduced to the system) or decreased by running the system at a low pressure for a longer period of time. The Van-Slyke method [8] was used to determine the quantity of the dissolved gasses. According to Peterson [9] and Arndt and Keller [10] the increase of the dissolved gasses is proportional to the increase of the cavitation nuclei content. The gas content ϕ was varied between 14.3 ± 0.5 mg_g/l_w (milligram of gas per liter of water) for low gas content to 48.5 ± 0.7 mg_g/l_w for high gas content (Table 10.1).

Figure 10.4 shows cavitation from the top view for the case of CLE (left) and ALE hydrofoils (right) at 13 m/s flow velocity, low gas content and cavitation number 2.3. The flow is from left to the right.

Table 10.1 Parameters of the experiments

Hydrofoil	σ	V (m/s)	ϕ (mg_g/l_w)
CLE	2.0	13	14.0
CLE	2.3	13	13.8
CLE	2.5	13	14.0
CLE	2.3	13	25.9
CLE	2.3	13	36.7
CLE	2.3	13	48.9
CLE	2.3	10	13.9
CLE	2.3	16	13.8
ALE	2.0	13	14.8
ALE	2.3	13	14.7
ALE	2.5	13	14.7

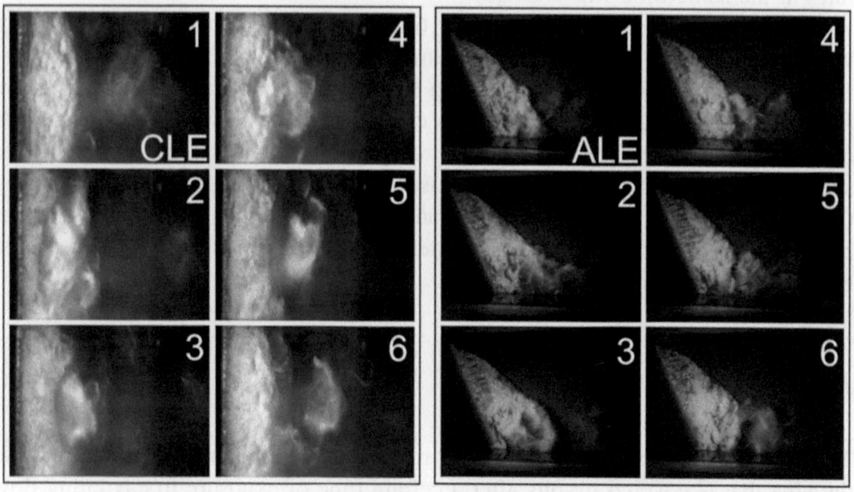

Fig. 10.4 *Top view* of cavitation on CLE (*left*) and ALE hydrofoils (*right*) at flow velocity 13 m/s, low gas content and cavitation number 2.3. The flow is from *left* to the *right*. Based on [7], reprinted with permission from Elsevier

Cavitation is approximately symmetrical in the case of CLE hydrofoil, which is not the case of swept leading edge of ALE hydrofoil. Significant fluctuations of cavitation region with separation of the cavitation cloud can be seen in the region where the ALE hydrofoil is the shortest, while cavitation at the rear wall (where the hydrofoil length is the longest remains nearly steady (with no cloud separation). A detailed experimental and numerical study of the phenomenon can be found in Dular et al. [11].

When one observes cavitation from the side view (Fig. 10.1) no significant difference between CLE and ALE hydrofoils can be seen.

10.2.2 Pure Copper Specimens and Mass Loss

The rotating disc test rig was developed to generate an aggressive type of cavitation (Fig. 10.5).

It consists of a closed water loop and the rotating disc with four holes where cavitation appears (right top picture in Fig. 10.5). Opposing to the rotating disc a stationary disc with specimen holders and observation window is positioned (right bottom picture in Fig. 10.5). Positions where the material specimens can be mounted are noted by p1–p6. For the present experiments specimens were mounted in positions 1–4 and the damage evaluation was only carried out on specimens in positions p1 and p2—specimen holders in positions p5 and p6 were empty. Asymmetrical distribution of specimens lead to different cavitation aggressiveness on evaluated specimens, hence performing tests at different erosion rates at the same time was enabled.

Specimens with size of 10 × 18 × 65 mm, made out of copper (99.96 %), were wet grinded and polished to 0.5 μm prior to the experiment.

Experiments at two different rotating frequencies (1,500 and 1,800 rpm) were performed. For easier comparison the cavitation number was held constant at $\sigma = 0.16$, where the velocity was the tangential velocity at the radius of the specimen/hole and the pressure was measured on the back side of the disc housing (behind the rotating disc), again at the radius of the specimen/hole. The water was neither aerated nor degassed. Gas content of 32.4 ± 1.2 mg_g/l_w was measured every time the test rig was stopped.

10.2.3 Pit-Counting Method

The pit-counting method is a simple approach that is based on the assumption that the area that is covered by pits after a certain time of exposure to cavitating flow gives a quantitative measure of the intensity of cavitation erosion. The erosion study is constricted to the incubation period, the period where damage (plastic deformations—pits) can already be detected, but there is no mass loss.

Pits have a diameter of magnitude order 10 μm, and can be distinguished only by sufficient magnification. Figure 10.6 shows an image of the copper coated surface of the hydrofoil before the exposure to cavitating flow (left) (0 % damaged surface) and after 1 h of exposure at $\sigma = 2.3$, $V = 13$ m/s (right) (4.98 % damaged surface).

A problem that has to be considered is the possibility of overlapping of the pits. Pit clusters are created by superposition of impacts during long tests, by collapse of a group of bubbles or by rebounds of a single bubble. The principle that is used for pit separation is that a single pit cannot form a concave shape. Hence concavely shaped dark regions are divided into a number of individual objects each having a circular (convex) shape. The separated objects are then enlarged to fill out the original object size [12]. If an image object is caused by overlapping pits it is possible that one pixel is shared by two or more pits (Fig. 10.7).

Fig. 10.5 Rotating disc test rig. The disc has a diameter of 500 mm and is driven by a 35 kW motor. The maximal rotating frequency is about 2,000 rpm. Based on [13], reprinted with permission from Elsevier

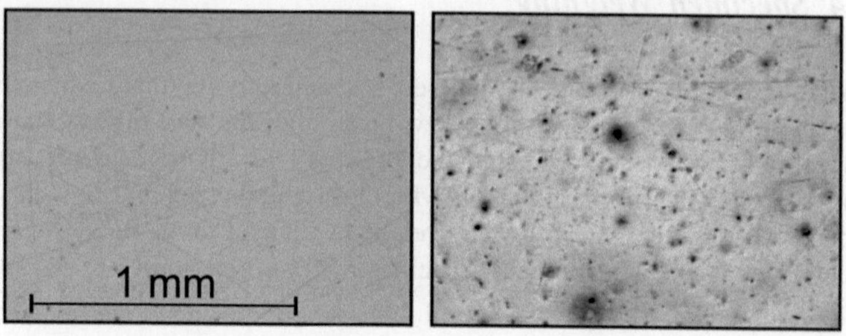

Fig. 10.6 Image of surface prior (*left*) and after (*right*) the exposure to the cavitating flow (copper coated hydrofoil at $\sigma = 2.3$, $V = 13$ m/s and low gas content). Based on [7], reprinted with permission from Elsevier

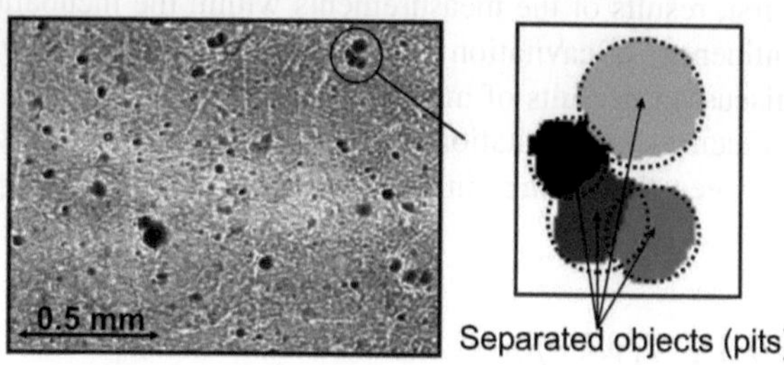

Fig. 10.7 Detection of pits on the surface by the pit—count method. Based on [7], reprinted with permission from Elsevier

About 1,000 images of the surface were taken under magnification and the area covered by pits on each image was measured. We can represent the results using pit diagrams where the local part of the damaged surface ES (on each of the 1,000 images) is presented. But we can also calculate the integral part of the damaged surface (ΣES) which represents the part of the whole evaluated surface that is covered by pits:

$$\Sigma\text{ES} = \frac{\sum(\text{ES} \times A)}{\sum A}, \tag{10.2}$$

where ES is the part (expressed in percent) of the surface that is covered by pits on one image after an exposure time of 1 h, and A is the surface of that image.

To evaluate the pit-count algorithm synthetic images with known area of the pits (ΣES) were created. No significant influence of pit overlapping could be seen until more than 15 % of the surface is covered by pits. From that point on, if the overlapping was not considered, the software measured a value which was smaller than the real one (for example, only 17.2 % of damaged surface was measured when in fact 20 % of the surface was covered by pits).

10.2.4 Specimen Weighing

For the case of copper specimens that were inserted into the rotating disc test rig, measurable mass loss occurred only a few hours after the start of the experiment. The specimens were removed from the rig every few hours and prepared for weighing by blowing all the water off with clean compressed air. The specimens with mass of approximately 95 g were weighed with a Sartorius BP301S precision balance scale. It has a maximum capacity of 303 g, readability of 0.1 mg and linearity of 0.3 mg.

10.3 Results and Discussion

First, results of the measurements within the incubation period are presented. The influences of cavitation dynamics and structures, flow velocity and gas content are discussed. Results of mass loss measurements follow, where we again discuss the influences of cavitation dynamics and flow velocity and also the relationship between the damage in incubation and in mass loss periods.

10.3.1 Hydrofoils

The distributions of magnitude of cavitation erosion on the surfaces of the hydrofoils are presented in Fig. 10.8. A sufficient number of pits was obtained after the hydrofoil was exposed to cavitating flow for a period of 1 h. Each contour diagram is a result of analyzing approximately 1,000 side by side microphotographs (like the one in Fig. 10.6) of the hydrofoil surface.

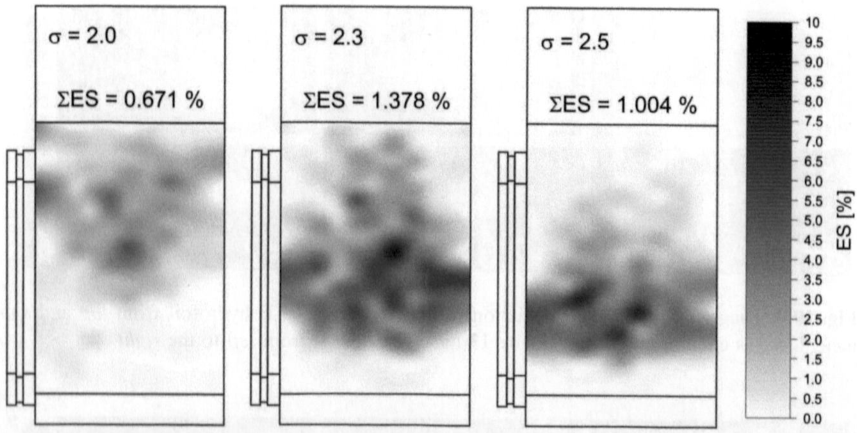

Fig. 10.8 Pit distribution for CLE hydrofoil, low gas content $\phi < 15$ mg_g/l_w. Exposure time is 1 h, flow velocity is 13 m/s. The flow is from *bottom* to *top*. Based on [7], reprinted with permission from Elsevier

We expected that the integral part of the damaged surface (ΣES) will increase as the cavitation number was lowered, due to the presence of larger vapor structures. But on the contrary it was the highest in the case of cavitation number 2.3. The region of greatest damage also moves away from the leading edge as the cavitation number is decreased since the structures grow to a larger size and cloud separation occurs further downstream. The easiest way to interpret the pit-counting results is to calculate the standard deviation of gray level of several hundred images of cavitation (like those in Fig. 10.4). Higher value of standard deviation of gray level corresponds to regions with greater dynamics of cavitation structures.

The standard deviation is high in the region where the cavitation clouds collapse. From the side view diagrams a thin strip, starting at the leading edge, with high value of deviation can also be seen—this can be related to small oscillations of the attached cavity size.

From the diagrams in Fig. 10.9 one can see that the maximum of standard deviation moves away from the leading edge and away from the surface of the hydrofoil as the cavitation number is decreased. This suggests that cavitation cloud collapses occur more downstream (even downstream of the copper foil in case of $\sigma = 2.0$) and away of the copper foil as the cavitation number is decreased. A result of this is the lower damage rate (lower value of ΣES) in case of $\sigma = 2.0$ since the amplitude of the emitted shock wave at cavitation could collapse is attenuated by the distance until it reaches the surface of the foil.

The cavitation structures in the cases with asymmetric hydrofoil (ALE) show dynamic cavitation behavior only in the region where the hydrofoil length is the shortest (Fig. 10.4). This can be nicely presented by the distribution of standard deviation (Fig. 10.10).

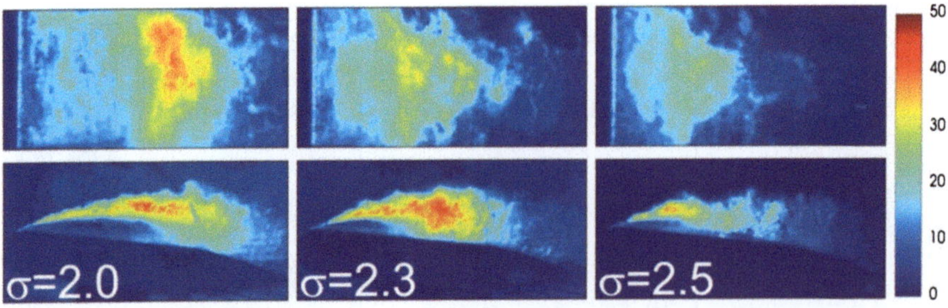

Fig. 10.9 Diagrams of standard deviation of gray level for CLE hydrofoil from *top* and *side view*, low gas content and flow velocity 13 m/s. The flow is from *left* to the *right*

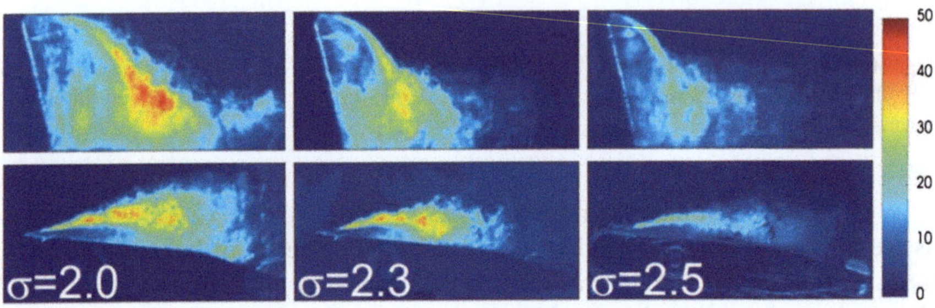

Fig. 10.10 Diagrams of standard deviation for ALE hydrofoil from *top* and *side view*, low gas content and flow velocity 13 m/s. The flow is from *left* to the *right*

We again see that the position of the cavitation cloud collapse can be related to high value of standard deviation—in the case of ALE hydrofoil this is the part of the hydrofoil where its length is the shortest. From the side view distributions, like in the case of CLE, we can see a thin strip, starting at the leading edge, with high value of deviation. In addition we see a similar region of higher standard deviation in the top view images (spreading from the point of the leading edge where the hydrofoil is the longest to the cavitation cloud collapse region)—this can again be related to small oscillations of the attached cavity size.

Again the region that sustains the most damage corresponds to the region where cavitation is the most dynamic, which confirms the hypothesis that the cavitation erosion is conditioned by the dynamic behavior of cavitation. Unlike in the case of CLE hydrofoil the ΣES value increases with decreasing cavitation number for the ALE hydrofoil (Fig. 10.11). The reason lies in a different flow pattern. The length of the cavitation structure does not exceed the length of the copper foil (even in the case with cavitation number 2.0), hence the cavitation cloud collapse occurs above the copper foil. Furthermore the distance of cloud implosion from the surface of the hydrofoil remains approximately constant for all the cases.

To investigate the influence of gas content we performed tests on a CLE hydrofoil at a constant flow velocity of 13 m/s, constant cavitation number

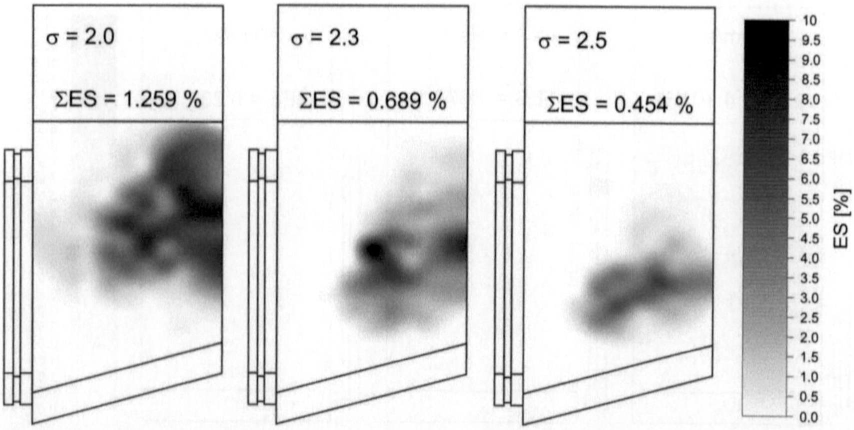

Fig. 10.11 Pit distribution for ALE hydrofoil, low gas content $\phi < 15\ \text{mg}_g/\text{l}_w$. Exposure time is 1 h, flow velocity is 13 m/s. The flow is from *bottom* to *top*. Based on [7], reprinted with permission from Elsevier

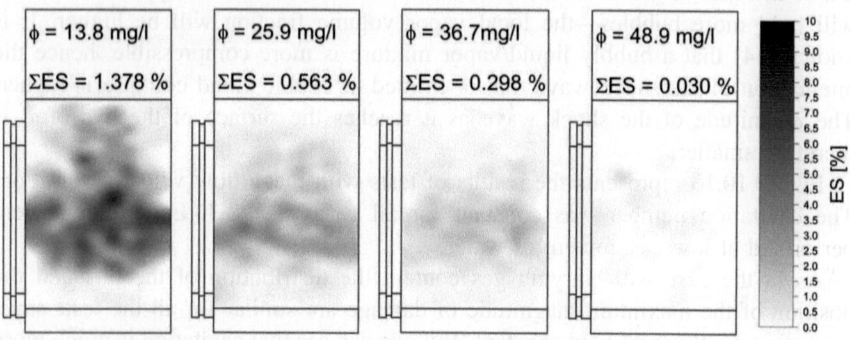

Fig. 10.12 Pit distribution for CLE hydrofoil, for different water gas contents ϕ at $\sigma = 2.3$ and flow velocity $V = 13$ m/s. Exposure time is 1 h. The flow is from *bottom* to *top*

$\sigma = 2.3$ but at different gas contents (Fig. 10.12). The gas content was measured at the beginning and at the end of each test. The gas content only slightly increased over the period of 1 h. The values given are the average values.

One can see that erosive aggressiveness decreases significantly when the gas content rises (Fig. 10.12). The surface sustains almost 50 times less damage in the case with high gas content (48.9 mg_g/l_w) than in the case with low gas content (13.8 mg_g/l_w). The position of maximal erosion magnitude and the distribution of pits on the hydrofoil remains almost the same for all cases, since the cavitation number was constant. The topology of cavitation structures practically did not change when the gas content was altered.

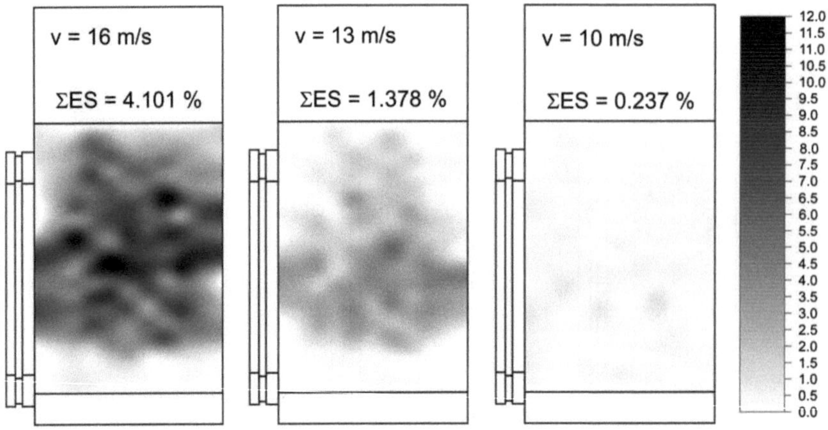

Fig. 10.13 Pit distribution for CLE hydrofoil, for different flow velocities at $\sigma = 2.3$ and low gas content $\phi < 15$ mg_g/l_w. The flow is from *bottom* to *top*

A possible reason for the less aggressive cavitation erosion (lower ΣES value) due to increase of gas content is that the cavitation cloud in strongly aerated water will hold more bubbles—the local vapor volume fraction will be higher. It is known [14] that a bubbly liquid/vapor mixture is more compressible, hence the attenuation of the shock wave that is emitted at bubble cloud collapse is higher. The magnitude of the shock wave as it reaches the surface of the hydrofoil is therefore smaller.

Figure 10.13 represents the results of tests with mean flow velocity variation. The cavitation number was constant for all cases $\sigma = 2.3$. Experiments were performed at low gas content of water.

As in the case with varying gas content the distribution of the pits and the position of the maximum magnitude of damage are similar for all the tests since cavitation number was held constant. But one can see that cavitation is much more aggressive at higher flow velocities. A power law with an exponent n = 5 to 8 is usually given in the literature [15]:

$$\frac{\sum ES_1}{\sum ES_2} = \left(\frac{V_1}{V_2}\right)^n. \tag{10.3}$$

It is common to use the mass loss as a measure of cavitation aggressiveness but due to limitations of our study we used the ΣES values instead (see Fig. 10.13). Nevertheless when we construct a trendline according to the power law (Eq. 10.3) we get a value of $n = 6.1$, which lies within the expected range (Fig. 10.14).

Possible reasons for the increase of cavitation aggressiveness due to the increase of velocity are:

- If the cavitation number remains constant and the flow velocity changes the pressure difference between operating and vapor pressure has to increase with

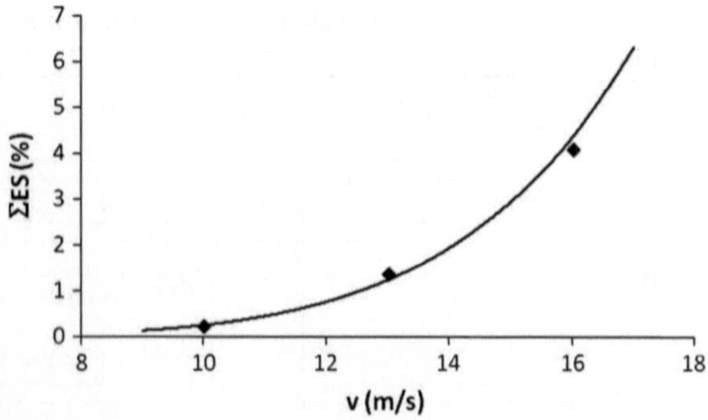

Fig. 10.14 Measured ΣES value as a function of flow velocity and a power law trendline (n = 6.1). CLE hydrofoil, at $\sigma = 2.3$ and low gas content $\phi < 15\ mg_g/l_w$

velocity squared (Eq. 10.1). Higher pressure difference could lead to higher amplitude of the shock wave, which is emitted at cavitation cloud collapse.

- Some studies [16] show that higher system pressure acts on the sonic velocity and compressibility of the bubbly mixture, that could result in smaller attenuation of the shock wave.
- Experiments on the same geometry [4] have shown that the cavitation cloud shedding frequency rises significantly (even with a power of two) when the flow velocity is increased. This results in more cavitation cloud collapses and consequently more shock waves, micro-jet impacts and finally pits.
- There is a finite number of bubbles, which have the potential to cause erosion, in the flow. Since the time of bubble implosion is much shorter than the time needed for transition of bubble over the foil, the probability for bubble to collapse and damage the surface does not change with velocity (this hypothesis was also confirmed by past experiments, for example Knapp et al. [17] write that only one, potentially erosive, bubble collapse in 30,000 will actually cause erosion). Hence when the velocity is increased, more bubbles collapse and cause erosion in a certain time period.

10.3.2 Pure Copper Specimens and Mass Loss

At the beginning of the experiment the specimens were removed from the test rig after 2, 5 and 9 min and images of the damaged surface were taken under a microscope. In all about 500 images were taken for the top and side surface of the specimen. Figure 10.15 shows the results of pit counting after 2, 5 and 9 min of exposure for the case of high rotating frequency (1,800 rpm).

Even after only 2 min of exposure, pits on the surface can be seen. The top surface sustained more damage. This is probably because it is oriented toward the upcoming cavitation cloud on the rotating disc and is consequently exposed for a long period of time—the exposure of the side surface, on the other hand, is brief (only when the cloud is directly above it). Although damage after 9 min seems to

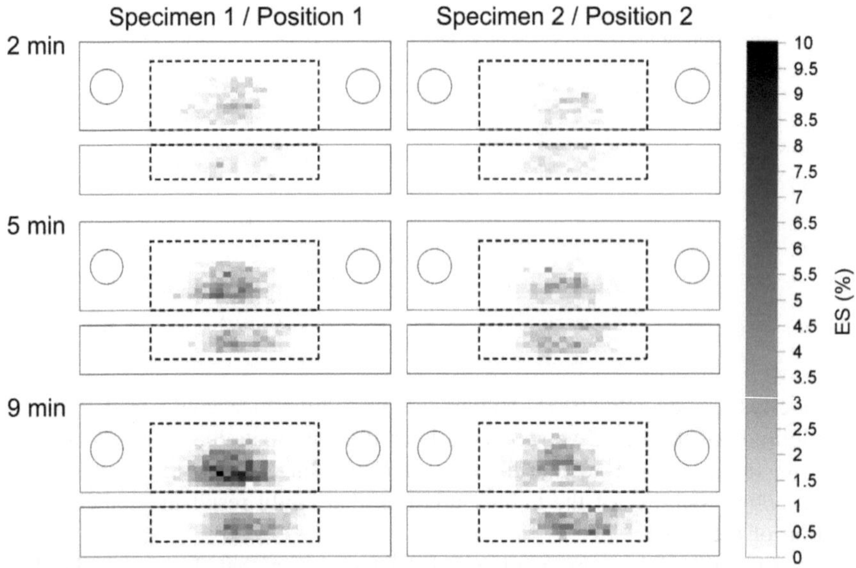

Fig. 10.15 Results of pit count after 2, 5 and 9 min of exposure to cavitation at 1,800 rpm. For each test, the top picture corresponds to the top surface of the specimen whereas the bottom picture corresponds to the side surface (see Fig. 10.5). Based on [13], reprinted with permission from Elsevier

be extensive, still no mass loss could be measured. Pursuing optical evaluation after longer exposure was meaningless since too many pits were overlapping and valid pit-count evaluation (quantification of results) could not be performed.

Obviously the specimen one receives more damage than the specimen two. The reason lies in the arrangement of the specimens on the circumference of the rotating disc. As already mentioned specimen holders in positions p5 and p6 (Fig. 10.5) were empty what gave cavitation enough space (180° of the disc rotation) to grow and severely damage the material specimen in position p1. Specimen in position p2 sustained less damage because cavitation did not have enough space to grow (only 60° of the disc rotation) and to become more aggressive. The difference in cavitation size and its unsteady nature at positions p1 and p2 is shown in Fig. 10.16.

The images were made through the observation window on the stationary disc (Fig. 10.5). One actually observes cavitation structures between positions 5 and 6 so the specimens needed to be repositioned to mimic the situations in positions 1 and 2. On the left in Fig. 10.16 a typical image of cavitation is shown—one sees a hole on the rotating disc, which rotates in the clockwise direction, and the cavitation trail behind it. On the right a diagram of standard deviation of the series of images of cavitation is presented—this can be related to the dynamics of cavitation. One can see that the size as well as the dynamics of the cavitation, characterized by the standard deviation, are much greater in front of position 1 than in front of position 2.

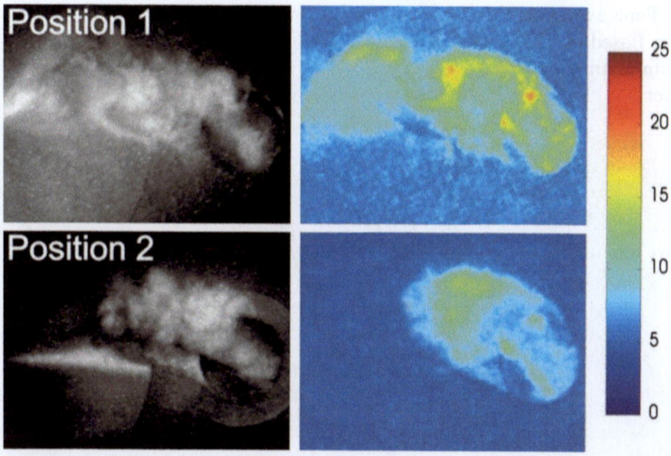

Fig. 10.16 Typical cavitation structure and the diagram of standard deviation just in front of position 1 *(up)* and position 2 *(bottom)*. Rotating frequency 1,800 rpm, $\sigma = 0.16$

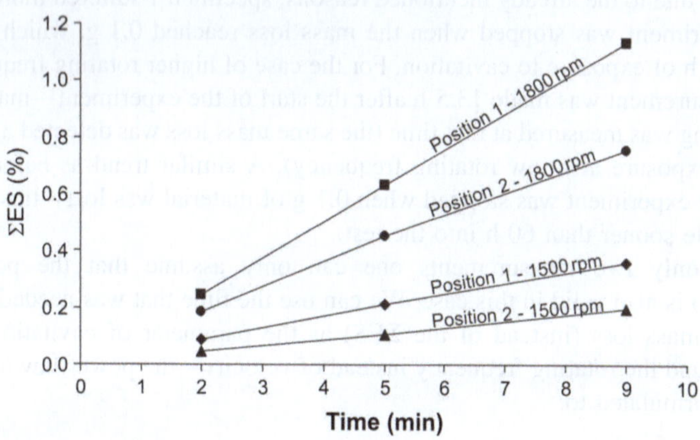

Fig. 10.17 Time evolution of the damaged surface area. The damaged surface increases linearly with time during the incubation period. Based on [13], reprinted with permission from Elsevier

Figure 10.17 shows the evolution of the damaged surface (surface covered by pits) as a function of time for several experiments. We see that the damage grows at a linear rate within the incubation. This means that pit-count method can be used already after a very short period of exposure. Confirmation of a constant pitting rate is also important for a possible prediction of erosion by CFD, where only very short periods of time can be simulated.

The periods between weighing of specimens varied during the experiment—from every half an hour at the beginning to as long as every 15 h at the end of the exposure to cavitation. Figure 10.18 shows results of mass loss measurements for low and high rotating frequency (for two specimens in each case).

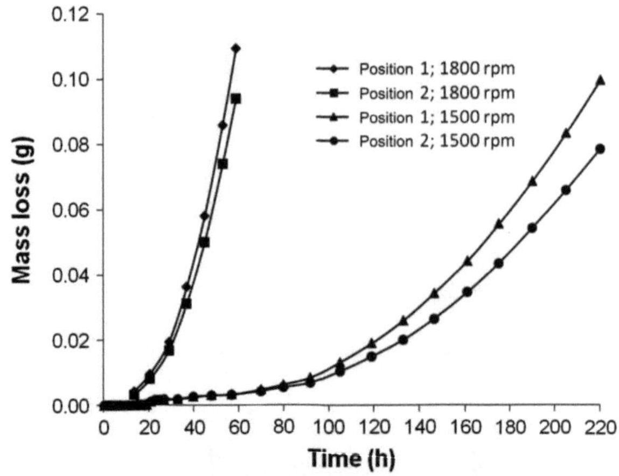

Fig. 10.18 Time evolution of mass loss. Based on [13], reprinted with permission from Elsevier

About 20 h of exposure were needed to detect mass loss (0.1 mg) for the case of low rotating frequency. After that the mass loss grew at an increasing rate. As expected, due to the already mentioned reasons, specimen 1 suffered more erosion. The experiment was stopped when the mass loss reached 0.1 g, which occurred after 220 h of exposure to cavitation. For the case of higher rotating frequency the first measurement was made 13.5 h after the start of the experiment—mass loss of about 4 mg was measured at that time (the same mass loss was detected after about 70 h of exposure at a low rotating frequency). A similar trend as before can be seen. The experiment was stopped when 0.1 g of material was lost—this occurred just a little sooner than 60 h into the test.

With only two measurements one can only assume that the power law (Eq. 10.3) is also valid in this case. We can use the time that was needed to detect 0.1 g of mass loss (instead of the ΣES) as the parameter of cavitation aggressiveness and the rotating frequency instead of velocity—the power law then needs to be reformulated to:

$$\frac{t_{1500}}{t_{1800}} = \left(\frac{f_{1800}}{f_{1500}}\right)^n. \tag{10.4}$$

A value of $n = 7.1$ is deduced, which again lies within the expected range.

10.4 Conclusions

A study of visually observable cavitation structures and erosion effects of cavitation on hydrofoils copper specimens was performed. A dependency of cavitation erosion on the dynamics of cavitation structures was found. The position and distribution of pits on the copper coated surface correlate to the distribution of the standard deviation of grey level, which can possibly be used as a measure of unsteadiness of cavitation. A hypothesis that the cavitation erosion is conditioned by the dynamic behavior of cavitation (cavitation cloud separation) was confirmed.

Additionally we employed the rotating disc test rig where the damage was studied during the incubation period and also during the mass loss period. By evaluating the pitted surface during the incubation period a linear increase of damage in time was determined. The linearity was independent of the specimen position or cavitation aggressiveness. This means that only very short tests can be conducted and the results can then easily be extrapolated until the end of the incubation period. Later on mass loss tests showed expected evolution of erosion.

The findings of this study could lead to considerable reduction of time needed to evaluate materials resistance to cavitation erosion and could contribute to cavitation erosion prediction models that are implemented into Computational Fluid Dynamics codes [18].

Acknowledgments The author would like to acknowledge the contributions of many colleagues with whom he worked in the past years. Among others: Brane Sirok, Bernd Stoffel, Bernd Bachert and Olivier Coutier-Delgosha. The presented work was performed at Technical University of Darmstadt (Germany) and University of Ljubljana (Slovenia).

References

1. Rayleigh L (1917) On the pressure developed in a liquid during the collapse of a spherical cavity. Phil Mag 34:94–98
2. Fujikawa S, Akamatsu T (1980) Effects of non-equilibrium condensation of vapor on the pressure wave produced by the collapse of a bubble in a liquid. J Fluid Mech 97:481–512
3. Dular M, Bachert R, Stoffel B, Širok B (2005) Experimental evaluation of numerical simulation of cavitating flow around hydrofoil. Eur J Mech B Fluids 24(4):522–538
4. Böhm R (1998) Erfassung und hydrodynamische Beeinflussung fortgeschrittener Kavitationsustände und ihrer Aggressivität. Ph.D. Thesis, Technische Universität Darmstadt, Darmstadt
5. Benjamin TB, Ellis AT (1966) The collapse of cavitation bubbles and the pressures thereby produced against solid boundaries. Philos Trans R Soc 260:221–240
6. Tong RP, Schiffers WP, Shaw SJ, Blake JR, Emmony DC (1999) The role of "splashing" in the collapse of a laser-generated cavity near a rigid boundary. J Fluid Mech 380:339–361
7. Dular M, Bachert B, Stoffel B, Širok B (2004) Relationship between cavitation structures and cavitation damage. Wear 257:1176–1184
8. Brandt F (1981) Ein physikalisches Verfahren zur Bestimmung von geloesten und ungeloesten Gasen in Wasser. Voith Forschung und Konstruktion, 27
9. Peterson FB (1972) Hydrodynamic cavitation and some considerations of the influence of free gas content. In: Proceedings of the 9th symposium on naval hydrodynamics, Paris, 1972
10. Arndt REA, Keller AP (1976) Free gas content effects on cavitation inception and noise in a free shear flow. IAHR Symposium. Two phase flow and cavitation in power generation systems, Grenoble, pp 3–16
11. Dular M, Bachert R, Schaad C, Stoffel B (2007) Investigation of a re-entrant jet reflection at an inclined cavity closure line. Eur J Mech B Fluids 26:688–705
12. Lohrberg H, Hofmann M, Ludwig G, Stoffel B (1999) Analysis of damaged surfaces: part II: pit counting by 2D optical techniques. In: Proceedings of the 3rd ASME/JSME joint fluids engineering conference, San Francisco, 1999

13. Osterman A, Bachert B, Širok B, Dular M (2009) Time dependant measurements of cavitation damage. Wear 266:945–951
14. Shamsborhan H, Coutier-Delgosha O, Caignaert G, Nour FA (2010) Experimental determination of the speed of sound in cavitating flows. Exp Fluids 49:1359–1373
15. Franc JP, Michel JM (2004) Fundamentals of cavitation. Kluwer Academic Publishers, Dordrecht
16. Gudmundsson JS, Celius HK (1999) Gas-liquid metering using pressure-pulse technology. Paper presented at the SPE annual technical conference and exhibition, Houston, 3–6 Oct 1999
17. Knapp RT, Daily JT, Hammit FG (1970) Cavitation. McGraw Hill, New York
18. Dular M, Coutier-Delgosha O (2009) Numerical modelling of cavitation erosion. Int J Numer Meth Fluid 61(12):1388–1410

Chapter 11
Recent Investigations on Cavitation Erosion at the University of Fukui

Shuji Hattori

Abstract This paper reviews the recent research achievements by the author on the construction of a cavitation erosion database, and on the data analysis of carbon steel, stainless steel, cast iron and nonferrous alloys. His studies on cavitation erosion mechanisms, temperature effects on cavitation erosion in liquid metals are also reviewed. Finally, an erosion prediction method based on impact load measurements is discussed. It was found that the erosion resistance of carbon steels, stainless steels, cast iron, aluminum alloys, copper alloys and titanium alloys can be estimated accurately from the material hardness for each series of materials and alloys. A cavitation erosion model is proposed for SUS304 and cobalt alloy ST6 based on observations via scanning electron microscopy. The temperature effect showed a similar tendency for both vibratory and cavitating liquid jet tests after defining a relative temperature (a percentage temperature between freezing and boiling points). Erosion rates in various liquids including liquid metals can be evaluated by a combination of material density and sound velocity for the test liquid and the specimen material. Impact load measurements were used to establish a prediction method for the incubation period.

11.1 Introduction

Cavitation often occurs in the contact area between solid and liquid in fluid machinery, pipes, ship propellers, and valves. Erosion is a phenomenon that damages the component surface manifested by deep local holes. Cavitation erosion causes serious problems that reduce the performance in apparatuses or the life by component failure. Three main objectives may be pursued for the control of cavitation erosion:

S. Hattori (✉)
University of Fukui, Fukui, Japan
e-mail: hattori@u-fukui.ac.jp

K.-H. Kim et al. (eds.), *Advanced Experimental and Numerical Techniques for Cavitation Erosion Prediction*, Fluid Mechanics and Its Applications 106,
DOI: 10.1007/978-94-017-8539-6_11,

- Maintain non-cavitating conditions
- Use a cavitation-erosion resistant material
- Predict the component life.

Cavitation erosion research is at the intersection of fluid engineering and material engineering. Material engineering research has focused mainly on the evaluation of cavitation erosion resistance of materials to predict component life.

The ASTM (American Society for Testing and Materials) maintains two standards for cavitation erosion: ASTM G32 and G134. The G32 [1] test method produces cavitation damage on a specimen through high frequency vibrations. The vibration induces the formation and collapse of cavities in the liquid. The collapsing cavities produce damage to the specimen surface. The ASTM G134 [2] test method includes a procedure that evaluates the cavitation erosion resistance of solid materials. A submerged cavitating jet, issuing from a nozzle, impinges on a test specimen placed in its path so that cavities collapse on its surface, thereby causing erosion. In our laboratory, cavitation erosion tests have been carried out for 42 years since 1970. Several papers have been published by the present author on various topics such as corrosion fatigue under cavitation erosion [3], observation of erosion particles [4], slurry wear versus cavitation erosion [5], and liquid impingement erosion versus cavitation erosion [6].

This study reviews recent research achievements [7–13] on the construction of a cavitation erosion database, and on the data analysis of carbon steel, stainless steel, cast iron and nonferrous alloys. The behavior of highly erosion resistant materials SUS304 and ST6 is discussed using vibratory and cavitating liquid jet apparatuses. The temperature effect on erosion in liquid metals is also discussed. Finally, a technique of cavitation erosion prediction through impact load measurements is presented.

11.2 Erosion Database

11.2.1 Construction of the Database

For the prediction of cavitation erosion an erosion database is required. In our laboratory, many cavitation erosion tests have been carried out, and cavitation erosion data has been documented for 42 years since 1970. A database from 1970 to 2007 has been constructed in the form of data sheets and Microsoft Excel data files [7–9]. The data are classified as: test material, chemical composition, heat treatment, mechanical properties, test method, and test condition.

Table 11.1 shows the number of tests under various test conditions. 143 types of materials were tested including iron, steel, cast iron, stainless steels, castings, Al alloys, Cu alloys, Ti alloys, Ni alloys, Co alloys, plastics, ceramics and others. The database includes a total of 990 tests, under various environmental conditions.

Table 11.1 Cavitation test configurations

Material			Test method			Frequency [kHz]			Amplitude [μm]				Test liquid			Standoff distance [mm]		
	Type	Total	Vibratory	Stationary	Others	14.7	19.5	Others	<30	40	50	>60	Deionized water	NaCl	Others	<0.5	1	Others
Iron & Steel	21	386	274	103	9	232	43	75	91	135	118	24	225	92	69	20	76	7
Cast iron	7	45	23	22	–	41	4	–	3	17	22	3	30	15	–	–	19	3
Stainless steel	21	172	74	98	–	60	103	9	17	18	126	11	145	19	8	–	94	4
Stainless steel castings	11	29	–	29	–	10	19	–	–	2	25	2	29	–	–	–	29	–
Al alloy	8	100	–	94	6	34	60	–	10	9	66	9	64	13	23	–	92	2
Cu alloy	3	77	10	51	16	12	36	13	16	10	33	2	48	7	22	4	46	1
Ti alloy	5	17	1	16	–	13	–	4	1	–	16	–	17	–	–	–	16	–
Ni alloy	2	4	2	2	–	1	3	–	–	–	4	–	4	–	–	–	2	–
Co alloy	5	5	–	5	–	1	4	–	–	–	5	–	5	–	–	–	5	–
Plastics	4	8	–	8	–	–	8	–	–	–	8	–	8	–	–	–	8	–
Ceramics	4	12	–	12	–	12	–	–	–	–	12	–	12	–	–	–	12	–
Others	52	135	2	133	–	33	80	22	7	3	123	2	129	6	–	–	133	–
Total	143	990	386	573	31	449	360	123	145	194	558	53	716	152	122	24	532	17

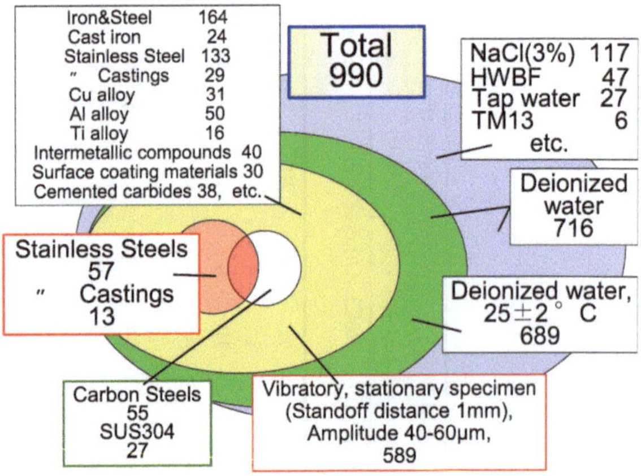

Fig. 11.1 Structure of cavitation erosion data. Based on [8], reprinted with permission from Elsevier. (*HWBF* High water-base fluid, *TM13* Pb–Bi liquid metal)

The cavitation erosion tests utilized a vibratory apparatus. Two procedures are employed for vibratory testing: (a) vibratory specimen method and (b) stationary specimen method. Most experiments in our laboratory have been carried out with vibratory apparatuses under various test conditions such as test liquid, liquid temperature, frequency, and peak-to-peak displacement amplitude. The data of the vibratory specimen and stationary specimen methods account for 40 and 50 % of the total data, respectively.

14.7 and 19.5 kHz frequencies are used primarily during vibratory testing. These two frequencies account for 50 and 30 % of the total data, respectively. The vibratory apparatuses have used an approach based on ASTM G32 [1], and since 1991 most of the cavitation erosion tests have been performed at a frequency of 19.5 kHz. In addition, a peak-to-peak displacement amplitude of 50 μm was used. The test liquid was deionized water at 25 °C, and the standoff distance between the vibrating disk and the specimen was 1 mm for the stationary specimen method.

Figure 11.1 shows the structure of the cavitation erosion test data. Again, the total number of tests is 990 including deionized water and various solutions. 689 of the tests used deionized water at 25 ± 2 °C as specified in ASTM G32 with 589 samples obtained under standard test conditions (frequency: 20 ± 0.5 kHz, standoff distance: 1 mm).

11.2.2 Carbon Steels and Stainless Steels

Heymann's results are shown in Fig. 11.2 [14]. These results are shown as a double logarithmic plot of the "normalized maximum instantaneous erosion resistance" N_e versus Vickers hardness. The erosion resistance is defined as the time to remove unit depth (hour/μm). The normalized erosion resistance of an arbitrary material X relative to that of a reference material *ref* with Vickers hardness HV_{ref} is defined [14] as,

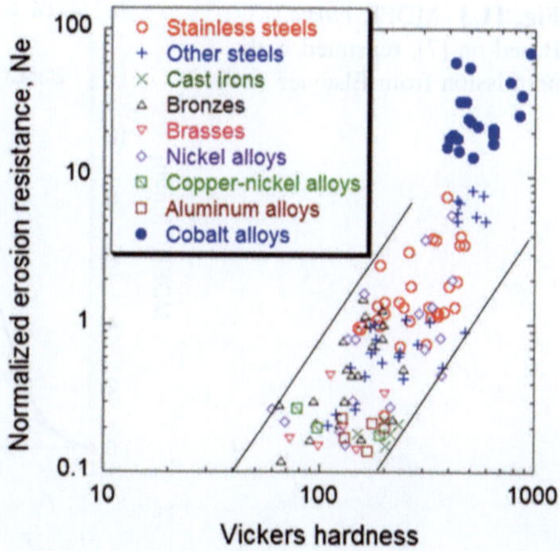

Fig. 11.2 Normalized erosion resistance versus hardness. Based on [7], reprinted with permission from Elsevier

$$Ne = \left(\frac{HV_{ref}}{170}\right)^2 \left(\frac{Erosion\ resistance\ of\ X}{Erosion\ resistance\ of\ ref}\right). \quad (11.1)$$

The reference material in Heymann's data is SUS304. The scatter of N_e for materials of the same hardness is virtually the same from low to high hardness. Therefore in this analysis, the scatter of Ne can be assumed to follow a normal distribution in the logarithmic scale. There was little difference in the mean and standard deviation between standard normal distribution and log-normal distribution [7].

For carbon steels, the data were analyzed for seven types of materials, SS400 (Rolled steel for general structure, tensile strength of approximately 400 MPa), S10C (0.10 % carbon steel), S15C, S35C, S45C, S55C and SK4 (Tool steel type 4) [7]. With the addition of stainless steel SUS304 eight types of materials were analyzed. The analysis of these materials produced 82 data points. The results of an erosion test are often expressed [1] as the "mean depth of erosion" (MDE), which is the mass loss divided by the material density and the eroded surface area. Another expression is the instantaneous "mean depth of erosion rate" (MDER), which is the slope of the MDE curve. For example, the MDER-time curve under the condition of 14.7 kHz for the vibratory specimen method for S35C is shown in Fig. 11.3. $MDER_{max}$, expressed as μm/h, is the "maximum of the mean depth of the erosion rate" which is the slope of the straight line that best approximates the steepest linear (or nearly linear) portion of the cumulative MDE-time curve.

Since many tests under the same conditions are required for a statistical analysis, $MDER_{max}$ under various conditions were converted [9] to $MDER_{max}$ under the standard test conditions specified by the ASTM G32 specification (amplitude: 50 μm, frequency: 19.5 kHz). The exponent of amplitude dependence on $MDER_{max}$ is 2, the ratio of vibratory/stationary is 4.7 and the ratio of 19.5 / 14.7 kHz is 1.2 [7], based on many experimental data.

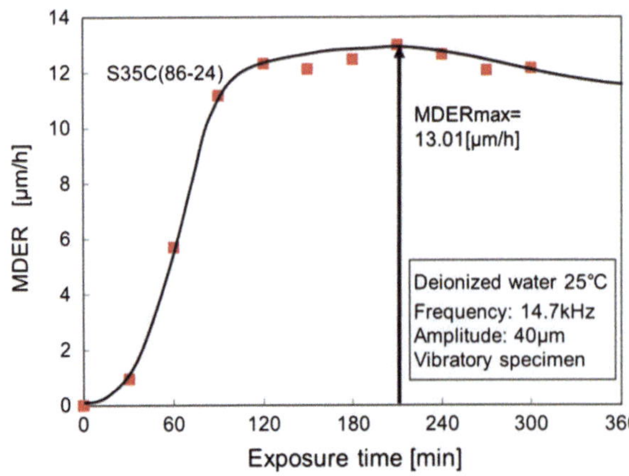

Fig. 11.3 MDER *curve*. Based on [7], reprinted with permission from Elsevier

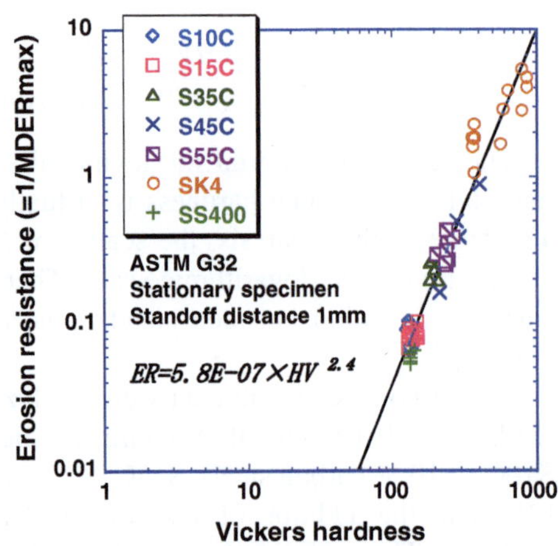

Fig. 11.4 Relation between hardness and erosion resistance (= 1/MDER$_{max}$) of carbon steels. Based on [7], reprinted with permission from Elsevier

After the data were converted into the standard test condition, the reciprocal value of MDER$_{max}$ was defined as the erosion resistance. The relation between Vickers hardness (HV) and erosion resistance is shown in Fig. 11.4. The erosion resistance approximately increases proportionally to the 2.4th power of HV.

As mentioned above, Heymann used the "normalized erosion resistance" (N_e) as the ordinate. For SUS304 of hardness 187 *HV* ($H_{ref} = 187$), the functional relation between *HV* and N_e became as follows (see Fig. 11.5):

$$Ne = 2.1 \times 10^{-6} HV^{2.4}. \tag{11.2}$$

Figure 11.6 shows the comparison of each of the 1–99 % probabilities in the relation between HV and N_e. When the mean *m* of log(N_e) at Vickers hardness HV and standard deviation σ of log(N_e) were obtained, $m - 2.32\ \sigma$ and $m + 2.32\ \sigma$ correspond to 1 and 99 % probability, respectively [15]. The 1 % erosion resistance means that when 100 materials with the same hardness are tested, only one

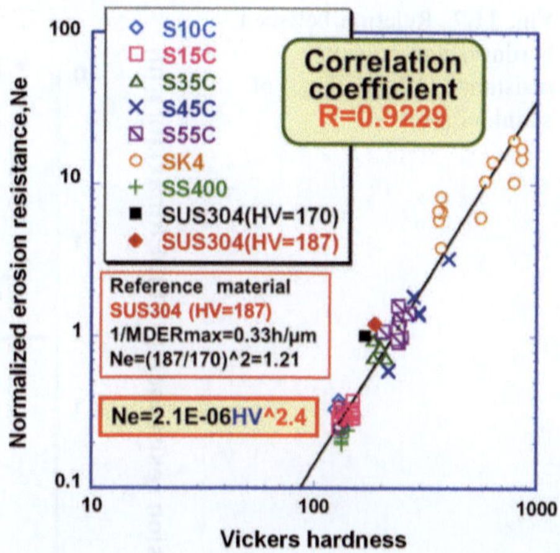

Fig. 11.5 Relation between hardness and normalized erosion resistance (Ne) of carbon steels. Based on [7], reprinted with permission from Elsevier

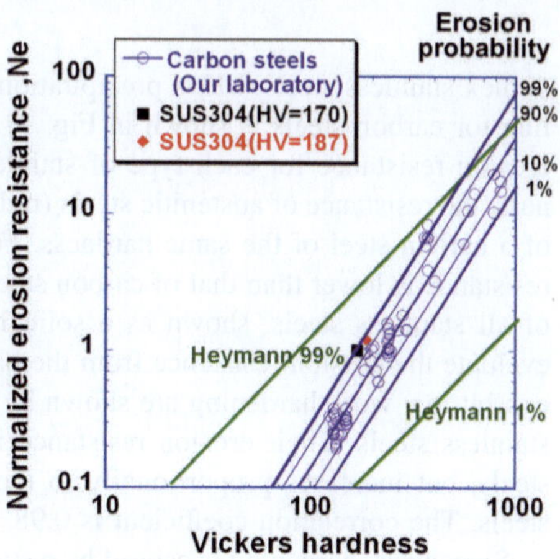

Fig. 11.6 Comparison of the data analysis of Heymann and the present study using erosion probability. Based on [7], reprinted with permission from Elsevier

material is lower than this erosion resistance [7]. The scatter of the "normalized erosion resistance" for the same hardness, between the erosion probability of 1 and 99 %, amounts to a factor of about 29 for Heymann's result which encompasses a wide range of metals. The scatter reduces to a factor of about 4.3 for the present study when limited to carbon steels. The latter is obviously much narrower, and shows that the erosion resistance for carbon steels is predictable from hardness with high reliability.

For stainless steels [8], the log–log relation between the Vickers hardness (HV) and the erosion resistance is shown in Fig. 11.7. All stainless steels were classified into austenitic (A), ferritic (F), martensitic (M), austenitic-ferritic (A + F) and

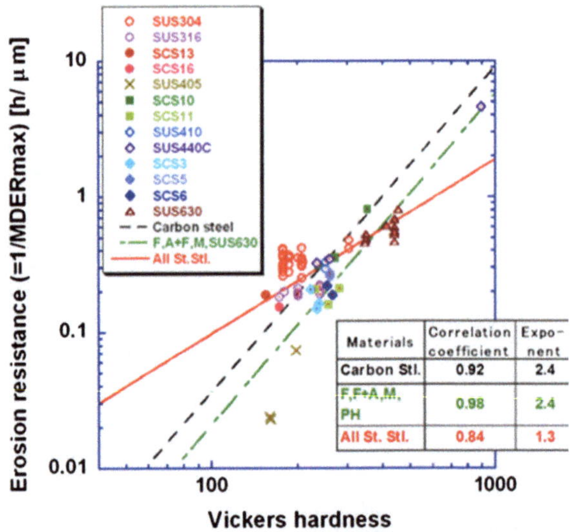

Fig. 11.7 Relation between hardness and erosion resistance ($1/MDER_{max}$) of stainless steels

duplex stainless steels. PH is precipitation hardened stainless steel. The reference line for carbon steels is shown in Fig. 11.4 as a black dashed line. Although the erosion resistance for each type of stainless steel increases with increased hardness, the resistance of austenitic steels (red marks) is 1.1–2.0 times higher than that of a carbon steel of the same hardness. For the other stainless steels, the erosion resistance is lower than that of carbon steels. Therefore, the correlation coefficient of all stainless steels, shown as a solid line, is only 0.84. It is then difficult to evaluate the erosion resistance from the hardness in this case. Stainless steels that exhibit low work hardening are shown by a green chain line, excluding austenitic stainless steels. Their erosion resistance is half the erosion resistance of carbon steels, but increases proportionally to the 2.4^{th} power of HV, similar to carbon steels. The correlation coefficient is 0.98.

Since work hardening is caused by a static tensile load in a tension test and by a compressive load in an erosion test, the amount of work hardening may be different between tension and erosion tests. The Vickers hardness was measured after each erosion test. Table 11.2 shows the mean value of the Vickers hardness at five locations on the original surface and the eroded surface after erosion tests of different stainless steels. Because severe erosion occurred, the hardness on the eroded surface was measured in the close vicinity of the eroded area. The increasing ratio of the hardness is between 1.4 and 1.7 for austenitic stainless steels, and the ratios for the other stainless steels are between 0.9 and 1.3.

To take into account the influence of work hardening, Fig. 11.8 shows the relation between the erosion resistance and the Vickers hardness after the erosion tests. The correlation coefficient is 0.98 for all stainless steels. The exponent of the erosion resistance versus hardness after erosion tests is 2.4, similar to that of carbon steels. Because it is difficult to measure the Vickers hardness after an erosion test, it seems advisable to simply correct the hardness with a material

Table 11.2 Vickers hardness of original and eroded surfaces. A, F, M, PH and A + F are austenitic, ferritic, martensitic, precipitation hardened and austenitic-ferritic stainless steels, respectively

Materials				HV (original surface)	HV (eroded surface)	Increasing ratio (material factor) Fmat
Stainless steels	Rolled	A	SUS304	178	296	1.7
			SUS316	180	250	1.4
		F	SUS405	160	164	1.0
		M	SUS410	251	290	1.2
			SUS440C	259	225	0.9
		PH	$SUS630_1$	345	363	1.1
			$SUS630_2$	411	430	1.0
	Cast	M	SCS3	238	260	1.1
			SCS5	260	296	1.1
			SCS6	255	283	1.1
		A + F	SCS10	270	355	1.3
			SCS11	231	231	1.0
		A	SCS13	155	251	1.6
			SCS16	173	244	1.4

Fig. 11.8 Relation between hardness after erosion test and erosion resistance ($1/MDER_{max}$) of stainless steels

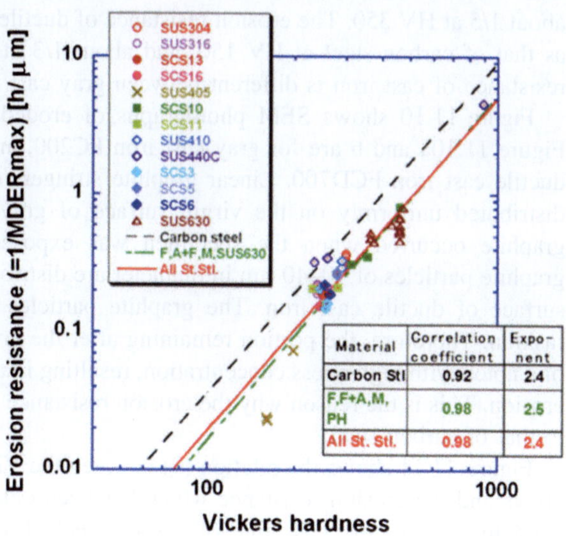

factor, F_{mat}, shown in Table 11.2. The erosion resistance (*ER*) of all stainless steels was obtained as follows:

$$ER = 2.6 \times 10^{-7}(HV \times F_{mat})^{2.4}. \tag{11.3}$$

Therefore, the erosion resistance of carbon and stainless steels can be estimated accurately from the hardness and the material factor.

Fig. 11.9 Relation between hardness and erosion resistance of cast iron

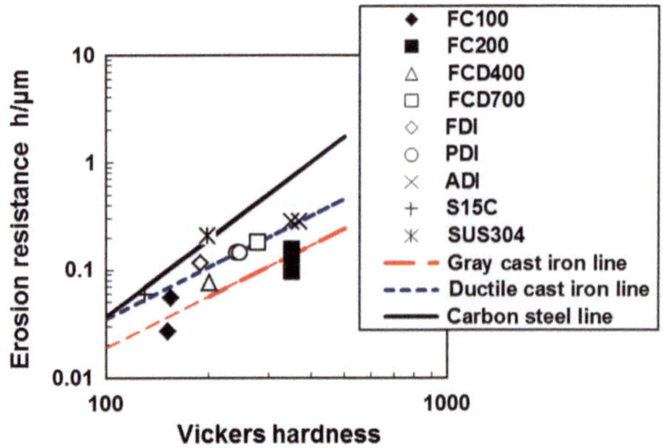

11.2.3 Cast Iron and Nonferrous Alloys

Figure 11.9 shows the relation [9] between the Vickers hardness of cast iron and the erosion resistance of carbon steels shown in Fig. 11.4. The solid line in this figure shows the base line of carbon steel. The erosion resistance of gray cast iron was compared to that of carbon steel. The resistance is about 1/3 at HV 150, and about 1/5 at HV 350. The erosion resistance of ductile cast iron is almost the same as that of carbon steel at HV 150, and about 1/3 at HV 350. Thus, the erosion resistance of cast iron is different between gray cast iron and ductile cast iron.

Figure 11.10 shows SEM photographs of eroded surfaces of cast iron [16]. Figure 11.10a and b are for gray cast iron FC200, and Fig. 11.10c and d are for ductile cast iron FCD700. Linear graphite stringers of 30–200 μm in length are distributed uniformly on the virgin surface of gray cast iron. The removal of graphite occurred when the specimen was exposed to cavitation. Spheroidal graphite particles of 10–40 μm in diameter are distributed uniformly on the virgin surface of ductile cast iron. The graphite particles were removed in a similar fashion. Therefore, the portion remaining after the graphite removal has the form of a notch with high stress concentration, resulting in a site favorable for cavitation erosion. This is the reason why the erosion resistance of cast iron is low compared to that of carbon steel.

Figure 11.11 shows the relation [9] between the Vickers hardness of aluminum alloys and the erosion resistance which has been obtained in our laboratory. The solid line shows the base line of carbon steel. The erosion resistance of pure aluminum at HV 100 or less cannot be compared with that of carbon steel, because the Vickers hardness HV of carbon steel is 100 or more. The erosion resistance of an aluminum alloy at HV100 or more compared to that of carbon steel is about a third lower at HV 100, and about a fifth lower at HV 150. This is because local overaging occurs in aluminum alloys [17]. The overaging is the local softening due to the increase in distance of precipitations by repeated stress.

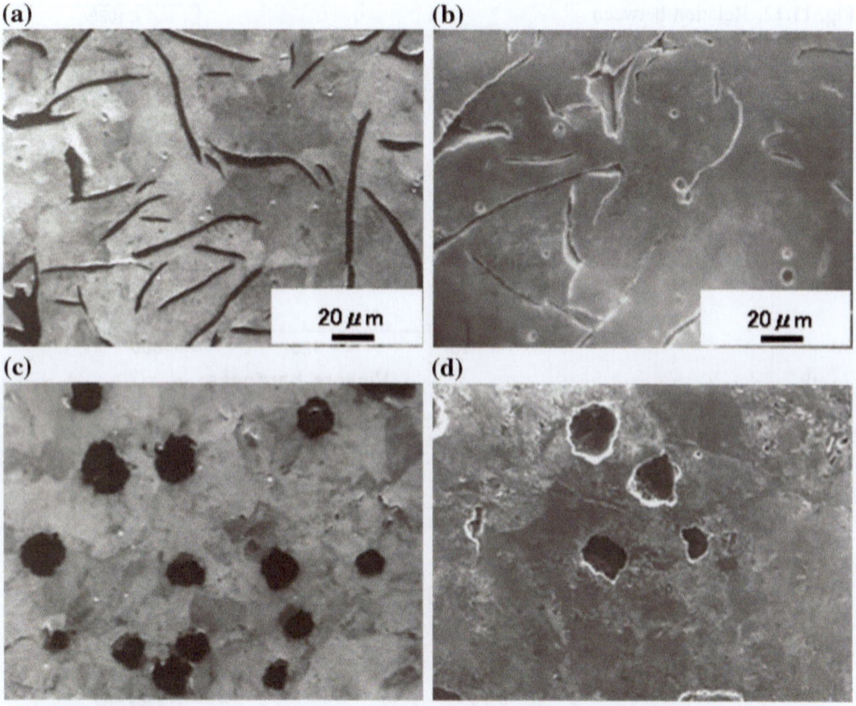

Fig. 11.10 Eroded surface of cast iron. Based on [16], reprinted with permission from Elsevier. **a** FC200 (0 min). **b** FC200 (5 min). **c** FCD700 (0 min). **d** FCD700 (0 min)

Fig. 11.11 Relation between hardness and erosion resistance of Al alloys. Based on [9], reprinted with permission from Elsevier

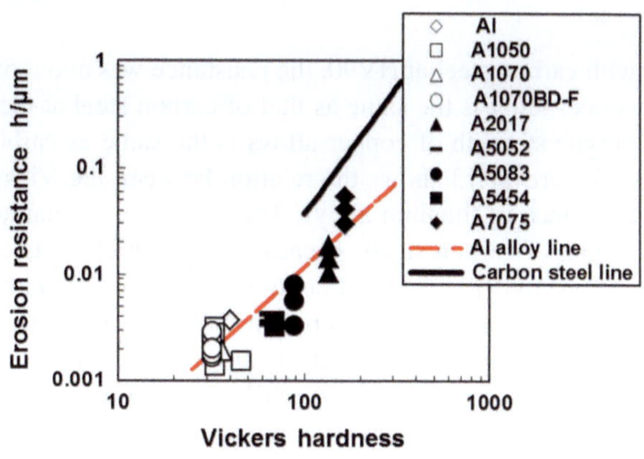

Figure 11.12 shows the relation between Vickers hardness and erosion resistance of copper alloys. The data of erosion resistance of pure copper are concentrated at HV 90. When the erosion resistance of pure copper was compared

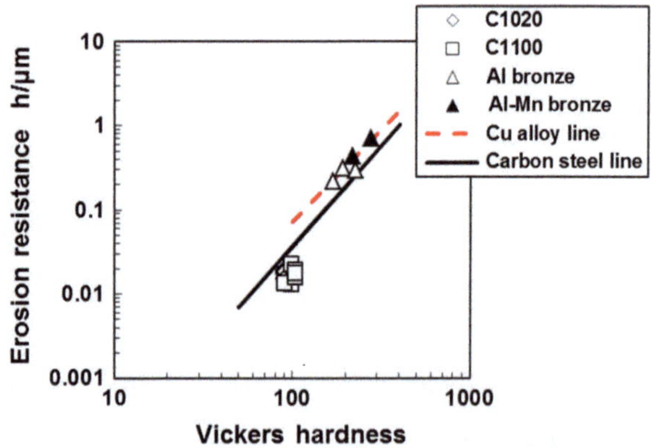

Fig. 11.12 Relation between hardness and erosion resistance of Cu alloys. Based on [9], reprinted with permission from Elsevier

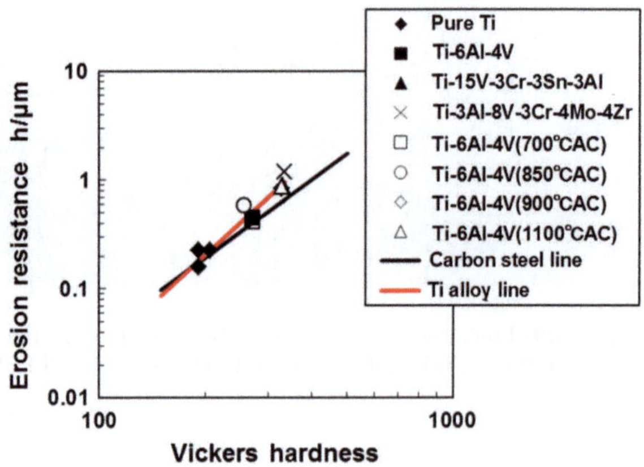

Fig. 11.13 Relation between hardness and erosion resistance of titanium alloys

with carbon steel at HV90, the resistance was about 50 % lower. The resistance of copper alloy is the same as that of carbon steel at the same hardness, because the fatigue strength of copper alloys is the same as carbon steel [18].

Figure 11.13 shows the relation between the Vickers hardness and the erosion resistance of titanium alloys. The resistance of titanium alloys is almost the same as that of carbon steels, because the notch effect is extremely small compared to cast iron [19], although titanium alloys have a microscopic dual phase.

We conclude that the erosion resistance of carbon steels, stainless steels, cast iron, aluminum alloys, copper alloys and titanium alloys can be estimated accurately from the hardness for each series of alloys.

11.3 Erosion Under Various Conditions

11.3.1 Erosion Mechanism of SUS304 and Co Alloys and the Comparison Between Vibratory and Cavitating Liquid Jet Methods

Cobalt (Co) alloys are known to have high resistance to cavitation erosion but the mechanism remains unclear. Eroded surfaces of ST6 and reference material SUS304 were observed by scanning electron microscopy (SEM) to clarify the erosion mechanism. Figure 11.14 shows SEM photographs of SUS304 eroded using the G32 vibratory apparatus [10]. After 45 min (Fig. 11.14a), plastic deformation occurred inside crystal grains by cavitation and accumulated at grain boundaries. After 3 h (Fig. 11.14b), crack-like grooves began to form in the highly deformed area near point A', along a grain boundary. After 4 h (Fig. 11.14c), material removal (erosion) occurred in the grain. After 5 h (Fig. 11.14d), the erosion extended across a wide area inside the grain.

Figure 11.15 is a schematic model of the erosion on the cross section A-A' based on the observation of the eroded surface. The original surface was repeatedly exposed to the collapses of cavitation bubbles, and plastic deformation of the material surface occurred by shock waves and micro jets generated by bubble collapses (Fig. 11.15a). Since bubble collapses repeatedly act on the deformed area, the area gradually expands, and plastic deformation accumulates at crystal grain boundaries as shown in Fig. 11.15b, which corresponds to the SEM photograph after 45 min (Fig. 11.14a). Since the material surface was plastically deformed, swelled parts appeared at the crystal grain boundaries. These swelled parts produce a step relative to the adjacent grain with less plastic deformation and cause a high stress concentration, resulting in crack initiation as in Fig. 11.15c. Figure 11.15d shows that the erosion easily occurred at the crack initiation site. Figure 11.15c and d correspond to the SEM photographs after 3–4 h (Fig. 11.14b and c), respectively. Figure 11.15e shows the development of erosion after 5 h (Fig. 11.14d).

Figure 11.16 presents the erosion model of ST6 based on SEM observations [10]. Figure 11.16a shows the virgin surface of the ST6 eutectic structure consisting of the Co matrix and carbide. Figure 11.16b illustrates that the Co matrix is softer than the carbide and plastic deformation occurs in the matrix. Thus, carbide plays the role of a grain boundary in the erosion mechanism of SUS304. Plastic deformation occurs in the matrix, and a swelled part appears near the carbide. The swelled part near the carbide has a high stress concentration which easily initiates cracks. Figure 11.16c shows that the erosion proceeds near the interface between the Co matrix and carbide, and then the carbides fall off, as shown in Fig. 11.16d. This is perhaps one mechanism which explains how the hard chrome carbide was removed by the bubble collapse pressures.

Figure 11.17 shows a comparison of the erosion resistance evaluated by vibratory and cavitating liquid jet methods. The erosion resistance was defined as

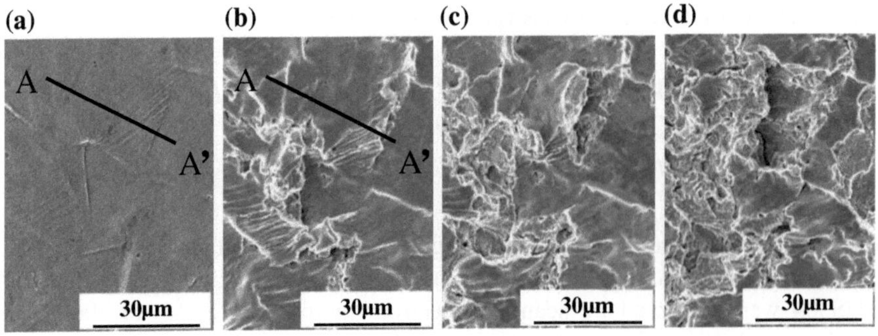

Fig. 11.14 SEM photographs of eroded surface of SUS304. Based on [10], reprinted with permission from Elsevier. **a** 45 min. **b** 3 h. **c** 4 h. **d** 5 h

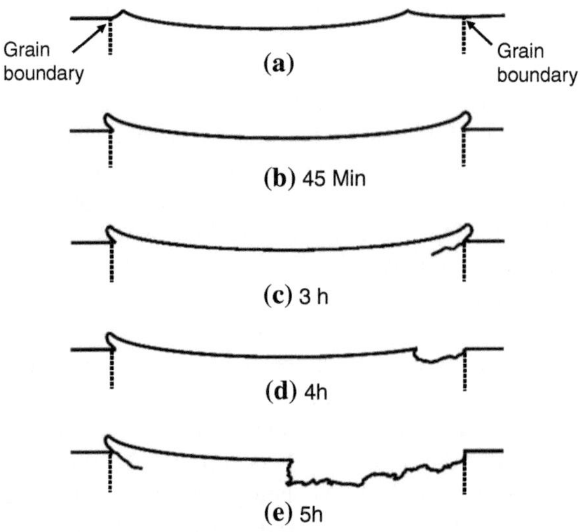

Fig. 11.15 Model of erosion process on SUS304 specimen. Based on [10], reprinted with permission from Elsevier

the reciprocal of MDER (mean depth of erosion rate) in the maximum rate stage. The vertical axis shows the erosion resistance for the cavitating liquid jet (upstream pressure 17.4 MPa and cavitation number 0.025), and the horizontal axis shows the erosion resistance for the vibratory method. Erosion resistance of ST6 materials is 10–20 times higher than that of SUS316. By taking the slope of the straight line, the test time of the vibratory method is seen to be about 3.5 times that of the cavitating liquid jet approach. This means the erosion resistance can be more quickly estimated using cavitating jets.

In this section, cavitation erosion models were proposed for SUS304 and cobalt alloy ST6 through observation by scanning electron microscopy. A relation for the erosion resistance was obtained between the vibratory method and the cavitating liquid jet method.

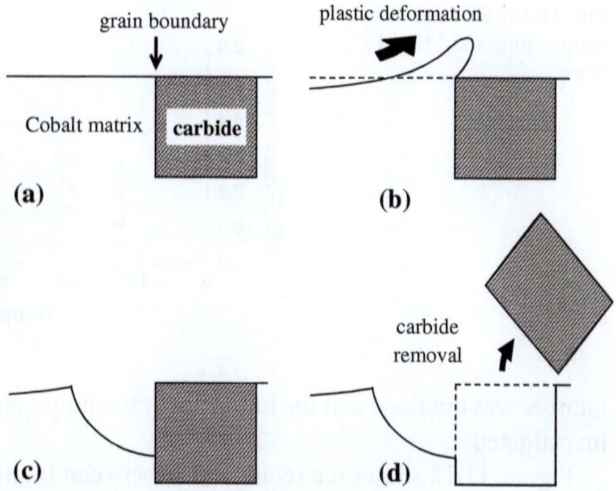

Fig. 11.16 Model of erosion process on ST6 specimen. Based on [10], reprinted with permission from Elsevier

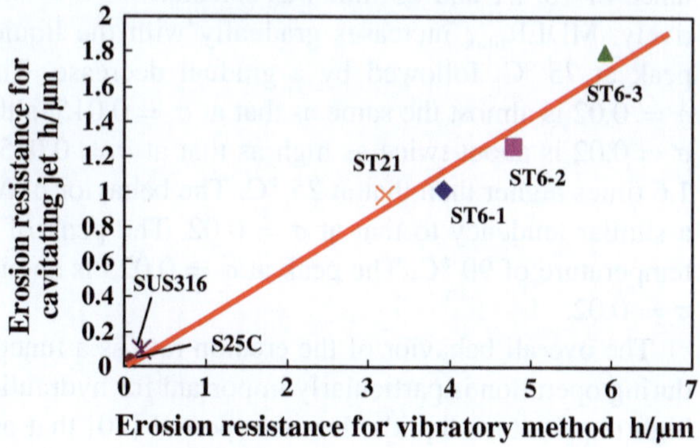

Fig. 11.17 Relation between erosion resistances for vibratory method and for cavitating liquid jet method. Based on [10], reprinted with permission from Elsevier

11.3.2 Effect of Temperature on Erosion

It is well known that cavitation erosion is affected by liquid parameters such as liquid temperature and vapor pressure. Hammitt [20], Plesset [21] and the present authors [22] have studied the influence of temperature on cavitation erosion. These studies concluded that erosion increases with increased temperature up to about 50 °C, followed by a decrease. However, all these experimental results were obtained by using a vibratory apparatus. A vibratory apparatus is unable to account for the effects of hydrodynamic parameters such as upstream pressure and downstream pressure. Electric power plants operate at high temperature and it is necessary to clarify the influence of the liquid temperature on the erosion under flow conditions. The cavitation number should be kept constant to isolate only the temperature effect. In this section, a heating system was added to the cavitating liquid jet apparatus specified in the ASTM G134 standard [2]. The optimum stand-off distance for each cavitation

Fig. 11.18 Relation between temperature and $MDER_{max}$ (Cu)

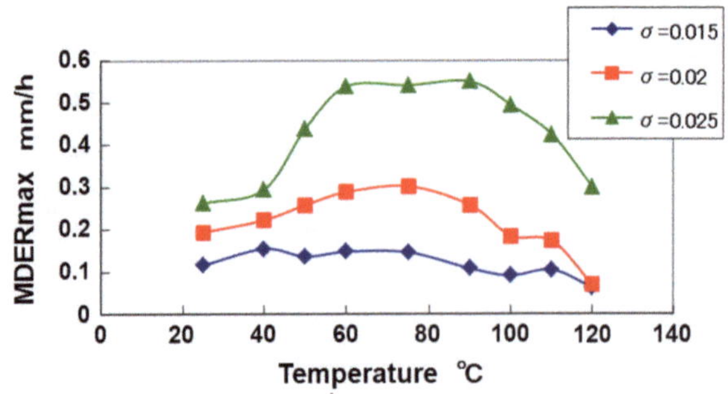

number was obtained and the influence of the liquid temperature on the erosion was investigated.

Figure 11.18 shows the relation [11] between liquid temperature and $MDER_{max}$ of pure copper at an upstream pressure of 17.4 MPa. An optimum standoff distance of 15, 21 and 25 mm was selected for $\sigma = 0.025$, 0.02 and 0.015, respectively. $MDER_{max}$ increases gradually with the liquid temperature and reaches a peak at 75 °C, followed by a gradual decrease. The behavior of $MDER_{max}$ at $\sigma = 0.02$ is almost the same as that at $\sigma = 0.015$, although the $MDER_{max}$ value at $\sigma = 0.02$ is about twice as high as that at $\sigma = 0.015$. The peak value at 75 °C is 1.6 times higher than that at 25 °C. The behavior of $MDER_{max}$ at $\sigma = 0.025$ shows a similar tendency to that at $\sigma = 0.02$. The peak of $MDER_{max}$ shifts to a higher temperature of 90 °C. The peak at $\sigma = 0.025$ is about 1.8 times larger than that at $\sigma = 0.02$.

The overall behavior of the erosion rate as a function of the liquid temperature during operation is particularly important for hydraulic components in contact with high temperature liquids. It was reported [20] that an important parameter is the location between freezing and boiling points based on test results using a vibratory apparatus. The relative temperature was defined as follows:

$$Relative\,Temperature = \frac{test\ temperature\ - freezing\ point}{boiling\ point\ - freezing\ point} \times 100. \tag{11.4}$$

The relative temperature is an index, which shows the location between freezing and boiling points.

In case of tap water, the vapor pressure p_v (MPa) at a test temperature T (K) is given by:

$$p_v = 107145\, e^{-\frac{5173}{T}}. \tag{11.5}$$

Therefore, if the downstream pressure p_d is given, the boiling point T_b is obtained by:

$$T_b = \frac{5173}{11.58 - \ln p_d}. \tag{11.6}$$

Fig. 11.19 Relation between relative temperature and normalized $MDER_{max}$

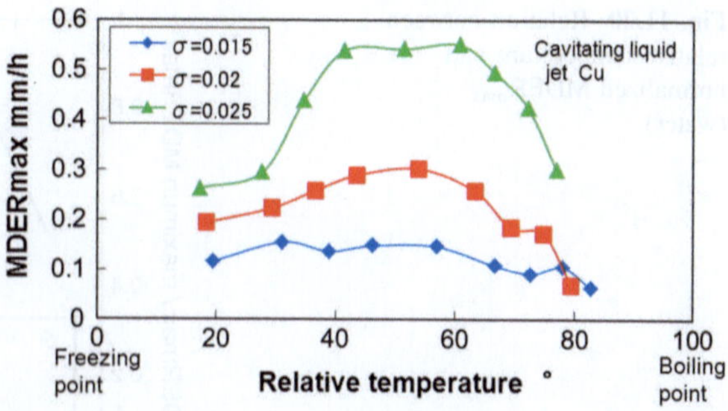

For example, in the case of $p_d = 0.46$ MPa ($\sigma = 0.025$), then the boiling point is obtained by Eq. (11.6) and is 419 K. When we carry out the erosion test at 347 K (75 °C), the relative temperature is $(347 - 273)/(419 - 273) \times 100 = 52°$.

Figure 11.19 shows the relation between the relative temperature and $MDER_{max}$. Test results using a cavitating liquid jet method at every cavitation number reach a peak at a relative temperature of 50°.

Figure 11.20 shows the relation between the relative temperature and normalized $MDER_{max}$. The test results in the vibratory apparatus are the results for an Al alloy given by Plesset [21]. $MDER_{max}$ reaches a peak at a relative temperature of 50° regardless of cavitation number and test apparatus. The erosion rate increases by 1 % between 5° and 50° of the relative temperature, and decreases by 2 % between 50° and 80°. As a result, the influence of temperature can be captured by a vibratory test if a relative temperature is used.

11.3.3 Erosion in Liquid Metals

Research on cavitation erosion in liquid metals is very important to confirm the safety of the fast breeder reactor using sodium coolant and to understand cavitation erosion of a liquid-mercury target in the spallation neutron source [23]. Cavitation erosion tests of SUS304 were carried out in liquid lead–bismuth alloy and deionized water using a vibratory apparatus [12]. We discuss the effect of liquid parameters and the temperature effect on the erosion rate.

Figure 11.21 shows the relation between $MDER_{max}$ of SUS304 and the relative temperature using a vibratory apparatus. The test results in sodium by Young and Johnston [24] and Thiruvengadam et al. [25, 26] are also included in Fig. 11.21. Young and Johnston [24] carried out vibratory erosion tests in sodium at three temperatures on a cobalt alloy (L605). The erosion tests of SUS316 were carried out only in sodium at 427 °C. The result was converted into the rate of SUS304 using a conversion coefficient of 1.5 [8]. Table 11.3 shows the converted results of SUS304 obtained from the cobalt alloy (L605). In Fig. 11.21 [12], cavitation erosion in deionized water depends on temperature with the peak at 45–50°, as

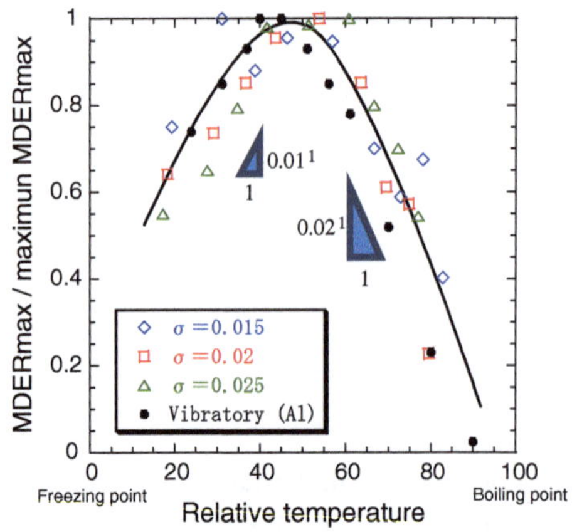

Fig. 11.20 Relation between relative temperature and normalized $MDER_{max}$ (water)

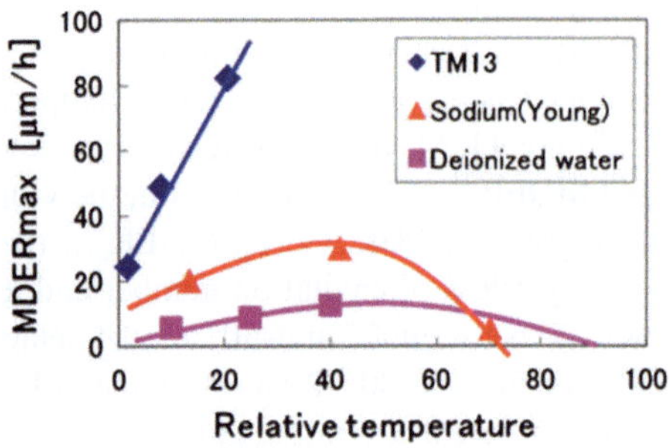

Fig. 11.21 Relation between relative temperature and $MDER_{max}$

Table 11.3 $MDER_{max}$ [μm/h]. Relative temperatures are shown in parenthesis

Material	Temperature [°C]		
	204 °C (13.5)	427 °C (42)	649 °C (70.3)
L605	11.1	25.6	4.3
SUS304	13.1[a]	29.9	5.0[a]

[a] Obtained by calculation

shown in the previous section. In the results of Young and Johnston [24], $MDER_{max}$ in sodium shows a similarly shaped curve to deionized water. At a relative temperature of 14°, the erosion rate was 10 times in lead–bismuth alloy, and 5 times in sodium, as compared with that in deionized water. In Fig. 11.21, the temperature dependence of $MDER_{max}$ increases 0.15 μm/h in deionized water, 0.59 μm/h in sodium and 3.1 μm/h in lead–bismuth alloy. Liquid metals have a higher temperature dependence compared to deionized water.

When a bubble collapses by inertia, the velocity of the bubble wall is given by the Rayleigh-Plesset equation [27]:

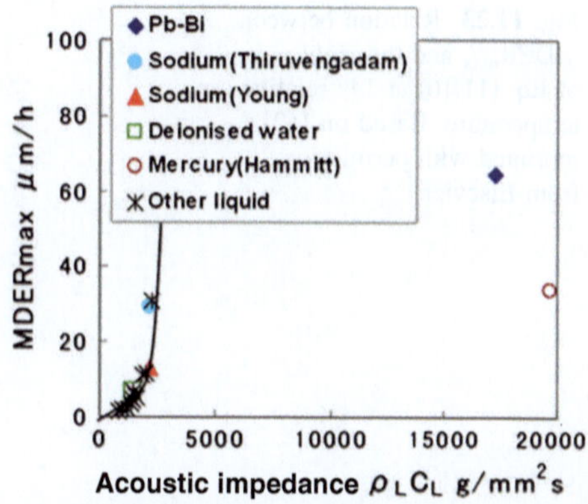

Fig. 11.22 Relation between acoustic impedance and MDER_{max}. Based on [12], reprinted with permission from Elsevier

$$V = \sqrt{\frac{2}{3}\frac{(p_\infty - p_v)}{\rho_L}\left[\left(\frac{R_0}{R}\right)^3 - 1\right] + \frac{2\gamma}{R_0\rho_L}\left(\frac{R_0}{R}\right)^3}, \tag{11.7}$$

where ρ_L is the liquid density and V the velocity of the bubble wall. p_∞, p_v, R_0, R and γ are downstream pressure, vapor pressure, initial value of the bubble radius, instantaneous value of the bubble radius and surface tension, respectively. On the other hand, the pressure p of a water hammer is given by [28]:

$$p = \frac{V}{\frac{1}{\rho_L C_L} + \frac{1}{\rho_S C_S}}, \tag{11.8}$$

where C_L is sound velocity in liquid, and ρ_S and C_S are solid material density and sound velocity in solid, respectively. Generally, the acoustic impedance of a solid is larger than that of a liquid. Eq. (11.6) can therefore be reduced to:

$$p \simeq \rho_L C_L V \tag{11.9}$$

Wilson and Graham [29] reported that the erosion rate in various liquids increases exponentially with the acoustic impedance $\rho_L C_L$. Therefore, the erosion rate in liquid metals was evaluated in terms of acoustic impedance $\rho_L C_L$. Figure 11.22 shows the relation between MDER_{max} and acoustic impedance $\rho_L C_L$. The symbol * marks the test results in heptane, butyl alcohol, aniline, anisole, benzene, ethylene glycol, trichloroethane, carbon tetrachloride, ethylene dibromide and bromoholm which were carried out by Wilson and Graham. The data in mercury were obtained by Garcia and Hammitt [30]. The data in sodium are located near Wilson's results. But the erosion rate in lead–bismuth alloy and mercury with high density cannot be evaluated in terms of $\rho_L C_L$. Since the acoustic impedance $\rho_L C_L$ of the lead–bismuth alloy or mercury is comparable to the acoustic impedance $\rho_S C_S$ of SUS304 test specimen, we have to use Eq. (11.8) instead of Eq. (11.9). When both the acoustic impedances of the test liquid and the

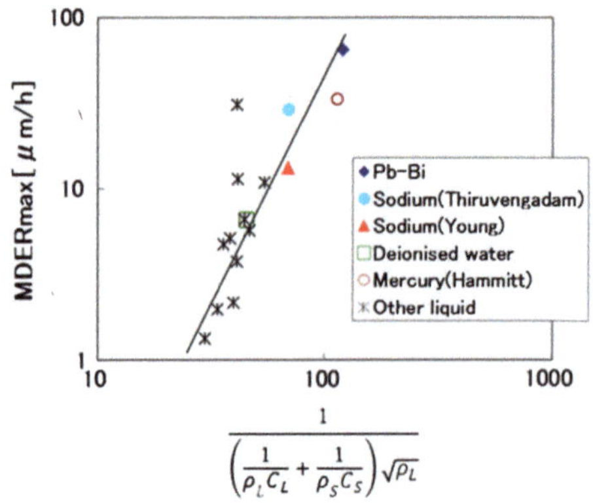

Fig. 11.23 Relation between MDER_{max} and the *right side* of Eq. (11.10) at 14° relative temperature. Based on [12], reprinted with permission from Elsevier

specimen are taken into account, the pressure p can be expressed as follows by considering Eqs. (11.7) and (11.8):

$$p \propto \frac{1}{\frac{1}{\rho_L C_L} + \frac{1}{\rho_S C_S}} \frac{1}{\sqrt{\rho_L}}. \tag{11.10}$$

Figure 11.23 shows the relation between MDER_{max} and the right hand side of Eq. (11.10) at 14° relative temperature. MDER_{max} has a linear relation in a log–log scale with the parameter in Eq. (11.10). As a result the erosion rate could be evaluated in terms of the parameter of Eq. (11.10).

11.4 Impact Load Measurement for Erosion Prediction

Cavitation erosion tests and impact load measurements were carried out using a pressure transducer made of piezoelectric ceramic developed in our laboratory [31]. These measurements are used to develop a method for the prediction of the incubation period under various cavitation conditions of a cavitating liquid jet apparatus.

Figure 11.24 shows the MDE (Mean Depth of Erosion) curves of S15C steel at a constant cavitation number of 0.025 and at flow velocities from 184 down to 80 m/s using a cavitating liquid jet apparatus [13]. All MDE curves exhibit an incubation period during which the erosion rate is zero or negligible compared to later stages and then they increase to reach a maximum rate period for each material. We discuss below a method to predict the incubation period.

Figure 11.25 shows the distribution of the bubble collapse impact loads, measured by the pressure transducer with a resonance frequency of 8 MHz located at an optimum standoff distance of 10 mm. Several tens of thousands counts/ minutes were detected having an impact load above a threshold of 0.2 N and several thousands were obtained above 1 N. Beyond that, lower numbers of

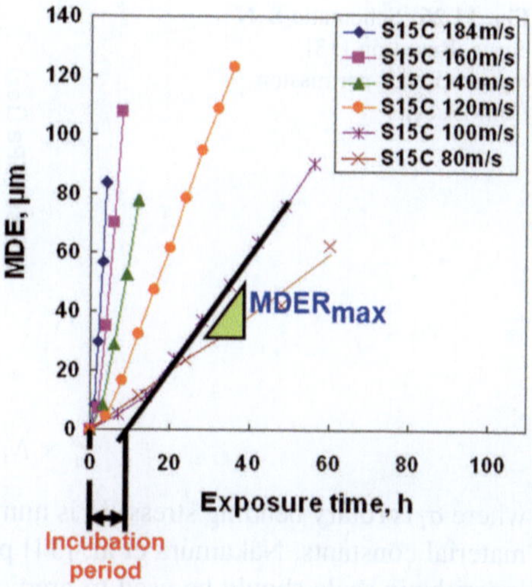

Fig. 11.24 MDE curves of S15C. Based on [13], reprinted by with from Elsevier

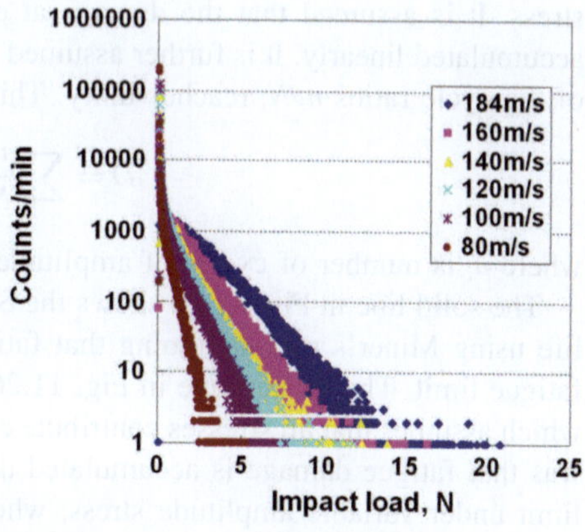

Fig. 11.25 Distribution of bubble collapse impact loads. Based on [13], reprinted with permission from Elsevier

counts/minutes at large impact loads were measured. The distributions at various flow velocities between 184 and 80 m/s are very similar, and the counts of the large impact loads are higher at higher flow velocities.

We previously proposed [32] a correlation between the parameter ΣF_i^2 (F_i: impact load) to predict the erosion rate. However, we found only a poor correlation between ΣF_i^2 and the incubation period. In this study, the damage is assumed to be based on the fatigue failure under variable amplitude stress. We therefore discuss the prediction method for cavitation erosion from the viewpoint of fatigue life prediction.

The linear cumulative damage rule is one of the prediction methods for the fatigue life under variable amplitude loading. Figure 11.26 shows the schematic S–N curves on a double logarithmic scale [33]. The slope part is given by:

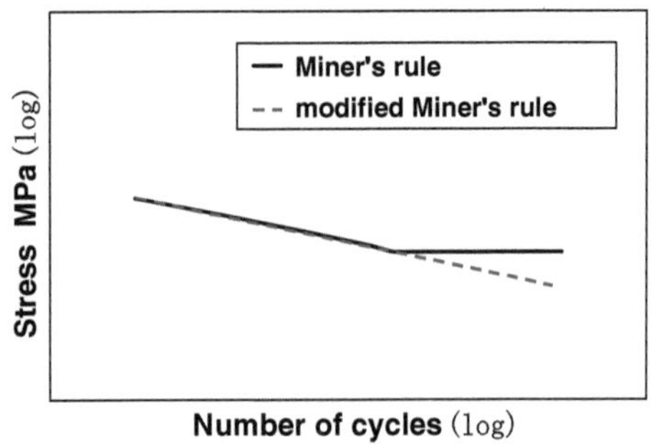

Fig. 11.26 Schematic S–N *curve*. Based on [13], reprinted with permission from Elsevier

$$\sigma_i^{\alpha} \times N_i = C, \tag{11.11}$$

where σ_i is rotary bending stress, N_i is number of cycles to fracture and α and C are material constants. Nakamura et al. [34] proposed that the S–N curve on a double logarithmic scale should be used to predict the fatigue life. The fatigue damage is given by the cycle ratio n_i/N_i when σ_i is repeated n_i times under variable amplitude stress. It is assumed that the damage at each stress level is independent and is accumulated linearly. It is further assumed that the material ruptures when the sum of the cycle ratios n_i/N_i reaches unity. This sum is given by:

$$D = \sum \frac{n_i}{N_i} = 1, \tag{11.12}$$

where n_i is number of cycles of amplitude σ_i.

The solid line in Fig. 11.26 shows the S–N curve in order to predict the fatigue life using Miner's rule, assuming that fatigue damage does not occur below the fatigue limit. The broken line in Fig. 11.26 is based on the modified Miner's rule which assumes that all stresses contribute equally to the fatigue damage. The result was that fatigue damage is accumulated due to the stress even below the fatigue limit under variable amplitude stress, when the stress is combined with stresses above the fatigue limit [33]. The modified Miner's rule is nowadays commonly used [33] to evaluate fatigue damage.

Since many impact loads at various intensities are measured in cavitation bubble collapses, the modified Miner's rule under the variable amplitude loading is applied to the prediction of cavitation erosion. The incubation period is defined as the point of intersection of the extended straight line of slope of the maximum rate period with the axis of exposure time [1], and the termination of incubation period is well assumed to coincide with macroscopic fatigue failure. The incubation period has been discussed based on the accumulation of fatigue damage.

The prediction method of cavitation erosion was constructed on the basis of the modified Miner's rule by using the relation between the impact load F_i and the number of impacts per unit time n_i obtained from the measurement of bubble collapse impact loads at the various flow velocities. The incubation periods were obtained from cavitation erosion tests at the various flow velocities. Since it is

Table 11.4 Values of constants α and C of each material

Material	α	C
S15C	1.64	1.29×10^8
SUS304	1.79	1.59×10^8
STPA24	1.11	8.89×10^7

impossible to measure the impacted area, we cannot convert the force into a stress. Therefore, the F–N curve was used. N is the impact number at the termination of the incubation period with constant impact load F. Since a test with a constant impact load F_i cannot be carried out for cavitation erosion, the impact number at the termination of incubation period N_i is basically unknown. But the F–N curve is derived using constant parameters α and C in:

$$F_i^{\alpha} \times N_i = C. \tag{11.13}$$

The parameter $\Sigma(n_i/N_i)$ is the cumulative damage per unit time. The incubation period t_d (the subscript d indicates the flow velocity) finishes when $t_d\Sigma(n_i/N_i)$ reaches unity. t_d is therefore given by:

$$t_d = \frac{1}{\Sigma \frac{n_i}{N_i}}. \tag{11.14}$$

To determine the constants α and C, the unknown quantity N_i in Eq. (11.13) is expressed by F_i and the constants α and C, and then substituted into Eq. (11.14). This procedure leads to the following equation:

$$\frac{1}{C}\Sigma\left(F_i^{\alpha} \times n_i\right) = \frac{1}{t_d}. \tag{11.15}$$

For a given material, the constants α and C are independent of the impact load F_i, the number of impacts n_i and the incubation period t_d. Therefore Eq. (11.15) with the data at a flow velocity of 184 m/s was divided by Eq. (11.15) with the data at a flow velocity of 160 m/s, which gives the following relation:

$$\frac{\Sigma\left(F_{i,184}^{\alpha} \times n_{i,184}\right)}{\Sigma\left(F_{i,160}^{\alpha} \times n_{i,160}\right)} = \frac{t_{160}}{t_{184}}. \tag{11.16}$$

Since t_d, F_i and n_i are already given, the constant α can now be determined with a trial & error method using Eq. (11.16). After α is determined, the constant C can be obtained from Eq. (11.15). Since the constants α and C are material dependent, it is necessary to find them for each material. Table 11.4 shows the values of α and C for the three materials.

Figure 11.27 shows the relation between the parameter $\Sigma(F_i^{\alpha} \times n_i)$ and the reciprocal of the incubation period obtained from cavitation erosion tests of three alloys at different flow velocities. Straight lines can be drawn passing through the origin.

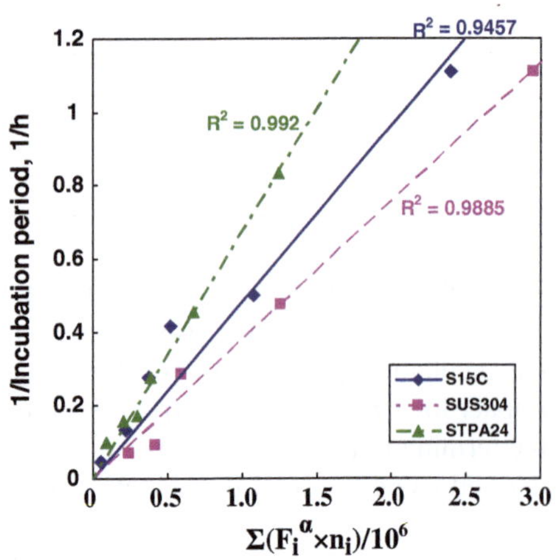

Fig. 11.27 Relation between $\Sigma(F_i^{\alpha} \times n_i)$ and reciprocal of the incubation period. Based on [13], reprinted with permission from Elsevier

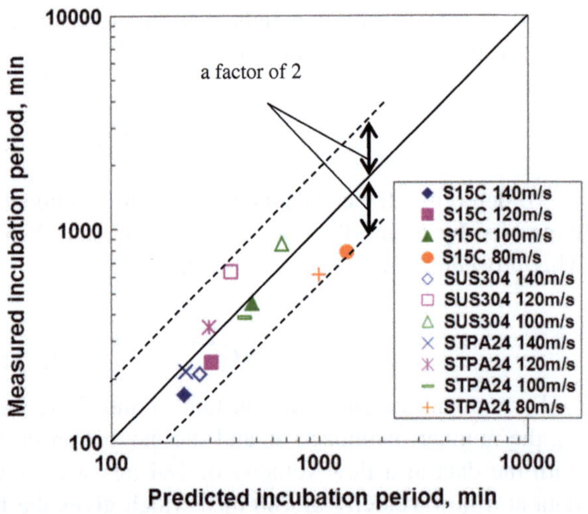

Fig. 11.28 Prediction accuracy at various flow velocities

The parameter $\Sigma(F_i^{\alpha} \times n_i)$ is suitable for the evaluation of cavitation erosion. By using the values of α and C obtained from two different cavitation conditions, we can predict the incubation period t_d by Eq. (11.15) using the values of F_i and n_i obtained from measurement of bubble collapse impact loads at each flow velocity. Figure 11.28 shows the relation between the predicted incubation periods and the measured incubation period for various flow velocities, plotted on a double logarithmic scale. A straight line with a slope of 45° on the double logarithmic scale was obtained. This shows that the measured incubation period can be obtained by the predicted incubation period with a factor of 2.

In this study, we showed that the incubation period can be predicted if the constants α and C are obtained from erosion test and the measurement of bubble collapse impact loads under two different cavitation conditions.

11.5 Conclusions

In this study, the recent research achievements by the author were reviewed. We draw the following conclusions.

1. The erosion resistance of carbon steels, stainless steels, cast iron, aluminum alloys, copper alloys and titanium alloys can be estimated accurately from the material hardness for each series of alloys.
2. We proposed a cavitation erosion model for SUS304 and cobalt alloy ST6 based on observations by scanning electron microscopy.
3. The temperature effect showed a similar tendency between vibratory and cavitating liquid jet tests after defining a relative temperature.
4. The erosion rate in various liquids including liquid metals can be evaluated on the basis of an ad hoc parameter combining density and sound velocity of both the test liquid and the eroded material.
5. Impact load measurements were made to establish a prediction method for the incubation period of erosion.

References

1. ASTM (ed) (2005) Annual book of ASTM standards. G32-03 106–122
2. ASTM (ed) (2005) Annual book of ASTM standards. G134-95 561–572
3. Okada T, Iwai Y, Hattori S (1984) Corrosion fatigue of steel under cavitation erosion generated intermittently in 3% Salt water (results of two-stage tests with and without cavitation erosion. Wear 96:85–98
4. Hattori S, Nakao E (2002) Cavitation erosion mechanisms and quantitative evaluation based on erosion particles. Wear 249:839–845
5. Sugiyama K, Nakahama S, Hattori S, Nakano K (2005) Slurry wear and cavitation erosion of thermal-sprayed cermet. Wear 258:768–775
6. Hattori S, Takinami M (2010) Comparison of cavitation erosion rate with liquid impingement erosion rate. Wear 269:31–316
7. Hattori S, Ishikura R, Zhang Q (2004) Construction of database on cavitation erosion and analyses of carbon steel data. Wear 257:1022–1029
8. Hattori S, Ishikura R (2010) Revision of cavitation erosion database and analysis of stainless steel data. Wear 268:109–116
9. Hattori S, Kitagawa T (2010) Analysis of cavitation erosion resistance of cast iron and nonferrous metals based on database and comparison with carbon steel data. Wear 269:443–448
10. Hattori S, Mikami N (2009) Cavitation erosion resistance of stellite alloy weld overlays. Wear 267:1954–1960
11. Hattori S, Goto Y, Fukuyama T (2006) Influence of temperature on erosion by a cavitating liquid jet. Wear 260:1217–1223
12. Hattori S, Inoue F, Watashi K, Hashimoto T (2008) Effect of liquid properties on cavitation erosion in liquid metals. Wear 265:1649–1654
13. Hattori S, Hirose T, Sugiyama K (2010) Prediction method for cavitation erosion based on measurement of bubble collapse impact loads. Wear 269:507–514

14. Heymann FJ (1970) Characterization and determination of erosion resistance, ASTM STP474. Am Soc Test Mater 212–222
15. Gibra IN (1973) Probability and statistical inference for scientists and engineers. Prentice-Hall, Englewood Cliffs
16. Okada T, Iwai Y, Yamamoto A (1983) A study of cavitation erosion of cast iron. Wear 84:297–312
17. The Society of Materials Science (1964) (ed) Fatigue of metals. Maruzen, Tokyo, pp 170–191
18. The Society of Materials Science (1981) (ed) Handbook of fatigue design for metals. Yokendo, Tokyo, pp 4–5
19. Hattori S, Maekawa N, Kawai Y (2000) Cavitation erosion of titanium alloys. Trans Jpn Soc Mech Eng 66A(648):1627–1633
20. Hammitt FG (1980) Cavitation and multiphase flow phenomena. McGraw-Hill International Book Company, New York, pp 250–251
21. Plesset MS (1972) Temperature effects in cavitation damage. Trans ASME J Basic Eng 94:559–566
22. Hattori S, Tanaka Y (2002) Influence of air content and vapor pressure of liquids on cavitation erosion. Trans Jpn Soc Mech Eng 68B:13–136
23. Futakawa et al (2003) Cavitation erosion in mercury target of spallation neutron source. Proceedings of CAV2003, Osaka GS-11-006
24. Young SG, Johnston JR (1970) Effect of temperature and pressure on cavitation damage in sodium ASTM STP 474. Am Soc Tst Mater 67–108
25. Thiruvengadam A, Preizer HS, Couchman CD (1964) Cavitation damage in liquid metals. NASA CR-54071 1-83
26. Thiruvengadam A, Preiser HS, Rudy SL (1965) Cavitation damage in liquid metal. NASA CR54459 1-10
27. Rayleigh LS (1917) On the pressure developed in a liquid during the collapse of a spherical cavity. Philos Mag 6(34):94, 200
28. Franc JP, Michel JM (2004) Fundamentals of cavitation. Kluwer Academic Publishers, New York, pp 281–291
29. Wilson ARW, Graham R (1957) Cavitation of metal surfaces in contact with lubricants. Conference on lubrication and wear, IME, London, pp 702–712
30. Garcia R, Hammitt FG (1967) Cavitation damage and correlations with material and fluid properties. Trans ASME J Basic Eng 89:753–763
31. Hattori S, Mori H, Okada T (1998) Quantitative evaluation of cavitation erosion. Trans ASME J Fluids Eng 120(1):179–185
32. Okada T, Iwai Y, Hattori S, Tanimura N (1995) Relation between impact load and the damage produced by cavitation bubble collapse. Wear 184:231–239
33. De Jonge JB, Nederveen A (1980) Effect of gust load alleviation on fatigue and crack growth in ALCLAD 2024–T3 ASTM STP. Am Soc Test Mater 714:170–184
34. Nakamura H (1981) Committee meeting on fatigue research. The Society of Materials Science, Tokyo No 158

Chapter 12
Effect of Nozzle Geometry on Aggressivity of Cavitating Jet for Cavitation Erosion Test and Applications

Satoshi Nishimura, Osamu Takakuwa and Hitoshi Soyama

Abstract In order to estimate cavitation erosion rate in hydraulic machinery the erosion resistance of materials can be investigated using a cavitating jet apparatus, standardized by ASTM G134. As the jet aggressivity is affected by the nozzle outlet geometry, this should also be considered to obtain reliable erosion tests. In this paper, we investigated the effects of the nozzle outlet geometry on the aggressivity of the cavitating jet by erosion tests, impact force measurements, and high-speed movie observations. The effect of Strouhal number, defined by the shedding frequency of the cavitation cloud, the width of the cavitating region, and the jet velocity, was also studied. The aggressivity of the cavitating jet peaked at a certain nozzle outlet bore, *D*, and outlet length, *L*. It was found that the Strouhal number, *St*, is 0.17 at the optimum *D* and *L*, even though the nozzle throat diameter *d* and injection pressure p_1 were different. It was also revealed that the frequency of the large cavitation impact is closely related to the shedding frequency of the cavitation cloud.

12.1 Introduction

Cavitation bubble collapse causes severe material damage in hydraulic machineries. Erosion resistance of materials for such applications can be investigated by using a cavitating jet apparatus, standardized by ASTM G134 [1]. Cavitating jets have been successfully applied for surface modification method [2–16] to improve fatigue strength by cavitation peening instead of shot peening. Further applications of cavitating jets for enhancement of chemical reactions were also reported [17, 18].

S. Nishimura · O. Takakuwa · H. Soyama (✉)
Department of Nanomechanics, Tohoku University, 6-6-01 Aoba, Aramaki, Aoba-ku Sendai 980-8579, Japan
e-mail: soyama@mm.mech.tohoku.ac.jp

K.-H. Kim et al. (eds.), *Advanced Experimental and Numerical Techniques for Cavitation Erosion Prediction*, Fluid Mechanics and Its Applications 106,
DOI: 10.1007/978-94-017-8539-6_12,

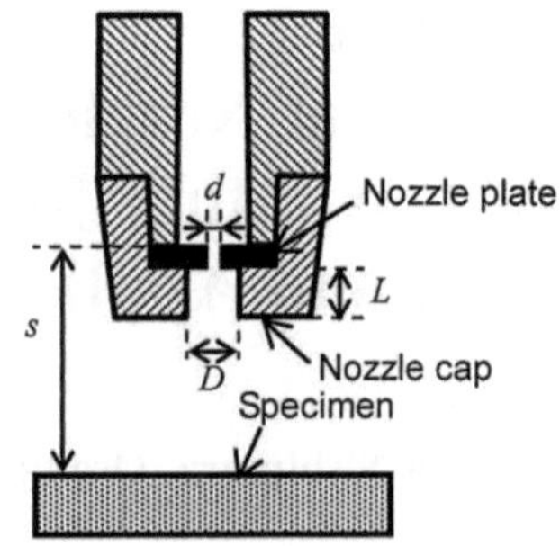

Fig. 12.1 Nozzle geometry for cavitating jet. Based on Soyama [29], reprinted with permission from ASME

It has been shown that the nozzle geometry affects the aggressivity of the cavitating jet [19–26]. By optimizing the geometry of the nozzle outlet, jet aggressivity was enhanced while avoiding nozzle damage [27–29]. The nozzle outlet geometry for the cavitating jet is shown in Fig. 12.1. The nozzle consists of a nozzle plate and nozzle cap, which has an outlet bore. The *d, D, L* and *S* are the nozzle throat diameter, bore diameter, bore length and standoff distance, respectively. The optimum geometry corresponds to $D/d = 8$ and $L/d = 8$, or *d, D, L* being 1, 8, 8 for various cavitating conditions [29]. The optimum nozzle outlet geometry was determined by considering the Strouhal number. This parameter is defined by the geometry of nozzle outlet, the cavitation cloud shedding frequency and the velocity of the jet, as the cavitating jet produces a periodical cavitation cloud shedding [30]. Thus, in order to obtain reliable erosion tests and advantages for other applications of cavitating jets, the relation between jet aggressivity and nozzle outlet geometries should be investigated in detail.

The aggressivity of the cavitating jet depends on the size of the cavitation cloud, since it can be assumed that a large sized cavitation cloud produces a large impact at its collapse. Small cavitation vortices at the nozzle exit merge to form a cavitation cloud. The merging process of the cavitation vortices affects the size and the shedding frequency of the cavitation clouds. Cavitation clouds take the form of cavity rings at high cavitation numbers [23]. The size of the cavitation cloud is also affected by the cavitation number, since low injection pressure and high ambient pressure inhibits the growth of the cavitation cloud. On the other hand, the aggressivity of the jet also depends on the shedding frequency of the cavitation cloud, since it affects the impact rate. Namely, the Strouhal number defined by the cavitation cloud shedding frequency, width and velocity of the jet at the nozzle exit might be optimum at the most aggressive condition.

In this paper, we investigated the effects of the nozzle outlet geometry on the aggressivity of the cavitating jet by erosion tests, impact force measurements and high-speed movie observations. Regarding the erosion tests, the mass loss of the pure aluminum specimen was considered as the measure of the aggressivity of the cavitating jet. The impact force measurement was realized by using a PVDF transducer [31], and the effect of the cavitation impact force and frequency on the cavitation erosion was discussed. The Strouhal number was obtained by high speed observations. It is noted that the Strouhal number at the optimum nozzle outlet geometry had a certain value for various cavitating conditions.

12.2 Experimental Apparatus and Procedure

12.2.1 Cavitating Jet Apparatus

The cavitating jet apparatus used in this paper is shown in Fig. 12.2. The high speed water jet is pressurized by a plunger pump that has a maximum pressure and flow rate of 35 MPa and 3.0×10^{-2} m^3/min, respectively, and is injected perpendicular to the specimen, through the nozzle. The injection pressure, p_1, is controlled by the velocity of the pump's inverter motor. The test section ambient pressure, p_2, is controlled by the downstream valve as shown in Fig. 12.2. The water temperature is controlled by a chiller, and kept constant at 300 ± 3 K.

The nozzle outlet geometry for the test is shown in Fig. 12.1. The nozzle outlet bore, D and the nozzle outlet length, L are varied by changing the nozzle cap. The nozzle throat diameter, d, is varied by changing the nozzle plate. The studied combinations of D and L are shown in Table 12.1. The thickness of the nozzle plate is kept constant at $3d$.

The main parameter of the cavitating jet is the cavitation number [1], σ, which is defined by the injection pressure p_1, ambient pressure p_2, and the vapor pressure p_v, as given by Eq. (12.1):

$$\sigma = \frac{p_2 - p_v}{p_1 - p_2}. \tag{12.1}$$

12.2.2 Erosion Test

The aggressivity of the cavitating jet was evaluated by measuring the mass loss, Δm, of a pure aluminum specimen (Japanese Industrial Standards JIS A1050). Conditions of the cavitating jet for the erosion test are shown in Table 12.2. Since the aggressivity of the cavitating jet has a maximum at $\sigma = 0.014$ [32], this σ value was chosen for all cavitating conditions. This was realized by controlling the downstream valve and the ambient pressure. As the mass loss Δm changes with the standoff distance, erosion tests were conducted at various standoff distance. Then, the maximum mass loss at the optimum standoff distance Δm_{opt} was obtained, for each nozzle outlet geometry and cavitating condition.

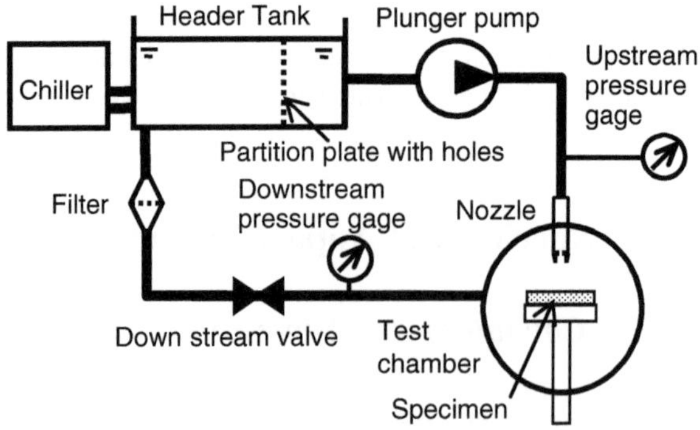

Fig. 12.2 Cavitating jet apparatus. Based on Soyama [29], reprinted with permission from ASME

Table 12.1 Tested combination of the nozzle outlet geometry

D/d	*L/d*				
	6	7	8	9	10
6	–	–	○	–	–
7	–	○	○	○	–
8	○	○	○	○	○
9	–	○	○	○	–
10	–	–	○	–	–

Table 12.2 Conditions of cavitating jet for the erosion tests

	Nozzle throat diameter *d* (mm)	Injection pressure p_1 (MPa)	Ambient pressure p_2 (MPa)	Cavitation number σ (-)	Erosion time *t*
Condition 1	1	30	0.42	0.014	60 s
Condition 2	2	15	0.21	0.014	2 min
Condition 3	2	30	0.42	0.014	30 s

12.2.3 Impact Force Measurement

The total aggressivity of the cavitating jet can be measured by the erosion test. However, the number and the intensity of the impacts caused by the cavitating jet cannot be evaluated. Soyama et al. [31, 33] developed a PVDF (Polyvinylidene Fluoride) transducer and a calibration method to measure the cavitation impacts. PVDF transducers are also useful for estimating the erosion rate in hydraulic machinery. The resistance of a material to cavitation erosion, which is a kind of fatigue strength, can be characterized by a threshold level and obtained using PVDF transducers and cavitation erosion tests [34].

In the present paper, in order to investigate the relation between the frequency of impacts and cavitation features such as the shedding frequency of the cavitation cloud, the cavitation impacts at Condition 1 were investigated using a PVDF

Fig. 12.3 PVDF transducer

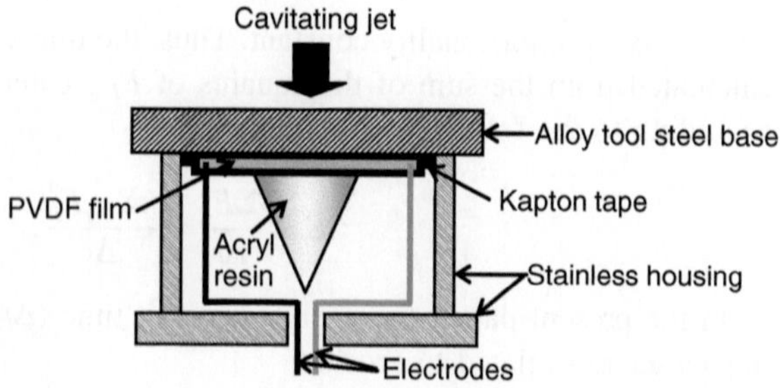

transducer. The schematic diagram of the PVDF transducer is shown in Fig. 12.3. The cavitating jet impacts the alloy tool steel (Japanese Industrial Standards JIS SKD11) base at the optimum standoff distance, to avoid the damage of the PVDF film. Elastic waves induced by the cavitation collapse propagate through the steel base and stimulate the PVDF film sealed by the Kapton (polyimide) tape. Transient elastic waves attenuate at the block of acryl resin. Output signals are analyzed by a cavitation impact counter [33], which counts the number of pulses having an amplitude over a certain threshold level, per unit time.

The impact energy induced by a single cavitation collapse, E_i, is calculated using the following formula:

$$E_i = I_i \tau_i A_i. \tag{12.2}$$

The I_i, τ_i and A_i denote the intensity of sound power, pulse duration and effective area, respectively. I_i can be calculated as follows:

$$I_i = \frac{P_i^2}{2\rho C}, \tag{12.3}$$

where P_i, ρ and C denote the sound pressure, density and sound speed, respectively. The sound pressure, P_i, can be calculated as follows:

$$P_i = \frac{F_i}{A_i}, \tag{12.4}$$

where F_i is the impact force, measured by the PVDF transducer. From Eqs. (12.2), (12.3) and (12.4) can be rewritten as follows:

$$E_i = \frac{F_i P_i \tau_i}{2\rho C}. \tag{12.5}$$

If it is assumed that P_i is proportional to F_i and τ_i is constant, E_i can be described as follows:

$$E_i = kF_i^2, \tag{12.6}$$

where k is a proportionality constant. Thus, the impact energy per unit time can be calculated from the sum of the squares of F_i greater than the threshold level F_{th} according to the following equation:

$$E = \frac{\Sigma E_i}{\Delta t} = \frac{k\Sigma_{F_{th}} F_i^2}{\Delta t}. \qquad (12.7)$$

In the present paper, the energy per unit time ($\Delta t = 1$), $\Sigma_{F_{th}} F_i^2$, is used as the impact energy related to erosion.

12.2.4 High-Speed Movie Observation

The cavitating jet was recorded with a high-speed video camera (Motion Pro HS) through the acrylic window of the test section. Two halogen lights and one xenon lamp were placed on the same side of the video camera, so that the jet would appear in white in the movie. In order to observe the jet clearly, an exposure time of 2 μs, a frame rate of 20,000 fps and frame size of 128 × 512 pixels were chosen.

The shedding frequency of the cavitation cloud was calculated from the fluctuation of the gray scale value at the analysis point in the movie. As shown in Fig. 12.4, the analysis point was chosen at $x/d = 10$, where x denotes the distance from the downstream edge of the nozzle plate, along the jet center, since this corresponds to the end of the potential core region [35]. A series of continuous data of time dependent gray scale values was transformed into a frequency spectrum using a Fourier transform program. The average of the frequencies of the 20 highest peaks in the spectrum were considered as the shedding frequency of the cavitation cloud, f_{shedd}. The standard deviation of the 20 biggest peaks frequency was considered as the measurement uncertainty.

The width of the cavitating region, w, was also studied. A composite picture was made by superimposing 2,000 frames of the high-speed movie, and the maximum width of the cavitating region was evaluated as w. The edge of w was determined by the stationary point of the gray scale value, as shown in Fig. 12.5.

In order to discuss the effect of the nozzle outlet geometry on the jet aggressivity, Strouhal number St was studied. As well known, cavitation number, σ, affects both shedding frequency of cavitation cloud and cavitating region. However, the conventional St based on the nozzle throat diameter, d, cannot consider the effect of σ on the cavitating region, since d is not a function of σ. Therefore, in the present study, the representative length of St was chosen as the width of the cavitating region, since it can consider the effect of σ on the cavitating region. St is given by Eq. (12.7), by using the frequency of the cavitation cloud, f_{shedd}, width of the cavitating region, w, and the velocity of the jet at the nozzle exit, U:

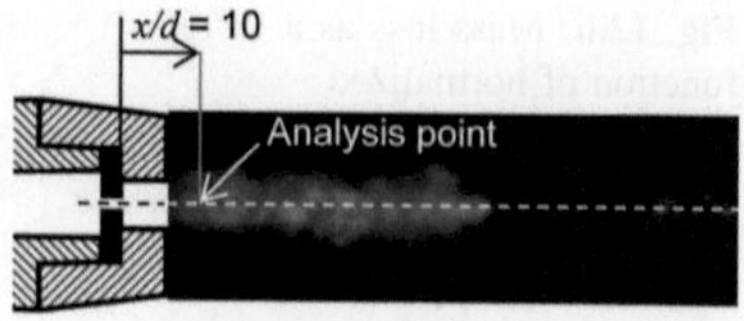

Fig. 12.4 Analysis point chosen as $x/d = 10$

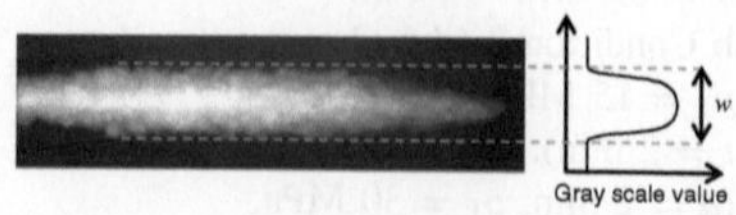

Fig. 12.5 Composite picture of the cavitating jet

$$St = \frac{f_{shedd}\, w}{U}, \tag{12.8}$$

where U is determined by the injection pressure, p_1, ambient pressure, p_2, and water density, ρ, as given by Eq. (12.8):

$$U = \sqrt{\frac{2(p_1 - p_2)}{\rho}}. \tag{12.9}$$

The calculated values of U are 244 m/s for Conditions 1 and 3, and 173 m/s for Condition 2.

12.3 Results

12.3.1 Optimum Standoff Distance

Figure 12.6 shows the mass loss, Δm, as a function of normalized standoff distance, s/d, for various nozzle outlet geometries and cavitating conditions [29]. It can be seen that the maximum value of Δm occurs at different s/d, for each nozzle outlet geometry. The optimum standoff distance, s_{opt}, was obtained by these results, as the standoff distance where the mass loss is at a maximum. The s_{opt} decreases with an increase in nozzle outlet bore, D, and on the contrary, increases with an increase in nozzle outlet length, L.

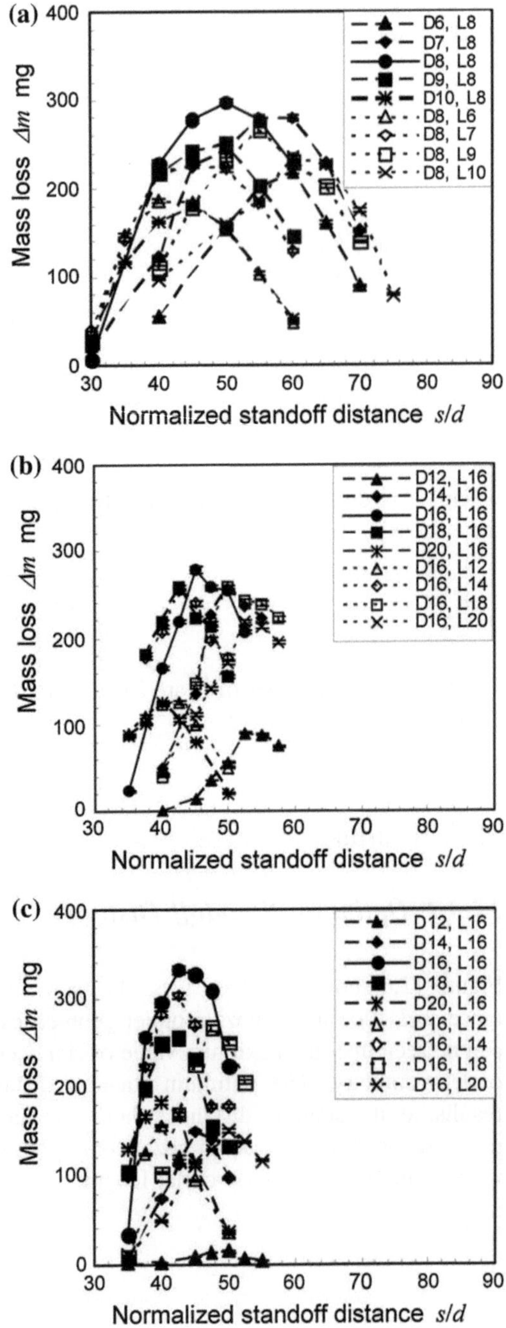

Fig. 12.6 Mass loss as a function of normalized standoff distance for various nozzle outlet geometries. Based on Soyama [29], reprinted with permission from ASME. **a** Condition 1 ($d = 1$ mm, $p_1 = 30$ MPa, $\sigma = 0.014$, $t = 60$ s). **b** Condition 2 ($d = 2$ mm, $p_1 = 15$ MPa, $\sigma = 0.014$, $t = 2$ min). **c** Condition 3 ($d = 2$ mm, $p_1 = 30$ MPa, $\sigma = 0.014$, $t = 30$ s)

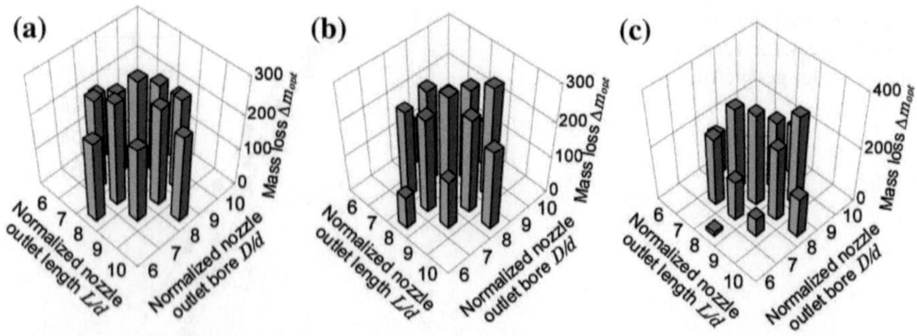

Fig. 12.7 Mass loss at the optimum standoff distance for various nozzle outlet geometries and cavitating conditions. Based on Soyama [29], reprinted with permission from ASME. **a** Condition 1 (d = 1 mm, p_1 = 30 MPa, σ = 0.014). **b** Condition 2 (d = 2 mm, p_1 = 15 MPa, σ = 0.014). **c** Condition 3 (d = 2 mm, p_1 = 30 MPa, σ = 0.014)

12.3.2 Influence of Nozzle Outlet Geometry on Mass Loss

Figure 12.7 shows the mass loss at the optimum standoff distance, Δm_{opt}, for various nozzle outlet geometries and cavitating conditions. The Δm_{opt} varied significantly and is at a maximum at certain values of normalized nozzle outlet bore, D/d, and length, L/d, i.e., the optimum nozzle outlet geometry. The optimum nozzle outlet geometry for Conditions 1 and 3 was $D/d = 8$, $L/d = 8$. For Condition 2, the optimum nozzle outlet geometry was $D/d = 9$, $L/d = 9$.

12.3.3 Influence of Nozzle Outlet Geometry on Impact Force

Figure 12.8 shows the distribution of the impact forces for various nozzle outlet geometries. This represents the relationship between the threshold level F_{th} and the frequency of cavitation impacts, f_i, larger than F_{th}. It can be seen that a large number of impacts of small amplitude are produced, and the number of impacts decreases monotonically with an increase in F_{th}. Obviously, the optimum nozzle outlet geometry produces more impacts of larger amplitude than the others.

Figure 12.9 shows the distribution of impact energy for various nozzle outlet geometries, calculated from the data of Fig. 12.8 by Eq. (12.6). The impact energy also decreases with an increase in amplitude F_{th}.

Figure 12.10 shows the impact energy as a function of erosion rate, obtained from the data of Figs. 12.7a and 12.9. The threshold level of the impact force was detrmined to be 273 N. The impact energy increased linearly with an increase in erosion rate. It can be concluded that the erosion rate is proportional to the impact energy.

In order to obtain a representative value of the impact force and the impact frequency, the impact force distribution was approximated by a combination of two individual Poisson distributions [36]. The Poisson distribution at the high

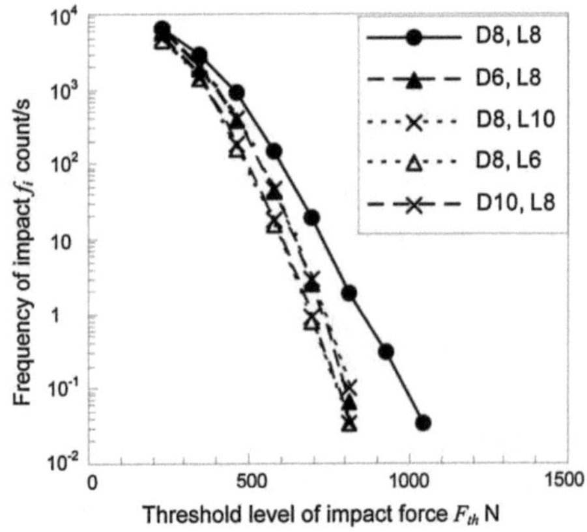

Fig. 12.8 Distribution of impact forces for various nozzle outlet geometries

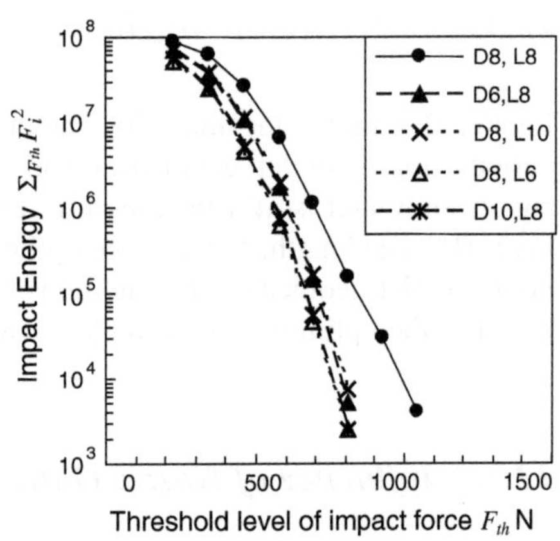

Fig. 12.9 Distribution of impact energy for various nozzle outlet geometries

frequency was named Poisson distribution 1, and the other one Poisson distribution 2, as shown in Fig. 12.11.

Figure 12.11 shows the approximation of the impact force distribution at the optimum nozzle outlet geometry. The average force and the peak frequency of Poisson distribution 1 were 273 N and 6.6 kHz, respectively. For Poisson distribution 2 these values were 371 N and 2.29 kHz. Thus, it can be assumed that the properties of Poisson distribution 2 are quite important.

Table 12.3 shows the effect of nozzle outlet geometry on the average impact force, F_{ave}, peak frequency of Poisson distribution 2, f_{peak}, and shedding frequency of the cavitation cloud, f_{shedd}. Obviously, f_{peak} corresponds to f_{shedd}, except the case of $(D, L) = (8, 6)$ and $(10, 8)$, where f_{peak} might correspond to the sub harmonics of f_{shedd}. Cavitation erosion is caused by the accumulated cavitation impacts [37, 38].

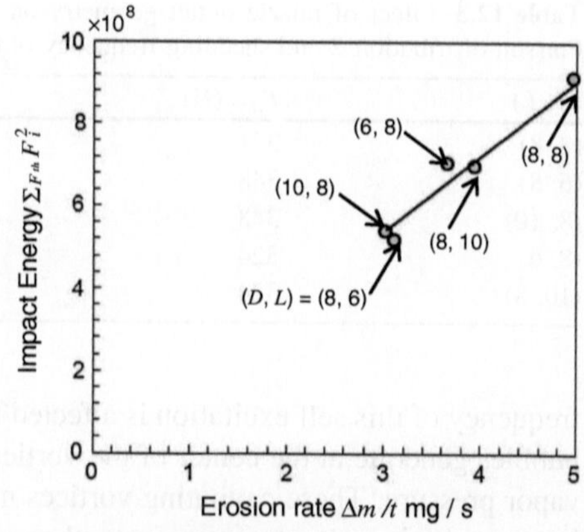

Fig. 12.10 Impact energy as a function of erosion rate (F_{th} = 232 N)

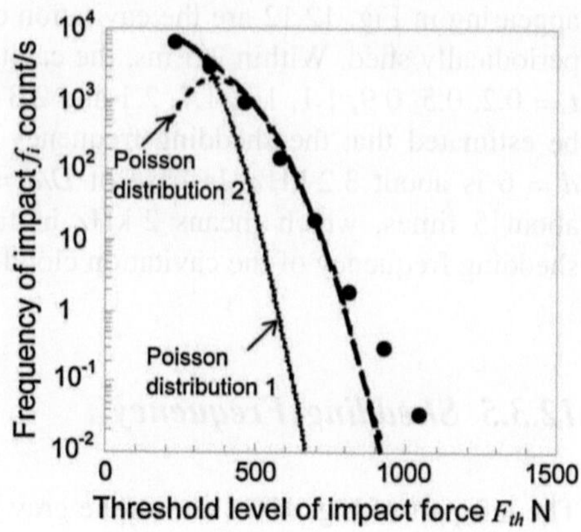

Fig. 12.11 Approximation of the impact force distribution at the optimum nozzle outlet geometry

Thus, the frequency of cavitation impacts is an important factor for cavitation erosion, and therefore, the shedding frequency of the cavitation cloud should be investigated in detail.

12.3.4 Visualizations of the Cavitating Jet

Figure 12.12 shows the appearance of the cavitation cloud shedding at Condition 1, for different nozzle outlet bores, *D*, and lengths, *L*. When a submerged water jet is produced, the jet self excites at a certain frequency and discrete vortices generate in the shear layer, around the nozzle exit. It is known that the

Table 12.3 Effect of nozzle outlet geometry on the average impact force, peak frequency of Poisson distribution 2 and shedding frequency of cavitation cloud

(*D*, *L*)	F_{ave} (N)	f_{peak} (kHz)	f_{shedd} (kHz)
(8, 8)	371	2.29	2.48 ± 0.10
(6, 8)	348	1.90	2.2 ± 0.06
(8, 10)	348	2.00	2.09 ± 0.04
(8, 6)	324	1.37	3.02 ± 0.29
(10, 8)	324	1.36	2.9 ± 0.16

frequency of this self excitation is affected by the nozzle geometry [23]. Cavitation bubbles generate at the center of the vortices, where the pressure is lower than the vapor pressure. These cavitating vortices merge with each other and form discrete large cavitating structures, namely the cavitation clouds. The cavitation clouds expand in size as they travel downstream and finally collapse. White masses appearing in Fig. 12.12 are the cavitation clouds, which are seen to discharge and periodically shed. Within 2.5 ms, the cavitating jet sheds a cloud about 8 times at $t = 0.2, 0.5, 0.9, 1.1, 1.4, 1.7, 2.1$ and 2.3 ms, for $D/d = 8$, $L/d = 6$. Thus, it can be estimated that the shedding frequency of the cavitation cloud at $D/d = 8$, $L/d = 6$ is about 3.2 kHz. In case of $D/d = 8$, $L/d = 10$, the cavitating jet sheds about 5 times, which means 2 kHz in frequency. Thus, it is obvious that the shedding frequency of the cavitation cloud depends on the nozzle outlet geometry.

12.3.5 Shedding Frequency

The left part of Fig. 12.13 shows the gray scale value fluctuation of the cavitating jet at the analysis point at Condition 1. Since the intensity of the gray scale value indicates the distribution of the cavitation clouds, the gray scale value fluctuations signify the cyclic transit of the cavitation clouds at the analysis point, i.e., the cyclic shedding of the cavitation clouds. The right part of Fig. 12.13 shows the frequency spectrum, obtained by the Fourier transform of the gray scale value fluctuation of the cavitating jet, shown on the left-hand side. It can be observed that the spectrum peak shifts with the nozzle outlet geometry.

Figure 12.14 shows the shedding frequency of the cavitation cloud, f_{shedd}, for various nozzle outlet geometries and cavitating conditions. For all cavitating conditions, f_{shedd} increases with the normalized nozzle outlet bore, D/d. On the other hand, f_{shedd} decreased with an increase in normalized nozzle outlet length, L/d. These results indicate that the nozzle outlet geometry affects the vortex structure around the jet near the nozzle exit, which results in differences of shedding frequency of the cavitation cloud [21, 23, 39]. For the same nozzle outlet geometry, f_{shedd} increased with an increase in the injection pressure, p_1, and decreased with an increase in the nozzle throat diameter, d.

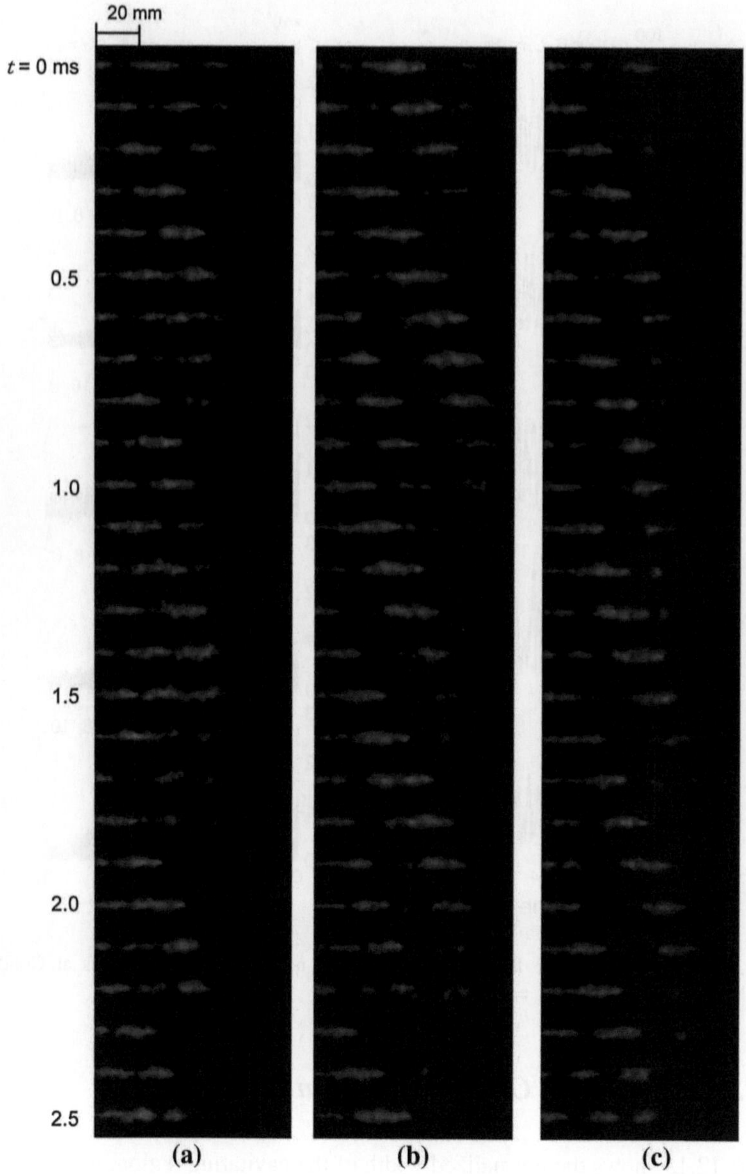

Fig. 12.12 Cavitation cloud shedding at Condition 1 (d = 1 mm, p_1 = 30 MPa, σ = 0.014). **a** D/d = 8, L/d = 6. **b** D/d = 8, L/d = 8. **c** D/d = 8, L/d = 10

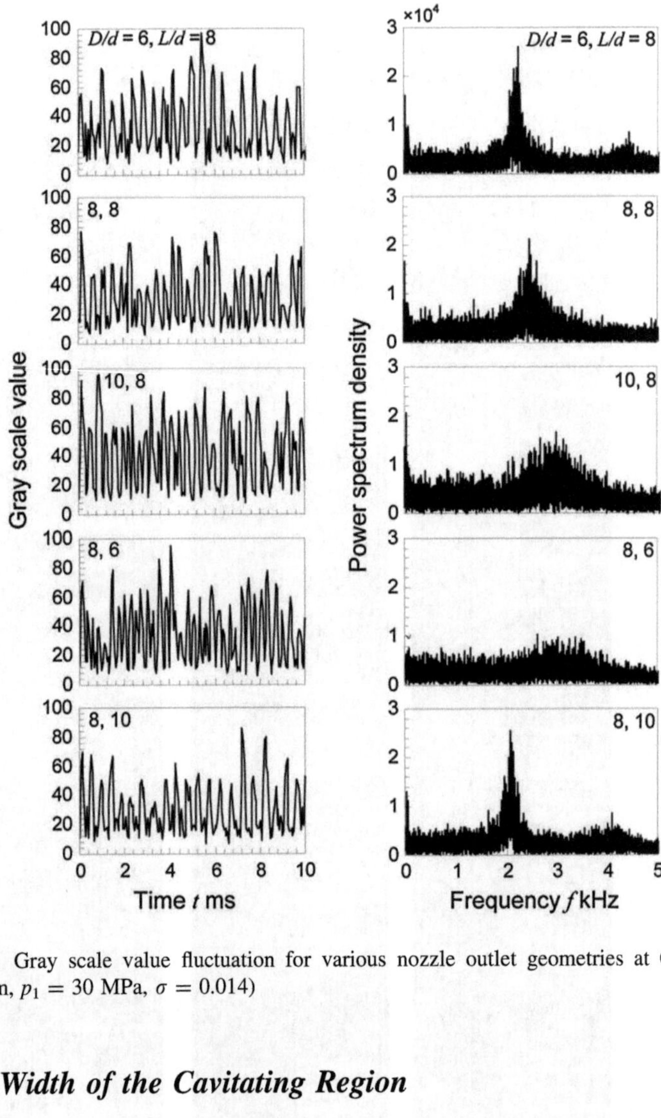

Fig. 12.13 Gray scale value fluctuation for various nozzle outlet geometries at Condition 1 ($d = 1$ mm, $p_1 = 30$ MPa, $\sigma = 0.014$)

12.3.6 Width of the Cavitating Region

Figure 12.15 shows the normalized width of the cavitating region, *w/d*, for various nozzle outlet geometries and cavitating conditions. The normalized width *w/d* slightly increased with an increase in normalized nozzle outlet bore, *D/d*. On the contrary, w/*d* slightly decreased with an increase in normalized nozzle outlet length, *L/d*. For the same normalized nozzle outlet geometry, the value of *w/d* was almost similar, among the three cavitating conditions. Thus, it can be concluded that *w/d* mainly depends on the cavitation number and the nozzle outlet geometry.

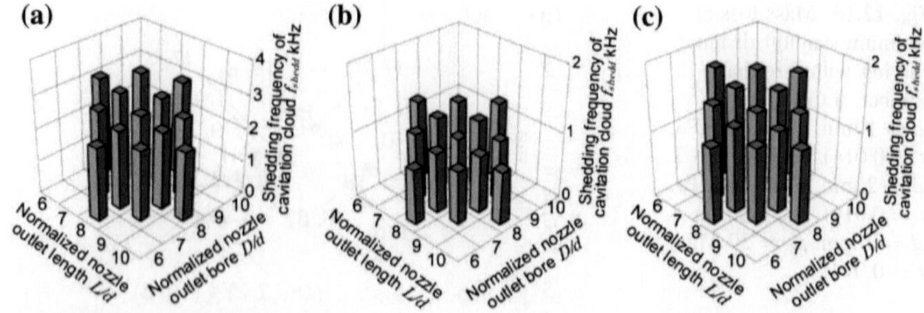

Fig. 12.14 Shedding frequency of cavitation cloud for various nozzle outlet geometries and cavitating conditions. **a** Condition 1 (d = 1 mm, p_1 = 30 MPa, σ = 0.014). **b** Condition 2 (d = 2 mm, p_1 = 15 MPa, σ = 0.014). **c** Condition 3 (d = 2 mm, p_1 = 30 MPa, σ = 0.014)

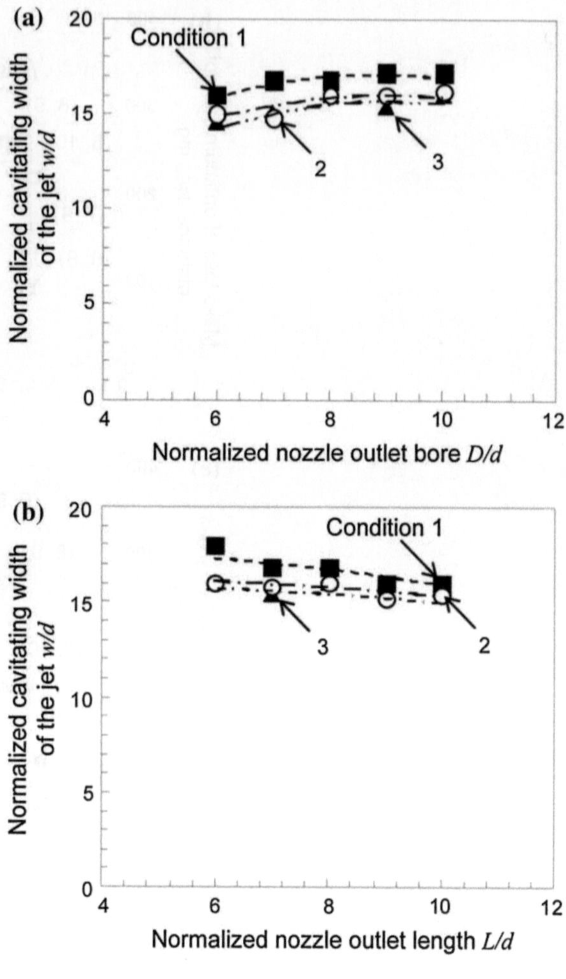

Fig. 12.15 Width of cavitating region for various nozzle outlet geometries and cavitating conditions. **a** Effect of nozzle outlet bore at $L/d = 8$. **b** Effect of nozzle outlet length at $D/d = 8$

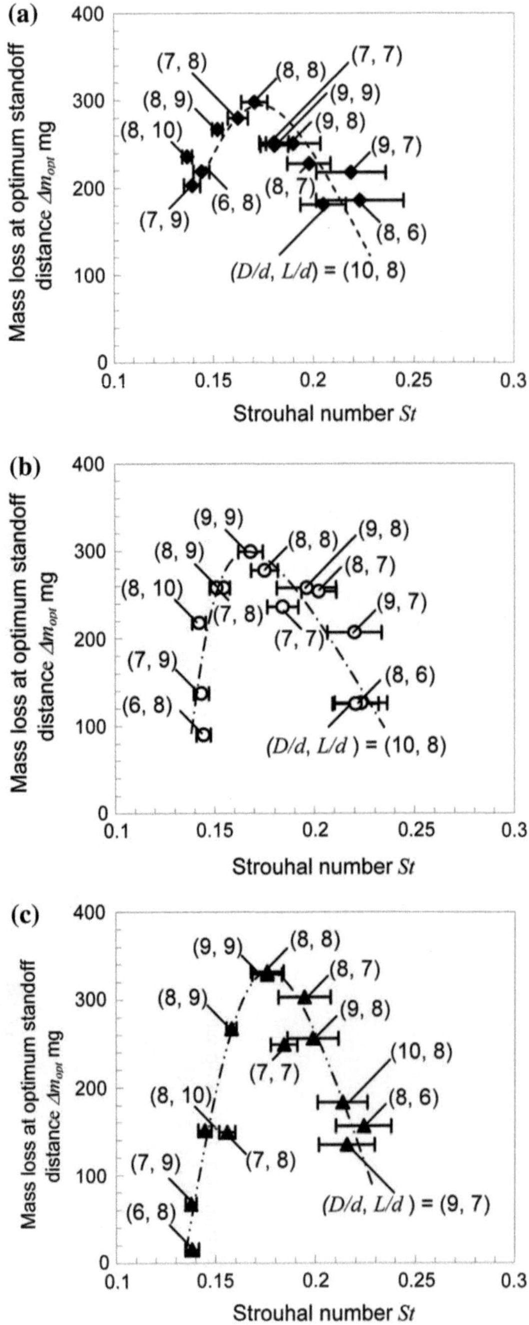

Fig. 12.16 Mass loss at optimum standoff distance varying with Strouhal number. **a** Condition 1 ($d = 1$ mm, $p_1 = 30$ MPa, $\sigma = 0.014$). **b** Condition 2 ($d = 2$ mm, $p_1 = 15$ MPa, $\sigma = 0.014$). **c** Condition 3 ($d = 2$ mm, $p_1 = 30$ MPa, $\sigma = 0.014$)

12.3.7 Optimum Strouhal Number

Figure 12.16 shows the mass loss at the optimum standoff distance, Δm_{opt}, as a function of Strouhal number, *St*. For all cavitating conditions, *St* varied with the nozzle outlet geometry, and the maximum values of Δm_{opt} occurred at a similar *St* value.

As expected, an optimum Strouhal number exists for the aggressivity of the cavitating jet. The optimum Strouhal number for all cavitating conditions was approximately 0.17, where Δm_{opt} is maximum.

The mechanism of the optimum nozzle outlet geometry can be explained as follows. The optimum nozzle outlet geometry affects the cavitating vortex merging, and makes the cavitation cloud shed at the optimum Strouhal number [21, 23, 39]. As mentioned in the previous section, the size of the cavitation cloud affects its single impact energy, and the shedding frequency of the cavitation cloud affects the impact number per unit time. At the optimum nozzle outlet geometry, and/or the optimum Strouhal number, the total impact energy per unit time is getting maximum, and therefore the mass loss is maximum. It is also noted that the optimum Strouhal number does not change with the nozzle throat diameter, *d*, and the injection pressure, p_1, as long as the cavitation number is kept constant.

12.4 Conclusions

The practical application of the cavitating jet for erosion tests and/or surface modification mandated further study on the aggressivity of the cavitating jet. The aggressivity of the cavitating jet was affected by the nozzle outlet geometry. There was an optimum nozzle outlet geometry and this was verified by the Strouhal number associated with the cavitation cloud shedding. This phenomenon was investigated via erosion tests, impact energy measurements, and image analysis of the cavitating jet at various nozzle throat diameters, *d*, outlet bores, *D*, and lengths, *L*. The obtained results were summarized as follows.

1. The aggressivity of the cavitating jet was affected by the nozzle outlet geometry, and an optimum nozzle outlet geometry was found.
2. The shedding frequency of the cavitation cloud and the width of the cavitating region were affected by the nozzle outlet geometry.
3. The frequency of the large impacts measured by the PVDF transducer corresponded to the shedding frequency of the cavitation cloud or its sub harmonics.
4. At the optimum nozzle outlet geometry, the cavitation cloud is shed at the optimum Strouhal number, $St = 0.17$, and this was considered as the reason for the high aggressivity.
5. Since the effect of the nozzle geometry is significant in determining the agressivity, erosion tests of materials should be conducted with standardized nozzle by ASTM G134.

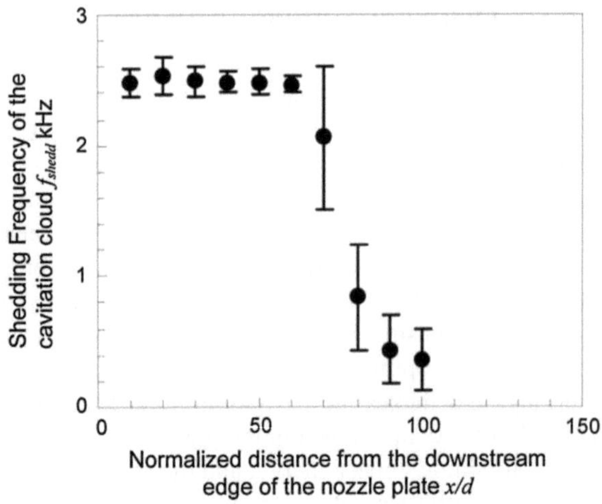

Fig. 12.A1 Influence of the detection point on the shedding frequency of the cavitating jet (Condition 1 $d = 1$ mm, $p_1 = 30$ MPa, $\sigma = 0.014$)

Acknowledgments This work was partially supported by the Cannon Foundation. The authors thank Mr. M. Mikami, technician, Tohoku University for his help in the experiment.

A.1 Appendix

Figure 12.A1 shows the influence of the detection point on the shedding frequency of the cavitating jet, at the optimum nozzle outlet geometry of Condition 1. The shedding frequency of the cavitating jet, f_{shedd}, is almost constant from $x/d = 10$ to $x/d = 60$, and then decreases with an increase in x/d. It can be assumed that the decrease of f_{shedd} is invoked by the collapse of the cavitation cloud, since the normalized optimum standoff distance, s_{opt}/d, is 50 (see Fig. 12.6a) at $D/d = 8$, $L/d = 8$, for which $x/d = 47$.

References

1. ASTM Designation G134-95 (2006) Standard test method for erosion of solid materials by a cavitating liquid jet. Annual book of ASTM standards 03.02, pp 559–571
2. Soyama H, Yamauchi Y, Ikohagi T, Oba R, Sato K, Shindo T, Oshima R (1996) Marked peening effects by high speed submerged-water-jets—residual stress change on SUS304. J Jet Flow Eng 13:25–32
3. Hirano K, Enomoto K, Hayashi E, Kurosawa K (1996) Effects of water jet peening on corrosion resistance and fatigue strength of type 304 stainless steel. J Soc Mater Sci Jpn 45:740–745
4. Soyama H (2000) Improvement in fatigue strength of silicon manganese steel SUP7 by using a cavitating jet. JSME Int J 43A:173–178

5. Soyama H, Park JD, Saka M (2000) Use of cavitating jet for introducing compressive residual stress. ASME J Manuf Sci Eng 122:83–89
6. Soyama H, Kusaka T, Saka M (2001) Peening by the use of cavitation impacts for the improvement of fatigue strength. J Mater Sci Lett 20:1263–1265
7. Soyama H, Saito K, Saka M (2002) Improvement of fatigue strength of aluminum alloy by cavitation shotless peening. ASME J Eng Mater Technol 124:135–139
8. Odhiambo D, Soyama H (2003) Cavitation shotless peening for improvement of fatigue strength of carbonized steel. Int J Fatigue 25:1217–1222
9. Soyama H, Macodiyo DO, Mall S (2004) Compressive residual stress into titanium alloy using cavitation shotless peening method. Tribol Lett 17:501–504
10. Soyama H (2004) Introduction of compressive residual stress using a cavitating jet in air. ASME J Eng Mater Technol 126:123–128
11. Qin M, Ju DY, Oba R (2006) Investigation of influence of incident angle on the process capability of water cavitation peening. Surf Coat Technol 201:1409–1413
12. Grinspan AS, Gnanamoorthy R (2007) Effect of nozzle traveling velocity on oil cavitation jet peening of aluminum alloy, AA6063-T6. ASME J Eng Mater Technol 129:609–614
13. Soyama H, Shimizu M, Hattori Y, Nagasawa Y (2008) Improving the fatigue strength of the elements of a steel belt for CVT by cavitation shotless peening. J Mater Sci 43:5028–5030
14. Soyama H, Sekine Y (2010) Sustainable surface modification using cavitation impact for enhancing fatigue strength demonstrated by a power circulating-type gear tester. Int J Sustain Eng 3:25–32
15. Takakuwa O, Ohmi T, Nishikawa M, Yokobori AT Jr, Soyama H (2011) Suppression of fatigue crack propagation with hydrogen embrittlement in stainless steel by cavitation peening. Strength Fract Complex 7:79–85
16. Saitou N, Enomoto K, Kurosawa K, Morinaka R, Ishikawa T, Yoshimura T (2003) Development of water jet peening technique for reactor internal components of nuclear power plant. J Jet Flow Eng 20:4–12
17. Sato K, Mizoguchi T, Fujita K, Ikohagi T (1999) Development of advanced aqua-environmental protection technology using high-speed cavitating water jets. In: Kobayashi R (ed) Proceedings of the international symposium new applications water jet technology, Ishinomaki Japan, pp 95–104
18. Kalumuck KM, Chahine GL (2000) The use of cavitating jets to oxidize organic compounds in water. ASME J Fluids Eng 122:465–470
19. Conn AF, Radtke RP (1978) CAVIJET™ augmented deep-hole drilling bits. ASME J Press Vessel Technol 100:52–59
20. Johnson Jr VE, Lindenmuth WT, Conn AF, Frederick GS (1981) Feasibility study of tuned-resonator, pulsating cavitating water jet for deep-hole drilling. Sandia National Laboratories contractor report SAND81-7126, pp 1–131
21. Johnson VE Jr, Chahine GL, Lindenmuth WT, Conn AF, Frederick GS, Giacchino GJ Jr (1984) Cavitating and structured jets for mechanical bits to increase drilling rate—Part I: theory and concepts. ASME J Energy Resour Technol 106:282–288
22. Yanaida K, Nakayama M, Eda K, Nishida N (1985) Water jet cavitation performance of submerged horn shaped nozzle. In: Proceedings 3rd U.S. water jet conference, Pittsburgh, pp 336–349
23. Chahine GL, Johnson Jr VE, Kalumuck KM, Perdue TO, Waxman DN, Frederick GS, Watson RE (1987) Internal and external acoustics and large structure dynamics of cavitating self-resonating water jets. Sandia National Laboratories contractor report SAND86-7176, pp 1–202
24. Vijay MM, Zou C, Tavoularis S (1991) A study of the characteristics of cavitating water jets by photography and erosion. In: Proceedings 10th international conference jet cutting technology, pp 37–67
25. Surjaatmadja JB, Howlett JJ Jr (1992) Surge enhanced cavitating jet. US Patent 5,125,582

26. Yamauchi Y, Soyama H, Adachi Y, Sato K, Shindo T, Oba R, Oshima R, Yamabe M (1995) Suitable region of high-speed submerged water jet for cutting and peening. JSME Int J 38B:31–38
27. Soyama H, Nishizawa K (2009) Enhancement of the impact force around a cavitating jet by changing the nozzle outlet geometry. Japanese J Multiph Flow, Prog Multiph Flow Res 4:77–85
28. Soyama H, Takakuwa O (2011) Enhancing the aggressive strength of a cavitating jet and its practical application. J Fluid Sci Technol 6:510–521
29. Soyama H (2011) Enhancing the aggressive intensity of a cavitating jet by means of the nozzle outlet geometry. ASME J Fluids Eng 133:101301 pp 1–11
30. Soyama H, Yamauchi Y, Adachi Y, Sato K, Shindo T, Oba R (1995) High-speed observations of the cavitation cloud around a high-speed submerged water jet. JSME Int J 38B:245–251
31. Soyama H, Sekine Y, Saito K (2011) Evaluation of the enhanced cavitation impact energy using a PVDF transducer with an acrylic resin backing. Measurement 44:1279–1283
32. Soyama H (1998) Material testing and surface modification by using cavitating jet. J Soc Mater Sci Jpn 47:381–387
33. Soyama H, Lichtarowicz A, Momma T, Williams EJ (1998) A new calibration method for dynamically loaded transducers and its application to cavitation impact measurement. ASME J Fluids Eng 120:712–718
34. Soyama H, Kumano H (2003) The fundamental threshold level—a new parameter for predicting cavitation erosion resistance. J Test Eval 30:423–431
35. Soyama H, Lichtarowicz A (1996) Cavitating jets—similarity correlations. J Jet Flow Eng 13:9–19
36. Soyama H, Kikuchi T, Nishikawa M, Takakuwa O (2011) Introduction of compressive residual stress into stainless steel by employing a cavitating jet in air. Surf Coat Technol 205:3167–3174
37. Soyama H, Futakawa M, Homma K (2005) Estimation of pitting damage induced by cavitation impacts. J Nucl Mat 343:116–122
38. Soyama H, Futakawa H (2006) Estimation of cavitation intensity from the time taken for bubbles to develop. Tribol Lett 23:23–26
39. Crow SC, Champagne FC (1971) Orderly structure in jet turbulence. J Fluid Mech 48:547–591

Chapter 13
Prediction of Cavitation Erosion and Residual Stress of Material Using Cavitating Flow Simulation with Bubble Flow Model

Masashi Fukaya

Abstract We have developed a numerical simulation code that can predict the cavitation erosion and the residual stress of a material, which are both closely related to plastic deformation. The shock wave generated at the cavitation bubble collapse hits the material surface and the impact energy causes plastic deformation which changes the residual stress. When the impact energy is high, mass loss (i.e., erosion) occurs after the plastic deformation stage. We numerically simulated impulsive bubble pressures that varied on the order of a microsecond in the cavitating flow with the 'bubble flow model'. The bubble flow model simulates the abrupt time-variations in the radius and inner pressure of bubbles based on the Rayleigh-Plesset equation. Although the shock wave propagation was not simulated, the impact energy was estimated based on the bubble pressure. The predicted impact energy was compared with the distribution of plastic deformation pits, which were observed on the impeller blade surface of a centrifugal pump. The predicted impact energy was also compared with the distribution of residual stress measured on a stainless steel plate after a cavitating jet impinged on the plate. The distribution of the impact energy corresponded qualitatively to that of the residual stress improvement caused by the plastic deformation. High correlation between the predicted impact energy and the plastic deformation of material was confirmed, and we found that our numerical method is relevant for the prediction of cavitation erosion and residual stress on a material surface.

M. Fukaya (✉)
Hitachi, Ltd., Hitachi Research Laboratory, Mechanical Engineering Research Center, 832-2, Horiguchi, Hitachinaka, Ibaraki 312-0034, Japan
e-mail: masashi.fukaya.kp@hitachi.com

K.-H. Kim et al. (eds.), *Advanced Experimental and Numerical Techniques for Cavitation Erosion Prediction*, Fluid Mechanics and Its Applications 106,
DOI: 10.1007/978-94-017-8539-6_13, © Springer Science+Business Media Dordrecht 2014

13.1 Introduction

Cavitation has both advantages and disadvantages for industrial products. On one side, cavitation causes performance reduction, vibration, noise, and erosion in turbomachinery such as pump, water turbine, and ship propeller. On the other side, cavitation is applied to apparatus for sterilization, purification, material cutting, and material modification as work hardening and residual stress improvement. Water Jet Peening (WJP) is a preventive maintenance technology for nuclear power plants [1, 2]. A cavitating jet is injected and impinges on the weld surfaces of structures in a nuclear reactor. The shock wave generated at the cavitation bubble collapse hits the material surface and the impact energy causes plastic deformation of the weld surface, and changes the residual stress from tensile to compressive. Compressive residual stress prevents the occurrence of stress corrosion cracking (SCC) on the weld surface.

CFD (Computational Fluid Dynamics) is commonly used to predict the influence of cavitation on industrial products. However, the prediction of cavitation erosion by CFD has only recently been attempted. Dular et al. pointed out the high correlation between the void fraction variation and pits on a thin hydrofoil [3]. The erosion area was predicted based on the numerically predicted void fraction [4]. Nohmi et al. proposed indexes for predicting cavitation erosion area. The indexes consist of void fraction, void fraction variation, pressure, and pressure variation [5, 6]. Miyabe and Maeda predicted the cavitation erosion area by using CFD and the indexes proposed by Nohmi et al. in a double-suction centrifugal feedpump [7]. Ochiai et al. numerically simulated the cavitating flow around a hydrofoil with a locally homogeneous model of a gas–liquid two phase medium [8]. The bubble dynamics was solved along the bubble moving path in the simulated cavitating flow field, and the bubble collapse pressure, frequency and energy were estimated for predicting the erosion area. Zima et al. numerically simulated the cavitating flow in a mixed-flow pump, and solved the bubble dynamics along the bubble moving path [9]. The predicted area of high cavitation intensity corresponded with the actual erosion area.

While the above mentioned studies focus on the prediction of cavitation erosion, the prediction of residual stress improvement has hardly been reported. We developed a simulation code with the bubble flow model to solve the bubble dynamics in the whole cavitating flow field. The volumetric and transitional bubble motion were modeled, and the distribution of bubble nuclei was also taken into account [10, 11]. The impact energy at the bubble collapse was estimated based on the bubble pressure and the bubble number density. In this paper, we report high correlation between the impact energy estimated by our simulation and the plastic deformation of material, which was caused by the bubble collapse. The simulation code was applied to predict the erosion [11] and the residual stress improvement [12], which are both closely related to plastic deformation of material.

13.2 Numerical Method

13.2.1 Governing Equations

We adopted the bubble flow model [10] in the simulation code to predict the cavitating flow. The following assumptions concerning bubbles were made in the model.

- The liquid phase is water, which is incompressible.
- The gas phase consists of spherical bubbles. The bubbles are filled with vapor and non-condensable gas. The evaporation and condensation on the bubble surface are modeled assuming the non-condensable gas follows an isothermal transformation during expansion and an adiabatic one during contraction [13]. Thus the gas phase is compressible.
- No collision, coalescence or breakup of bubbles occurs.
- The density and momentum of the gas phase are sufficiently small to be negligible.

The governing equations are described in generalized coordinates as follows [10, 11]. The momentum conservation equation of bubble flow is:

$$\frac{\partial(\rho_L f_L u_{Li})}{\partial t} + J\frac{\partial(\rho_L f_L u_{Li} U_{Li}/J)}{\partial \xi_j}$$
$$= -\nabla_i p_L + \nabla_j(\mu \nabla_j u_{Li}) + \frac{1}{3}(\mu \nabla_k u_{Lk}) - \rho_L\{2\varepsilon_{ijk}\Omega_j u_{Lk} + \varepsilon_{ijk}\Omega_j(\varepsilon_{klm}\Omega_l r_m)\}, \tag{13.1}$$

where ρ_L (= 1,000 kg/m^3) and f_L are the density and the volume fraction of the water. u_L is the water velocity and U_L is the contravariant velocity of the water. p_L and μ (= 1.0×10^{-3} Pa s) are the static pressure and the viscosity of the water. J is the Jacobian and ε is the Eddington's epsilon.

In Eq. (13.1), the water velocity represents the relative velocity in a rotating coordinate system defined by the rotation speed vector Ω. In the case $\Omega = 0$, this represents the absolute velocity in a static coordinate system. The fourth term on the right hand side includes the Coriolis force and the centrifugal force. Here, r is the distance from the rotating axis.

The conservation of the bubble number density, n_G, is:

$$\frac{\partial n_G}{\partial t} + J\frac{\partial(n_G U_{Gi}/J)}{\partial \xi_j} = 0. \tag{13.2}$$

The pressure equation is:

$$\frac{1}{c^2}\frac{\partial p_L}{\partial t} + \nabla_j f_L u_{Lj} + \nabla_j f_G u_{Gj} - 4\pi r_G^2 n_G \frac{D r_G}{Dt} = 0, \tag{13.3}$$

where c is the coefficient for pseudo-compressibility, f_G is the void fraction, u_G is the bubble velocity, r_G is the bubble radius, and D/Dt is the substantial derivative.

Equation (13.3) is based on a pseudo-compressibility method, which is derived from the conservation of volumetric fractions ($f_L + f_G = 1$) and the conservation of bubble number density [10]. The water pressure is influenced by the bubble behavior through the fourth term on the left hand side.

The volumetric motion of an isolated bubble is described by the Rayleigh-Plesset equation [14],

$$r_G \frac{D^2 r_G}{Dt^2} + \frac{3}{2}\left(\frac{Dr_G}{Dt}\right)^2 = \frac{p_B - p_L}{\rho_L} + \frac{1}{4}(u_{Li} - u_{Gi})^2, \tag{13.4}$$

where p_B is the bubble pressure. Here, i means the x, y and z directions. The second term on the right hand side accounts for the effect of the reduction of the surrounding pressure of the bubble due to the velocity difference between the bubble and its surrounding water. The bubble pressure is:

$$p_B = p_G + p_v - \frac{2T}{r_G} - 4\mu \frac{1}{r_G}\frac{Dr_G}{Dt}, \tag{13.5}$$

with:

$$p_G r_G^3 = const.\ \text{if}\ \frac{Dr_G}{Dt} > 0, \tag{13.6}$$

or:

$$p_G r_G^{3\kappa} = const.\ \text{if}\ \frac{Dr_G}{Dt} < 0, \tag{13.7}$$

where p_G is the non-condensable gas pressure. Both the surface tension, T (= 0.072 N/m) and the vapor pressure, p_v (= 2,300 Pa) are constant. The viscosity, μ is assumed to be the same as that of water ($\mu = 1.0 \times 10^{-3}$ Pa s). The specific heat ratio, κ, is 1.4. Eqs. (13.6) and (13.7) are simplified equations deduced from a more precise simulation [13]. In Ref. [13], the motion of one point-symmetric bubble in incompressible liquid was simulated including the heat exchange and the phase change at the interface, the temperature distribution, the diffusion between vapor and non-condensable gas, the homogeneous nucleation of mist and its growth within the bubble. It was concluded that the cavitation bubble behaves isothermally when expanding and adiabatically when shrinking.

The validation of the present simplified model is shown in Fig. 13.1. The line labeled "DNS" was the result of the precise simulation [13]. Other lines in Fig. 13.1 were the results of the Rayleigh-Plesset equation with different models of non-condensable gas. The line labeled "Switched" corresponded to the present model and follows the DNS line well.

The translational motion of a bubble is solved by taking into consideration the force balance of the bubble,

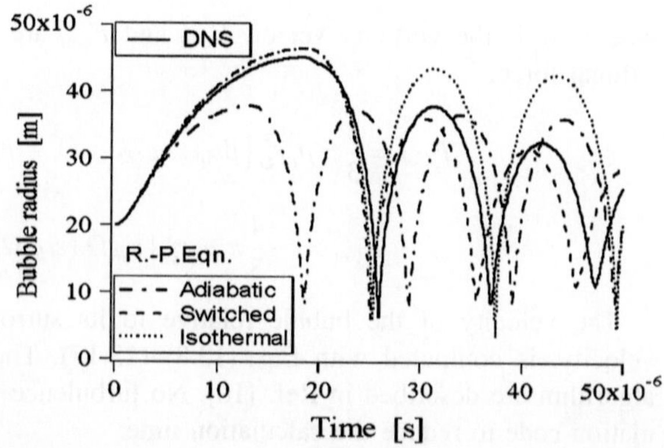

Fig. 13.1 Bubble dynamics models

$$F_{Ai} + F_{Pi} + F_{Di} + F_{Li} + F_{Coi} + F_{Cei} = 0, \tag{13.8}$$

where F_{Ai} is the added mass force,

$$\begin{aligned} F_{Ai} = \frac{4}{3}\pi\beta\left\{\left(\frac{\partial\left(\rho_L\, r_G^3\, u_{Gi}\right)}{\partial t} + U_{Gj}\frac{\partial\left(\rho_L\, r_G^3\, u_{Gi}\right)}{\partial \xi_j}\right)\right\} \\ - \left\{\left(\frac{\partial\left(\rho_L\, r_G^3\, u_{Li}\right)}{\partial t} + U_{Lj}\frac{\partial\left(\rho_L\, r_G^3\, u_{Li}\right)}{\partial \xi_j}\right)\right\}, \end{aligned} \tag{13.9}$$

Here, β is a constant of 0.5 for a spherical bubble, U_G is the contravariant velocity of the bubble. F_{Pi} is the force of the acceleration of the surrounding fluid,

$$F_{Pi} = -\frac{4}{3}\pi\,\rho_L r_G^3\left(\frac{\partial\, u_{Li}}{\partial t} + U_{Lj}\frac{\partial\, u_{Li}}{\partial \xi_j}\right). \tag{13.10}$$

F_{Di} and F_{Li} are the drag and lift forces [15, 16],

$$F_{Di} = \frac{1}{2}\pi\,\rho_L\, r_G^2\, C_D |u_G - u_L|(u_{Gi} - u_{Li}), \tag{13.11}$$

$$F_{Li} = \frac{1}{2}\pi\,\rho_L r_G^2\, C_L\,|u_G - u_L|\varepsilon_{ijk}\,\omega_{Lk}(u_{Gj} - u_{Lj})/|\omega_L|, \tag{13.12}$$

$$C_D = \frac{24}{\mathrm{Re}_{bub}}(1 + 0.15\,\mathrm{Re}_{bub}^{0.687}), \tag{13.13}$$

$$\mathrm{Re}_{bub} = \frac{2\, r_G\, \rho_L\, |u_G - u_L|}{\mu_L}, \tag{13.14}$$

$$C_L = 0.59\left(\frac{|\omega_L| r_G}{|u_G - u_L|}\right)^{0.25}, \tag{13.15}$$

where ω_L is the vorticity vector. F_{Coi} and F_{Cei} are the Coriolis force and centrifugal force,

$$F_{Coi} = \frac{8}{3}\pi\, \rho_L r_G^3 \left\{ \beta \varepsilon_{ijk} \Omega_j u_{Gk} - (1+\beta)\varepsilon_{ijk}\Omega_j u_{Lk} \right\}, \tag{13.16}$$

$$F_{Cei} = -\frac{4}{3}\pi\, \rho_L r_G^3 \left\{ \varepsilon_{ijk} \Omega_j (\varepsilon_{klm} \Omega_l r_m) \right\}. \tag{13.17}$$

The velocity of the bubble relative to its surrounding water, i.e., the slip velocity, is computed with Eqs. (13.8)–(13.17). The details on the calculation algorithm are described in Ref. [10]. No turbulence model was used in the simulation code to reduce the calculation time.

The void fraction in this simulation is calculated from the bubble radius and the bubble number density by:

$$f_G = \frac{4}{3}\pi r_G^3 n_G. \tag{13.18}$$

Equation (13.18) means that the void fraction increases when the bubble expands due to the pressure difference at the liquid–gas interface or where the bubble nuclei have accumulated.

13.2.2 Cavitation Impact Energy

In the actual bubble collapse phenomenon, a bubble breaks up into minute bubbles after an abrupt shrinking of the bubble, and high energy is released from the bubble in the form of high pressure. The energy released at the bubble collapse was evaluated in the accelerated cavitation erosion tests with a shock sensor as:

$$E = k \sum_{i=1,2,3\ldots} p_{ci}^2, \tag{13.19}$$

where p_c is the measured bubble collapse pressure and k is a proportionality constant [17, 18]. Although the collapse pressure was not calculated in the present simulation, the abrupt shrinking of the bubble and the impulsive bubble pressure were calculated. We assumed that the predicted maximum bubble pressure was qualitatively related to the bubble collapse pressure. We therefore numerically investigated the cavitation impact energy by defining:

$$E = k \sum_{i=1,2,3\ldots} p_{Bi}^2 (n_G \Delta V). \tag{13.20}$$

The bubble collapse pressure in Eq. (13.19) is replaced with the bubble pressure, p_B, in Eq. (13.20). The $n_G \Delta V$ represents the number of bubbles in each numerical cell.

13.2.3 Simulated Objects and Calculation Conditions

There are two simulated objects in this paper. We predicted the cavitation erosion area in a centrifugal pump and the residual stress improvement by a cavitating jet.

Figure 13.2 shows the numerical mesh and boundary conditions in the centrifugal pump. The impeller had six blades and a shroud, which was made of aluminum for the accelerated cavitation erosion test. The maximum diameter of the impeller was 302 mm, and the hub diameter was 78 mm. The region between the pressure and suction sides of the impeller blades was investigated using periodical boundaries.

A cylindrical suction channel, which was 180 mm long axially, was connected to the region including the impeller blades. The flow rate of the water, Q, was 4.56 m^3/min, and $Q/Q_{\eta max}$ was 0.60 where $Q_{\eta max}$ is the flow rate at the highest efficiency. The liquid velocity at the inlet boundary was uniform and equal to 2.12 m/s. This pump had a volute-type casing positioned downstream of the impeller. However, a fan-shaped discharge channel, which was 50 mm long radially, was connected downstream of the impeller blades region instead of the volute-type casing. The static pressure at the outlet boundary was varied to change the *NPSH* conditions.

The initial void fraction was 0.001. The initial bubble radius was 1.0×10^{-5} m. The initial bubble number density and the bubble number density at the inlet boundary were 2.39×10^{11} m^{-3} based on Eq. (13.18).

No-slip conditions were assumed for the solid surface of the blade, hub, and shroud of the impeller. No-slip conditions were also assumed for the wall surface of the suction and discharge channels.

Figure 13.3 shows the numerical mesh and boundary conditions of the cavitating jet. The simulated region was limited between the nozzle and the flat plate, where the submerged cavitating jet from the nozzle impinged vertically on the plate. We assumed an axisymmetric flow field and conducted calculations within a range of 2 degrees in the θ direction using periodic boundary conditions since the nozzle and the flat plate shapes were axisymmetric. The nozzle had a cylindrical throat and a horn-shaped flow passage downstream of the throat. The nozzle throat diameter was 2 mm. The length and the expansion angle of the horn-shaped flow passage were 11 mm and 30 degrees. The cartesian mesh was prepared in the r–z plane except for the horn-shaped flow passage.

The pressure boundary conditions are plotted on the dotted line in Fig. 13.3, and a constant pressure of 3.0×10^5 Pa or 1.0×10^5 Pa was assumed. When WJP is conducted in a nuclear reactor, the water depth above the weld and the nozzle is over 20 m in some cases. Then, the boundary pressure was 3.0×10^5 Pa. On the other hand, the pressure boundary condition was fixed at 1.0×10^5 Pa since the water depth at the plate was below 1 m in the laboratory experiment.

There are uniform velocity conditions at the nozzle inlet. The inlet velocity and the distance between the nozzle edge and the flat plate, i.e., the standoff distance, were changed as listed in Table 13.1. The inlet velocity and the standoff distance

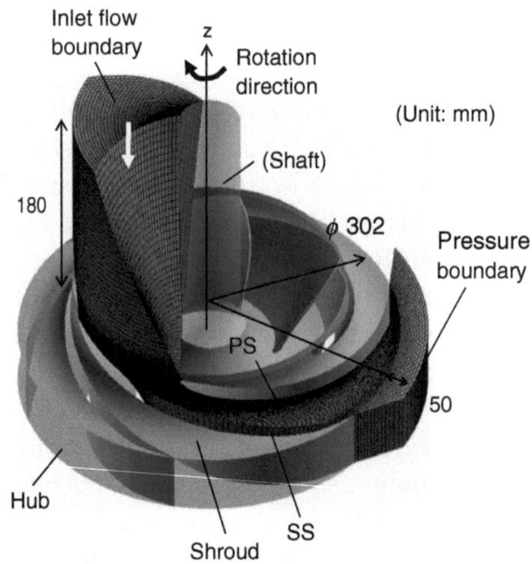

Fig. 13.2 Simulated region and boundary conditions for pump. Based on [11], reprinted with permission from JSME

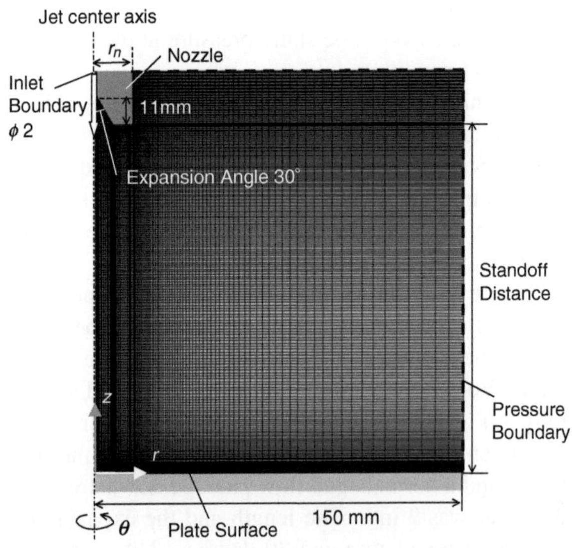

Fig. 13.3 Simulated region and boundary conditions for cavitating jet. Based on [12], reprinted with permission from ASME

were different in Cases 2–5 while the pressure boundary condition was fixed at 1.0×10^5 Pa. There are no-slip velocity conditions on the solid surfaces of the nozzle and the flat plate.

The initial void fraction was 0.001. The initial condition of the bubble radius was assumed to be 1.0×10^{-5} m under the pressure of 1.0×10^5 Pa, and the initial bubble radius of 7.0×10^{-6} m was estimated under the pressure of 3.0×10^5 Pa. The boundary and initial conditions are summarized in Table 13.1.

We carried out the simulation in two stages to reduce the calculation time. First, the flow field was computed using local time stepping. At this stage the bubbles

Table 13.1 Boundary and initial conditions

Case	Inlet velocity (m/s)	Stand-off distance (mm)	Boundary pressure (Pa)	Initial bubble radius (m)
1	254.6	140	3.0×10^5	7.0×10^{-6}
2	254.6	140	1.0×10^5	1.0×10^{-5}
3	254.6	100	1.0×10^5	1.0×10^{-5}
4	197.3	100	1.0×10^5	1.0×10^{-5}
5	98.7	100	1.0×10^5	1.0×10^{-5}

Based on [12], reprinted with permission from ASME

were assumed in quasi-static equilibrium at the local water pressure to avoid instability in the calculation. Next, we started to calculate the bubble behavior in detail, without assuming a quasi-equilibrium condition. The time step in the second stage of the simulation was 0.02 microseconds in the pump simulation, and 0.08 microseconds in the cavitating jet simulation.

13.3 Results and Discussion

13.3.1 Centrifugal Pump

13.3.1.1 Cavitation Performance

Figure 13.4 compares the calculated and experimental cavitation performance of the centrifugal pump under the flow rate condition $Q/Q_{\eta max} = 0.60$. The *NPSH* (Net Positive Suction Head) and total head were divided by $U_t^2/(2g)$ to make them dimensionless: *NPSH*' and ψ, respectively. The *NPSH* was obtained from the total pressure at the outlet of the suction pipe. The total head was equivalent to the total pressure increase between the outlet of the suction pipe and the outlet of the impeller.

The experimental and predicted total heads remained nearly constant when *NPSH*' was high, and both decreased when *NPSH*' was below a certain value. $NPSH_R$ is the 'Required *NPSH*' at the 3 % drop of total head. While the experimental $NPSH_R$' was 0.094, the predicted $NPSH_R$' was 0.068 having a prediction error of –27 %. The underestimation of $NPSH_R$ was due to an underestimation of the cavitation volume since, for high vapor volume fractions, bubble coalescence and sheet cavitation were not sufficiently modeled in the simulation.

The total head in the experiment included the effect of loss caused by flow in the volute-type casing; however, the flow in the volute-type casing was not calculated in the present simulation. Therefore, the predicted total head exceeded the experimental one, and the predicted total head had an error of +28 %. The predicted cavitation performance was not in quantitative agreement with that in the experiments;

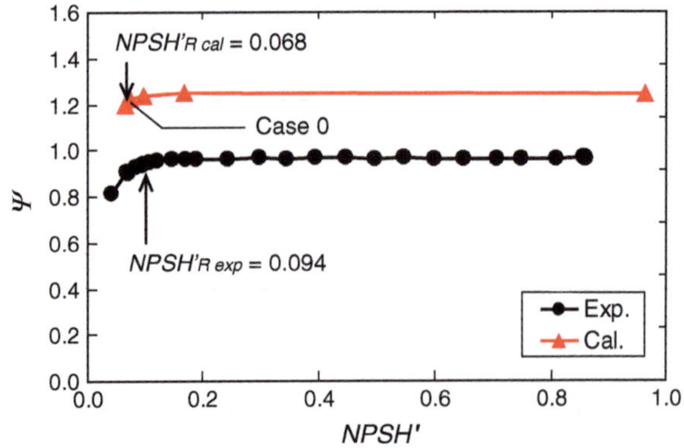

Fig. 13.4 Cavitation performance of centrifugal pump ($Q/Q_{\eta max} = 0.60$). Based on [11], reprinted with permission from JSME

however, the pressure and velocity variations caused by cavitation in the impeller were qualitatively simulated from a macroscopic viewpoint.

The discussion in the following sections is based on the numerical results from Case 0 ($NPSH_R' = 0.073$), shown in Fig. 13.4.

13.3.1.2 Static Pressure, Void Fraction and Bubble Number Density

Figure 13.5 shows the predicted instantaneous distributions of the static pressure of water, the void fraction, and the bubble number density close to the impeller blade. Figure 13.5a shows the static pressure distribution. The average pressure decreased inside the region surrounded by the dotted line on the suction side of the impeller blade near the leading edge (LE). High pressure regions locally occurred as spots in the low-pressure region since the static pressure was affected by the bubble behavior, i.e., the bubble radius and the bubble number density through the fourth term of Eq. (13.3).

Figure 13.5b shows the distribution of bubble number density. The bubble nuclei are accumulated in the low-pressure region in Fig. 13.5a. Flow separation near the blade surface occurred in the low-pressure region, and the bubble nuclei were trapped there. Bubble residence also appeared near the hub in the low energy flow.

Figure 13.5c shows the distribution of the void fraction. The high void fraction means the occurrence of cavitation. The void fraction depended on the bubble radius and the bubble number density based on Eq. (13.18). The void fraction increased in the same region as the low-pressure region in Fig. 13.5a since bubble expansion was caused by the decrease in the static pressure. The void fraction also increased in the region of bubble nuclei accumulation shown in Fig. 13.5b.

Figure 13.6 compares the distributions of bubble number density and void fraction close to the impeller blade at two different times. Both of these were obtained along the line between the leading edge (LE) and the trailing edge (TE) of the impeller blade shown in Fig. 13.5c. The line was located between the shroud and the mid-span of the blade. The distance, s, from the LE was

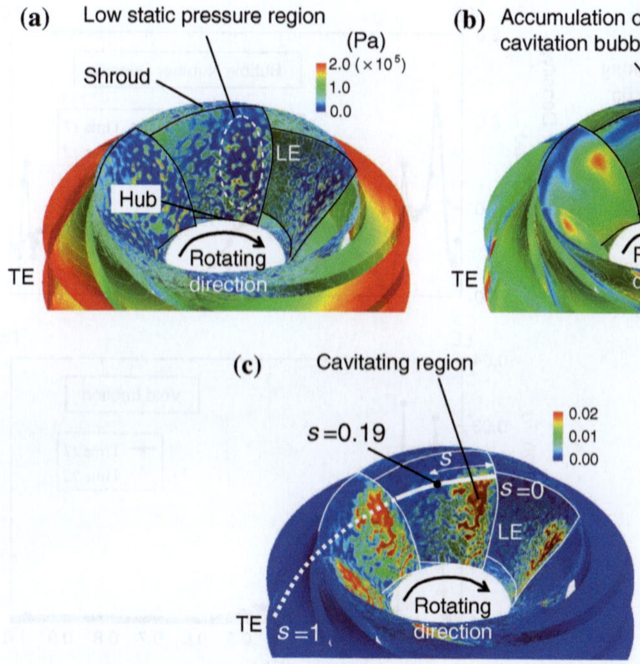

Fig. 13.5 Distributions of static pressure of water, bubble number density and void fraction close to impeller blade, **a.** static pressure, **b.** bubble number density, **c.** void fraction. Based on [11], reprinted with permission from JSME

nondimensionalized by the total line length between the leading and trailing edges ($s = 0.0$ at LE and $s = 1.0$ at TE). The bubble number density was obtained as the ratio against the density at the inlet boundary, n_{G0}.

The bubble number density distributions at the different times were nearly same. This result means that the unsteadiness of the bubble number density distribution was not strong. The bubble nuclei accumulated in a range from $s = 0.05$ to 0.15 at times t_1 and t_2. The void fraction also increased in a range from $s = 0.05$ to 0.15; however, the distributions were different at the two times. This was because the bubble radius varied as will be explained by Fig. 13.7 in the next section. The high dependence of the void fraction on bubble accumulation and bubble expansion and contraction was numerically confirmed.

13.3.1.3 Bubble Radius and Bubble Pressure

Figure 13.7 compares the evolutions of the bubble radius and pressure close to the impeller blade at the same times t_1 and t_2 as in Fig. 13.6. The bubble radius and pressure were calculated in each cell along the line shown in Fig. 13.5c. The number of cells between the leading edge (LE) and the trailing edge (TE) of the

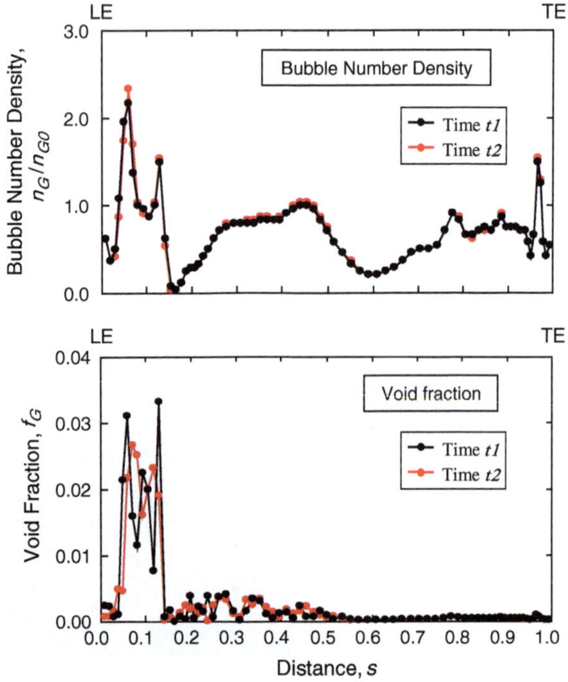

Fig. 13.6 Bubble number density and void fraction close to impeller blade along line in Fig. 13.5c. Based on [11], reprinted with permission from JSME

impeller blade was 76. The bubble radius was obtained as the ratio against the initial value at the inlet boundary, r_{G0}.

While the bubble radius increased on the whole from the leading edge ($s = 0$) to location $s = 0.1$ due to the local reduction of water pressure, it decreased beyond $s = 0.1$ since pressure was increased by impeller rotation. Figure 13.7 shows the irregular variation of bubble radius between the leading edge and location $s = 0.6$. This behavior caused the non-uniform distribution in the void fraction visible in Fig. 13.5c. At several locations between $s = 0.1$ and $s = 0.4$, the bubble radius locally decreased, and the bubble pressure distribution exhibited simultaneously peaks of large amplitude due to the abrupt contraction of the bubbles. Figure 13.7 shows that location and height of the bubble pressure peaks changed unsteadily. The transient behavior of the bubble radius caused the different distributions of void fraction visible in Fig. 13.6.

13.3.1.4 Transient Bubble Pressure

Figure 13.8 shows the instantaneous bubble pressure close to the impeller blade. Scattered local high-pressure regions appeared especially between LE and the throat, which is the narrowest location between the impeller blades. These high-pressure spots show the locations of the bubble collapse. The location of the high-pressure regions unsteadily changed. Figure 13.9 plots the transient bubble

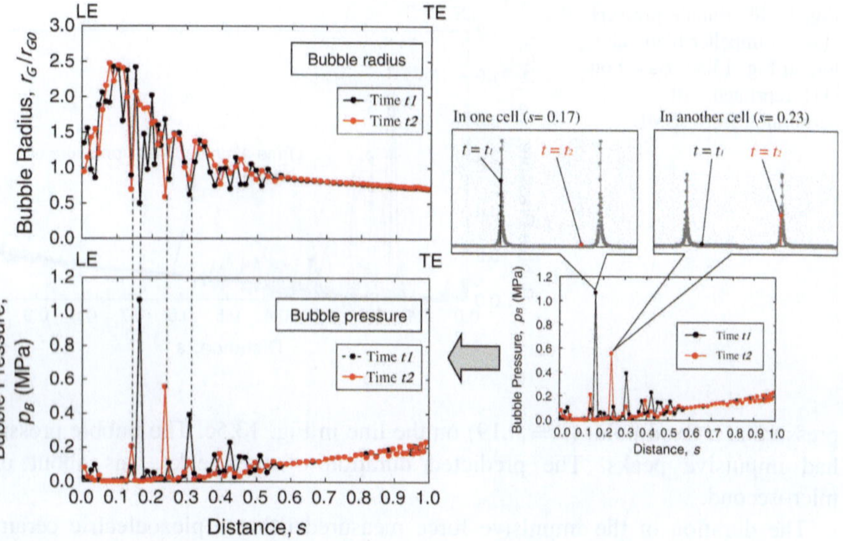

Fig. 13.7 Bubble radius and bubble pressure close to impeller blade along line in Fig. 13.5c. Based on [11], reprinted with permission from JSME

Fig. 13.8 Bubble pressure close to impeller blade. Based on [11], reprinted with permission from JSME

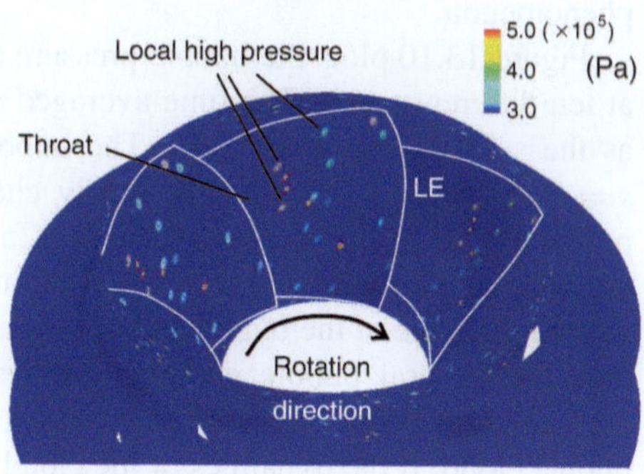

Fig. 13.9 Transient bubble pressure at a fixed point ($s = 0.19$) along line in Fig. 13.5c. Based on [11], reprinted with permission from JSME

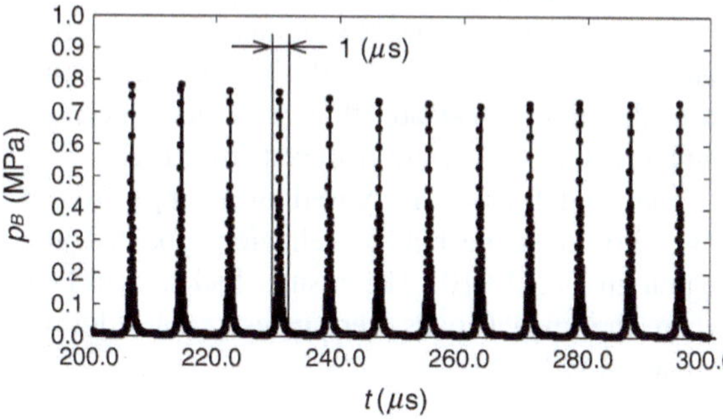

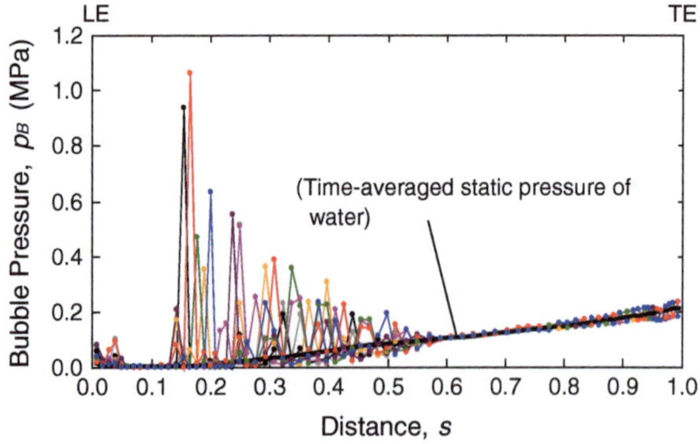

Fig. 13.10 Bubble pressure close to impeller blade along line in Fig. 13.5c. Based on [11], reprinted with permission from JSME

pressure at a fixed point ($s = 0.19$) on the line in Fig. 13.5c. The bubble pressure had impulsive peaks. The predicted duration of the peaks was about one microsecond.

The duration of the impulsive force measured using a piezoelectric ceramic sensor attached to the impeller blade surface was actually below one microsecond [19]. In the present study, the breakup of bubbles and the final stage of the bubble collapse were not simulated. However, bubble motion was simulated, which changed on the order of microseconds, in the same manner as the actual phenomenon.

Figure 13.10 plots the bubble pressure distributions along the line in Fig. 13.5c at ten different times. The time-averaged static pressure of water was also shown as the solid line in Fig. 13.10. The bubble pressure had peaks mainly between $s = 0.15$ and $s = 0.5$ while it hardly changed and was equivalent to the static pressure in ranges from $s = 0.05$ to 0.13, and $s > 0.6$. The bubble pressure became maximum near the location of $s = 0.15$, and decreased with the increase in s. The maximum value of the bubble pressure was about 1 MPa, which was much smaller than the general bubble collapse pressure of over 10 MPa [20]. The collapse pressure of bubbles increased in turn as a chain reaction toward the center of the bubble cloud. If the dynamics of the cloud were included in the simulation, bubble pressure may reach 1 GPa as calculated in Ref. [21]. The bubble pressure was underestimated mainly because the behavior of the bubble cloud was not modeled in the simulation. A strong unsteady flow with vortices was supposed to occur in the region where the bubbles collapsed in the present model pump under the flow rate condition of $Q/Q_{\eta max} = 0.60$. If the large static pressure fluctuations resulting from the highly unsteady flow were accurately simulated, it is expected that the bubble collapse pressure deduced from Eqs. (13.3) to (13.5) would be larger.

The void fraction increased in a range from $s = 0.05$ to 0.15, as shown in Fig. 13.6. Unstable bubble behavior appeared just downstream of the cavitating region in Fig. 13.10. The results had a correlation with the experimental data where the impact loads were measured at a location where the cavitation bubble collapsed [19].

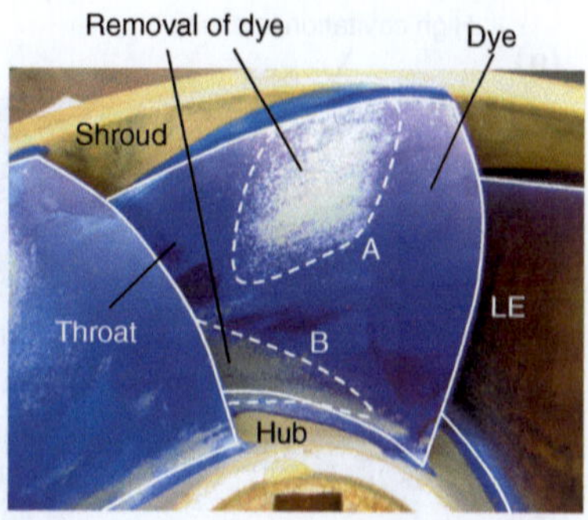

Fig. 13.11 Visualized cavitation erosion area in experiment. Based on [11], reprinted with permission from JSME

13.3.1.5 Cavitation Intensity

The cavitation intensity, I, was defined by the following equation,

$$I_c = \sum_{\substack{p_{Bi} \geq p_{Bth} \\ i=1,2,3\ldots}} \frac{p_{Bi}^2 n_G \Delta V \Delta t}{2\rho_L C} \Big/ \Delta T, \tag{13.21}$$

where p_B is the bubble pressure, n_G is the bubble number density, ΔV is the volume of the numerical cell, ρ_L (= 1,000 kg/m^3) is the water density, C (= 1,500 m/s) is the sound velocity in water, Δt is the time step, and ΔT is the elapsed time in the simulation. Equation (13.21) was derived from Eq. (13.20). The bubble pressure in Eq. (13.21) represents the bubble collapse pressure, and $n_G \Delta V$ means the number of bubbles in each numerical cell.

Material deformation and mass loss only occurs when the impact force applied to the material surface caused by the bubble collapse pressure exceeds a threshold value [17]. A threshold of the bubble pressure, p_{Bth}, was thus also assumed in Eq. (13.21). Since the appropriate value of p_{Bth} has not been investigated, $p_{Bth} = 0.6$ MPa or $p_{Bth} = 0.2$ MPa was temporarily assumed based on the numerical results in Figs. 13.9 and 13.10. The cavitation intensity was only estimated when the bubble pressure exceeded the threshold pressure, p_{Bth}.

Figure 13.11 shows the visualized cavitation erosion area. A blue dye was painted on the suction surface of the impeller blade. After the pump had been operated for four hours, the dye was removed due to cavitation bubble collapse in two different areas. A great deal of plastic deformations in the same areas, i.e., tiny pits, was observed on the aluminum blade surface. Cavitation erosion area A was distributed between LE and the throat of the blade, and between the shroud and mid-span of the blade. The other cavitation erosion area B was distributed between LE and the throat, and between the hub and mid-span of the blade. The dye was more severely peeled off in area A than in area B. The surface roughness caused by the plastic deformations in the area A was also remarkably higher than that in the area B.

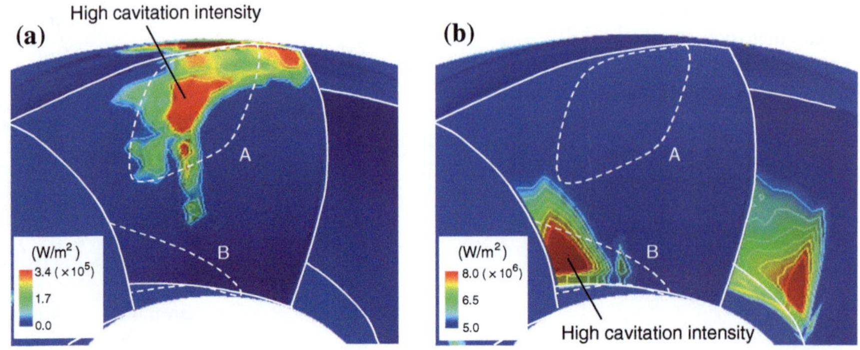

Fig. 13.12 Predicted area of high cavitation intensity. **a.** Cavitation intensity distribution ($p_{Bth} = 0.6$ MPa), **b.** Cavitation intensity distribution ($p_{Bth} = 0.2$ MPa). Based on [11], reprinted with permission from JSME

Figure 13.12 shows the predicted cavitation intensity under different conditions of p_{Bth}. Figure 13.12a shows the results at $p_{Bth} = 0.6$ MPa. The high cavitation intensity area agreed with area A. The high cavitation intensity area at $p_{Bth} = 0.2$ MPa agreed with area B, as seen in Fig. 13.12b.

The peak value of the bubble pressure in area A was larger than that in area B; however, the bubble number density in area B was larger, as seen in Fig. 13.5b. This means that there is a small number of strong collapses in area A, and in contrast, a large number of weak collapses in area B. The numerical results corresponded qualitatively to the above experimental states for the dye removal and surface roughness.

The cavitation intensity in area A was lower than that in area B. This is because the bubble pressure was not sufficiently high, as explained in Sect. 13.3.1.4, and the bubble number density was dominant in the calculation of Eq. (13.21). The cavitation intensity in area A can be larger than that in area B when a higher bubble pressure is simulated and an appropriate threshold bubble pressure is set up.

Although the threshold of the bubble pressure, p_{Bth}, should be investigated in detail in future studies, the present simulation and the method of estimating the impact energy and the cavitation intensity effectively predicted cavitation erosion area around the impeller of a centrifugal pump.

13.3.2 Cavitating Jet

13.3.2.1 Flow Pattern and Bubble Behavior

The flow pattern and the bubble behavior under the condition of Case 1 in Table 13.1 are explained in this section by using Figs. (13.13)–(13.16). The velocity difference between the liquid and the bubble, i.e., the slip velocity, was

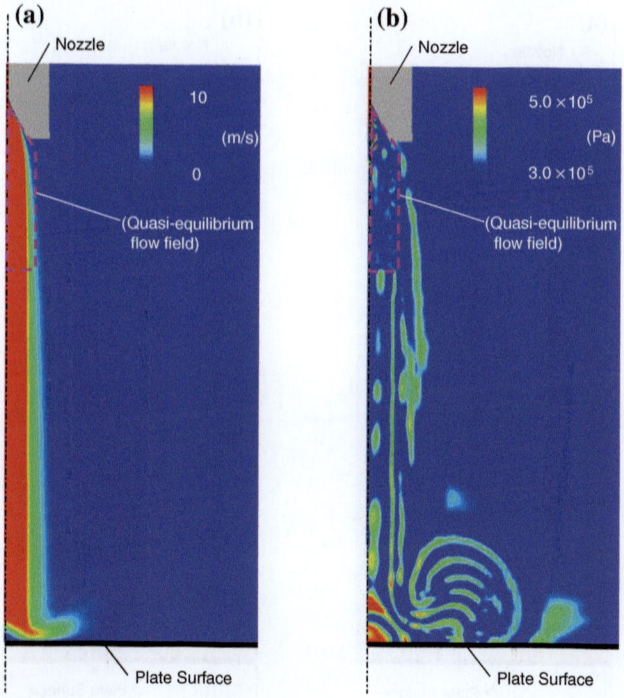

Fig. 13.13 Velocity and static pressure of water in Case 1, **a.** velocity, **b.** pressure. Based on [12], reprinted with permission from ASME

not taken into account in this cavitating jet simulation. The flow pattern and the bubble behavior were unsteady; however, Figs. (13.13)–(13.16) show the instantaneous results. The calculation of bubble pressure diverged around the nozzle. We set a quasi-equilibrium region surrounded by the dotted line in Figs. (13.13)–(13.16) to avoid the divergence of calculation. In the numerically unstable flow field, the quasi-equilibrium between the bubble pressure and its surrounding pressure of water was assumed instead of solving Eqs. (13.4)–(13.7).

Figure 13.13 shows the velocity and the static pressure of water in Case 1. The water jet injected from the nozzle impinged on the flat plate, and the water flowed along the plate while rolling up. The unsteadiness was not strong since the local vortex structure formed in the region where the shear stress is high was not sufficiently resolved in this simulation. The static pressure varied with a strong unsteadiness affected by the bubble behavior through the fourth term of Eq. (13.3). This result was similar with Fig. 13.5a.

Figure 13.14 shows the bubble number density and the void fraction. The bubble nuclei were distributed in the jet and near the flat plate within a radial range from the jet center axis. The void fraction depends on the distribution of the bubble number density and the bubble radius, as shown in Eq. (13.18). The void fraction increased in the jet, which means that a cavitating jet was generated.

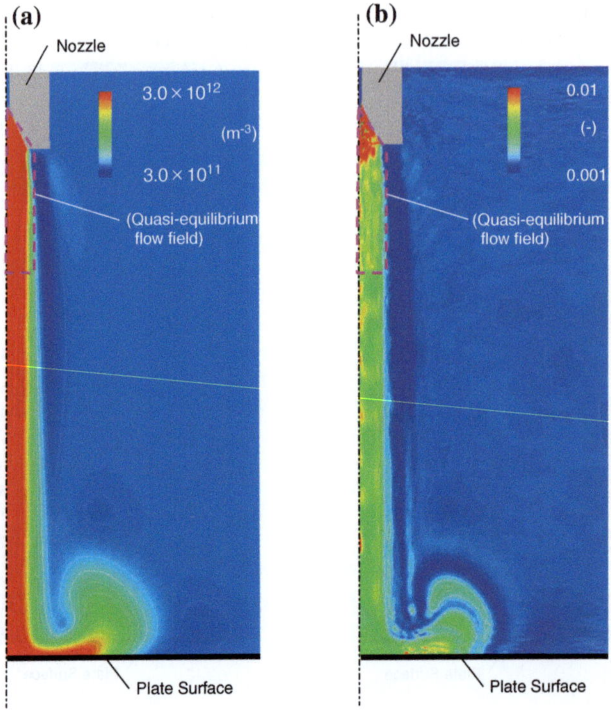

Fig. 13.14 Bubble number density and void fraction in Case 1, **a.** bubble number density, **b.** void fraction. Based on [12], reprinted with permission from ASME

Figure 13.15 shows the bubble pressure abruptly fluctuated in and around the main flow. Figure 13.15b shows the transient bubble pressure near the flat plate at point A in Fig. 13.15a. The bubble pressure impulsively increased in about 5 microseconds during the bubble collapse.

Sato et al. observed bubble cloud behavior in a cavitating jet impinging vertically on a wall, which was captured with a high-speed video camera [22]. The measured collapse time of the bubble cloud near the wall surface was about 50–150 microseconds, which was analyzed from successive image frames. The predicted time for the bubbles to shrink was shorter than the measured cloud collapse time since (i) we simulated isolated bubble behavior, and (ii) the jet velocity and the surrounding water pressure were higher than those in the experiment conducted by Sato et al. However, the time scale of microseconds for the bubble pressure fluctuations in the simulation corresponded with the experimental data.

The duration of the impulsive peak in Fig. 13.9 was shorter than that in Fig. 13.15b since the expansion and contraction of the bubble became quicker due to the pressure gradient caused by the impeller rotation.

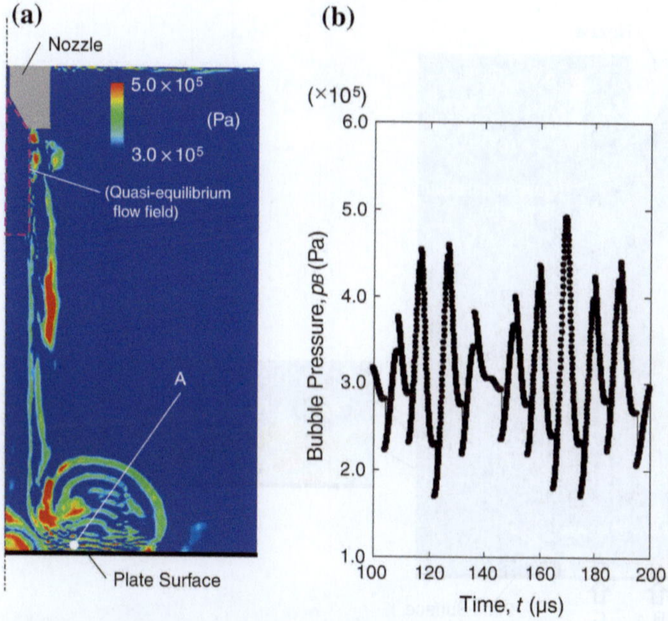

Fig. 13.15 Bubble pressure in Case 1, **a.** bubble pressure, **b.** time variation in bubble pressure at point A. Based on [12], reprinted with permission from ASME

Figure 13.16 shows the density of the cavitation impact energy obtained from Eq. (13.20), i.e., $E/\Delta V$ where ΔV is the volume of numerical cell, and k is assumed to be 1.0. The cavitation energy was high in the region (indicated by arrow B) around the jet center axis and in the other peripheral region (indicated by arrow C) away from the jet center axis near the plate surface. When the cavitating jet impinged vertically on the flat plate, the occurrence of ring-like erosion has been observed on the material surface (e.g., in Ref. [22]). The peripheral region indicated by arrow C has a correlation with the ring-like erosion.

13.3.2.2 Cavitation Impact Energy and Compressive Residual Stress

The correlation between the estimated cavitation impact energy and the measured compressive residual stress on the flat plate after WJP was investigated in Cases 2–5. The inlet velocity and the standoff distance were changed in Cases 2–5 while the pressure boundary condition was fixed at 1.0×10^5 Pa. We used a stainless steel plate, on which tensile residual stress was introduced by grinding before WJP. The X-ray residual stress measurement was conducted on the flat plate before and after WJP.

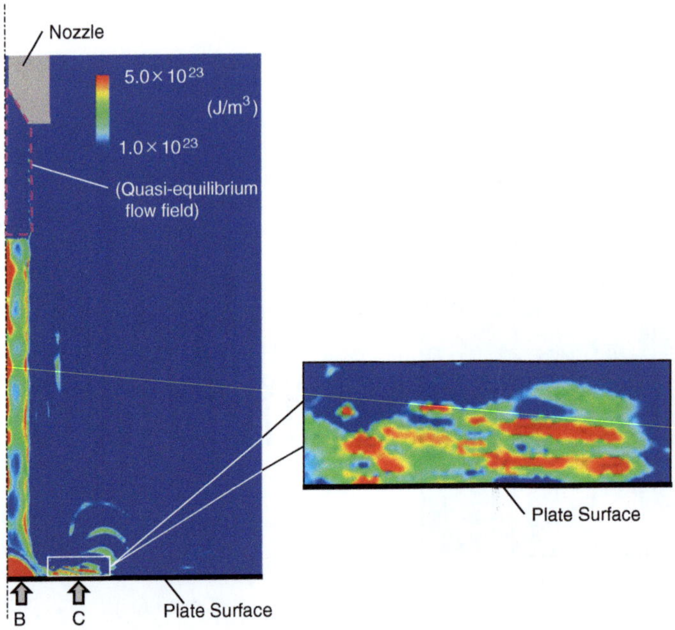

Fig. 13.16 Cavitation impact energy in Case 1. Based on [12], reprinted with permission from ASME

Although the bubble does not release the impact energy only in one direction at the collapse, the impact energy that acted vertically on the plate surface was estimated in the present study as follows:

$$E_{surf} = \alpha \frac{E}{\Delta S}\left(\frac{T_{\exp}}{\Delta T}\right) = \sum_{j=1,2,3\ldots} \sum_{\substack{p_{Bi} \geq p_{Bth} \\ i=1,2,3\ldots}} \alpha \left(p_{Bi}^2 n_G \Delta z\right)_j \left(\frac{T_{\exp}}{\Delta T}\right), \tag{13.22}$$

where E is the impact energy defined in Eq. (13.20) where k is assumed to be 1.0, p_B is the bubble pressure, n_G is the bubble number density, ΔT is the elapsed time in the simulation, and T_{exp} is the jet injection time. T_{exp} was 2 min without moving the nozzle in the experiment. ΔS is the cross-section of each numerical cell parallel to the plate surface, and Δz is the height of each cell ($\Delta z = \Delta V/\Delta$S). Equation (13.22) gives the energy per unit cross-section, which was summed up by j in the direction of the jet center axis, i.e., in the z direction since the cartesian mesh in the r–z plane was used. The α is a damping coefficient of the energy released at the bubble collapse, which is transferred as the pressure wave [23]:

$$\alpha = e^{-\left(\frac{4}{3}\right)\left(\frac{f^2}{\rho C'^3}\right)\mu d_j}, \tag{13.23}$$

where f is the pressure wave frequency, C' is the sonic velocity, i.e., the velocity of a pressure wave in water including bubbles, and d is the vertical distance between the bubble collapse location and the plate surface. When the pressure wave is transferred with high frequency and low speed in a media with small density and high viscosity, the energy damping is large. The energy damping also depends on the distance from the collapse location. Equation (13.23) was adopted since the collapse pressure propagation from the bubble was not numerically simulated in the present study. $f = 1.0 \times 10^6$ Hz and $C' = 30$ m/s were temporarily assumed; then, the cavitation impact energy was damped to about 10 % when d was 50 mm.

The adding up in Eq. (13.22) was conducted when the bubble pressure exceeded a threshold value since only high impact energy caused plastic deformation of the material surface [17]. The bubble pressure threshold was temporarily assumed to be 1.02×10^5 Pa in the simulation.

Figure 13.17 compares the cavitation impact energy calculated from Eq. (13.22) and the measured residual stress in Cases 2–5. The radial location, r, and the residual stress, σ, were simultaneously nondimensionalized by the nozzle radius, r_n, and the absolute value of the minimum residual stress in Case 2, $|\sigma_{2min}|$. Compressible residual stress was introduced into the stainless steel plate by WJP, and this was distributed within a radial range from the jet center axis. The distribution of compressible residual stress had a peak, indicated by arrow D, away from the jet center axis. The peak indicates the strong impacts caused by cavitation bubble collapses and the correlation with ring-like erosion (e.g., in Ref. [22]). The other peak indicated by arrow E was caused not only by bubble collapse impacts but also by the impingement of water. No peak appeared in Case 5 since the jet velocity was too low.

The distribution of impact energy also had a peak indicated by arrow *F* similar to that indicated by arrow D. Figure 13.18 compares the radial locations and the values of peaks indicated by arrows D and F in Fig. 13.17. The radial peak location of compressive residual stress in Case 5 was closer to the jet center axis compared with the other Cases 2–4. The peak value of compressive residual stress was the lowest in Case 4. The cavitation impact energy also had similar tendencies.

The radial range of compressive residual stress from the jet center axis, R_{exp}, is one of the most important measures of performance of WJP. WJP can cover the wider area of the weld surface of structures and is more efficient when R_{exp} is larger. R_{exp} ranged from 1.5 to 3.7, as shown in Fig. 13.17. Figure 13.19 compares R_{exp} and R_{cal}. R_{cal} is the radial range of the cavitation impact energy in Fig. 13.17, and ranged from 1.6 to 3.0. R_{exp} and R_{cal} decreased when the standoff distance was shorter in Cases 2 and 3. R_{exp} and R_{cal} also decreased when the inlet velocity was decreased in Cases 3–5. R_{cal} corresponded to R_{exp} with a prediction error of $\pm$ 20 % in Cases 2–5.

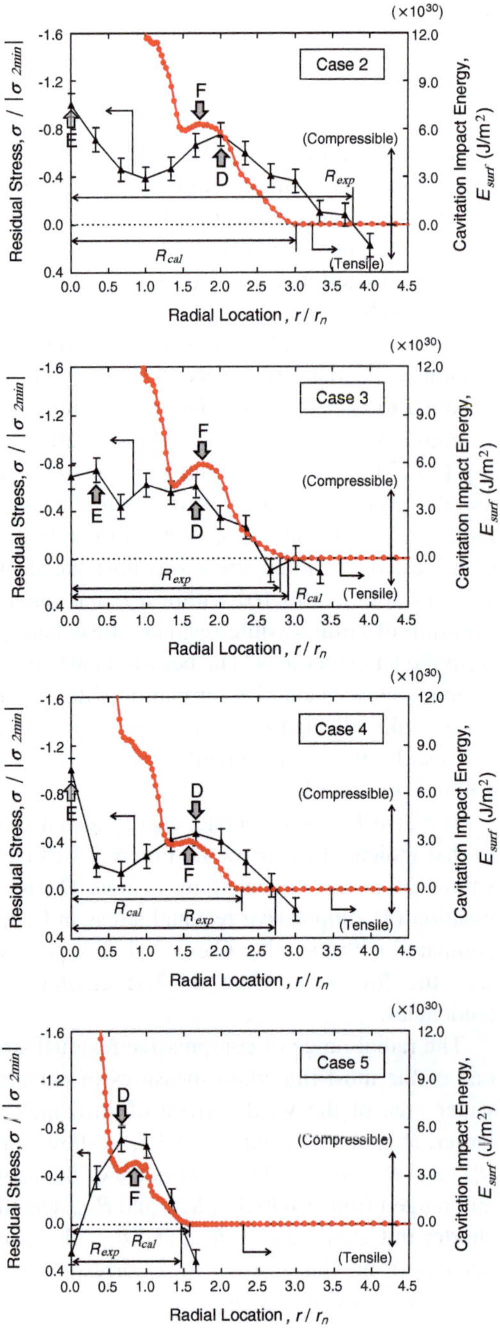

Fig. 13.17 Comparison between predicted cavitation impact energy (*red symbols*) and measured residual stress (*black symbols*) in Cases 2–5. Based on [12], reprinted with permission from ASME

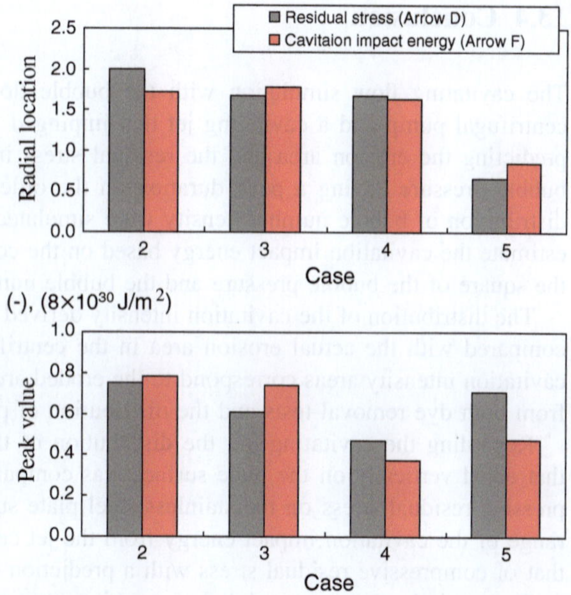

Fig. 13.18 Radial location and value of distribution peak of compressive residual stress and cavitation impact energy in Fig. 13.17. Based on [12], reprinted with permission from ASME

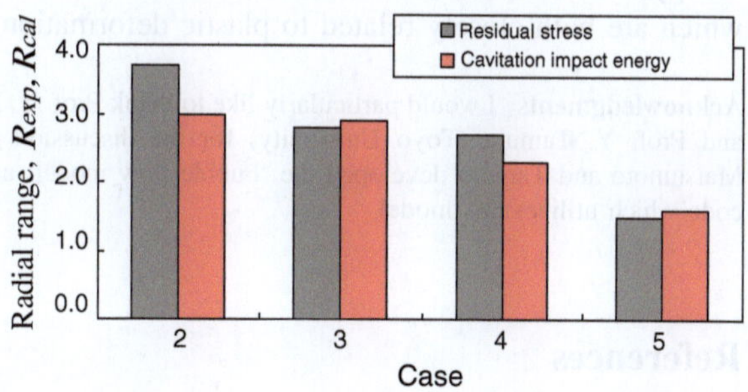

Fig. 13.19 Radial range of compressive residual stress and cavitation impact energy in Fig. 13.17. Based on [12], reprinted with permission from ASME

The simulation however could not predict the values of peaks indicated by arrows E in Fig. 13.17. The cavitation impact energy was exceedingly overestimated when the radial location was below about 1.4 in Cases 2–4 or below 0.6 in Case 5. The main cause is the following; the static pressure of water around the flow stagnation point on the flat plate and the jet center axis was much too high since turbulent flow diffusion was not taken into account in the simulation. The overestimated water pressure surrounding the bubble enormously increased the bubble pressure through Eqs. (13.3)–(13.7).

The above results demonstrated that the numerical method we developed for predicting the region of compressive residual stress after WJP was valid except near the jet center axis.

13.4 Conclusion

The cavitating flow simulation with the bubble flow model was applied to a centrifugal pump and a cavitating jet that impinged vertically on a flat plate for predicting the erosion area and the residual stress improvement. The impulsive bubble pressure having a peak duration on the order of a microsecond and the distribution of bubble number density were simulated. We proposed a method to estimate the cavitation impact energy based on the computation of the product of the square of the bubble pressure and the bubble number density, i.e., $p_B^2\, n_G$.

The distribution of the cavitation intensity derived from the impact energy was compared with the actual erosion area in the centrifugal pump. Computed high cavitation intensity areas correspond to the eroded areas identified experimentally from both dye removal tests and the distribution of plastic deformation pits.

Regarding the cavitating jet, the distribution of the cavitation impact energy that acted vertically on the plate surface was compared with the measured compressive residual stress on the stainless steel plate surface after WJP. The radial range of the cavitation impact energy from the jet center axis corresponded with that of compressive residual stress with a prediction error of ± 20 %.

To conclude, the proposed technique of computation of the cavitation impact energy accurately reflects the distributions of erosion damage and residual stress, which are both closely related to plastic deformation of the material.

Acknowledgments I would particularly like to thank Prof. Y. Matsumoto (University of Tokyo) and Prof. Y. Tamura (Toyo University) for the discussion of my present study. Professors Matsumoto and Tamura developed the 'bubble flow model' and the cavitating flow simulation code which utilizes this model.

References

1. Mochizuki M, Enomoto K, Sakata S, Kurosawa K, Saito H, Tsujimura H, Ichie K (1993) A study on residual stress improvement by water jet peening. In: Proceedings of the 5th international conference on shot peening, Oxford, pp 247–256, 1993
2. Anzai H, Sagawa W, Morinaka R, Hirano K, Hayashi E, Enomoto K, Ishida T, Gotoh H (1999) Residual stress improvement of BWR core shroud for IGSCC mitigation using water jet peening technology. In: PLIM + PLEX conference, Madrid, 1999
3. Dular M, Sirok B, Stoffel B (2006) Experimental and numerical modeling of cavitation erosion. In: Proceedings of sixth international symposium on cavitation, CAV2006, Paper No. 55, Wageningen
4. Dular M, Coutier-Delgosha O (2008) Numerical modelling of cavitation erosion. In: Proceedings of the ASME fluids engineering division summer meeting, FEDSM2008-55034, Jacksonville
5. Nohmi M, Iga Y, Ikohagi T (2008) Numerical prediction method of cavitation erosion. In: Proceedings of the ASME fluids engineering division summer meeting, FEDSM2008-55126, Jacksonville
6. Nohmi M, Ochiai N, Iga Y, Ikohagi T (2010) Detailed observation of hydrofoil cavitation and a proposal for improving cavitation model. In: Proceedings of the ASME 2010 3rd joint US-

European fluids engineering summer meeting and 8th international conference on nanochannels, microchannels, and minichannels, FEDSM-ICNMM2010-30435, Montreal
7. Miyabe M, Maeda H (2011) Numerical prediction of pump performance drop and erosion area due to cavitation in a double-suction centrifugal feedpump. In: Proceedings of the ASME-JSME-KSME joint fluids engineering conference 2011, AJK2011-06023, Hamamatsu
8. Ochiai N, Iga Y, Nohmi M, Ikohagi T (2010) Numerical prediction of cavitation erosion intensity in cavitating flows around a Clark Y 11.7 % hydrofoil. J Fluid Sci Technol 5(3):416–431
9. Zima P, Sedlář M, Maršík F (2009) Modeling collapse aggressiveness of cavitation bubbles in hydromachinery. In: Proceedings of the 7th international symposium on cavitation, CAV2009, Paper No. 131, Ann Arbor
10. Tamura Y, Fukaya M, Matsumoto Y (2002) Numerical method for cavitating flow simulations and its application to axial flow pumps. In: Proceedings of the 9th international symposium on transport phenomena and dynamics of rotating machinery (ISROMAC-9), FD-ABS-129, Honolulu, 2002
11. Fukaya M, Tamura Y, Matsumoto Y (2010) Prediction of cavitation intensity and erosion area in centrifugal pump by using cavitating flow simulation with bubble flow model. J Fluid Sci Technol 5(2):305–316
12. Fukaya M, Morinaka R, Saitou N, Hatou H, Tamura Y, Matsumoto Y (2010) Prediction of residual stress improvement by water jet peening using cavitating jet simulation with bubble flow model. In: Proceedings of the ASME 2010 3rd joint US-European fluids engineering summer meeting and 8th international conference on nanochannels, microchannels, and minichannels, FEDSM-ICNMM2010-30419, Montreal
13. Takemura F, Matsumoto Y (1994) Internal phenomena in bubble motion: bubble dynamics and interface phenomena. Kluwer Academic Publishers, pp 467–474
14. Plesset MS (1954) On the stability of fluid flows with spherical symmetry. J Appl Phys 25:96–98
15. Schiller L, Nauman AZ (1933) Uber die Grundlegenden Berechnungen bei der Schwerkraftaufbereitung. Ver Deut Ing 77:318–320
16. Sridhar G, Katz J (1995) Drag and lift forces on microscopic bubbles entrained by a vortex. Phys Fluids 7(2):389–399
17. Soyama H, Kumano H (2002) The fundamental threshold level: a new parameter for predicting cavitation erosion resistance. J Test Eval, ASTM international, 30(5):421–431
18. Hattori S et al (2010) Prediction method for cavitation erosion based on measurement of bubble collapse impact loads. Wear 269(7–8):507–514
19. Maeda M, Yamada I, Kawakami T (1992) The prediction of cavitation Erosion on pump impeller by measuring of cavitation bubble collapse impact loads. In: Proceedings of the 16th symposium of IAHR, section of hydraulic machinery and cavitation, San Paulo, pp 681–690
20. Kato H et al (1999) Cavitation. Maki Shoten, ISBN4-8375-0658-5, pp 203–204 (in Japanese)
21. Shimada M et al (1999) Dynamics of the cloud cavitation and cavitation erosion. Trans JSME B 65(634):1934–1941 (in Japanese)
22. Sato K, Sugimoto Y, Ohjimi S (2009) Pressure-wave formation and collapses of cavitation clouds impinging on solid wall in a submerged water jet. In: Proceedings of the 7th international symposium on cavitation, CAV2009, Paper No. 66, Ann Arbor
23. Dular M et al (2006) Development of a cavitation erosion model. Wear 261(5–6):642–655

Chapter 14
Assessment of Erosion Sensitive Areas via Compressible Simulation of Unsteady Cavitating Flows

Steffen J. Schmidt, Michael S. Mihatsch, Matthias Thalhamer and Nikolaus A. Adams

Abstract The objective of this paper is the assessment of the numerical predictability of erosive events arising in cavitating flows. First, a numerical method and an efficient thermodynamic model for the simulation of cavitating flows are briefly described. The prediction of typical flow details is evaluated by simulating the 3-D flow around a quasi 2-D NACA hydrofoil. We find that the maximum length of the attached cavity, the Strouhal number, and the average diameter of detached clouds are essentially grid independent. Scale enrichment and enhanced 3-D flow details are observed on refined grids. Even delicate flow features, such as cavitating vortices and irregular 3-D break-up patterns, are reproduced, provided that the spatial resolution is sufficiently high. The simulation of cloud collapses and resulting instantaneous peak pressures is assessed in a second investigation. Here, we analyze the effect of the computational grid resolution with respect to typical collapse characteristics, such as the collapse duration, and the instantaneous maximum pressure within the flow field and at walls. The proposed methodology is confirmed by a third investigation, where an experimental setup to investigate cavitation erosion is simulated, and regions of experimentally observed cavitation damage are compared with numerical predictions of strong collapses. The excellent agreement of numerically predicted collapse positions and experimentally observed damage justifies the proposed methodology.

14.1 Introduction

The numerical prediction of erosive events arising in cavitating flows is a serious challenge due to the large range of involved spatial and temporal scales. The extent of a partial cavity at the suction side of a propeller blade might be four to

---

S. J. Schmidt (✉) · M. S. Mihatsch · M. Thalhamer · N. A. Adams
Institute of Aerodynamics and Fluid Mechanics, Technische Universität München, Munich, Germany
e-mail: Steffen.Schmidt@aer.mw.tum.de

K.-H. Kim et al. (eds.), *Advanced Experimental and Numerical Techniques for Cavitation Erosion Prediction*, Fluid Mechanics and Its Applications 106,
DOI: 10.1007/978-94-017-8539-6_14,

five orders of magnitude larger than the diameter of a collapsing bubble or of a nucleus [1–5]. Similarly for temporal scales, the duration of a peak pressure due to a shock impinging on a material surface is in the order of a few microseconds, whereas the duration of a typical cloud-shedding process might be several milliseconds [6, 7]. Moreover, the process of cavitation erosion might take place over the lifetime of the propeller. Except for a few academic cases, current computational capabilities do not allow the resolution of all scales arising in typical cavitating flows. Therefore, an accurate prediction of medium and large scale dynamics, such as cloud formation and cloud collapses, is highly desired.

One approach for removing unresolvable small scales is to filter the continuous governing equations using spatial low-pass filters [8]. Filtering not only removes unresolvable small-scale kinematics, it additionally replaces unresolvable two-phase structures with locally homogeneous mixtures. It can be shown that finite-volume methods provide approximate solutions of the top-hat filtered governing equations without explicit filtering [9]. The filter width of a finite volume discretization is related to the computational mesh, which constitutes the lower bound of resolvable spatial scales [10]. In most cases of practical relevance, the computational grid is far too coarse to resolve single bubbles or small-scale kinematics in the range where viscous dissipation is dominant, i.e. the grid is insufficient for direct numerical simulations. However, the investigations summarized within this paper provide a strong indication that a considerable amount of relevant detail for the prediction of cavitation and cavitation erosion are much larger than nuclei or single bubbles.

The time discretization method needs to ensure sufficient resolution of the dynamic processes as well. Here, the predominant physical constraint is the fastest signal speed, which is in the order of the speed of sound in the liquid: e.g., assuming that the spatial resolution is 0.1 mm and the speed of sound is 1,500 m/s, the numerical time step must be on the order of 60–70 ns to resolve collapse-induced peak pressures that are known to be a driving mechanism of cavitation erosion.

14.2 Numerical Method

In the following sections we present investigations of cavitating flows as predicted by our flow simulation code CATUM (CAvitation Technische Universität München [11, 12]), which is a density-based finite volume method employing a low Mach-number consistent flux function and an explicit time marching procedure. Since the objective of this investigation is the prediction of erosive collapses, we focus on inertia-driven dynamics. Thus, the governing equations are the 3-D unsteady Euler equations resulting from mass, momentum and energy balances. We denote that the applied computational grids are far too coarse to resolve boundary layers or kinematics with wave numbers close to the viscous sub-range. Therefore, we neglect viscous effects on purpose and apply a nonlinear discretization scheme that implicitly provides a sufficient amount of numerical regularization [13].

The spatial reconstruction of the velocity field follows a WENO-3 procedure [14]. Density and internal energy are reconstructed by monotonic TVD limiters ("minmod") [15]. Time marching is performed by adopting an explicit low-storage four-step Runge–Kutta method with optimized stability region [11, 12, 16]. The combination of both methods results in a second-order numerical approach in space and time for smooth flow, and it ensures sharp representation of discontinuous flow features, such as shocks and contact waves.

In order to allow for the simulation of shock formation and wave propagation, the compressibility of the fluids (liquid and vapor) is taken into account. Consequently, the numerical time step is proportional to the ratio of the smallest length scale (minimum grid size) and the fastest signal speed ($\sim$ speed of sound in the liquid). The fluids can be characterized by closed form equations of state, or by look-up tables for complex fluid models, e.g. for fluids where closed form equations are not available or computationally expensive. For this investigation we use closed form equations for the simulation of the flow around a hydrofoil, and more efficient look-up tables for simulation of the cloud collapse and the nozzle-target flow. Both thermodynamic models provide high quality approximations of the IAPWS database [17] of water, vapor, and liquid-vapor equilibrium.

The phase-transition model is based on local equilibrium assumptions for pressure, temperature and specific Gibbs functions [18]. Therefore, two-phase regions are modeled as saturated mixtures (liquid-vapor equilibrium with vapor volume fraction α), where the resulting pressure is the temperature-dependent vapor pressure [19]. For each computational cell, and at each instant in time, the average density, the average momentum, and the average total energy are given by the governing equations. These quantities allow for a cell-wise evaluation of the average internal energy. Using the previously specified local equilibrium assumptions together with suitable fluid-models (closed form equations or look-up tables) all unknown thermodynamic properties can be obtained from density and internal energy. Thereby, the effects of latent heat are fully taken into account and the method is inherently conservative. Moreover, the speed of sound is the equilibrium speed of sound that leads to a well-posed mathematical problem [20].

We consider it advantageous that the equilibrium model does not require user-specified and potentially unknown parameters, such as initial nuclei distributions or evaporation/condensation rates. Surely, equilibrium assumptions might be questionable in case of cavitation inception where meta-stable states (pressures below the vapor pressure or even tension [21, 22]) may occur. However, our observations indicate that equilibrium assumptions may lead to a sufficiently accurate model for developed cavitating flows, such as those investigated within this paper.

14.3 Hydrofoil Cavitation and Predictability of Large and Medium Scale Dynamics

In this section we assess the effect of the chosen computational grid on typical cavitation phenomena around a quasi 2-D hydrofoil. The investigation represents a brief summary of Ref. [23]. We simulate the cavitating flow around a 2-D NACA 0015 hydrofoil (angle of attack 6°, cord length $l_{cord} = 0.13$ m, span width $l_{span} = 0.3$ m), which is placed in the middle of a rectangular test section (height 0.3 m, depth 0.3 m, length 0.9 m). The walls of the test section and the surface of the hydrofoil are modeled as inviscid adiabatic walls, since the grid resolution is insufficient to resolve boundary layers. At the inlet of the numerical domain the velocity $u_{in} = 30$ m/s and the static temperature $T_{in} = 293$ K of the liquid inflow are prescribed. Since the inlet conditions correspond to a pure liquid state, the initial vapor volume fraction is zero. At the outlet the static pressure of $p_{exit} = 4.5$ bar is prescribed. The resulting cavitation number is $\sigma_{ref} = 2(p_{exit}\text{-}p_{sat}(T_{in}))/(\rho_{liquid}\, u_{in}^2) = 1.0$.

The numerical domain is discretized by 2×10^5 cells (coarse grid) and 2.4×10^7 cells (fine grid), respectively. First, a simulation is performed on the coarse grid until the typical periodic shedding of the cavity is statistically developed. This may require 2–5 shedding processes where the maximum length of the sheet might be initially overestimated. To determine the shedding frequency, the integrated vapor volume content V_{vap} [%] within the computational domain is analyzed. The simulated time interval of 0.15 s corresponds to nearly 17 shedding cycles. As depicted in Fig. 14.1—left, the attached cavity is mainly two-dimensional, while spanwise variations of the vapor volume are observed for the detached cloud. The results obtained with the coarse grid are now compared to those obtained by using the refined counterpart with 2.4×10^7 cells, see Fig. 14.1—right. Here, the simulated time interval of 0.11 s corresponds to 12

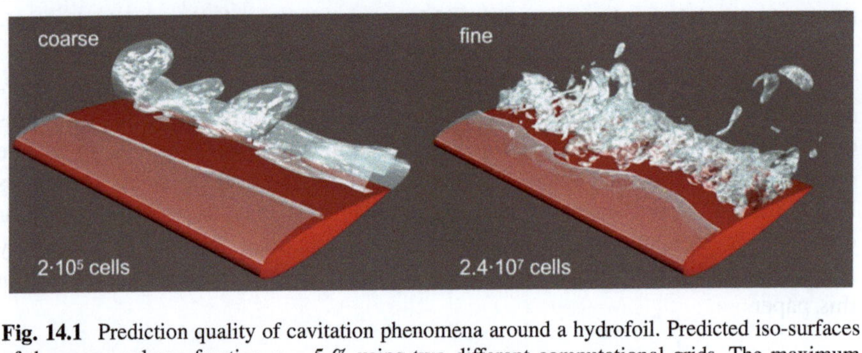

Fig. 14.1 Prediction quality of cavitation phenomena around a hydrofoil. Predicted iso-surfaces of the vapor volume fraction $\alpha = 5$ % using two different computational grids. The maximum length l_{max}/l_{cord} of the sheet cavity is 0.75 (*coarse grid*) and 0.8 (*fine grid*). The shedding frequency is approximately 110 Hz (±5 Hz) on both grids. The typical diameter of the detached cloud close to the trailing edge (as shown in this figure) is approximately $l_{cord}/3$ for both grids

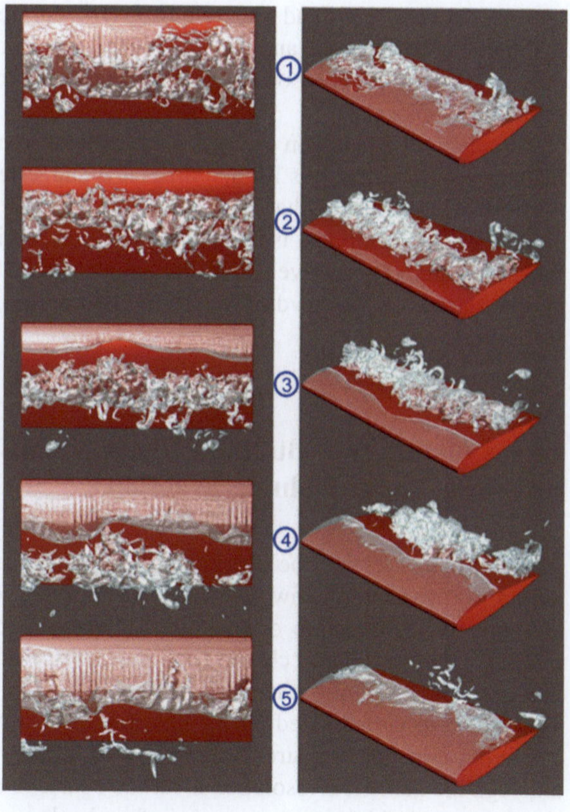

Fig.14.2 A typical shedding cycle with f $\approx$ 110 Hz using the fine grid. Iso-surfaces of $\alpha = 5$ % at 5 consecutive instants in time showing the break-up of the sheet cavity, the formation of cavitating vortical structures and the disintegration and collapse of two-phase regions. On the *left* the *top* view is presented, while on the *right perspective* views at the same instants in time are shown

shedding cycles. The shedding frequency is approximately 110 Hz on both grids. The numerical time step for the simulation on the fine grid is $\Delta t_{CFD,fine} = 8.5 \times 10^{-8}$ s. We observe that the maximum length of the sheet cavity, the typical diameter of the cloud close to the trailing edge, and the shedding frequency compare well for both spatial resolutions. This is a strong indication that large-scale dynamics of cavitating flows can be predicted even if the spatial resolution is relatively coarse. Grid refinement enables the computation of medium-scale phenomena, such as cavitating vortices in the streamwise direction and enhanced cloud fragmentation. All predicted disturbances develop as a result of Rayleigh-Taylor and Kelvin-Helmholtz instabilities initiated by the re-entrant flow. As analyzed in [23], the flow field develops perfectly two-dimensional in case of pure liquid flow ($\sigma_{ref} \gg 1$) where the numerical method predicts a stationary solution. Hence, numerical round-off errors are insignificant, although they might provide initial disturbances. It is thus reasonable to denote the observed instabilities as intrinsic instabilities [3] that seem to be mainly inertia controlled [23].

Figure 14.2 shows a time series of iso-surfaces of the vapor volume fraction $\alpha = 5$ %. On the left the top view is presented, while on the right perspective views at the same instants in time are shown. Due to different illumination on the left and on the right hand side, slightly different details of the iso-surfaces are highlighted. The position and intensity of the streaks observed on the left side vary from cycle to cycle. Their occurrence seems to be related with the break-up pattern of the previous cycle.

Although the prediction of large-scale dynamics might be sufficient to quantify integral properties such as lift and drag, it is suggested that a sufficient amount of medium-scale dynamics has to be resolved in order to capture at least a certain amount of flow detail that is related to cavitation erosion. In particular, we suggest that it is required to resolve collapsing clouds and collapse-induced peak pressures at the surface of the hydrofoil. These issues are addressed in the following sections.

14.4 Collapse of a Bubble Cluster Versus Collapse of a Vapor Volume Fraction

In the previous section the prediction of medium-scale phenomena, such as cavitating vortices in streamwise direction and enhanced cloud fragmentation was assessed on the example of hydrofoil cavitation. In this section we assess the prediction of the collapse characteristics of a resolved bubble cluster and compare them with the collapse characteristics predicted by under-resolved simulations. This test case is motivated by the observation that our numerical simulations of cavitating flows predict strong shock waves in case of collapsing vapor regions, although the numerical resolution is far too coarse to resolve single bubbles. These shocks result in high instantaneous surface-loads as soon as they impinge onto a material surface. Depending on the flow conditions, the loads last for a few microseconds and can reach several thousand bars. In order to allow for a reliable forecast of erosive collapses, the influence of the numerical resolution on the predicted loads needs to be evaluated.

We generate a random distribution of spherical bubbles with the following properties: The "numerical cloud" consists of 125 spherical vapor bubbles with radii between $R_{min} \approx 0.70$ mm and $R_{max} \approx 1.65$ mm. The average radius is $R_{av} \approx 0.95$ mm and the minimum distance between two bubbles is 0.2 mm. The cloud is located within a liquid-filled cubic domain of $20 \times 20 \times 20$ mm^3, which will be referred to as the "inner domain". This inner domain is embedded into a larger "outer domain" of $4 \times 4 \times 2$ m^3. The inner domain is discretized with uniform cubic cells while stretched cells are applied to discretize the outer domain. The bottom faces of both domains are coplanar and impermeable for representing solid walls. The other faces of the outer domain correspond to far-field boundaries. The total vapor volume fraction is 5.8 % with respect to the inner domain. Since we apply a thermodynamic equilibrium approach, we simplify the test case by neglecting surface tension and non-condensable gas. Furthermore, the following properties are enforced: the initial pressure within the surrounding liquid satisfies a

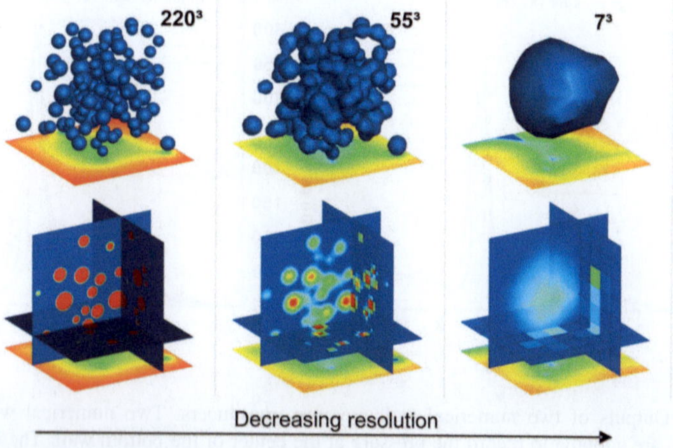

Fig. 14.3 Discrete representation of a bubble cluster using different grids. Bubbles are fully resolved by using 220^3 cells, partial resolution is achieved with 55^3 cells. Bubbles are represented by a locally homogeneous mixture using 7^3 cells

Laplace equation $\Delta p = 0$, where the pressure sufficiently far away from the bubbles is $p_\infty = \text{const.} = 40$ bar, and the pressure inside the bubbles is equal to the vapor pressure $p_v = 2{,}340$ Pa. Thereby we ensure that the initial pressure field does not contain spurious acoustics. It is further assumed that the velocity field is initially at rest and the initial temperature is $T = 293$ K. Viscous effects are neglected.

A series of 6 grids with 7^3, 14^3, 28^3, 55^3, 110^3, and 220^3 cubic cells is used to discretize the inner domain. On the finest computational grid (220^3 cells), the smallest bubbles are resolved by at least 2000 cells each, and each of the largest bubbles is resolved by about 25,000 cells. On the coarsest grid (7^3 cells), the bubbles are no longer resolvable but their vapor volume content contributes to the specified cell-averaged void fraction. Three of the initial fields are shown in Fig. 14.3. One horizontal cut plane shows the initial vapor volume fraction by using continuous coloring and the remaining cut planes show the vapor volume fraction by using cell-center coloring. The colored bottom-plane indicates the initial pressure field. The shapes of the bubbles are sharply represented by the fine grids (220^3, 110^3 not shown), while they are not resolved by the coarse grids (7^3, 14^3 not shown). The medium grids (55^3, 24^3 not shown) allow for partial resolution of the bubble shapes. The finest grid (220^3) requires a time-step size of 1.95×10^{-8} s, while for the coarsest grid (7^3) the resulting time-step size is 6.24×10^{-7} s.

For each computational grid we adopt two "numerical wall pressure transducers" located at the center of the bottom wall. The larger transducer records the average pressure on an area of 1×1 cm^2, while the smaller one covers an area of 1×1 mm^2. Using two sensors with different areas allows for a grid independent

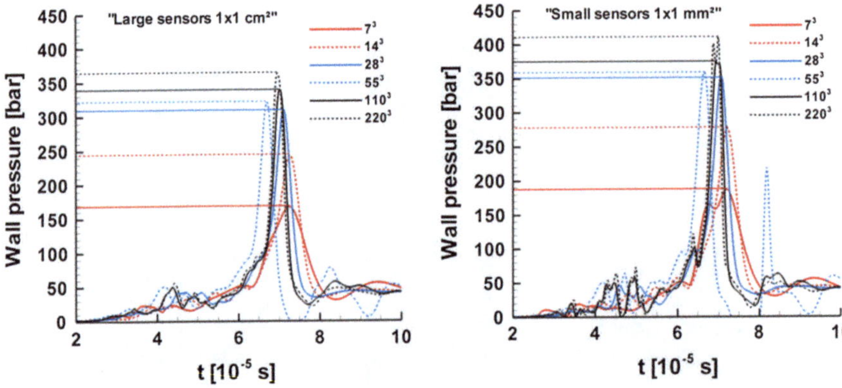

Fig. 14.4 Outputs of two numerical wall pressure transducers. Two numerical wall pressure transducers are adopted to record the pressure at the center of the bottom wall. The large sensor covers an area of 1×1 cm^2 and the small one covers an area of 1×1 mm^2. The figures show the obtained signals for the applied numerical resolutions. It is observed that the signals obtained with the large sensors (*left*) are slightly smeared and the amplitudes are slightly reduced, compared to the signals obtained with the small sensors (*right*)

evaluation of the pressure at the wall. Both transducers store the pressure for each instant in time, resulting in a maximum sampling frequency of 5.1×10^7 Hz (using 220^3 cells). The signals obtained by the large sensor are shown in Fig. 14.4—left, and the corresponding outputs of the small sensor are shown on the right hand side. As expected, the large sensor provides slightly smoother outputs with reduced amplitudes. The numerical sampling frequencies vary from 51 MHz (using 220^3 cells) to 1.6 MHz (using 7^3 cells). The peaks are more pronounced when computing on the fine grid and with high temporal resolution. In the present case, the duration of the shock-induced peak pressure is only about 5 μs for the run using 220^3 cells. On the coarse grid (run using 7^3 cells), the amplitude of the peak is reduced, and the peak is smeared. The small sensor detects a secondary collapse for the simulation using 55^3 cells (dashed blue line). This secondary event is actually the collapse of a small vapor pocket after rebound that occurs in this case close to the surface of the small sensor.

Aside from the "wall pressure transducers", the maximum pressure within the complete flow field is monitored during each simulation. The resulting data are provided in Table 14.1 and visualized in Fig. 14.5—left.

Before we analyze the obtained data, we investigate the following thought experiment: We consider the temporal evolution of a weak spherical shock front with given radius $R(t_2)$ and pressure amplitude $A(t_2)$, where t_2 is a given instant in time and $R(t_2)$ is the distance of the front to its origin. Since the shock is assumed to be weak, we may apply the linear theory [24] and obtain $A(t_2) \times R(t_2) = A(t_1) \times R(t_1)$. If we assume that the pressure amplitude and the radius are known at time $t_2 > t_1$, we can compute the amplitude $A(t_1)$ for any radius $R(t_1) > 0$ as long as the linear theory remains valid. With respect to numerical computations,

Table 14.1 Comparison of recorded maximum pressure within the domain and at the wall (small sensor and large sensor) and comparison of collapse durations with respect to six different numerical resolutions

Grid	p_{max} (domain) (Pa)	p_{max} (wall) small sensor (Pa)	p_{max} (wall) large sensor (Pa)	Collapse duration (s)
220^3	3.14×10^9	4.10×10^7	3.63×10^7	6.5×10^{-5}
110^3	1.17×10^9	3.74×10^7	3.40×10^7	6.6×10^{-5}
55^3	0.63×10^9	3.59×10^7	3.21×10^7	6.3×10^{-5}
28^3	0.34×10^9	3.53×10^7	3.10×10^7	6.7×10^{-5}
14^3	0.12×10^9	2.76×10^7	2.44×10^7	6.9×10^{-5}
7^3	0.07×10^9	1.87×10^7	1.68×10^7	7.0×10^{-5}

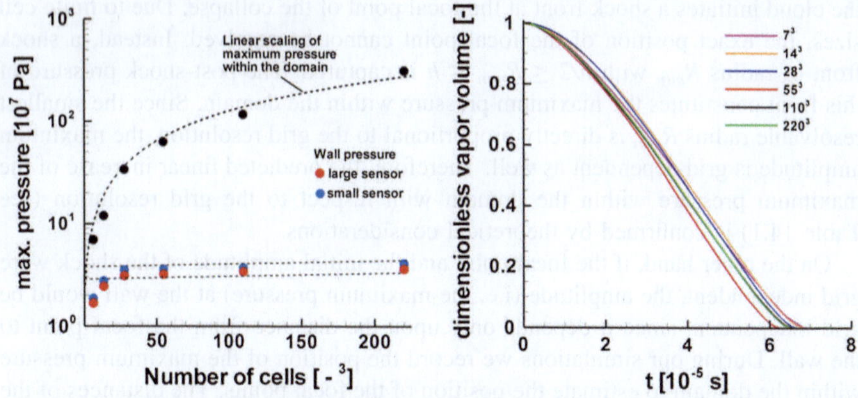

Fig. 14.5 Comparison of maximum pressures within the domain and at the wall (*left*) and collapse durations (*right*) using different numerical resolutions. On the left, the predicted maximum pressures (*logarithmic scale*) within the domain and at the bottom wall are plotted against the applied grid resolution (*number of cells*). One observes an approximately linear increase of the maximum pressure within the domain, and a very weak increase of the maximum pressures at the wall with increased grid resolution. On the right, the temporal evolution of the dimensionless vapor volume as predicted by different grids is shown. It can be concluded that the collapse durations compare well for all numerical resolutions

the smallest resolvable radius of a shock front is limited by the resolution of the applied computational grid. By assuming that the origin of the front is located at the center of a computational cell of size $h \times h \times h$, the smallest resolvable radius R_{min} is approximately $h/2$. Therefore, the maximum resolvable amplitude Amax is directly proportional to the resolution of the applied computational grid. Finally, one obtains $A_{max} = A(t_2) \times R(t_2) \,/\, R_{min} \approx 2 \times A(t_2) \times R(t_2) \,/\, h$, provided that the origin of the front is located at the center of a computational cell. If the origin of the front is not located at the cell center, the smallest resolvable radius is slightly larger ($h/2 \leq R_{min} \leq h$), and the corresponding maximum amplitude is smaller as well.

Table 14.2 Distance of the focal points to the wall as predicted by six different numerical resolutions

Grid	Wall distance of focal point (mm)	Cell size h (mm)
7^3	10 ± 1.43	2.86
14^3	9.3 ± 0.71	1.43
28^3	8.2 ± 0.36	0.71
55^3	7.8 ± 0.18	0.36
110^3	7.5 ± 0.09	0.18
220^3	7.4 ± 0.05	0.09

The result of this thought experiment allows for an illustrative interpretation of the data provided in Table 14.1 and visualized in Fig. 14.5—left: The collapse of the cloud initiates a shock front at the focal point of the collapse. Due to finite cell sizes, the exact position of the focal point cannot be resolved. Instead, a shock front of radius R_{min} with $h/2 \leq R_{min} \leq h$ is captured. The post-shock pressure of this front constitutes the maximum pressure within the domain. Since the smallest resolvable radius R_{min} is directly proportional to the grid resolution, the maximum amplitude is grid dependent as well. Therefore, the predicted linear increase of the maximum pressure within the domain with respect to the grid resolution (see Table 14.1) is confirmed by theoretical considerations.

On the other hand, if the focal point and the initial amplitude of the shock were grid independent, the amplitude (i.e. the maximum pressure) at the wall would be grid independent since it depends only upon the distance from the focal point to the wall. During our simulations we record the position of the maximum pressure within the domain to estimate the position of the focal points. The distances of the focal points to the wall are summarized in Table 14.2. Note that due to the grid dependent cell size h, an uncertainty of $h/2$ of the position of the focal point is assumed. We notice a slight displacement of the focal point towards the wall with increasing resolution. However, convergence towards the final position is clearly visible. It is supposed that the slight increase of the wall pressure is caused by the displacement of the focal point towards the wall, and by an improved resolution of the shock front. By using at least 28^3 computational cells the predicted wall pressures differ only by about 15 %. Therefore, the investigation demonstrates that the predicted wall pressure is only weakly grid dependent, provided that the position of the focal point is accurately captured.

Furthermore, we analyze the time history of the total vapor volume within the domain. Figure 14.5—right shows the time history of the vapor volume as predicted by the simulations. One observes that the predicted collapse durations agree well for all resolutions. These results confirm that the numerical methodology provides reasonable approximations of collapsing bubble clusters, even if the applied computational grid is far too coarse to resolve individual bubbles. However, this statement might not apply to dilute clusters, where the interaction of the involved bubbles is weak. More details on this investigation are given in Ref. [25].

14.5 Prediction of Erosion-Sensitive Areas in an Axisymmetric Nozzle-Target Flow

We simulate an experimentally investigated nozzle-target flow [26, 27] to evaluate the numerical predictability of erosion-sensitive areas. The experimental setup results in an axisymmetric stagnation flow, as shown in Fig. 14.6—left. At the exit of the nozzle the fluid accelerates along a small radius and forms a toroidal cavitation pocket. Collapse-induced erosion is quantified experimentally within an annulus at the surface of the target disc. The remaining parts are manufactured out of highly cavitation resistant material. Figure 14.6—right shows a cut through the computational domain. In accordance with the experiment, water at approximately room temperature enters the nozzle with an inlet velocity of $u_{in} = 31$ m/s. All observed flow details develop on their own without imposition of additional disturbances, such as random noise. An asymptotic pressure boundary condition $p_e = 10.1$ bar is imposed at the exit of a circular reservoir attached at a radial distance of 100 mm. All solid boundaries are treated as inviscid adiabatic walls, since the applied grid resolution is too coarse to resolve boundary layers. The computational grid consists of 5×10^5 body-fitted hexahedra.

We observe the onset of sheet cavitation at the rounded edge of the nozzle outlet. The fragmentation of the sheet and its transient shedding are strongly non-uniform in circumferential direction. The collapse of the vapor structures is generated by a positive pressure gradient in the radial direction. The part of the simulation analyzed here corresponds to a physical time interval of 7×10^{-2} s, and consists of 2.5×10^6 time steps with a time-step size of $\Delta t \approx 2.8 \times 10^{-8}$ s. Figure 14.7 shows a perspective view of two consecutive time instants with an increment $\Delta t = 2.8 \times 10^{-5}$ s. At the top, vapor structures (marked in red) are observed at a radial position of $\approx 2.6 \times 10^{-2}$ m. At the bottom, most of these vapor structures have already collapsed. At the focal point of the collapse a shock front is initiated that leads to a strong increase of the pressure (Fig. 14.7—bottom).

The duration of the complete collapse is about 2.8×10^{-5} s, which is resolved by approximately 1,000 time steps. Figure 14.8—right shows a photograph (from an angle) of an eroded target disc. The damage exhibits an almost perfect circular shape. Most pits are found at a radial distance between $r = 19$ and $r = 32$ mm measured from the nozzle axis [26]. In [27] the authors present pressure measurements for a series of operating points. They apply a flush-mounted sensor whose sensitive area has a diameter of 3.6 mm and the specified natural frequency is larger than 250 kHz. Pressures up to 400 bar are reported [27] for comparable operating conditions. However, the authors clearly state that the impact pressure can only be estimated if the impacted surface area is known, which is not the case. Figure 14.8—left shows the maximum pressure for each computational cell on the target disc recorded during the analysis interval of 7×10^{-2} s. Only pressures about one order of magnitude higher than the stagnation pressure are shown. The highest recorded value at the wall is 1,560 bar, but values between 250 bar and 700 bar seem to be predominant. The small discrepancy between the

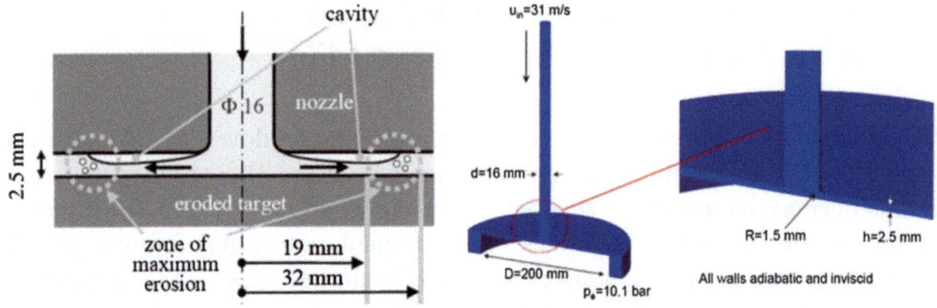

Fig. 14.6 Schematic and numerical model of the experimental setup. Based on [26], reproduced with permission of ASME

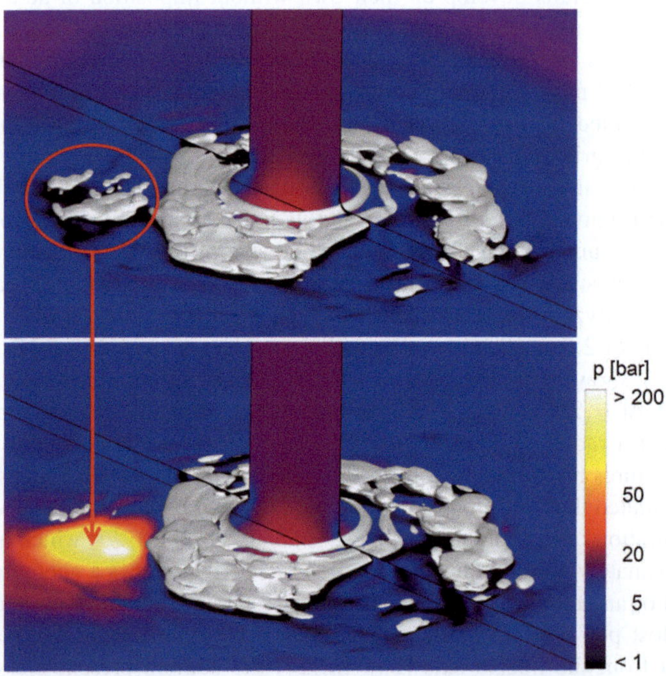

Fig. 14.7 Collapse of vapor structures and resulting shock wave. Iso-surfaces of a vapor volume fraction $\alpha = 10\ \%$ and pressure at the target disc and on a cut-plane at two consecutive instants in time. Top: Fragments of vapor structures are advected into regions of increased pressure. Due to the pressure gradient between the vapor structures and the surrounding liquid an acceleration of liquid towards the center of the vapor pockets is initiated and re-condensation occurs. Bottom: At the final stage of the collapse the inertia of the accelerated liquid results in the formation of a shock ("*water hammer*")

experimentally obtained pressure pulses and our numerical prediction might be related to the different sensitive areas. The sensitive area of the transducer applied in [27] is approximately 10 mm^2, while the areas of the "numerical transducers" are about 0.4–0.6 mm^2. We observe a convincing similarity of the maximum-pressure distribution and the areas of erosion damage as detected experimentally. However, further investigations using much larger analysis intervals are required to improve the statistical relevance of the prediction.

The "foot-print" of maximum pressures on the target disc already provides information about erosion-sensitive areas, but it does not provide information about the frequency of collapses or of the number of collapses above a certain threshold-pressure. To overcome these drawbacks, a numerical "collapse detector" was developed [28]. Cells where the vapor content has condensed entirely during the previous time step are denoted as "collapse candidates". If the surrounding cells of a "collapse candidate" contain only liquid, an isolated collapse is detected. Once a collapse is detected, the maximum pressure is recorded at that instant in time when the divergence of the velocity field changes its sign. The strength of the collapse is characterized by the maximum (negative) divergence of the velocity field and, in particular, by its maximum pressure. The main advantages of the collapse detector as compared to the maximum pressure approach are as follows: The collapse detector automatically distinguishes between collapse-induced maximum pressures and high pressures at stagnation points. Furthermore, the number of collapse events, as well as their position and their strength, provide important information about the stress profile the material is exposed to. This information can be used to estimate erosion rates. A potential drawback of the collapse detector could be the lack of information on the collapse intensity at material surfaces. As proposed in [28], an efficient projection method based on the linear decay law of spherical waves may be used to estimate the surface loads.

Figure 14.9 shows the collapses detected within the gap between the nozzle and the target disc during the analyzed part of the simulation. The collapses are indicated as spheres plotted at the position of collapse occurrence. The diameter of each sphere and its color represent the collapse intensity. On the left side, the collapse pressure is visualized, whereas the collapse pressure projected onto the target wall using the linear decay law of spherical waves is depicted on the right side, respectively. In both cases, most of the collapses are found within the marked area (red circles) where pits were detected experimentally. However, the thickness of the band where the strongest pitting was observed (dark band in Fig. 14.8—right) is narrower than our numerical prediction. Still, both numerical indicators (maximum pressure and detected collapses) might be suitable surrogates for the prediction of cavitation erosion.

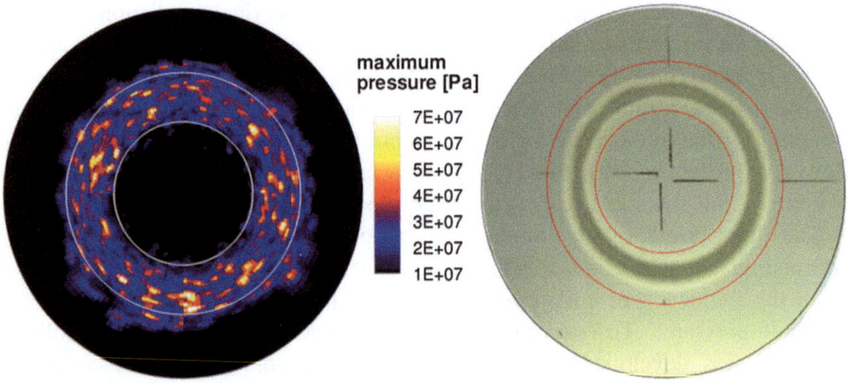

Fig. 14.8 Predicted maximum pressures at the surface of the target disc (*left*) and damaged target (*right*). Regions where the majority of pits occurred are indicated by circles. Based on [26], reproduced with permission of ASME

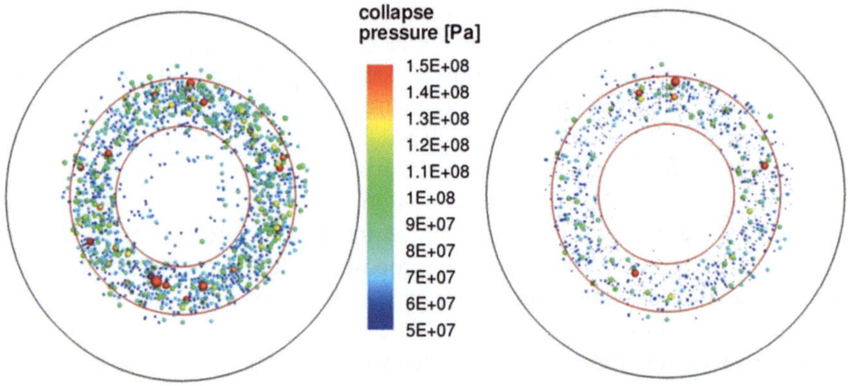

Fig. 14.9 Detected collapses within the gap between the nozzle and the target. Spheres represent detected collapses. Size and color of the spheres indicate the collapse intensity. On the left side, the intensity is the collapse pressure. On the right side, the collapse pressure is weighted by the inverse distance from the collapse location to the target as proposed in [28]. The application of this projection method slightly improves the prediction since the collapses within the inner circle disappear

14.6 Conclusions and Outlook

We assessed the numerical predictability of erosive events, such as collapsing vapor clouds, and resulting shocks. It is demonstrated that at least a certain range of large and medium-scale two-phase structures are represented correctly by our numerical simulations. In particular, typical features of hydrofoil cavitation, such as crescent-shaped regions [29] and irregular break-up patterns are well predicted.

By comparison of the collapse-characteristics of a numerically-resolved bubble cluster with the collapse of a locally homogeneous two-phase mixture it is demonstrated that important information about the collapse intensity and its duration can be obtained even if the numerical resolution is far too coarse to resolve individual bubbles. The ability to predict areas where intense collapse events damage the material is evaluated by the simulation of an experimentally investigated nozzle-target flow. We note a convincing agreement of numerically predicted collapse positions with experimentally observed pitting. Further investigations using state of the art Large Eddy Simulation techniques [10] are planned.

References

1. Knapp RT, Daily JW, Hammitt FG (1970) Cavitation. McGraw-Hill, New York
2. Brennen CE (1995) Cavitation and bubble dynamics. Oxford University Press, Oxford
3. Franc JP, Michel JM (2004) Fundamentals of cavitation. Springer, New York
4. Lecoffre Y (1999) Cavitation Bubble Trackers. Balkema, New York
5. D'Agostino L, Salvetti MV (2007) Fluid dynamics of cavitation and cavitating turbopumps. Springer, New York
6. Lauterborn W (1980) Cavitation and inhomogeneities in underwater acoustics. Springer, Berlin
7. Kendrinskii VK (2005) Hydrodynamics of explosion. Springer, Berlin
8. Gullbrand J, Chow FK (2003) The effect of numerical errors and turbulence models in large-eddy simulations of channel flow, with and without explicit filtering. J Fluid Mech 495:323–341
9. Garnier E, Adams N, Sagaut P (2009) Large eddy simulation for compressible flows. Springer, Berlin
10. Hickel S, Mihatsch M, Schmidt SJ (2011) Implicit large eddy simulation of cavitation in micro channel flows. In: Proceedings of WIMRC cavitation forum 2011, Warwick, UK, e-publication, 4–6 July 2011
11. Schnerr GH, Sezal IH, Schmidt SJ (2008) Numerical investigation of three-dimensional cloud cavitation with special emphasis on collapse induced shock dynamics. Phys Fluids 20(4):040703
12. Schmidt SJ, Sezal IH, Schnerr GH, Thalhamer M (2008) Riemann Techniques for the Simulation of Compressible Liquid Flows with Phase-transition at all Mach numbers—Shock and Wave Dynamics in Cavitating 3-D Micro and Macro Systems. In: 46th AIAA Aerospace Sciences Meeting and Exhibit, 7–10 January 2008, Reno, Nevada, AIAA paper 2008–1238
13. Grinstein FF, Margolin LG, Rider WJ (2007) Implicit large eddy simulation: computing turbulent fluid dynamics. Cambridge University Press, NewYork
14. Jiang GS, Shu CW (1996) Efficient implementation of weighted ENO schemes. J Comput Phys,126:202–228
15. Toro EF (1999) Riemann solvers and numerical methods for fluid dynamics. Springer, Berlin
16. Kennedy CA, Carpenter MH, Lewis RM (2000) Low-storage, explicit Runge-Kutta schemes for the compressible Navier-Stokes equations. Appl Numer Math 35:177–219
17. The International Association for the Properties of Water and Steam. http://www.iapws.org/
18. Hirschfelder JO, Curtiss CF, Bird RB (1954) Molecular theory of gases and liquids. Wiley, New York
19. Moran MJ, Shapiro HN, Boettner DD, Bailey MB (2011) Fundamentals of engineering thermodynamics. Wiley, New Jersey

20. Menikoff R, Plohr BJ (1989) The riemann problem for fluid flow of real materials. Rev Mod Phys 61:75–130
21. Trevena DH (1984) Cavitation and the generation of tension in liquids. J Phys D 17:2139–2164
22. Andersen A, Mørch KA (2011) In situ measurement of the tensile strength of water. In: Proceedings of WIMRC cavitation forum 2011, Warwick, UK, e-publication, 4–6 July 2011
23. Schmidt SJ, Thalhamer M, Schnerr GH (2009) Inertia controlled instability and small scale structures of sheet and cloud cavitation. In: Proceedings of 7th CAV 2009—7th international symposium on cavitation, Ann Arbor, Michigan, USA, 16.8–21.8.2009, paper 17, CD-ROM publication
24. Witham GB (1999) Linear and nonlinear waves. Wiley, New Jersey
25. Schmidt SJ, Mihatsch M, Thalhamer M, Adams NA (2011) Assessment of the prediction capability of a thermodynamic cavitation model for the collapse characteristics of a vapor-bubble cloud. In: Proceedings of WIMRC cavitation forum 2011, Warwick, UK, e-publication, 4–6 July 2011
26. Franc JP, Riondet M (2006) Incubation time and cavitation erosion rate of work-hardening materials. In: Proceedings of CAV2006—6th international symposium on cavitation, Wageningen, The Netherlands, 11–15 September 2006, CD-ROM publication
27. Franc JP, Riondet M, Karimi A, Chahine GL (2011) Impact load measurement in erosive cavitating flow. J Fluids Eng 133:121301–121303
28. Mihatsch M, Schmidt SJ, Thalhamer M, Adams NA (2011) Numerical prediction of erosive collapse events in unsteady compressible cavitating flows. In: Proceedings of marine 2011, International conference on computational methods in marine engineering, Barcelona, 2011
29. Reisman GE, Wang YC, Brennen CE (1998) Observations of shock waves in cloud cavitation. J Fluid Mech 355:255–283

Chapter 15
Scaling of Cavitation Bubble Cloud Dynamics on Propellers

Georges L. Chahine, Chao-Tsung Hsiao and Reni Raju

Abstract This paper addresses the issue of transposing laboratory scale results of cavitation on a propeller to other geometrically similar propellers of larger sizes. A particular emphasis is placed on nuclei dynamics and similarities and differences in the behavior of the bubbles at the different scales. Considering a realistic nuclei size distribution, an Eulerian–Lagrangian method is used to numerically model propeller flow and the motion, dynamics, and collection of the bubbles for a set of propeller sizes. Strong scaling effects are found for propellers operating under the same cavitation number and advance coefficient and in waters, which have the same nuclei density distribution. Inclusion of strong bubble interactions in the simulations is required for future cavitation erosion modeling efforts. A multi-bubble interaction model is introduced in the second part of the paper and a preliminary study of potential collective effects on the cavitation impulsive loads is presented.

15.1 Introduction

A major objective in cavitation erosion prediction studies is the development of tools and procedures to enable prediction of full-scale performance based on model-scale results. This requires the characterization of the cavitation intensity level at both scales and can be achieved through determination of the distribution and characteristics of the cavitation impulsive pressures at each scale (i.e. number, amplitude, duration, and spatial extent). One of the typical applications for such

G. L. Chahine (✉) · C.-T. Hsiao · R. Raju
DYNAFLOW, INC., 10621-J Iron Bridge Road, Jessup 20794 MD, USA
e-mail: glchahine@dynaflow-inc.com

K.-H. Kim et al. (eds.), *Advanced Experimental and Numerical Techniques for Cavitation Erosion Prediction*, Fluid Mechanics and Its Applications 106,
DOI: 10.1007/978-94-017-8539-6_15,

scaling and predictions is marine propellers, where little is known about full scale impulsive loads. A computational tool appears at this time as having the most potential for usefulness. This study presents some scaling results based on an Eulerian–Lagrangian approach. On-going and future efforts will test the validity of these predictions and will improve on some aspects of the present work.

Although various types of cavitation are known to occur on and around a propeller, the most erosive form is known to be cloud cavitation, where bubble clouds are shed from unsteady cavitation sheets [1–3]. These clouds have been reported to contain travelling bubbles interacting with each other and with the rest of the flow or being trapped in local micro-vortices. These clouds can behave collectively to generate extremely high pressures, which can in turn drive the collapse of micro-cavities on the propeller surface and result in damage. A cavitation *cascading effect* results from the following succession of events: large macroscopic flow pressure fluctuations, bubble cloud formation, very high cloud collapse pressures and shock waves generation, small bubbles driven by the cloud pressures, and impulsive loads leading to damage. Therefore, it is of great interest to model the cavitation impulsive loads due to this type of cavitation and deduce from it general scaling laws.

The dynamics of sheet-cloud cavitation formation on a rotating propeller is modeled here with an Eulerian viscous flow solver coupled with a Lagrangian bubble tracking model. These enable the study of the dynamics of bubble nuclei, their growth, collection into areas of high void fraction, and break-up of these regions back into micro-bubbles, which form bubble clouds and vortical structures propagating downstream. This approach offers a first-order approximation for computing scaling effects, deducing how cavitation impulsive loads change for different propeller sizes, and deriving some scaling laws.

15.2 Numerical Models

The numerical modeling of the cavitating flow field and the bubble dynamics considered in this paper is briefly described below and involves:

(a) a variable density finite volume method Navier–Stokes equations flow solver, 3DynaFS-Vis© [4–6], for modeling complex flows such as around a propeller,
(b) a Lagrangian particle tracking and bubble dynamics model, 3DynaFS-Dsm© [6, 7], and
(c) a bubble interaction model, PhantomCloud©, for bubble clouds modeling [8, 9].

We present below some details of the models (a) and (b), which are used for the propeller scaling study. Description of the model (c) is given at the beginning of Sect. 15.4 in order to present the cloud cavitation modeling study separately.

15.2.1 Viscous Flow Solver

3DYNAFS-VIS© is a Navier–Stokes flow solver based on a finite volume scheme [4–6]. The code solves three-dimensional unsteady incompressible continuity and momentum equations:

$$\frac{\partial u_i}{\partial x_i} = 0, \tag{15.1}$$

$$\frac{\partial u_i}{\partial t} + u_j \frac{\partial u_i}{\partial x_j} = -\frac{\partial p}{\partial x_i} + \frac{1}{R_e} \frac{\partial \tau_{ij}}{\partial x_j}, \tag{15.2}$$

where $u_i = (u, v, w)$ are the Cartesian components of the velocity, $x_i = (x, y, z)$ are the Cartesian coordinates, p is the pressure, $R_e = \rho\, u^* L^* / \mu$ is the Reynolds number, u^* and L^* are the characteristic velocity and length, ρ is the liquid density, and μ is its dynamic viscosity. The effective stress tensor τ_{ij} is given by:

$$\tau_{ij} = \left(\frac{\partial u_i}{\partial x_j} + \frac{\partial u_j}{\partial x_i} \right) - \frac{2}{3} \delta_{ij} \frac{\partial u_k}{\partial x_k}, \tag{15.3}$$

where δ_{ij} is the Kronecker delta.

The solver is based on the artificial-compressibility method [10] in which a time derivative of the pressure divided by an artificial-compressibility factor, β, is added to the continuity equation as

$$\frac{1}{\beta} \frac{\partial p}{\partial \tau} + \nabla \cdot \mathbf{u} = 0, \tag{15.4}$$

where τ is a pseudo-time for iteration and convergence of Eq. (15.4) to a divergence free velocity.

As a consequence, a hyperbolic system of equations is formed and can be solved using a time marching scheme. The solution is marched in pseudo-time, τ, to reach a steady-state solution. To obtain a time-dependent solution, a Newton iterative procedure is performed at each physical time step in order to satisfy the continuity equation.

This solver is capable of applying a level-set scheme to handle large free surface deformations and breakup [11]. It also uses dynamic regridding schemes and moving overset grid schemes to capture bubble dynamics [4, 12] and includes visco-elastic models for non-Newtonian flows [5]. 3DYNAFS-VIS© can be run in a coupled mode with 3DYNAFS-DSM© using an Eulerian/Lagrangian coupling approach for simulation of bubble/liquid two phase flows [4, 7, 11] and/or coupled with DYNA3D to solve fluid–structure interaction problems [13, 14].

15.2.2 Bubble Tracking and Dynamics Model

3DynaFS-Dsm© models the dynamics and tracks the motion of discrete bubbles. The bubble dynamics is simulated by solving either a modified Rayleigh-Plesset equation or a modified Keller-Herring equation [15], which apply the Surface Averaged Pressure (SAP) method [5–7]:

$$\begin{aligned}&\left(1-\frac{\dot{R}}{c}\right)R\ddot{R}+\frac{3}{2}\left(1-\frac{\dot{R}}{3c}\right)\dot{R}=\frac{(\mathbf{u}_{enc}-\mathbf{u}_b)^2}{4}\\&+\frac{1}{\rho}\left(1+\frac{\dot{R}}{c}+\frac{R}{c}\frac{d}{dt}\right)\left[p_v+p_g+p_{enc}-\frac{2\gamma}{R}-4\mu\frac{\dot{R}}{R}\right],\end{aligned} \tag{15.5}$$

where c is the sound speed, ρ is liquid density, p_v is the vapor pressure, p_g is the gas pressure, γ is the surface tension and μ is the viscosity. With the SAP method, Eq. (15.5) accounts for any slip velocity between the bubble and the host liquid, and also for a non-uniform pressure field along the bubble surface. The encounter pressure, p_{enc}, and velocity, $\mathbf{u}_{enc}$, are defined respectively as the average of the liquid pressures and velocities over the bubble surface.

The use of p_{enc} results in a major improvement over the classical spherical bubble model which uses the pressure at the bubble center in its absence [4, 5]. The bubble trajectory is obtained using the following bubble motion equation [4, 5]:

$$\begin{aligned}\frac{d\mathbf{u}_b}{dt}&=-\frac{3}{\rho}\nabla P-2g+\frac{3}{4}\frac{C_D}{R}(\mathbf{u}_{enc}-\mathbf{u}_b)|\mathbf{u}_{enc}-\mathbf{u}_b|\\&+\frac{3}{2\pi}\frac{C_L}{R}\sqrt{\frac{\mu}{\rho}}\frac{(\mathbf{u}_{enc}-\mathbf{u}_b)\times\boldsymbol{\omega}}{\sqrt{|\boldsymbol{\omega}|}}+\frac{3}{R}(\mathbf{u}_{enc}-\mathbf{u}_b)\dot{R},\end{aligned} \tag{15.6}$$

where $\mathbf{u}_b$ is the bubble velocity, g is the acceleration of gravity, C_L is the lift coefficient, $\boldsymbol{\omega}$ is the local vorticity, and C_D is the drag coefficient given by an empirical equation such as that of Haberman and Morton [16]:

$$C_D=\frac{24}{R_{eb}}\left(1+0.2R_{eb}^{0.63}+2.6\times10^{-4}R_{eb}^{1.38}\right),\ R_{eb}=\frac{2\rho R|\mathbf{u}_{enc}-\mathbf{u}_b|}{\mu}. \tag{15.7}$$

The last term on the right hand side of Eq. (15.6) is the force due to the bubble volume variations, which is obtained by solving Eq. (15.5).

3DynaFS-Dsm© was developed to simulate complex flows involving particles and/or bubbles, vortices, or other singularities. It can be coupled with 3DynaFS-Vis© through an Eulerian/Lagrangian coupling approach [5, 7, 11] and with 3DynaFS-Bem© through a Lagrangian/Lagrangian coupling approach [13].

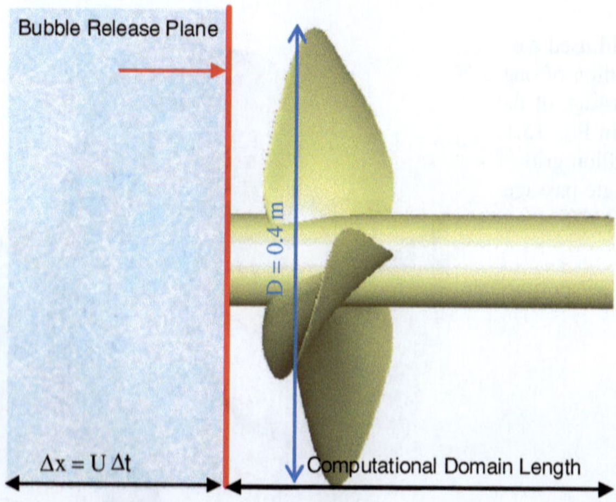

Fig. 15.1 Geometry of the NSWCCD open propeller 5530 used for the simulations shown below and illustration of the fictitious volume upstream of the propeller and the computational domain, which was used to feed randomly distributed nuclei with a given size distribution into the inlet of the computational domain. The length of the fictitious domain, Δx, was determined by the incoming upstream velocity, U, and Δt, the desired physical duration of the computation. Its radius is the same as that of the computational domain

15.3 Cavitation Dynamics and Pressure Loading on a Rotating Propeller

15.3.1 Computational Domain and Grid

The basic propeller chosen for the simulations shown below is the NSWCCD three-bladed open propeller 5530, which has a diameter D = 2R = 0.4 m as illustrated in Fig. 15.1. 3DYNAFS-VIS© was used to solve the unsteady Reynolds-Averaged Navier–Stokes (RANS) equations in a frame of reference rotating with the blades. Axial symmetry and periodic boundary conditions were used on both the suction and pressure sides of the blade to compute the flow field in one blade-to-blade passage domain.

Following the work of Hsiao and Pauley [17] an H–H grid with 2.1 million grid points was used for the RANS calculations. The grid followed the incoming flow angle forming a spiral domain as shown in Fig. 15.2, with the inflow located at 1R and the outflow at 3R from the propeller mid-plane. The computational domain extended radially to a distance 2R from the hub, and a free stream boundary condition was imposed on the external domain edge while the hub was represented as a rigid boundary. Care was taken to ensure that the grid points closest to the blade surface were located at distances of at least $y^+ \sim 2$. Turbulence was modeled using the one-equation Baldwin Lomax model [18].

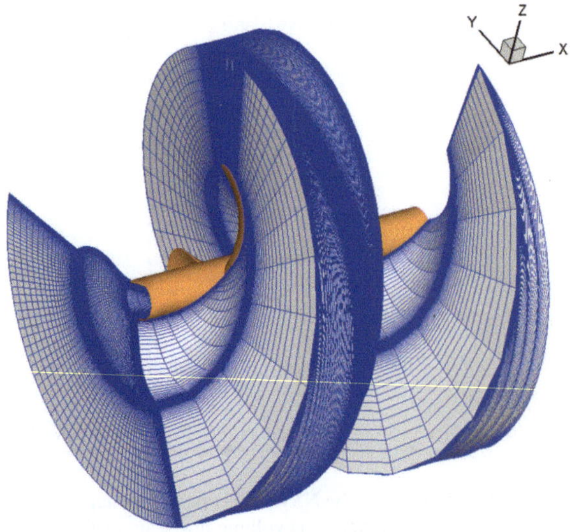

Fig. 15.2 H-H computational grid used for the RANS simulation of one blade-to-blade passage of the propeller shown in Fig. 15.1. A total of 2.1 million grid points for the single passage was used in the RANS calculations

15.3.2 Pressure Distribution on Propeller

We present below two sets of RANS computations (liquid phase only) for a uniform inflow and two values of the advance coefficient, J, and the cavitation number, σ,

$$J = \frac{U_\infty}{nD}, \tag{15.8}$$

$$\sigma = \frac{P_\infty - P_v}{\frac{1}{2}\rho U_\infty^2}, \tag{15.9}$$

where n is the number of propeller revolution per unit time, D is the propeller diameter, and U_∞ is the advance speed of the propeller or the free stream velocity. P_∞ is the free stream pressure at the same depth as the propeller axis. The two cases are selected for illustration purposes as they represent two very different cavitation configurations. For both cases n was kept at 1,600 rpm, while two values of U_∞ were considered: 9.6 m/s and 11.9 m/s.

The resulting pressure distributions for these two cases are shown in Fig. 15.3, which presents contours of the pressure coefficient, $C_p = (p - P_\infty)/(0.5\rho U_\infty^2)$, on the suction-side surfaces of the propeller blades and in a plane perpendicular to the propeller axis right behind the propeller tip trailing edge. The contours for the two flow conditions show that the low pressure region covers the mid-chord region of the blade for $J = 1.1$ and $\sigma = 1.4$, while it is concentrated along the leading edge and the blade tip for $J = 0.9$ and $\sigma = 4.0$. This can be understood by considering the flow-blade effective angle of attack under each condition.

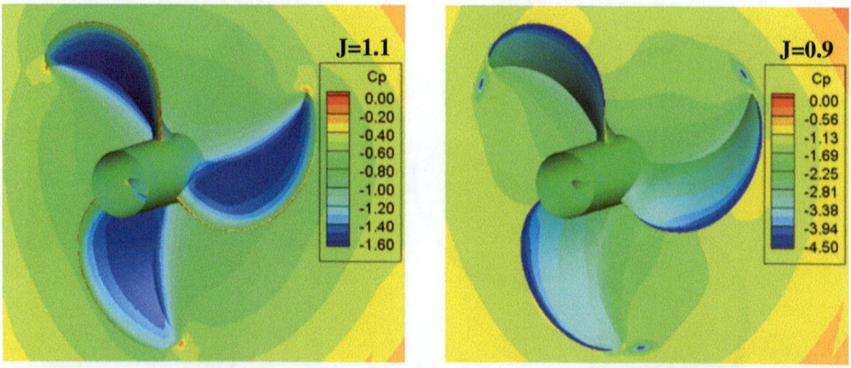

Fig. 15.3 Contours of the pressure coefficient, C_p, on the suction-side surfaces of the 5530 open propeller blades and on a plane perpendicular to the propeller axis right behind the propeller tip trailing edge. *Left* $n = 1{,}600$ rpm, $U_\infty = 11.9$ m/s, J = 1.1, $\sigma = 1.4$. *Right* $n = 1{,}600$ rpm, $U_\infty = 9.6$ m/s, J = 0.9, $\sigma = 4.0$

Fig. 15.4 3DynaFS© Eulerian–Lagrangian simulations of cavitation on the open propeller 5530. $U_\infty = 11.9$ m/s, J = 1.1, $\sigma = 1.4$, $n = 1{,}600$ rpm. Contours show pressure distributions. Simulations indicate formation of traveling bubble cavitation and sheet to cloud cavitation starting from the mid-section of the blade

This implies that two very different cavitation patterns will occur in each case as seen in Figs. 15.4 and 15.5. Cavitation inception occurs much earlier for the $J = 0.9$ case because c_{pmin} is much lower. In addition, a well-defined tip vortex cavitation is expected for the $J = 0.9$ case as the cavitation number decreases.

Fig. 15.5 3DYNAFS© Eulerian–Lagrangian simulations of cavitation on the open propeller 5530. $U_\infty = 9.6$ m/s, J = 0.9, $\sigma = 4.0$, $n = 1{,}600$ rpm. Contours show pressure distributions. Simulations indicate formation of traveling bubble cavitation at the leading edge and strong tip vortex cavitation

15.3.3 Eulerian–Lagrangian Cavitation Simulations

In order to simulate realistic water conditions, a nuclei size distribution was selected and used to seed the water. The bubble sizes ranged from 10 to 200 μm and the distribution followed a classical power law of coefficient 4, $N(R) \sim R^{-4}$ [19–22] and the void fraction was 2.9 x 10^{-6}. The nuclei were distributed randomly in a fictitious supply volume, which fed the inlet surface of the computational domain at the free stream velocity ($U_\infty = 11.9$ m/s for $J = 1.1$, $U_\infty = 9.6$ m/s for $J = 0.9$). The supply volume size was determined by multiplying the release area by the desired physical duration of the simulation and the characteristic inlet velocity as illustrated in Fig. 15.1. Computations of bubble dynamics, trajectories, and resulting flow and pressure fields were conducted.

Figures 15.4 and 15.5 show the distribution (location and size) of the visible bubbles and bubble clouds for $J = 1.1$ and $\sigma = 1.4$, and for $J = 0.9$ and $\sigma = 4.0$, respectively. Note that no attempt was made to model sheet cavitation and sheet-to-cloud cavitation in these simulations. Instead, by concentrating on travelling cavities, detecting when they grow explosively (e.g. when the volume of a bubble increases by a factor, say, larger than 100) and examining their collection in clouds, we hope to deduce scaling laws for different propeller sizes. Both front and side views of the propeller field are shown in the figures at selected instants to highlight the concentration of cavitation occurring on the propeller blade surfaces, as well as in the trailing tip vortex.

By comparing the two cases, one can easily see two very different cavitation patterns. For the relatively high advance coefficient, $J = 1.1$, and low σ, only partial sheet-to-cloud type cavitation is observed at the mid-chord of the blade,

Table 15.1 Simulation conditions and the corresponding number of nuclei used for keeping the void fraction α the same

Propeller diameter (m)	Scaling factor (λ)	No. of nuclei (λ^3) (thousands)
0.4	1	2
0.6	1.5	6.75
0.8	2	16
1.2	3	54
1.6	4	128
2.0	5	250

This allows varying λ only, separating its effects from the effects of α

while both very short sheet-like cavitation (formed by the collection of all the bubbles) and tip vortex cavitation appear at the low advance coefficient, $J = 0.9$. Furthermore, for the $J = 0.9$ case sheet-like cavitation appears near the leading edge while tip vortex cavitation starts in the blade tip region near the trailing edge where more intensive bubble growth and collapse events can be observed. This implies that the tip and leading edge areas are subject to more intensive cavitation erosion when the propeller is operated under this higher loading condition. We note again that the computations presented here do not include bubble interactions and actual collective effects. These are the subject of ongoing studies at DynaFlow.

15.3.4 Scaling of Propeller Cavitation Dynamics

In order to study the effect of propeller scale on the cavitation dynamics six different propeller sizes were selected and the 3DynaFS© cavitation computations conducted. The propeller diameters ranged from 0.4 to 2.0 m, as shown in Table 15.1, resulting in a geometric scaling from $\lambda = 1$ to $\lambda = 5$. The same random nuclei distribution with a selected size distribution and the same initial void fraction, $\alpha = 2.9 \times 10^{-6}$, were considered for all cases. As a result the total number of nuclei involved in the computations increased as λ^3 while the initial nuclei size and distribution were kept the same. The actual number used in each simulation is shown in Table 15.1.

Figure 15.6 illustrates a time history of a bubble growing explosively then collapsing strongly in the propeller flow field. Shown are the bubble radius (blue curve) and the emitted field pressure (red curve) for a 50 μm radius bubble. The blue curve shows the bubble radius versus time when the bubble crosses a low pressure region and then implodes in a pressure recovery region of the propeller blade (the encountered pressures are further explained in Fig. 15.8). The bubble is seen to grow explosively from 50 μm to about 3.5 mm before executing multiple collapses and rebounds due to its interaction with the propeller local flow field. As shown by the red curve, this results in a pressure on the blade emitted by the bubble collapse, which involves multiple impulsive peaks. The total pressure on

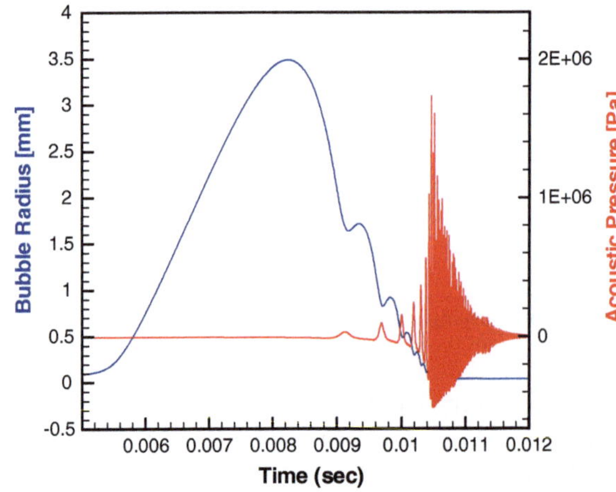

Fig. 15.6 Illustration of the dynamics of a 50 μm radius bubble moving through the 0.4 m diameter open propeller 5530 flow field. *Blue curve* Bubble radius versus time. *Red curve* Emitted pressure versus time at the blade surface. $U_\infty = 9.6$ m/s, J = 0.9, $\sigma = 4.0$, n = 1,600 rpm. 3DynaFS© Eulerian–Lagrangian simulations

the propeller blade surface will be a combination of the emitted pressures by all cavitating bubbles in the flow field. Note that the pressures illustrated in this figure are high for sound emission (cavitation noise) but are extremely low for erosion.

15.3.4.1 Scaling Study Framework

Model propeller experiments in the water tunnel are generally conducted at the same cavitation number σ, and the same advance coefficient J, for different scale propellers. In most water tunnels, the rotational speed, n, is varied while keeping the tunnel speed, U_∞, the same in order to maintain the same J values for different propeller diameters.

In the scaling studies presented below, we also conserve σ and J. This results in conserving the incoming velocity U_∞, and varying the rotational speed, n. In addition, we will also conserve the initial void fraction and the initial bubble sizes and initial bubble size distribution for all scales.

15.3.4.2 Bubble Size Scaling

In order to understand how the bubble dynamics is affected by changes in the geometric scales, we compare for the various scales the bubble dynamics along the same scaled flow trajectories (i.e. for bubbles having the same normalized initial positions and following the same normalized trajectories relative to the propeller).

Figure 15.7a illustrates the bubble radius versus time for all simulated scales. Figure 15.7b shows bubble radius versus time with both radii and times normalized by the geometrical scaling factor λ. Doing so accounts for the fact that bubbles growing explosively reach an asymptotically linear growth speed and grow λ times larger if exposed to the same pressure field for a λ times longer

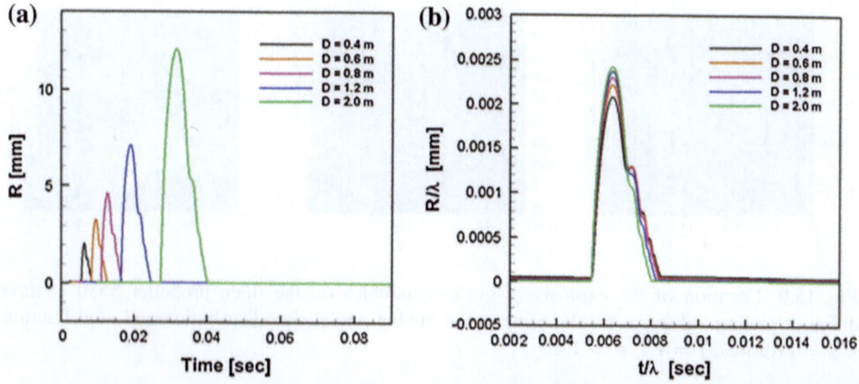

Fig. 15.7 Illustration of the behavior of a 50 μm radius bubble in the flow field of the open propeller 5530 at five different scaled propeller diameters, D. **a** Bubble size variations and **b** normalized curves, where both radii and times have been divided by the geometric size ratios, λ. $U_\infty = 11.9$ m/s, J = 1.1, $\sigma = 1.4$

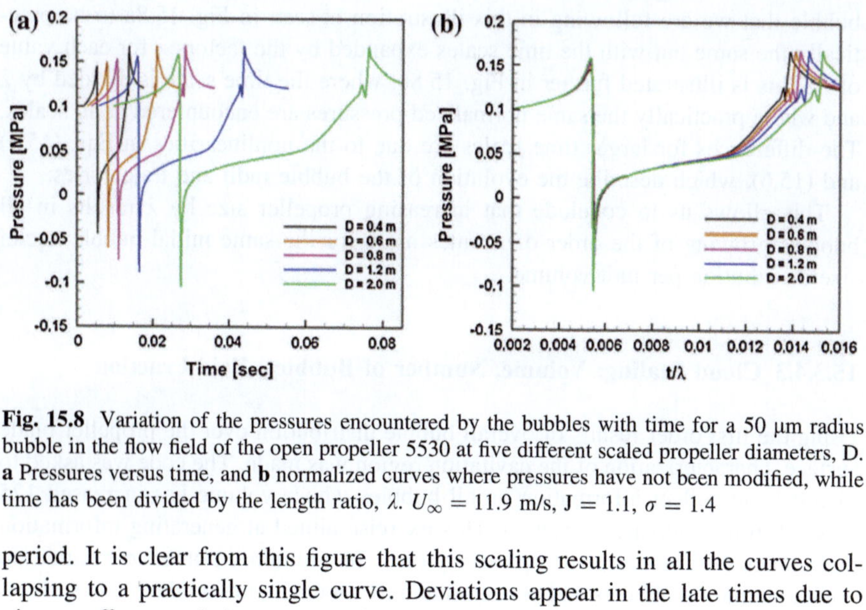

Fig. 15.8 Variation of the pressures encountered by the bubbles with time for a 50 μm radius bubble in the flow field of the open propeller 5530 at five different scaled propeller diameters, D. **a** Pressures versus time, and **b** normalized curves where pressures have not been modified, while time has been divided by the length ratio, λ. $U_\infty = 11.9$ m/s, J = 1.1, $\sigma = 1.4$

period. It is clear from this figure that this scaling results in all the curves collapsing to a practically single curve. Deviations appear in the late times due to viscous effects not being conserved at the various scales. At larger scales increased viscous drag slows down the bubbles relatively more than at the smaller scales. This increases the duration that bubbles are exposed to the propeller pressure field, resulting in further increase in size and deviation from perfect scaling.

As the propeller size increases, the bubble grows to a λ times larger size even though the initial radius is the same. Since the propellers are geometrically scaled, the pressures seen by the bubbles at the scaled locations are the same (this is illustrated in Fig. 15.8b). On the other hand, when the propeller is λ times larger

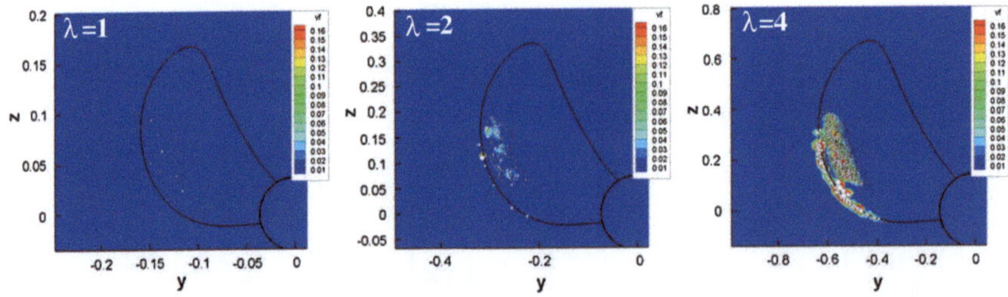

Fig. 15.9 Location of the explosively growing bubbles on the open propeller 5530 at three different scaling values, λ, for the same initial bubble nuclei size distribution and void fraction. $U_{\infty} = 11.9$ m/s, J = 1.1, $\sigma = 1.4$

the duration of the bubble exposure to the propeller pressure is also λ times larger for the same incoming velocity.

Since the incoming velocity U_{∞} is conserved for different scales under the framework described in the previous section, increasing the propeller size by λ results in stretching the time scale by a factor λ. The pressure encountered by the bubble that we are following in this illustration is seen in Fig. 15.8a to be practically the same but with the time scales expanded by the factors λ for each value of λ. This is illustrated further in Fig. 15.8b, where the time scale is divided by λ and where practically the same normalized pressures are encountered at all scales. The differences for larger time scales are due to the nonlinearities in Eqs. (15.5) and (15.6), which describe the evolution of the bubble radii and trajectories.

This allows us to conclude that increasing propeller size by λ results in all bubbles growing of the order of λ times more for the same initial bubble nuclei size distribution per unit volume.

15.3.4.3 Cloud Scaling: Volume, Number of Bubbles, Void Fraction

Using the first order results of overall bubble distribution over the propeller blade surface, characterization of the cavitation region was made. The code was asked to identify and collect information on all bubbles whose volume has increased 125 times (5 times increase in radius). This exercise aimed at generating information on bubbles that have grown explosively only. In addition, in order to exclude tip vortex cavitation bubbles and concentrate only on cavitation on the blade surface, additional constraints were placed on the location of the bubbles to be counted. This automated screening was intended to reflect experimental evaluations of cavitation extent, which resemble the comparisons for different λ in Fig. 15.9. This figure shows a characteristic void fraction distribution at selected times for each scale (λ = 1, 2, and 4). As the scale increases, the number of bubbles in the field increases for the same initial free field void fraction (same free field nuclei concentration). Also, as the scale increases, the bubbles grow to larger sizes thus

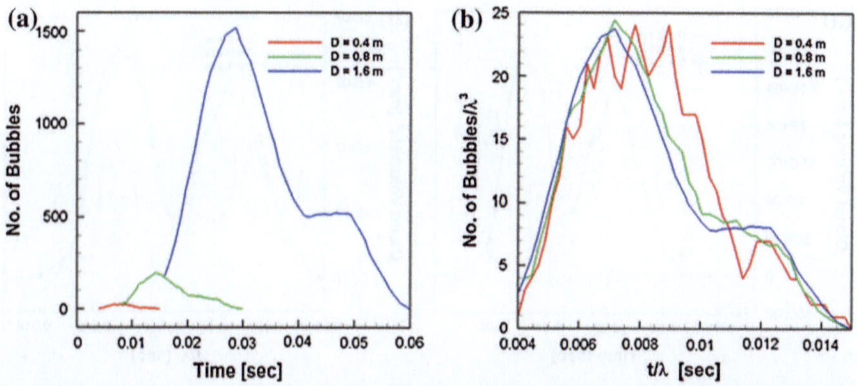

Fig. 15.10 Comparison of the number of explosively growing bubbles over the blade for three different propeller diameters geometrically scaled by λ. **a** Actual number of cavitation activated bubbles versus time. **b** Number of activated bubbles divided by λ^3 and the times divided by λ. $U_\infty = 11.9$ m/s, J = 1.1, $\sigma = 1.4$

increasing the local void fractions. It is clear from this figure that cavitation scaling effects are very important and need to be carefully studied. Even though the cavitation number and the advance coefficient are the same for various scales, and the initial nuclei size distribution and void fraction are the same, as propeller size increases, cavitation extent or advancement far exceeds pure geometric scaling. The three figures in Fig. 15.9 actually have qualitatively the appearance of the same propeller when reducing the cavitation number.

The results in Fig. 15.9 are further expanded in Figs. 15.10 and 15.11. The number of *activated* bubbles versus time is seen in Fig. 15.10a for the three scales. Good collapse of the curves is seen in Fig. 15.10b when the bubble number is divided by λ^3 and the time is divided by λ. The smallest scale $\lambda = 1$ shows significant fluctuations because of the smaller number of nuclei involved in that computation (low statistics). The smoothness of the curve is expected to significantly improve with increasing number of nuclei (and computational times).

Figure 15.11 compares the sum of the volumes versus time of the bubbles identified as growing explosively between three scales. As expected, the largest scales show much higher "bubble cloud" volumes due to a combination of increased number of bubbles in the field and enhanced bubble size growth. From pure scaling reasoning the number of cavitation *activated* bubbles (i.e. the number of bubbles whose volumes increase for example 125 times) should be proportional to λ^3, as the number of nuclei involved in the computations. Since the bubble radii scale with λ, the total volume of all bubbles is expected to scale as λ^6. On the other hand the volume of the cloud, i.e. volume of the region where the growing bubbles is located scales only as λ^3 (see summary in Table 15.2). Therefore the void fraction in the cloud, ratio of total bubbles volume to cloud volume is expected to scale as λ^3.

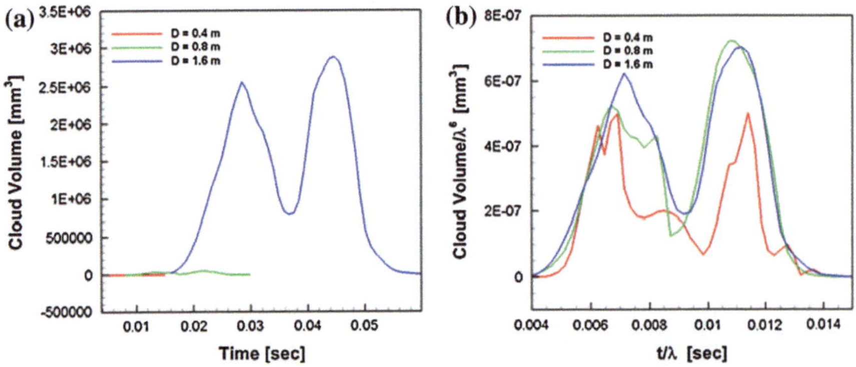

Fig. 15.11 Comparison of the total volume of the bubbles explosively growing over the blade for three different propeller diameters geometrically scaled by λ. **a** Actual volume of cavitation activated bubbles versus time. **b** Volume of activated bubbles divided by λ^6 and the times divided by λ. $U_\infty = 11.9$ m/s, J = 1.1, $\sigma = 1.4$

Figure 15.11a shows the total volume of activated bubbles (bubble cloud) versus time. The difference of the dimensional values of the volumes between the scales is extremely large. In order to better compare the various cases, the volumes are divided by λ^6 while the times are divided by λ. This shows a better comparison with the curves not fully collapsed to a single curve for the same reason mentioned above, i.e. due to low statistics for the smaller scales.

15.3.4.4 Pressure Scaling

As discussed in Sect. 15.4, actual cloud cavitation pressures are the result of collective bubble dynamics, which are not captured yet by the approach presented in this section where bubble interactions were not considered, and where cloud cavities emanate mostly from the breakup and shedding of bubble clouds from receding sheet cavities. Such computations require additional developments of ongoing modeling efforts at DynaFlow and elsewhere. Here, we consider simply the summation of the pressures due to the dynamics and collapse of the traveling bubbles. This is meant as just an indication of the potential scaling of the pressures driving the individual bubble collapses, which are actually responsible for cavitation erosion.

Figure 15.12 compares for various scales the cumulative pressure generated by the bubbles at a given location on the propeller blade after dividing times by the scale λ. Pressures are seen to increase with the propeller size, but the derivation of a simple power law scaling of λ^α is more difficult. This is probably due to the fact that while the bubble number and size increases with λ (thus increasing the pressure, p, with λ), also the distance of these bubbles from the surface increases with λ, which results in an opposite effect (reduction of p with λ). Note the

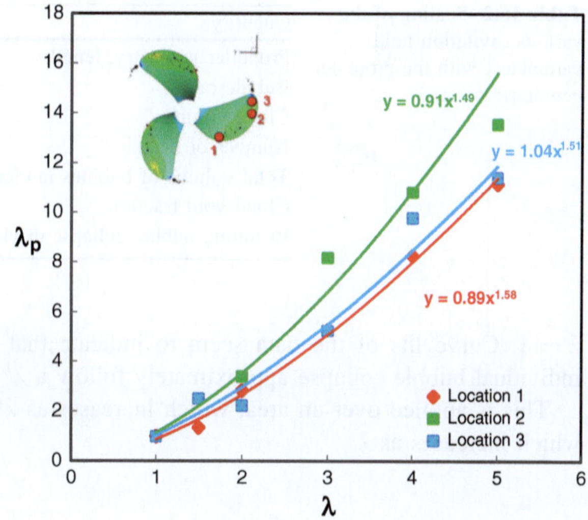

Fig. 15.13 Ratio, λ_p, between the peak pressure value for a given propeller size (geometrical scale parameter, λ) and that for $\lambda = 1$, as a function of λ from computations at three locations on the propeller blade shown in the *upper left* insert. $U_\infty = 11.9$ m/s, J = 1.1, $\sigma = 1.4$

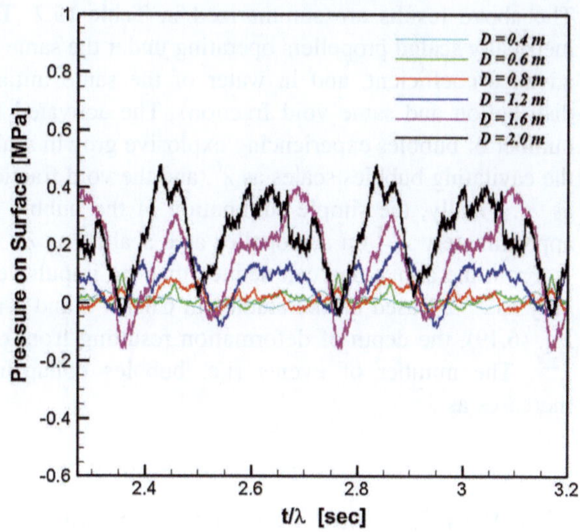

Fig. 15.12 Comparison of the pressure versus time at a point on the blade (location 1 in Fig. 15.13) for the six different propeller diameters, D. $U_\infty = 11.9$ m/s, J = 1.1, $\sigma = 1.4$

relatively low frequency repeated oscillations in Fig. 15.12, which correspond to a much lower frequency than the bubble oscillations frequency. These low frequency oscillations correspond to the overall "cloud" volume oscillations and shedding.

In order to analyze this further, the peak pressures at three different locations on the blade are compared between the scales. Plotted in Fig. 15.13 are the ratios, λ_p, between the peak pressure value for a given λ (different propeller size) and that for

Table 15.2 Scaling of the various cavitation field parameters with the propeller geometric scale, λ

Quantity	Scaling
Propeller geometry, lengths	λ
Bubble radii	λ
Cloud volume	λ^3
Number of nuclei	λ^3
Total volume of bubbles in cloud	λ^6
Cloud void fraction	λ^3
Resulting bubble collapse driving pressures	$\lambda^{1.5}$

$\lambda = 1$. Curve fits of the data seem to indicate that the peak pressures driving individual bubble collapse approximately follow a $\lambda^{1.5}$ power law.

This is applied over an area, which increases as λ^2, and over a time duration, which increases as λ.

15.3.4.5 Scaling Summary

The above results are summarized in Table 15.2. These results apply for geometrically scaled propellers operating under the same cavitation number, the same advance coefficient, and in water of the same initial quality (same nuclei size distribution and same void fraction). The activated bubble sizes scale as λ, the number of bubbles experiencing explosive growth scales as λ^3, the total volume of the cavitating bubbles scales as λ^6, and the void fraction in the cloud region scales as λ^3. Finally, the simple summation of the bubble collapse pressures scales as approximately $\lambda^{1.5}$ on a propeller area scaling as λ^2.

From the above one can deduce that the impulsive loads (pressure times area) vary as $\lambda^{3.5}$. Based on the studies in Chaps. 3 and 6 in the first part of this book, Eq. (6.19), the depth of deformation resulting from one impact load increases as $\lambda^{2.5}$. The number of events (i.e. bubbles collapsing near the blade surface) increases as λ^2.

15.4 Bubble Cloud Cavitation Dynamics

The collapse of clouds of cavitating bubbles is presently recognized as the most aggressive and erosive form of cavitation and has been the subject of many studies starting with the pioneering works of van Wijngaarden [23], Mørch [24], d'Agostino and Brennen [25], and Chahine [26]. These studies have shown that bubble collective effects in the cloud result in much enhanced collapse pressures exerted over longer periods of time. To enable simulation of bubble cloud dynamics and deduce the scaling of bubble clouds, a module of 3DynaFS© addressing multi-bubble interaction with a large number of bubbles, PhantomCloud© is used.

15.4.1 Cloud Cavitation Modeling

PhantomCloud© stems from earlier work [26–32] on bubble cloud modeling and from asymptotic expansions modeling of bubble dynamics and deformations [33–35]. It also includes the more recently introduced concept of the Surface Averaged Pressure (SAP) scheme in 3DynaFS© [5, 6]. PhantomCloud© extends these earlier concepts to strong interactions (i.e. interactions appear at the leading order) of multiple bubbles with each other and with the surrounding flow field without considering the effects on the shape of the individual bubbles (i.e. including only the effects of the interaction on bubble volume and bubble motion but not on bubble shape). The cloud itself behaves in a fully three-dimensional fashion. Each bubble can be represented by a series of singularities; e.g. a source/sink to represent volume change, a dipole to represent slip velocity relative to the surrounding liquid, and higher order terms for bubble deformations…etc.

The volume change source term can be obtained through a spherical bubble equation by solving a modified Keller-Herring-Rayleigh-Plesset (KHRP) equation, which uses the SAP approach to determine the relevant liquid pressure and velocity, and which accounts for the presence of all the other bubbles and compressibility of the bubbly medium. In the case of N bubbles in the computational domain, the modified KHRP equation can be written as follows for each bubble:

$$\left(1-\frac{\dot{R}_i}{c}\right)R_i\ddot{R}_i+\frac{3}{2}\left(1-\frac{\dot{R}_i}{3c}\right)\dot{R}_i=\frac{(\mathbf{u}_{enc,i}-\mathbf{u}_{bi})^2}{4}+\frac{1}{\rho}\left(1+\frac{\dot{R}_i}{c}+\frac{R_i}{c}\frac{d}{dt}\right)\times\left[p_v+p_{gi}-p_{enc,i}-\frac{2\gamma}{R_i}-4\mu\frac{\dot{R}_i}{R_i}\right]\quad i=1,2,\ldots,N. \tag{15.10}$$

Here R_i is the instantaneous equivalent bubble radius of index i, c is the sound speed, p_v and p_{gi} are the vapor and the gas pressure inside the bubble i. γ is the surface tension, and ρ and μ are the liquid density and dynamic viscosity. The first term in the right hand side of the equation is a pressure term due to the slip velocity between the bubble, whose center of gravity translates at the velocity, $\mathbf{u}_{bi}$, and the surrounding liquid of bubble surface average velocity $\mathbf{u}_{enc,i}$. The pressure term, $p_{enc,i}$, is fundamental in this approach and includes the contributions of all the other bubbles. In addition, it also uses the SAP scheme in which the instantaneous pressures and velocities are averaged over the bubble surface to account for non-uniformities of the pressure and velocity fields, and deduce the pressures and velocities 'felt' by each bubble.

The translation velocity of each bubble is taken into account through a dipole term, which expresses the flow around the approximated sphere. The equations for the size and motion for each bubble i can be expressed as functions of the pressure and the flow field around each bubble [5, 6], which account for any existing basic flow field (e.g. due to a nearby body motion) as well as the flow due to all other bubbles, and the acceleration of gravity, $\mathbf{g}$:

$$\frac{d\mathbf{u}_{bi}}{dt} = -2\mathbf{g} + \frac{3}{4}(\mathbf{u}_{enc,i} - \mathbf{u}_{bi})\frac{|\mathbf{u}_{enc,i} - \mathbf{u}_{bi}|C_d}{R_i} + 3\frac{\dot{R}_i}{R_i}(\mathbf{u}_{enc,i} - \mathbf{u}_{bi}) \\ - \frac{3}{\rho}\nabla p_{enc,i} + 12.88R_i^2\sqrt{\rho\mu}\frac{(\mathbf{u}_{enc,i} - \mathbf{u}_{b,i}) \times \mathbf{\Omega}}{\sqrt{|\mathbf{\Omega}|}}, \tag{15.11}$$

where C_d is a bubble drag coefficient, $\mathbf{\Omega}$ is the local vorticity, and i is the index of the particular bubble.

At a point $\mathbf{X}$, the influence from bubble j of radius R_j and radial velocity $\dot{R}_i$, expressed in terms of the velocity potential, ϕ_j, is [9, 26–30]:

$$\phi_j = -\frac{R_j^2 \dot{R}_j}{r_j} - \frac{R_j^3}{2}\frac{1}{r_j^2}\mathbf{V}_{Bj} \cdot \mathbf{e}_{rj}, \tag{15.12}$$

where

$$\mathbf{e}_{rj} = \frac{\mathbf{X} - \mathbf{X}_{Bj}}{r_j}, \tag{15.13}$$

is the unit vector from the bubble center $\mathbf{X}_{Bj}$ to the point $\mathbf{X}$, and r_j is the radial distance from the center of bubble B_j to the point $\mathbf{X}$. The interaction among the bubbles is realized by introducing the encounter velocity potential due to sources other than the considered bubble, ϕ_{enc}, and the corresponding velocity, $\boldsymbol{U}_{enc}$, and pressure, P_{enc}, at the location of a given bubble i.

The various flow quantities are added up linearly or as vectors as follows:

$$\phi_{enc,i} = \sum_{j \neq i}^{j=1,N} \phi_j, \quad \mathbf{U}_{enc,i} = \sum_{j \neq i}^{j=1,N} \nabla\phi_j, \tag{15.14}$$

where j represents the surrounding bubbles indices, and ϕ_j is the velocity potential due to the source term of the other bubbles dynamics.

As a result, the encounter pressure each bubble feels at any instant, t, can be expressed by:

$$\frac{p_{enc,i}}{\rho} = \frac{P_{amb,i}}{\rho} - \sum_{j \neq i}^{j=1,N} \frac{\partial\phi_j}{\partial t} - \frac{1}{2}\left(\sum_{j \neq i}^{j=1,N} \nabla\phi_j\right)^2, \tag{15.15}$$

and the pressure gradient is calculated by:

$$\frac{\nabla p_{enc,i}}{\rho} = \frac{\nabla P_{amb,i}}{\rho} + \nabla\left[-\sum_{j \neq i}^{j=1,N} \frac{\partial\phi_j}{\partial t} - \frac{1}{2}\left(\sum_{j \neq i}^{j=1,N} \nabla\phi_j\right)^2\right] \\ = \frac{\nabla P_{amb,i}}{\rho} - \sum_{j \neq i}^{j=1,N} \frac{\partial\nabla\phi_j}{\partial t} - \left(\sum_{j \neq i}^{j=1,N} \nabla\nabla\phi_j\right) \cdot \left(\sum_{j \neq i}^{j=1,N} \nabla\phi_j\right), \tag{15.16}$$

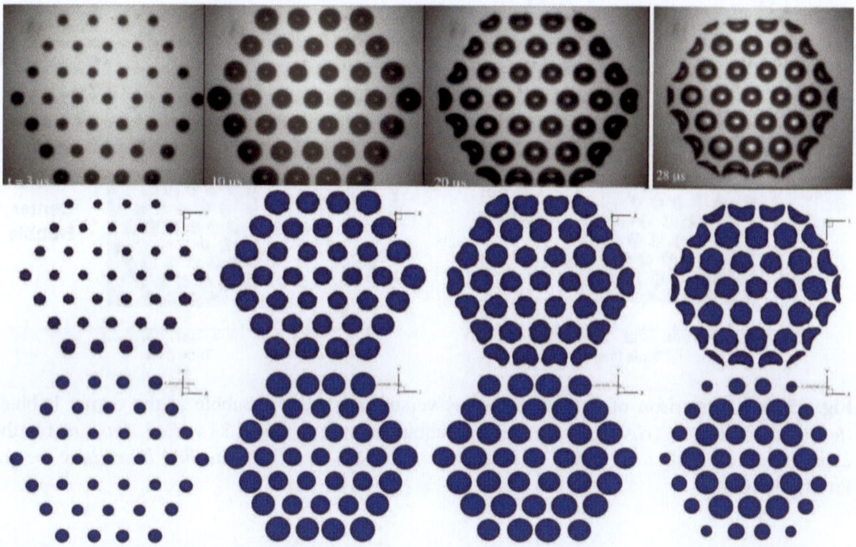

Fig. 15.14 Comparison of bubble outline time evolution during the interaction of a set of 37 bubbles for PhantomCloud© (3rd row), 3DynaFS-Bem© (2nd row) and the experiments of Bremond et al. [36] (1st row)

where P_{amb} is the ambient pressure due to the hydrostatic pressure and any other motions not directly connected to the other bubbles motion.

PhantomCloud© computes all above quantities and solves Eqs. (15.10) and (15.11) at each time step. This then provides the updated positions and sizes of all bubbles. An example to illustrate results obtained with the method can be seen in Fig. 15.14, which simulates the controlled multibubble experiments by Bremond et al. [36]. In the experiment a hydrophobic surface patterned with 37 micro-cavities was used to control surface cavitation. The expansion of the nuclei in the micro-cavities was triggered by a fast drop of the liquid pressure. Figure 15.14 shows the comparison of the cavitation bubble outlines at different time frames between the experimental time sequence pictures, the results of a full 3D computation with 3DynaFS-Bem© and those obtained with the much faster PhantomCloud© approximation.

It can be seen that bubbles located at the edge of the cloud complete their oscillation cycle earlier than bubbles at inner locations. The edge bubbles complete their growth and collapse forming re-entrant jets directed toward the center of the cloud. The inner bubbles have a much longer period and appear shielded by the surrounding bubbles. This was observed earlier by Chahine using an asymptotic approach [26] and then through numerical modeling using 3DynaFS-Bem© [30, 37]. Quantitative comparisons of the equivalent bubble radii for an edge and central bubble are shown in Fig. 15.15.

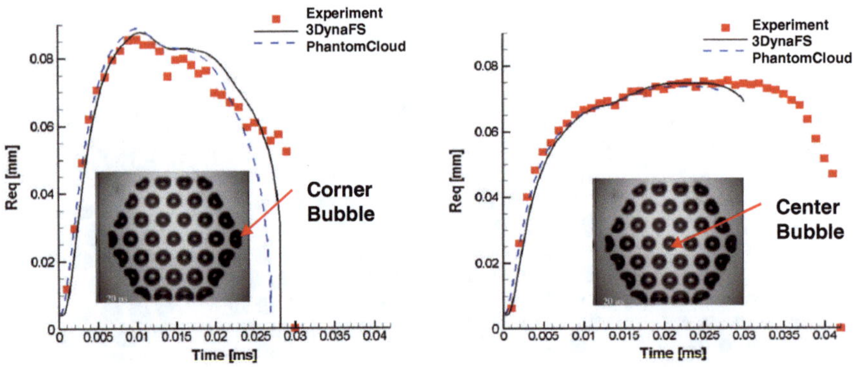

Fig. 15.15 Comparison of equivalent radius versus time for the bubble at the corner bubbles (*left*) and for *center* (*right*) for the set of bubbles shown in Fig. 15.14. *Red dots* are for the experiments of [36], *solid black lines* are for 3DynaFS-Bem© and *dashed blue lines* are for PhantomCloud©

Another illustration is made by comparing PhantomCloud© results with the multi-bubble results presented by Schmidt et al. in this book [38]. They solved 3-D unsteady compressible Euler equations resulting from mass, momentum, and energy balance using a density-based finite volume method. The effects of the grid resolution for a cloud of 125 bubbles on the pressure signal were studied. All bubbles had an initial radius of 0.95 mm and an initial pressure of 2,340 Pa (vapor pressure) in a fluid domain of 20 × 20 × 20 mm^3. The bottom face of the domain is considered impermeable and represents a solid wall. An ambient pressure of 40 MPa was imposed on the bubble cloud and the pressure at the wall was obtained using a fictitious 1 × 1 cm^2 transducer at the wall through averaging of the pressure over the transducer surface.

Reducing the number of grids in those computations resulted in merging bubbles, i.e. reducing the equivalent number of bubbles, while keeping the same void fraction. The same procedure was mimicked using PhantomCloud© with the assumption of a smooth distribution between the number of bubbles and the Schmidt number of grids as illustrated in Fig. 15.16. In both methods the total bubble volume is conserved, while their apparent number changes with the approximation (gridding finiteness) level.

Results of the two methods became very close when the sound speed in the medium in Eq. (15.10) was selected close to the sound speed used in the compressible computation, i.e. about 170 m/s. Figure 15.17 shows a comparison of nondimensional vapor volume versus time between the two methods. The correspondence between the two for the large number of bubbles is remarkable given the differences in the approaches between the two methods. Figure 15.18 illustrates the computed average pressures using PhantomCloud© at the nearby wall center at collapse for different number of bubbles (with the same initial total volume) and Fig. 15.19 compares the pressure peak and period of bubble collapse

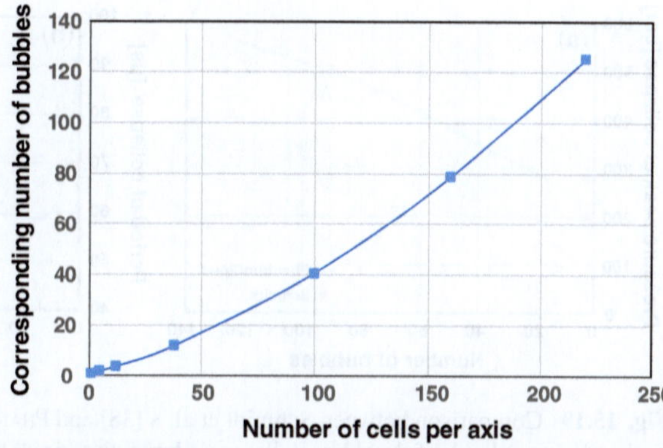

Fig. 15.16 Assumed equivalence between the number of cells per axis in the Schmidt computations [38] and the number of bubbles in PhantomCloud©

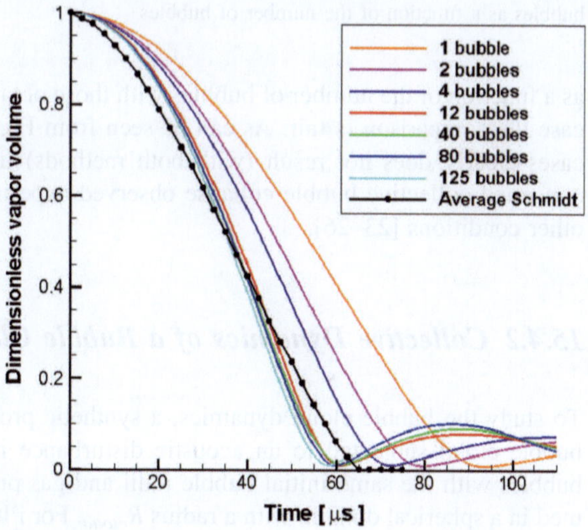

Fig. 15.17 Dimensionless vapour volume versus time when a cloud of bubbles collapses near a rigid wall. Influence of the number of bubbles for PhantomCloud© results compared to an average curve of Schmidt et al.'s results ([38] Fig. 15.5)

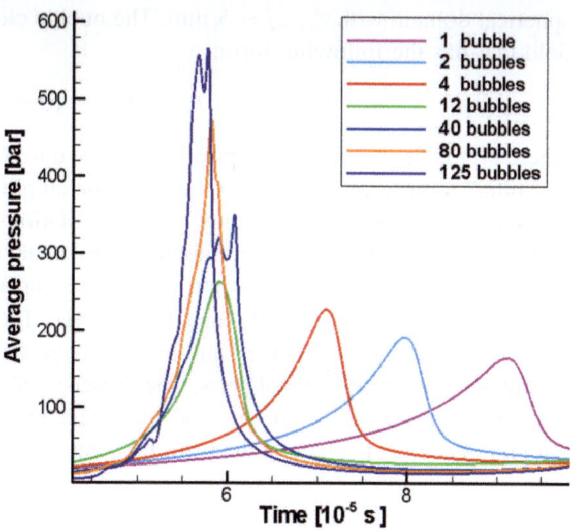

Fig. 15.18 Average pressure versus time when a cloud of bubbles in PhantomCloud© collapses near a rigid wall for different number of bubbles

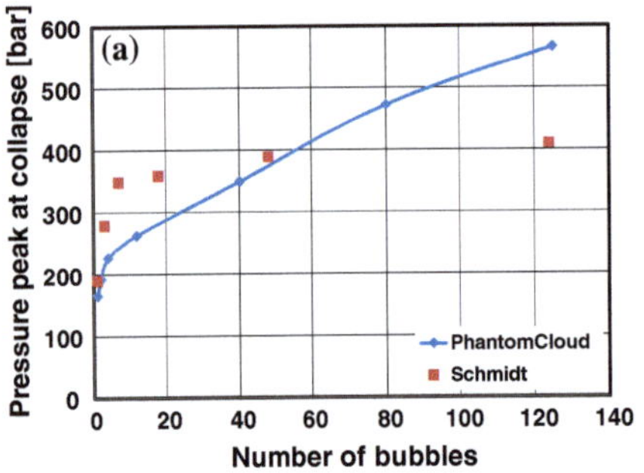

Fig. 15.19 Comparison between Schmidt et al.'s [38] and PhantomCloud© results for **a** Pressure peak when a cloud of bubbles collapses above the rigid wall at the bottom edge of the computational domain as a function of the number of bubbles. **b** Period of collapse of a cloud of bubbles as a function of the number of bubbles

as a function of the number of bubbles with those obtained by Schmidt et al. In this case the comparison is fair. As can be seen from Fig. 15.19, however, this set of cases clearly does not result (with both methods) in very large pressures as in concerted collective bubble collapse observed experimentally or computed under other conditions [23–26].

15.4.2 Collective Dynamics of a Bubble Cloud

To study the bubble cloud dynamics, a synthetic problem involving an idealized bubble cloud subjected to an acoustic disturbance is simulated. For this study, bubbles with the same initial bubble radii and gas pressure are randomly distributed in a spherical domain with a radius R_{cloud}. For illustration, Fig. 15.20 shows a bubble cloud with 400 bubbles having the same initial radius of 200 μm in a spherical domain with $R_{cloud} = 5$ mm. The bubble cloud is subjected to a pressure field that has the following form:

$$p(t) = P_\infty + P_0 \cos\,(2\pi f t), \tag{15.17}$$

where $p(t)$ is the instantaneous pressure imposed to the fluid around the cloud, P_∞ the initial ambient pressure in the fluid, P_0 the amplitude of the oscillations of the pressure field, and f the frequency of the oscillations.

Figure 15.21 illustrates the bubble cloud behaviour at different time steps within one oscillation cycle when excited by an acoustic pressure wave with $P_\infty = 10^5$ Pa, $P_0 = 2 \times 10^5$ Pa, and $f = 500$ Hz. The figure shows only half the cloud to represent a hemispherical cloud collapse on a rigid flat wall, which cuts the full sphere into two equal parts. The presence of the wall was accounted for in the simulations by using the method of images and adding symmetrical bubbles with reference to the plane of the wall. It is seen that due to bubble interactions, the

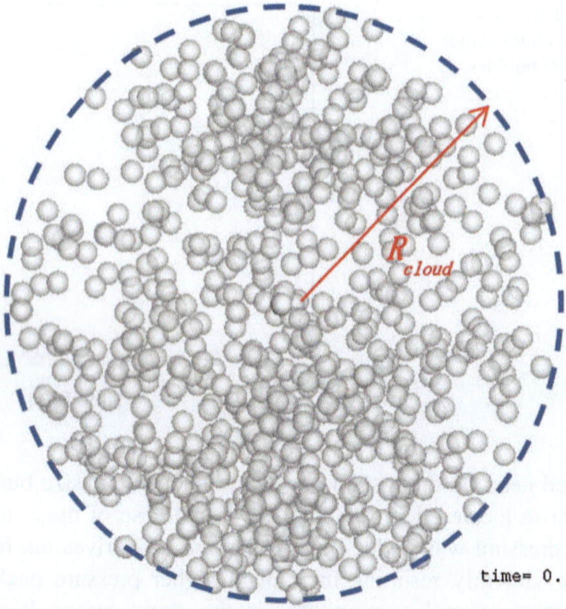

Fig. 15.20 A cloud of bubbles with the same initial radius of 200 μm randomly distributed in a spherical domain with a radius of R_{cloud}

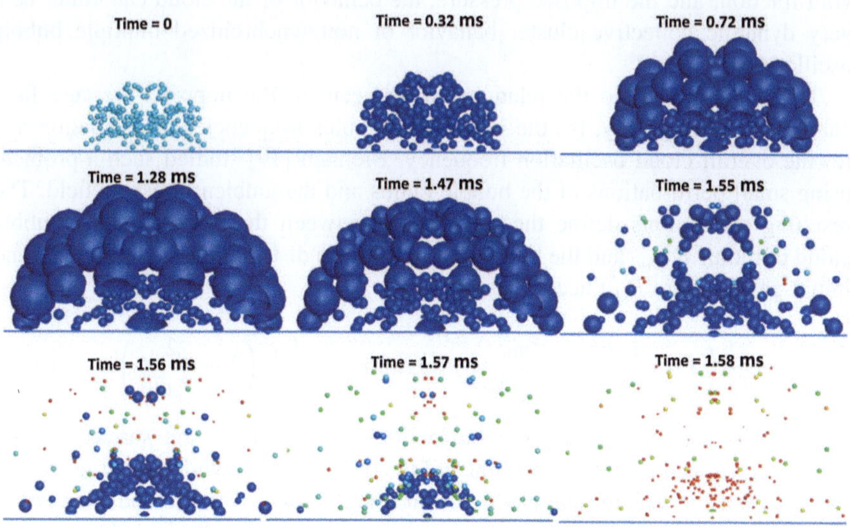

Fig. 15.21 Bubble cloud behavior at different time steps within one oscillation cycle when excited by an acoustic pressure wave. $P_\infty = 10^5$ Pa, $P_0 = 2\times10^5$ Pa and $f = 500$ Hz, $R_{cloud} = 5$ mm, 400 bubbles, $R_0 = 200$ μm. The bubbles are *colored* according to their resulting internal pressure with *red* being very high pressures and *blue* very low pressures

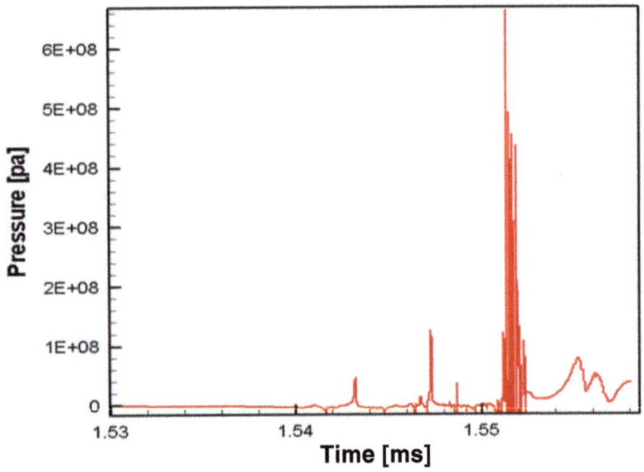

Fig. 15.22 Field pressure monitored at the cloud center of Fig. 15.21. 400 bubbles, $P_\infty = 1$ atm, $P_0 = 2$ atm, and $f = 500$ Hz. $R_{cloud} = 5$ mm, $R_0 = 200$ μm

bubbles located near the outer shell grow to a much larger size but collapse much earlier than those located near the center. The collapse of the outer bubbles generates a high pressure wave which moves inward and drives the inner bubbles to collapse more violently resulting in a much higher pressure peak. Figure 15.22 shows the pressure signal as monitored at the cloud center. It is seen that the pressure generated by the cloud dynamics reaches its maximum as the central bubbles collapse.

Depending on the driving frequency f, the bubble radii, the cloud radius, the void fraction, and the imposed pressure, the behavior of the cloud can either be a very dynamic collective cluster behavior or non-synchronized multiple bubble oscillations.

This is connected to the relationship between (a) the imposed pressure field characteristic frequency, (b) the individual bubbles frequencies of oscillation and (c) the overall cloud oscillation frequency. Brennen [19] studied such a problem using small perturbations of the bubble radius and the ambient pressure field. The resulting oscillations define the relationship between the pulsation of a bubble cloud of radius R_{cloud} and the bubbles in it (uniform distribution with radius R_o and initial gas pressure P_{g0}) according to:

$$\omega_1 = \omega_0 \left(1 + \frac{4}{3\pi^2} \frac{R_{cloud}^2}{R_0^2} \frac{\alpha_0}{1-\alpha_0}\right)^{-1/2} \text{ with } \omega_0^2 = \frac{P_{g0}}{\rho R_0^2}\left(3 - \frac{2\gamma}{P_{g0} R_0}\right), \quad (15.18)$$

where α_0 is the initial void fraction in the cloud, and γ is the surface tension parameter. ω_1 is the angular frequency of the overall cloud, while ω_o is the angular frequency of the isolated bubble. Resonance and strong overall cloud oscillations are expected to occur when the driving frequency matches the cloud frequency. This corresponds then to a "tuned" system.

Figures 15.23 and 15.24 illustrate the differences in behavior for tuned and detuned bubble cloud systems. Figure 15.23 shows the bubble radii versus time for a large set of bubbles for two initial uniform size bubble clouds: $R_0 = 10$ and

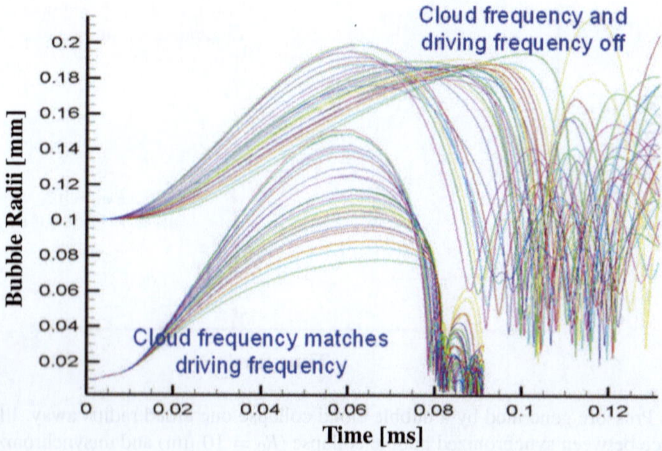

Fig. 15.23 Bubble radius versus time of the various bubbles in the cloud (each *color line* corresponds to an individual bubble). Illustration of the difference between collective bubble effects ($R_0 = 10$ μm) and just many bubble collapse ($R_0 = 100$ μm). 400 bubbles, $R_{cloud} = 0.5$ mm, $f = 10$ kHz

100 μm. In both cases the radius of the overall bubble cloud is selected to be $R_{cloud} = 0.5$ mm. This one order of magnitude difference in the initial bubble radius (and three order of magnitude difference in the cloud initial void fraction) results in major differences when the imposed pressure fluctuations are the same in the two cases, $f = 10$ kHz. While the 100 μm bubbles appear to be non-synchronized and grow and collapse at different times, the 10 μm bubbles appear to align their collapse almost simultaneously, even though they all achieve different maximum radii at very different times as previously illustrated in Fig. 15.21. This "collective" collapse is such that the high pressure field acts as a driving pressure for the collapse of the next successive sets of bubbles until a very high pressure is generated at the collective collapse time.

This is illustrated in Fig. 15.24, where we can see more than two orders of magnitude higher pressures for the case of the synchronized collapse. It is probably behaviors of this type that result in the very large pressures required to erode materials, as deduced from the two-way fluid–structure interaction analysis presented in Chap. 6 in the first part of this book.

15.5 Summary

The study presented in this paper has shown that scaling effects on a propeller (i.e. lack of similitude in the results between geometrically scaled different propeller sizes) can be very significant due to the effect of changing scale on the growth, collection, and collapse of bubble clouds. Conserving the cavitation number, the

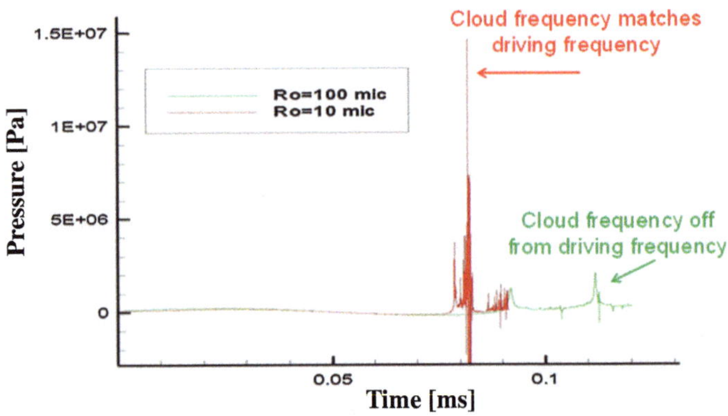

Fig. 15.24 Pressure generated by a bubble cloud collapse one cloud radius away. Illustration of the difference between synchronized bubble collapse (R_0 = 10 μm) and unsynchronized collapse (R_0 = 100 μm). 400 bubbles, R_{cloud} = 0.5 mm, f = 10 kHz, R_0 = 10 μm (*red*) and 100 μm (*green*)

advance coefficient, and the bubble size distribution between two scales results in major differences in the generated bubble clouds. Bubble sizes, void fraction distributions, and resulting impulsive loads are much larger at the larger scales. This implies much higher cavitation intensities at full scale than in the laboratory. Such a scaling cannot be uncovered with models considering the liquid phase only. Many aspects of the scaling require, however, further investigation. These include modelling and scaling of sheet cavitation, sheet-to-cloud breakup, and bubble cloud collective collapse. The present study has indicated that the tuning between the bubble cloud characteristics and the pressure field is essential to generating very high cavitation impulsive loads. Future studies will need to consider this aspect, together with the statistics of bubble cloud formation and their interaction with the flow field generated pressures.

Acknowledgments This study was supported by the US Office of Naval Research under contract N00014-11-C-0378, monitored by Dr. Ki-Han Kim. This support is acknowledged with gratitude. Many colleagues at DYNAFLOW and most particularly Dr. Jin-Keun Choi, Mr. Sowmitra Singh, Mr. Arvind Jayaprakash, and Ms. Tiffany Fourmeau have contributed significantly to the study presented here as well as to the various chapters in Part I of this book. Their efforts are much appreciated.

References

1. Shen Y, Petersen FB (1978) Unsteady cavitation on an oscillating hydrofoil. In: Proceedings of the 12th symposium on naval hydrodynamics, pp 362–384
2. de Lange DF, de Bruin GJ (1997) Sheet cavitation and cloud cavitation, re-entrant jet and three-dimensionality. Appl Sci Res 58(1–4):91–114. doi:10.1023/A:1000763130780

3. Bark G, Bensow RE (2014) Hydrodynamic processes controlling cavitation erosion. In: Kim KH, Chahine GL, Franc J-P, Karimi A (eds) Advanced experimental and numerical techniques for cavitation erosion prediction. Series fluid mechanics and its applications, Springer, New York
4. Hsiao C-T, Chahine GL (2004) Prediction of vortex cavitation inception using coupled spherical and non-spherical models and Navier-Stokes computations. J Mar Sci Technol 125(3):99–108
5. Chahine GL (2009) Numerical simulation of bubble flow interactions. J Hydrodyn 21(3):316–332
6. Hsiao C-T, Chahine GL, Liu H-L (2003) Scaling effects on prediction of cavitation inception in a line vortex flow. J Fluids Eng 125:53–60
7. Hsiao C-T, Chahine GL (2012) Effect of gas diffusion on bubble entrainment and dynamics around a propeller. J Hydrodyn 24(6):809–882
8. Chahine GL, Choi J-K, Lu X (2008) Modeling of air gun generated underwater bubble and pressure field using GEMINI and PHANTOMCLOUD©. 79th SAVIAC shock and vibration symposium, Orlando, October 2008
9. Hsiao C-T, Chahine GL, Singh S (2011) Basic model for PHANTOMCLOUD© simulation of single and multiple airgun source models. DYNAFLOW, INC. Technical Report 2M11022-2
10. Chorin AJ (1967) A numerical method for solving incompressible viscous flow problems. J Comput Phys 2:12–26
11. Hsiao C-T, Ma J, Chahine GL (2012) Numerical and experimental study of bubble entrainment due to a plunging jet. 29th symposium on naval hydrodynamics, Gothenburg, 26–31 August 2012
12. Choi J, Hsiao C-T, Chahine GL, Ceccio S (2009) Growth oscillation and collapse of vortex cavitation bubbles. J Fluid Mech 624:255–279
13. Chahine GL, Hsiao C-T (2012) Modeling microbubble dynamics in biomedical applications. J Hydrodyn 24(2):169–183
14. Hsiao C-T, Lu X, Chahine GL (2010) Three-dimensional modeling of the dynamics of therapeutic ultrasound contrast agent. Ultrasound Med Biol 36(12):2065–2079
15. Keller JB, Kolodner II (1956) Damping of underwater explosion bubble oscillations. J Appl Phys 27:1152–1161
16. Haberman WL, Morton RK (1953) An experimental investigation of the drag and shape of air bubbles rising in various liquids. DTMB Report 802
17. Hsiao C-T, Pauley L (1998) Numerical study of the steady-state tip vortex flows over a finite-span hydrofoil. J Fluids Eng 120:345–353
18. Baldwin BS Barth TJ (1990) A one-equation turbulence transport model for high reynolds number wall-bounded flows. NASA TM 102847
19. Brennen CE (1995) Cavitation and bubble dynamics. Oxford engineering sciences series 44. Oxford University Press, New York
20. Billet ML (1985) Cavitation nuclei measurements: a review. ASME cavitation and multiphase flow forum, FED-vol 23, June 1985
21. Breitz N, Medwin H (1989) Instrumentation for in situ acoustical measurements of bubble spectra under breaking waves. J Acoust Soc Am 86:739–743
22. Duraiswami R, Prabhukumar S, Chahine GL (1998) Bubble counting using an inverse acoustic scattering method. J Acoust Soc Am 104(5):2699–2717
23. van Wijngaarden L (1964) On the collective collapse of a large number of gas bubbles in water. In: Proceedings of the 11th international conference on applied mechanics, Springer-Verlag, Berlin, pp 854–861
24. Mørch KA (1979) Dynamics of cavitation bubbles and cavitating liquids. Treatise on Mater Sci Technol 16:309–355
25. d'Agostino L, Brennen CE (1983) On the acoustical dynamics of bubble clouds. ASME cavitation and polyphase flow forum, Houston, pp 72–76

26. Chahine GL (1983) Cavitation cloud theory. In: Proceedings of the 14th symposium on naval hydrodynamics, Ann Arbor, Michigan, National Academy Press, Washington, D.C, pp 165–194
27. Chahine GL, Liu HL (1984) Collective effects on the growth of vapour bubbles in a superheated liquid. J Fluids Eng 106:486–491
28. Chahine GL, Liu HL (1985) A singular perturbation theory of the growth of a bubble cluster in a superheated liquid. J Fluid Mech 156:257–279
29. Chahine GL, Kalumuck KM (1987) The influence of gas diffusion on the growth of a bubble cloud. ASME cavitation and multiphase flow forum, Cincinnati, Ohio 50:17–21
30. Chahine GL (1982) Experimental and asymptotic study of nonspherical bubble collapse. Appl Sci Res 38:187–197
31. Chahine GL, Duraiswami R (1992) Dynamical interaction in a multi-bubble cloud. J Fluids Eng 114(4):680–686
32. Chahine GL (2005) Numerical studies of the interaction of multiple underwater explosion bubbles. Critical technologies in shock and vibrations, Walter Pilkey (ed) 3(1):1–13
33. Chahine GL, Bovis AG (1980) Oscillation and collapse of a cavitation bubble in the vicinity of a two-liquid interface. Cavitation and inhomogeneities in underwater acoustics. Springer-Verlag, New York, pp 23–29
34. Bovis AG, Chahine GL (1981) Etude asymptotique de l'interaction d'une bulle oscillante avec une surface libre voisine. J de Mécanique 20(3):537–556
35. Chahine GL, Bovis AG (1983) Pressure field generated by nonspherical bubble collapse. J Fluids Eng 105(3):356–364
36. Bremond N, Arora M, Ohl C-D, Lohse D (2006) Controlled multibubble surface cavitation. Phys Rev Lett 96:224501
37. Chahine GL (1991) Dynamics of the interaction of non-spherical cavities. In: Miloh T (ed) Mathematical approaches in hydrodynamics. SIAM, Philadelphia
38. Schmidt SJ, Mihatsch MS, Thalhamer M, Adams NA (2014) Assessment of erosion sensitive areas via compressible simulation of unsteady cavitating flows. In: Kim KH, Chahine GL, Franc J-P, Karimi A (eds) Advanced experimental and numerical techniques for cavitation erosion prediction, Series Fluid mechanics and its applications. Springer, New York

Chapter 16
Numerical Simulations of Shock Emission by Bubble Collapse Near a Rigid Surface

With Applications to Shock-Wave Lithotripsy (SWL) and the Spallation Neutron Source

Eric Johnsen

Abstract The event central to cavitation erosion is bubble collapse. Yet, the detailed physics of the process are not well characterized. In recent years, direct simulations of the Euler equations have been used to study the collapse of a single bubble and the subsequent shock emission, in contexts ranging from naval engineering to biomedical applications. In the present work, shock-induced collapse and Rayleigh collapse of a single gas bubble are examined, with emphasis on cavitation damage in biomedical applications and the Spallation Neutron Source (SNS). After an overview following Johnsen [32], the non-spherical bubble dynamics are considered in detail for problems in which the collapse time is on the order of the time scale of shock propagation through the bubble. The pressures generated by bubble collapse near a solid surface are measured, and it is shown that the resulting wall pressure may be larger than that of the incoming shock. The current work is then applied to shock-wave lithotripsy, a procedure developed to treat renal calculi. In particular, one-way coupled fluid and elastic wave propagation simulations are used to investigate damage to kidney stones. Two stone comminution mechanisms are proposed: shock-induced bubble collapse and spallation due to shocks emitted by bubble collapse. The results suggest that the coupling between the fluid and solid mechanics are important in terms of understanding cavitation erosion. In the context of the SNS, the effect of confinement on the bubble dynamics and the pressure generated by collapse are examined.

E. Johnsen (✉)
Mechanical Engineering Department, University of Michigan, Ann Arbor, USA
e-mail: ejohnsen@umich.edu

K.-H. Kim et al. (eds.), *Advanced Experimental and Numerical Techniques for Cavitation Erosion Prediction*, Fluid Mechanics and Its Applications 106,
DOI: 10.1007/978-94-017-8539-6_16,

16.1 Cavitation Erosion Across Disciplines

Cavitation is the phenomenon in which vapor bubbles grow in liquids due to reductions in pressure, which are typically caused by two factors [41]: local variations due to the hydrodynamics of the flow field or the application of an acoustic field. *Hydrodynamic cavitation* is characterized by the fact that pressure drops are caused by increases in the average velocity field; this phenomenon frequently occurs along propellers and in turbomachines. In *acoustic cavitation*, acoustic or shock waves produce a time-varying pressure field that drive the cavitation dynamics; this process is exploited in several biomedical applications, specifically in therapeutic ultrasound. Regardless of the means by which cavitation is produced, one of the most severe consequences of cavitation is the material damage caused to neighboring solid surfaces [7].

16.1.1 Undesirable Cavitation Damage

By and large, hydrodynamically generated cavitation has negative outcomes. Initially, the focus of cavitation research lay in the need to understand to what extent bubbles affected the operation of naval structures. Thornycroft and Barnaby [65] were the first to report on the reduced performance and structural damage caused by cavitation. Naval engineering remains one of the main areas of cavitation research as collapsing bubbles and the resulting damage may be observed along propellers, rudders, struts, and even hulls [38]. In underwater explosions, a gas bubble containing the products of the explosion rapidly expands to a large size and collapses violently, thus loading the target and prompting it to recoil. This process, along with reflections of the initial shock and the dynamics of the large bubble, create regions of low pressure [72], thus providing conditions favorable to small-scale cavitation and subsequent damage. Damage caused by bubble collapse is observed over a wide range of scales. A spectacular example is the cavitation erosion sustained by spillways or the outlet of dams [1]. Liquid flows through rotating machines (turbomachinery) constitute another field where cavitation plays a deleterious role. The operation of valves and nozzles, e.g., fuel injectors [20], may also be affected by cavitation, due to changes in the velocity of the liquid passing through them [38]. In the biomedical field, damage to red blood cells in artifical heart valves [73] or trauma due to impact head injuries [43] have been related to cavitation. Recently, cavitation erosion has been observed in the target of the Spallation Neutron Source; this problem will be discussed in more detail below.

16.1.2 Desirable Cavitation Damage

In several recent applications, cavitation erosion is employed in a deliberate fashion to clean or break materials. In these instances, cavitation is usually controlled by acoustic means. In ultrasonic cleaning, the flow induced by a bubble oscillating stably under the action of an acoustic field, i.e., microstreaming [18], generates shear stresses that remove particulates from surfaces. Wet finishing, where textile materials are treated in an attempt to modify their properties [45], and silicon wafer cleaning [9] are examples of this process. Ohl et al. [51] recently suggested that the main contributions to the shear flow in ultrasonic cleaning emanate from the spreading of the liquid jet generated during bubble collapse along surfaces. In biomedical engineering, several procedures take advantage of cavitation damage. In dentistry, plaque is removed using an ultrasonically vibrating probe [70]. Such research has translated into the mainstream sonic toothbrush and, more recently, into its ultrasonic sibling [46]. In cataract surgery, the eye's natural lens is emulsified and removed [52]. In a recent medical procedure developed at the University of Michigan, focused shock waves are used to ablate pathogenic tissue in a procedure called histotripsy [62] in which cavitation occurs directly in tissue and exhibits different responses depending on the relevant viscoelastic properties [28]. This procedure originated from shock wave lithotripsy (SWL), described in more detail below, which is the most common treatment for kidney stone disease in the USA [12].

16.2 Single-Bubble Collapse and Cavitation Erosion

The event central to cavitation erosion is bubble collapse. From its maximum radius to collapse, the volume of a cavitation bubble decreases rapidly by several orders of magnitude. As this compression occurs, the pressure and temperature within the cavity reach high values, and shock waves are emitted. In fact, Suslick [64] remarks that by using conventional ultrasound equipment to drive a bubble to oscillate near its natural frequency: "One can create the temperature of the sun's surface, the pressure of deep oceanic trenches, and the cooling rate of molten metal splatted onto a liquid-helium-cooled surface!"

16.2.1 Past Theoretical and Experimental Work

Starting with the research of Lord Rayleigh [60], single-bubble collapse has been studied for almost a century. Rayleigh was the first to fully appreciate the engineering implications of bubbles repeatedly collapsing along propeller blades, and his studies set the basis for cavitation research henceforth [24]. The high pressures achieved during collapse and the impact of shock waves thereby emitted were the first proposed cavitation damage mechanisms [19, 26, 60]. In studies of traveling cavitation bubbles, Plesset [56] included viscous and surface tension effects into what is now known as the Rayleigh-Plesset equation. Further extensions, e.g., including liquid compressibility and thermal effects, and studies can be found in Plesset and Prosperetti [58] and Brennen [7]. Kornfeld and Suvorov [39] first predicted that the asymmetry in the flow field caused by the presence of a solid surface near a bubble would lead a re-entrant jet to form during the collapse. This phenomenon was later confirmed by the experiments and analysis of Naude and Ellis [48] and Benjamin and Ellis [3]. During the collapse, the far end of the bubble involutes, such that a high-speed re-entrant jet directed towards the solid surface forms and penetrates the bubble [42]. If the bubble is sufficiently close to the wall, the jet may directly impact it and generate a large water-hammer pressure [57].

Most early experiments employed a spark discharge method to vaporize a small portion of liquid and generate a bubble [3, 37, 48]. Using this technique, Tomita and Shima [67] studied the collapse of isolated cavitation bubbles near a solid surface and measured wall pressures up to 12 MPa. They found that bubbles near the wall migrate towards it and further argued that cavitation damage was caused by the collapse of tiny bubbles generated during the collapse of the main bubble. One important difficulty of such studies is that electrodes interfere with the flow field, and the bubbles thereby generated are not always spherical. Lauterborn and Bolle [40] used a laser to vaporize water, thus creating a bubble. The combination of this technique with high-speed photography has provided many advances in understanding bubble collapse near a wall and cavitation erosion. Vogel et al. [69] investigated the formation of the jet and the effect of the wall on the bubble dynamics. Philipp and Lauterborn [54] determined that bubbles in direct contact with solid surfaces were responsible for the largest part of the erosion. Lindau and Lauterborn [42] visualized shock waves emitted during collapse and studied the properties of the counterjet formed after the collapse. However, it is unclear whether optically produced cavitation behaves in a fashion similar to hydrodynamically or acoustically generated bubbles.

The collapse of a bubble near a rigid wall is equivalent to the simultaneous collapse of two identical neighboring bubbles, a configuration that breaks the spherical symmetry. Another type of asymmetry consists of the presence of an external force field, such as gravity. In such problems, a re-entrant jet forms in the direction of the buoyancy force [71]. The passage of a shock wave over a bubble also leads to the formation of a re-entrant jet in the direction of propagation of the shock [67], e.g., in underwater explosions and shock-wave lithotripsy. In cloud cavitation, shock waves may be generated by the collapse of individual bubbles, as explained previously, or by the coherent collapse of the cloud [25]. Two-dimensional shock-bubble interactions have been studied in the framework of shock propagation in non-uniform media: Bourne and Field [6] considered the impact of shock waves onto cavities in a water/gelatin mixture to better understand ignition and propagation of explosive reactions, while Haas and Sturtevant [22] investigated the impingement of shock waves onto gaseous cylinders.

16.2.2 Past Computational Work

Because of the complexity of the physics and the wide range of spatial and temporal scales present in cavitation dynamics, it is difficult to perform precise measurements in experiments. Numerical simulations have therefore emerged as a reliable tool to complement analytical and experimental studies. However, direct simulations of such flow phenomena are challenging because of high computational costs and difficulties in developing algorithms. Prior simulations of single-bubble collapse typically fall into one of two categories: the method accurately solves a problem in which the physics are highly restrictive, or the physics of the problem are more adequately represented at the expense of accuracy. Methods based on potential theory were first used to compute the collapse of a bubble near a wall, e.g., the Particle In Cell Method of Plesset and Chapman [57], the Boundary Integral Method of Blake et al. [5], Zhang et al. [74] and Best [4], and the Boundary Element Method of Chahine and Duraiswami [10] and Klaseboer et al. [36]. Popinet and Zaleski [59] used an axisymmetric front-tracking method to investigate the effect of viscosity for Rayleigh collapse near a solid surface. These methods capture the initial bubble behavior and the formation of the re-entrant jet. However, during the last stages of collapse, the bubble wall velocity may become comparable to the liquid sound speed, so that the assumption of liquid incompressibility breaks down. Thus, shock waves cannot be represented and the pressure inside the bubble is no longer uniform [34].

Compressibility accounts for a large part of the energy dissipation in cavitation bubble collapse [68], which is important for shock emission upon collapse and is therefore relevant to the damage potential. Recent efforts have therefore been devoted to developing methods capable of simulating the full compressible equations of motion in order to capture any shock generated during the process. The main difficulty consists of numerically treating shock waves and interfaces in a robust and stable fashion. Ball et al. [2] used a two-dimensional second-order accurate Free-Lagrange Method to study shock-bubble interactions based on the experiments of Bourne and Field [6]. The gas was observed to heat substantially due to internal reflections of the shock wave. However, the complexities of the mesh are not very practical for intricate bubble shapes in three dimensions. Nourgaliev et al. [49] employed adaptive characteristics-based matching, and Hu et al. [27] used a level set method with corrections to account for conservation losses to simulate the same problem and two-dimensional underwater explosions, with similar results; these methods are not conservative near interfaces since the numerical flux into a cell is not necessarily equal to that out of the neighboring cell. Chang and Liou [11] implemented a volume fraction formulation in the AUSM$^+$-up methodology to simulate similar problems. The main drawback of this method is that several problem-dependent constants must be adjusted. The Rayleigh collapse of a bubble was studied by Nagrath et al. [47], who combined a finite element and a level set method to simulate three-dimensional collapse of a spherical bubble initially following the Rayleigh-Plesset equation for a single set of parameters. Jamaluddin [29], using an axisymmetric implementation of the Free-Lagrange Method of Ball et al. [2], and Johnsen and Colonius [34] studied shock-bubble interaction. The shock wave resulting from the impact of the jet onto the distal side was captured, and predictions for jet speed and wall pressure were made for certain specific parameters.

16.3 Advances in Numerical Simulations of Single-Bubble Collapse

Recent advances in the numerical simulation of single-bubble collapse are discussed. First, the Rayleigh collapse and shock-induced collapse of a gas bubble are examined. The results are then interpreted in the context of shock-wave lithotripsy and the Spallation Neutron Source.

16.3.1 A Study of Rayleigh Collapse and Shock-Induced Collapse

16.3.1.1 Numerical Framework

Modern numerical methods used to study single-bubble dynamics are usually based on direct simulations of the Euler equations. Through this approach, the entire density, velocity and pressure fields are computed at all times. The present work focuses on the methodology of Johnsen and Colonius [31]. Since interactions between waves and interfaces constitute the dominant flow features of shock bubble collapse, compressible flows of multiple fluid components are considered, in which the components are assumed immiscible, and diffusion, surface tension and phase change are neglected. The gas is assumed to be perfect, and a stiffened equation of state is used to model water; this stiffened equation of state allows for negative pressure. No cavitation (phase change) model is used. Material interfaces are then represented by discontinuities in the fluid properties, so that advection equations for these quantities supplement the Euler system. A high-order accurate, quasi-conservative shock- and interface-capturing scheme is used to solve the equations of motion. The time marching is carried out using a third-order accurate TVD Runge–Kutta method, and the spatial discretization employs a fifth-order accurate finite volume WENO scheme [30] in which the average primitive variables are reconstructed and for which the HLLC solver [66] is modified to solve advection equations.

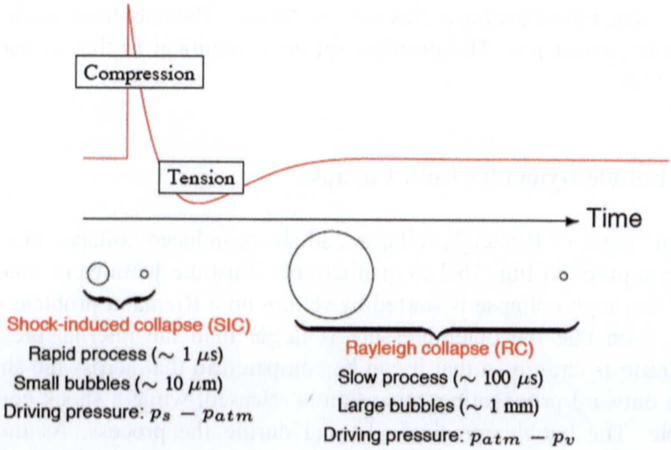

Fig. 16.1 Schematic of a lithotripter pulse and bubble collapse that occurs during shock-wave lithotripsy [32]

16.3.1.2 Rayleigh Collapse and Shock-Induced Collapse

The distinction between shock-induced collapse and Rayleigh collapse of a bubble can be understood by considering shock-wave lithotripsy (Fig. 16.1). In lithotripsy, a pre-existing bubble or nucleus is subjected to a sudden pressure increase (shock), thus collapsing it rapidly. The shock is immediately followed by a long expansion, which produces a negative pressure. Due to this tension, the bubble grows to its maximum size, at which points Rayleigh collapse occurs. In shock-induced collapse, a shock wave interacts with a pre-existing bubble; the pressure across the shock drives the collapse in a transient process as the shock passes through the bubble. Rayleigh collapse denotes the collapse driven by the ratio between the pressure of the surroundings (e.g., atmospheric pressure) and that of the bubble (e.g., vapor pressure); it is the archetype of cavitation bubble collapse, which starts when the bubble reaches its maximum radius. The limiting case of shock-induced collapse for which the shock speed in the liquid is infinite corresponds to a Rayleigh collapse. This latter problem is equivalent to raising the pressure of the water instantaneously by a given amount, while keeping the bubble pressure at atmospheric value, and is thus a generalization of the first problem for which the pressure ratio can be made arbitrary.

In the present work, a spherical air bubble of radius, R_o, is initially located at a distance, H_o, from a rigid surface, which may be placed infinitely far from the bubble. When the collapse is driven by a shock, the incoming shock propagates in the direction normal to the wall, so that the problem is axisymmetric; thus, cylindrical coordinates with azimuthal symmetry are used. The pressure ratio across the shock is $p_s/p_o = 353$, where p_s is the pressure downstream of the shock and p_o is the upstream pressure. This value is representative of shock wave lithotripsy. The initial bubble radius is $R_o = 50$ μm. The bubbles contain only non-condensable perfect gas. The problem set-up is identical to that of Johnsen and Colonius [34].

16.3.1.3 Bubble Dynamics Until Collapse

The initial stages of Rayleigh collapse and shock-induced collapse in a free field are first compared in Fig. 16.2 to qualitatively illustrate features of non-spherical collapse. Rayleigh collapse is started by setting up a Riemann problem across the interface, such that the outer pressure is larger than the internal pressure. The pressure ratio is chosen so that it can be compared to that across the shock. As a result, an outward-propagating expansion is released, while a shock converges in the bubble. The bubble remains spherical during the process. As the collapse progresses, the pressure in the liquid just outside the bubble becomes higher than that inside the bubble for a short time, as observed in Rayleigh-Plesset calculations. In shock-induced collapse, upon the impact of the incoming shock, the bubble starts its collapse. An expansion wave is reflected, a shock is transmitted into the bubble, and the incoming shock diffracts around the bubble. During the collapse, a re-entrant jet forms in the direction of propagation of the shock. The increased jet velocity leads to a high local pressure in the liquid just behind it; this high pressure is a consequence of the high velocity, not a cause for jet formation [34]. Clearly, shock-induced collapse is a non-spherical phenomenon, in which the detailed dynamics (e.g., jet formation) are important factors in the shock emission.

To highlight certain key events in non-spherical collapse, the history of the jet velocity, v_j, and that of the distal side, v_d, are shown in Fig. 16.3. The left figure shows these velocities based on the absolute time origin for shock-induced collapse. In the right figure, these velocities are shown for Rayleigh collapse and for shock-induced collapse, for which the time origin has been shifted by the time it takes for the shock to reach the bubble for v_j, and by the time it takes for the shock to diffract to the distal side for v_d; in addition, a shift in the jet velocity is added to "remove" the instantaneous increase due to the shock. The interface is stationary until the shock reaches the proximal side, from which the jet eventually forms. Then v_j increases impulsively due to the passage of the shock ($t \approx 1.4$); note that the time is non-dimensionalized using the liquid sound speed c_L and the initial radius R_o. Until $t \approx 6.0$, the velocity follows an approximately constant and small acceleration (*slow regime*); then, for $7.5 < t < 9.2$, the jet velocity obeys an approximately constant and larger acceleration (*fast regime*). The large deceleration is then due to the high compression within the bubble and the impact upon the distal side. The distal side starts its motion at a later time, once the shock has diffracted around the bubble ($t \approx 4.0$). Similarly to that of the proximal side, the velocity of the distal side initially obeys an approximately constant and small acceleration, but does not enter the fast regime due to jet impact. After impact, the two sides are very close to each other and move together downstream. The interaction of the shock with the bubble is clearly a transient process: the proximal

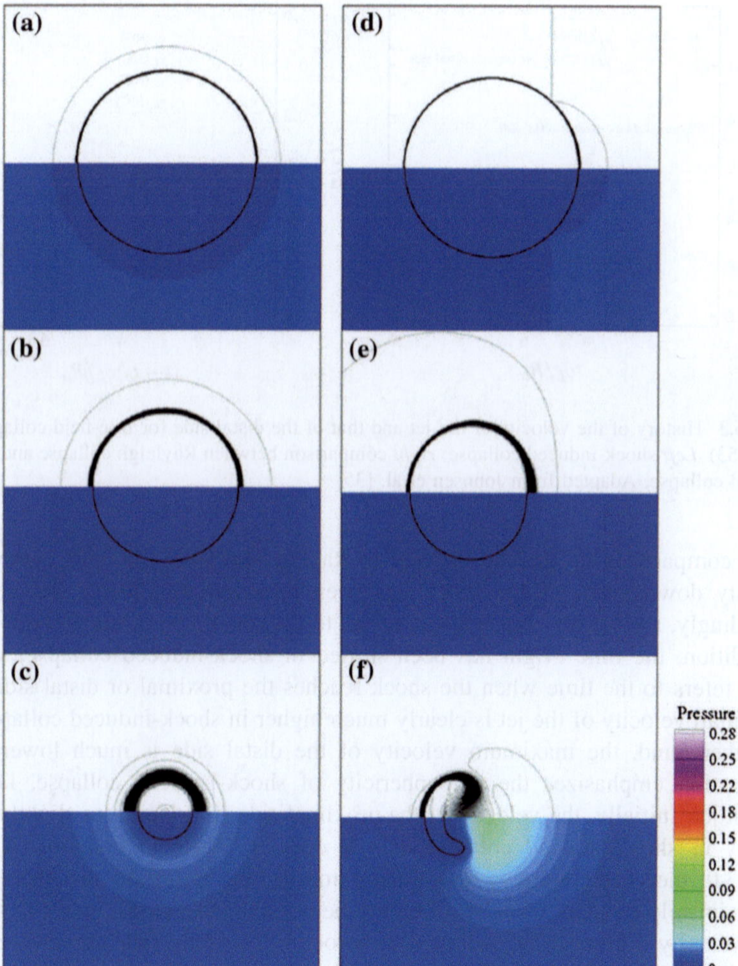

Fig. 16.2 Initial stages of the Rayleigh collapse (*left*) at times 0.265, 3.98 and 6.98, and shock-induced collapse (*right*) at times 0.280, 5.79 and 8.02, of an air bubble. The initial pressure ratio across the interface/shock is 353. The pressure is non-dimensionalized by the water density and sound speed, and time by the sound speed and initial radius. *Top half* density lines; *bottom half* pressure contours, with the interface highlighted. Reproduced from Johnsen and Colonius [34] with permission from Cambridge University Press

side is subjected to the shock sooner than the distal side. In other words, for sufficiently large shock amplitude, the collapse time is on the order of the time it takes for the shock to diffract about the bubble. Thus, assuming that a shock-induced collapse by a Rayleigh collapse in which the surrounding pressure is instantaneously increased by the same amount, e.g., in a Rayleigh-Plesset analysis, becomes a poor approximation [35].

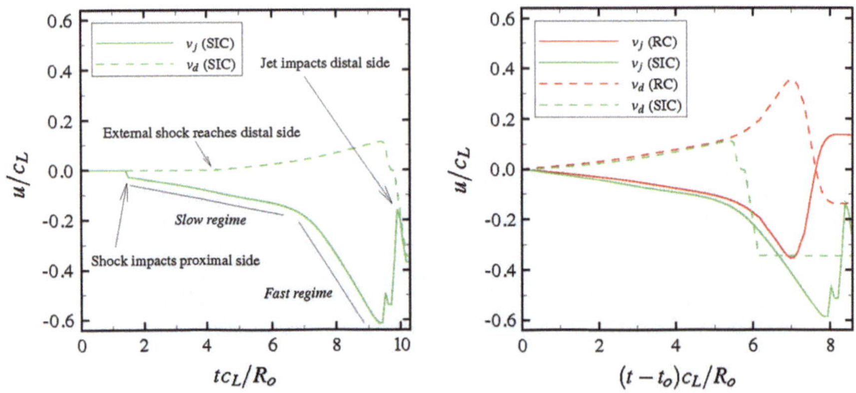

Fig. 16.3 History of the velocity of the jet and that of the distal side for free-field collapse ($p_s/p_o = 353$). *Left* shock-induced collapse; *right* comparison between Rayleigh collapse and shock-induced collapse. Adapted from Johnsen et al. [35]

To compare shock-induced collapse to Rayleigh collapse, the initial impulsive velocity downstream of the shock has been removed by shifting the velocity accordingly, so that the zero velocity refers to the initial conditions in both cases; in addition, the time origin has been shifted in shock-induced collapse, so that $t = 0$ refers to the time when the shock reaches the proximal or distal side. The maximum velocity of the jet is clearly much higher in shock-induced collapse; on the other hand, the maximum velocity of the distal side is much lower. This observation emphasizes the non-sphericity of shock-induced collapse, i.e., re-entrant jet. Initially, the velocity of the proximal side (jet) increases slightly more rapidly in shock-induced collapse than it does in Rayleigh collapse. After t-$t_o \approx 3.0$, the acceleration is nearly the same in both cases, so that there is an approximately constant velocity difference between the two cases, until t-$t_o \approx 5.0$. This velocity difference is attributed to baroclinic vorticity (Richtmyer-Meshkov instability, [8]) due to the misalignment of the pressure gradient (across the shock) and the density gradient (across the interface), which induces a velocity on the base flow. Shock-bubble interaction has been studied mainly in the context of a shock in air interacting with a helium cylinder [22], in which case baroclinic vorticity is the jet formation mechanism. In the present situation, the pressure ratio across the shock is much larger than in studies with gases—even though the shock Mach number is close to unity – so that changes in bubble volume are important. After this stage, the proximal side enters the fast regime; the higher velocity in shock-induced collapse allows the proximal side to enter the fast regime sooner than in Rayleigh collapse. The velocities of the distal side are almost identical in the two cases. However, because of the delay due to the finite speed of propagation of the shock in water, the distal side is still in the slow regime when the jet impacts it. The proximal side accelerates in the fast regime while the distal side is still contracting in the slow regime because of the finite time of propagation of the shock; the velocity of the proximal side accelerates at the given (fast) rate and can penetrate the bubble to a greater amount because nothing is there to arrest its motion, until impact with the distal side. In the spherical case, the collapse is arrested sooner, because the trajectory of the distal side is the exact opposite of that of the proximal side. Similar behavior is observed for sufficiently large pressure ratios (data not shown), as long as the characteristic time scale of the shock is on the order of the collapse time.

Although spherical bubble dynamics are strictly not applicable to shock-induced collapse, they may be used to understand the *slow* and *fast* regimes. For a gas bubble initially in equilibrium with its surroundings at pressure, p_o, and subjected to an instantaneous pressure rise, p_s/p_o, the Rayleigh-Plesset equation can be integrated to yield an expression for the interface velocity [7]:

$$\dot{R}^2 = \frac{2}{3}\frac{p_o}{\rho_L}\left[\frac{1}{\gamma-1}\left(\frac{R_o}{R}\right)^{3\gamma} - \frac{p_s}{p_o} + \left(\frac{p_s}{p_o} - \frac{1}{\gamma-1}\right)\left(\frac{R_o}{R}\right)^3\right],$$

where R is the bubble radius, R_o is the initial bubble radius, ρ_L is the density of the liquid and γ is the ratio of specific heats of the gas. Initially, while $R(t) \approx R_o(1\text{-}\varepsilon)$, where ε is small, the velocity obeys the following *slow* asymptotic growth:

$$\dot{R} \to -\sqrt{2\frac{p_o}{\rho_L}\left(\frac{p_s}{p_o} - 1\right)\varepsilon} + O(\varepsilon^2).$$

On the other hand, for $R \ll R_o$, the velocity asymptotes to the following *fast* growth [7]:

$$\dot{R} \to -\left(\frac{R_o}{R}\right)^{3/2}\sqrt{\frac{2}{3}\frac{p_o}{\rho_L}\left[\frac{p_s}{p_o} - \frac{1}{\gamma-1}\left(\frac{R_o}{R}\right)^{3(\gamma-1)}\right]}.$$

These asymptotic regimes are recovered in the simulations, as shown in the phase diagram of Fig. 16.4. Due to the limitations in grid resolution and to the fact that the present simulations represent additional physics (shock propagation inside the bubble), the current Rayleigh collapse matches the Rayleigh-Plesset solution only at early times. The maximum velocities of the jet and distal side are important quantities, because the jet formation and impact onto the distal play important roles in the shock emission process. This water-hammer shock generated by the non-spherical collapse propagates outward and impacts neighboring surfaces, as explained in the next section; the amplitude of this shock is directly proportional to the velocity difference between the jet and the distal side.

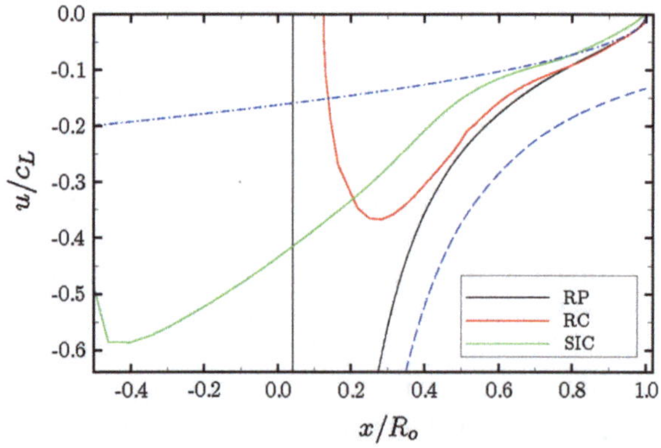

Fig. 16.4 Phase diagram showing velocity of the proximal side as a function of interface position for shock-induced collapse (SIC) and Rayleigh collapse (RC) for $p_s/p_o = 353$ [32]. The corresponding Rayleigh-Plesset solution (RP) is shown, as well as the slow asymptotic growth (*dash-dotted line*) and fast growth (*dashed line*)

16.3.1.4 Shock Emission Upon Collapse

To give a qualitative understanding of pressure and the shock emitted at collapse, the pressure history measured along the neighboring rigid surface in shock-induced collapse near a wall is shown in Fig. 16.5 for two different radial distances from the centerline, s. First, the incoming shock impacts the wall at $t \approx 7$. Due to the presence of the bubble, the shock lags slightly along the centerline. After reflection off the wall, the shock interacts with the bubble again. Because of the impedance mismatch, the reflected wave is an expansion wave, so that a dip is observed in the profile at $t \approx 9$. Thereafter, the pressure increases significantly due to the shock emitted upon bubble collapse. The pressure is smaller farther away from the centerline, because the shock is propagating radially outward. After reflection off the wall, this shock interacts with the bubble and inverts its amplitude due to the impedance mismatch. Thus, regions of tension (i.e., negative pressure) are observed. Multiple reflections then occur between the bubble and the wall.

Figure 16.6 shows the shock dynamics during the last stages of collapse. Two shocks are emitted at collapse [34]: a weak precursor shock, due to the piston-like motion of the jet, and a strong water-hammer shock, due to the impact of the jet on the distal side of the bubble. These shocks merge into one shortly after their emission into the liquid. The resulting shock propagates radially outward and is stronger in the direction of propagation of the jet. The bubble becomes a vortex ring [4] and convects toward the wall.

To quantify the potential damage of this process, the maximum wall pressure, $p_{wall,max}$, generated over the course of the simulation is measured and plotted as a function of the location of the bubble at collapse, H_c, in Fig. 16.7. This pressure is not the maximum pressure over the course of the simulation, but rather the pressure measured after the shock emitted at collapse has propagate to and impacted the wall; since the shock propagates radially outward, the amplitude of the shock decreases with distance from the origin. H_c is determined by calculating the distance between the bubble centroid and the wall. Very high local pressures are achieved (up to 2.3 GPa) for bubbles initially close to the wall. As expected

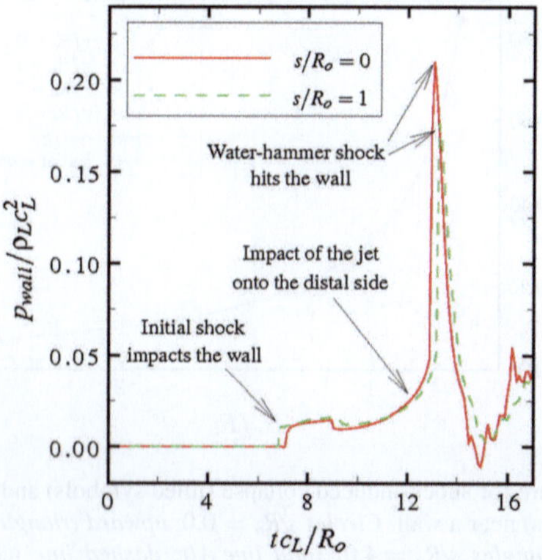

Fig. 16.5 History of the wall pressure at different locations along the wall for shock-induced collapse near a wall with $p_s/p_o = 353$, $H_o/R_o = 2.0$ [35]. Different distances s from the centerline are considered

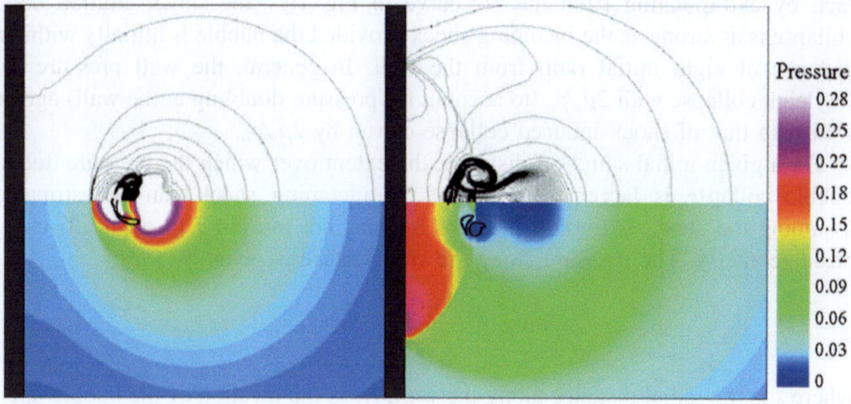

Fig. 16.6 Collapse and shock propagation in the shock-induced collapse of a gas bubble near a wall ($p_s/p_o = 353$, $H_o/R_o = 2.0$) at $t = 7.93$ (*left*) and $t = 9.42$ (*right*). *Top half* density lines; *bottom half* pressure contours. The solid *black* rectangles on the *left* of each frame denote the neighboring rigid wall. Adapted from Johnsen and Colonius [34] with permission from Cambridge University Press

[26], the shock strength decreases as $1/r$ with distance from the origin of the shock. At a farther radial distance along the wall ($s/R_o = 1$), the behavior of the pressure is different, in that the pressure is smaller for bubbles closer to the wall. This discrepancy can be explained by the fact that the shock emitted upon collapse propagates radially outward and reflects off the wall at an acute angle, thus leading to a lower pressure than expected in normal shock reflection. For bubbles

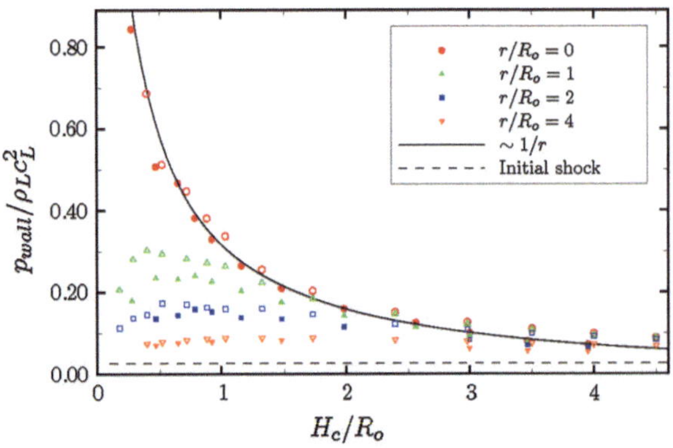

Fig. 16.7 Wall pressure for shock-induced collapse (filled symbols) and the equivalent Rayleigh collapse (open symbols) near a wall. *Circles* $s/R_o = 0.0$; *upward triangles* $s/R_o = 1.0$; *squares* $s/R_o = 2.0$; *inverted triangles* $s/R_o = 4.0$; *solid line* $1/r$; *dashed line* incoming shock. Adapted from Johnsen and Colonius [34] with permission from Cambridge University Press

sufficiently close to the wall, the shock emitted upon collapse is stronger than the incoming shock; in non-dimensional units, the pulse amplitude is 1.29×10^{-2}. In fact, by extrapolating from the $1/r$ curve in Fig. 16.7 the shock emitted upon collapse is as strong as the incoming shock provided the bubble is initially within a distance of eight initial radii from the wall. In general, the wall pressure for Rayleigh collapse with $2p_s/p_o$ (to account for pressure doubling at the wall) agrees well with that of shock-induced collapse driven by $2p_s/p_o$.

For a given initial stand-off distance, the extent over which the pressure due to bubble collapse is larger than that of the incoming shock can be estimated assuming the shock is normal (i.e., for bubbles not too close to the wall) using basic geometry. The pressure along the wall surface is given by [33]:

$$p_{wall} = \frac{c_1}{\sqrt{H_c^2 + s^2}} + c_2,$$

where s is the radial distance along the wall, H_c is the location of the bubble upon collapse for a given initial stand-off distance, H_o, and c_1 and c_2 are constant that can be determined if at least two measurements are known. For the case, $H_o/R_o = 2.0$, the distance along the wall for which the pressure is as large as that due to the incoming shock is $10R_o$. The current wall pressure results show good agreement with experimental data, provided that the acoustic properties of the solid wall is accounted for [34]: the acoustic impedance of the neighboring surface may not be much larger than that of water, such that shock transmission into the solid must be included, in addition to reflection.

16.3.2 Application to Shock-Wave Lithotripsy

Shock wave lithotripsy (SWL) is a non-invasive medical procedure in which shocks are focused on kidney stones in an attempt to break them [12]. A lithotripter pulse consists of a sharp compressive front, followed by a long expansion tail that has a tensile component (Fig. 16.1). Since kidney stones typically reside in urine and pooled blood, cavitation bubbles grow from nuclei after the passage of the tensile part of the pulse. The exact mechanism responsible for stone comminution has not yet been fully determined; however, two main mechanisms are thought to play an important role: wave propagation within the stone [13, 75] and cavitation erosion along the stone surface [15, 16]. Three failure modes are associated with wave propagation within the stone: shear waves and surface waves interfere constructively to form regions of high stresses [13], squeezing due to the pressure wave in the liquid acts as a compressive hoop stress [17], and spallation due to the inversion of the pulse amplitude upon reflection off the distal side of the stone and interaction with the incoming tensile pressure [21]. After the passage of the tensile part of the pulse, cavitation bubbles grow and gather as a cloud near the stone where they collapse as a whole, thereby generating high pressures [55]. The impact of the re-entrant jet onto the surface has been regarded as the primary cause of cavitation erosion in SWL [15, 16]. The individual effect of wave propagation within the stone or cavitation erosion is difficult to assess. Experiments show that the most effective comminution rate is achieved when both mechanisms work in synergy [75]: spallation due to the high tension created by the reflected pulse breaks the stones into smaller fragments, which are further eroded by the bubble collapse. Until now, the process thought to be responsible for the wave propagation within the stone is the lithotripter pulse; the present work shows that shocks are generated by bubble collapse and that the propagation of these waves may be a source of additional damage within the stone.

The incoming lithotripter pulse may interact with pre-existing bubbles or gas nuclei [50, 53, 63]. The pressures measured in shock-induced collapse exceed those estimated by Philipp and Lauterborn [54] for Rayleigh collapse. Recently, numerical simulations of shock-induced collapse of a gas bubble have shown that pressures on the order of 1 GPa are generated along the stone surface [33], which is much larger than the maximum compressive strength of kidney stones (on the order of 10 MPa, [14]. The reason that shock-induced collapse leads to higher pressures than Rayleigh collapse is that the pressure driving the bubble collapse is much larger in the former (350:1 in lithotripsy compared to 35:1). Since the pressures are so large, they may have the potential to cause failure by mechanisms similar to the direct action of the lithotripter pulses on the stone.

Three important observations made from prior studies of shock-induced collapse [33, 34] form the basis of the present analysis of structural damage within the stone: (i) An outward propagating shock wave is generated upon the non-spherical bubble collapse (ii) For bubbles close enough to the stone surface, the pressure may be far greater than that of the pulse, and (iii) The amplitude of the emitted shock wave decays as $1/r$. As the shock emitted upon bubble collapse interacts with the stone, part of the shock is transmitted into the stone and part is reflected. Thus effects similar to those due to the propagation of the lithotripter pulse through the stone can be anticipated. Because of the decay in pressure with distance, such an effect is expected to play a role for small stones of geometries favorable to shock focusing. To verify this hypothesis, the bubble dynamics simulations are prescribed as an initial condition for the propagation of elastic waves within an idealized stone, using the elastic-wave propagation code of Cleveland and Sapozhnikov [13]. In this model, the kidney stone is assumed to behave as a linear, isotropic, elastic solid, with the resulting elasto-dynamics equations solved using a time-domain finite difference method. The computational domain consists of a kidney stone immersed in water. For the water, the density is 1,000 kg/m^3 and the sound speed in water is 1,500 m/s, and for the stone, the density is 1,700 kg/m^3, the compressional wave speed is 2,635 m/s and the shear wave speed is 1,328 m/s. The pressure along the stone surface is given from the bubble dynamics simulations as a time-varying boundary condition, which is transformed into an initial condition in the water for the elastic waves simulations of the form, $f(t-x/c_L)$. The coupling between the hydrodynamics and the wave propagation in the stone is one-way only. To illustrate the propagation of the shock emitted during shock-induced collapse within the stone, pressure contours are plotted in Fig. 16.8 for a cylindrical stone (radius: $R_{stone} = 3.5$ mm; length: $L_{stone} = 7.5$ mm). The bubble is aligned with the axis of the cylinder and is initially located at a distance $H_o/R_o = 1.5$ below the stone, where $R_o = 100$ μm. The outline of the stone is included.

The shock generated by bubble collapse is partly transmitted as a shock into the stone and partly reflected as a shock back into the water; because of the higher sound speed in the stone, the wave propagates faster therein ($t = 1$ μs). The initial shock propagates radially outward, so that it reflects off the sides of the stone and then intersects along the centerline ($t = 2.5$ μs). In the meantime, the shock in water diffracts around the edge of the stone and leads to the formation of shear waves. The main front reflects off the distal side and inverts its amplitude since the end of the stone acts as a pressure-release surface, thus the reflected wave leads with a tensile phase ($t = 4$ μs). Furthermore, the different types of waves (e.g., reflected, shear, surface) interact with each other within the stone to produce other localized regions of tension ($t = 5$ μs). It should be noted that the amplitude of the stresses becomes smaller in each frame, as expected since the initial shock generated by the bubble collapse decreases as $1/r$. Hence, higher tensile stresses are expected to be generated in small stones. In addition, a geometry that favors shock focusing within the stone (e.g., a sphere) is expected to generate higher stresses.

To understand the implications of stone size and geometry on the tensile stress inside the stone, the principal stresses were calculated throughout the simulation and the peak tensile stress recorded at each location in the stone. Figure 16.9 shows the distribution of the peak tensile stress for a small ($R_{stone} = 1$ mm) and larger ($R_{stone} = 2$ mm) spherical stone. The units for the stresses are MPa, and the white line denotes the stone outline. Again, the bubble is initially located at $H_o/R_o = 1.5$ below the stone, with $R_o = 100$ μm. On the proximal surface, both cases

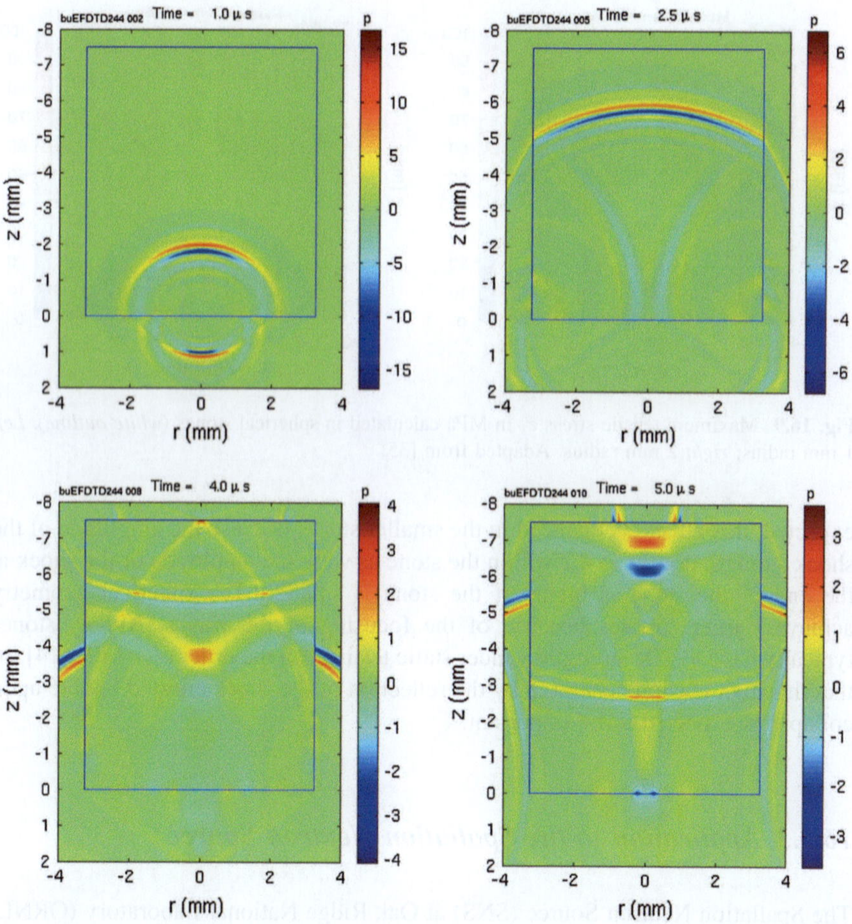

Fig. 16.8 Pressure contours showing the propagation of the shock generated by the collapse of a bubble near the *bottom* of a cylindrical stone (*radius* 3.5 mm; *length* 7.5 mm). The color scale is in MPa and is adjusted on each frame to the peak pressure. Adapted from [35]

show similar regions of high tension. This is expected, because the initial conditions based on the bubble collapse are identical; the only difference is the curvature of the stone. The reason for the high tension (as opposed to the previously discussed high compression) is as follows. The shock generated upon bubble collapse reflects off the stone back onto the bubble. As it subsequently reflects off the bubble, the amplitude of the shock inverts, thereby generating the large tension in the water on the proximal side of the stone. The stress contours within the stone are different in the two cases, because of the different geometry. The shock generated upon collapse propagates through the stone. Because of the shock reflection off the distal side and of the interference of the shear and surface waves, a high tension (approximately 100 MPa) is observed near the distal side of the stone. As

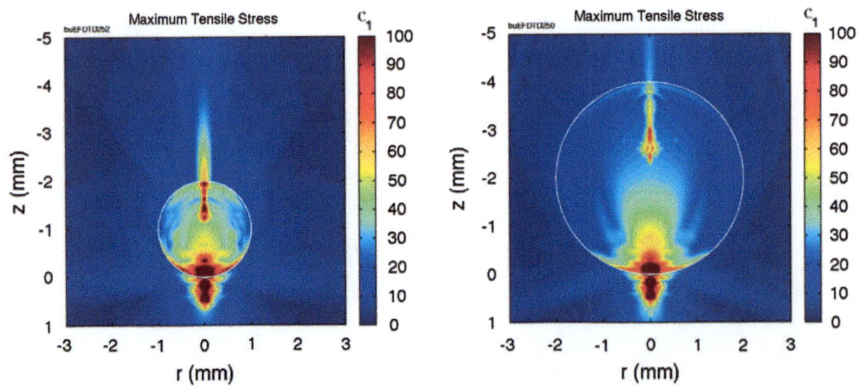

Fig. 16.9 Maximum tensile stress c_1 in MPa calculated in spherical stones (*white outline*). *Left* 1 mm radius; *right* 2 mm radius. Adapted from [35]

expected, this effect is enhanced in the smaller stone; because the amplitude of the shock strength decays as $1/r$ within the stone as well, the amplitude of the shock at the end of the stone is higher if the stone is smaller. The spherical geometry achieves larger stresses because of the focusing of the waves. Kidney stones typically have tensile strengths (under static testing) on the order of 1 MPa [14], so that this high tension generated by the reflection of the shock emitted bubble upon collapse is expected to be important.

16.3.3 Application to the Spallation Neutron Source

The Spallation Neutron Source (SNS) at Oak Ridge National Laboratory (ORNL) generates the most intense pulsed neutron beams in the world. During the operation of the SNS, the intense heating of the liquid mercury target by a proton beam causes large changes in pressure that in turn lead to the formation of cavitation bubbles. The violent collapse of these bubbles and the subsequent emission of shock and acoustic waves have been shown to generate damage on the container walls and therefore have the potential to significantly reduce its lifetime [23]. However, the dynamics of the cavitation bubbles and shock waves in the target are not well understood, thus preventing an effective elaboration of mitigation strategies. The goal of this study is to investigate the bubble dynamics in the small outer cooling channels, as opposed to the bulk mercury [61]. The channel is approximately 2 mm wide and the flow velocity is 2.5 m/s. The erosion pits observed in the channel exhibit elongated patterns in the direction of the flow [44].

First, to illustrate the destructive potential of cavitation in mercury, a simple and idealized Rayleigh-Plesset analysis reveals that the pressure generated at collapse by a cavitation bubble in mercury (Rayleigh collapse) is several orders of

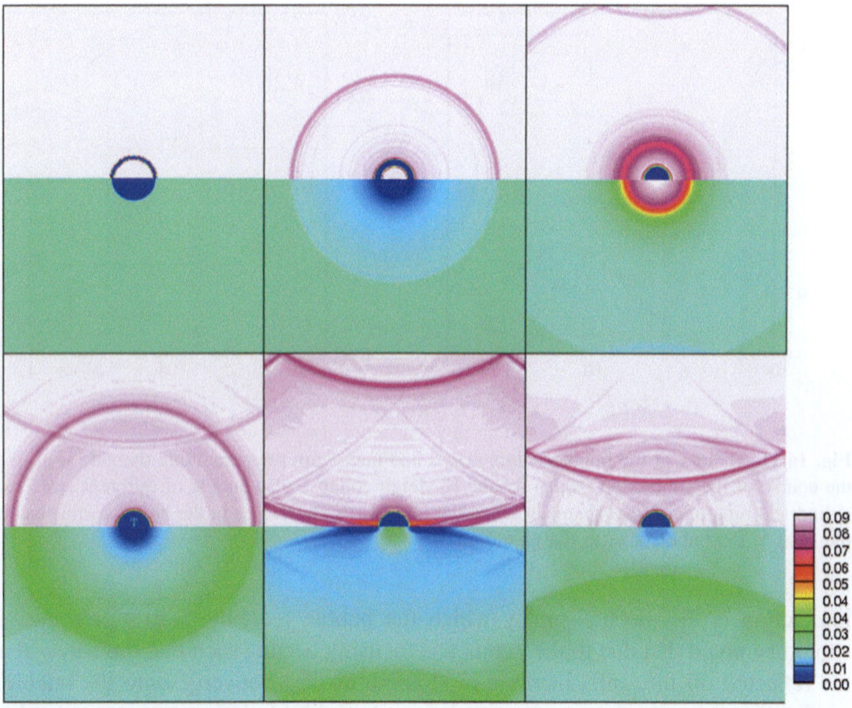

Fig. 16.10 Bubble collapse in a cylindrical pipe at times 0.0, 3.61, 7.38, 11.1, 14.8, 18.5. *Top* numerical Schlieren contours; *bottom* pressure contours. The initial pressure ratio across the interface is 1,400

magnitude larger than that produced by a bubble in water, due to the very low vapor pressure of mercury, which leads to a much larger pressure ratio driving the collapse, and the tenfold increase in liquid density. The flow in the target, particularly in the exterior channels, is significantly influenced by the presence of solid boundaries. To determine the effect of this confinement, direct simulations are conducted to study the axisymmetric collapse of a single air bubble in mercury along the centerline of a cylindrical vessel of diameter 2 mm. Although the geometry is slightly different from the experiments, the influence of the walls is expected to be captured. The bubble collapse is similar to the Rayleigh collapse described previously. Figure 16.10 shows sections of numerical Schlieren and pressure contours of this phenomenon, for a 100 micron bubble with an initial pressure ratio of 1,400. The cylinder walls are located at the top (and bottom); non-reflecting conditions are applied along the sides, effectively making the cylinder infinitely long. As observed previously, an expansion is initially released and propagates radially outward. In the third frame, collapse has just occurred, and the shock emitted during this process is starting to propagate radially outward; the expansion reflects off the cylinder walls. In the fifth frame, the shock reflects off the cylinder and the reflected expansion interacts with the bubble. This latter

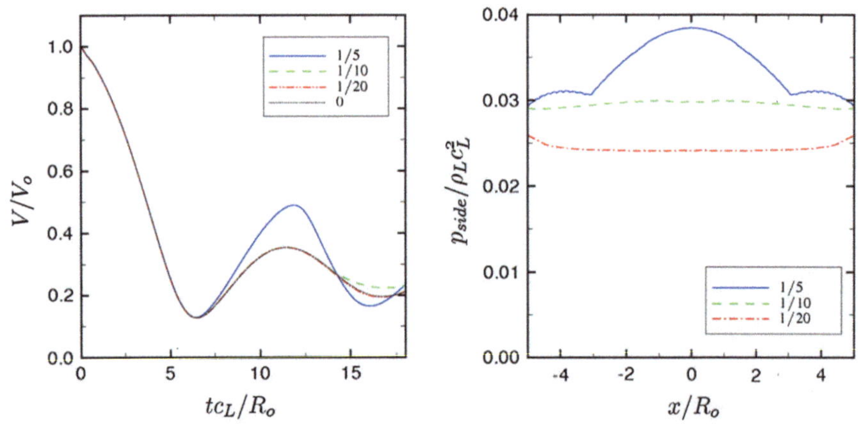

Fig. 16.11 History of the bubble volume (*left*) and maximum pressure along the side wall over the course of the simulation (*right*) for the Rayleigh collapse of a bubble of different sizes in a cylindrical pipe of constant diameter filled with mercury. The values in the boxes correspond to the ratio of the bubble size to the pipe diameter

interaction is the mechanism by which the bubble feels the wall's presence. It should be noted that due to the geometry, focusing occurs, such that the waves that are reflected off the walls increase in strength as they converge onto the bubble.

The dependence of the wall pressure on the initial bubble size is examined in Fig. 16.11. The following bubble radii for this problem are considered: 200, 100 and 50 μm. Here, the problem is set up such that the number of computational cells in the bubble is the same for all sizes. The first collapse is unaffected by the confinement, because the collapse time is smaller than the time it takes for the initial expansion to propagate to the wall and back to the bubble. However, the rebound and second collapse are affected by the proximity of the wall, especially for larger bubbles; at these later times, the reflected expansion and shock generated at collapse interact with the bubble. In the right plot, the maximum pressure measured over the course of the simulation along the wall is plotted. For a large bubble compared to the channel size, the emitted shock is clearly spherical and exhibits the rounded shaped shown in the figure. For smaller bubbles, i.e., bubbles located farther away, the emitted shock essentially looks planar, as expected. The small increase along the edges is due to boundary effects (corners), where the reflected shock is emphasized.

In prior experimental work, the damage on the walls exhibited asymmetries [44]. It was further postulated that bubble translation may lead to such effects. To test this hypothesis, simulations were carried out for different flow velocities (0.001, 0.01 and 0.1 non-dimensionalized using the sound speed in mercury) in the channel in Fig. 16.12. The maximum pressure results along the side walls over the course of the simulation yielded little dependence on the velocity. However, the present work demonstrated that another mechanism may lead to asymmetric behavior in this application: the shock-induced collapse of a bubble, e.g., by the

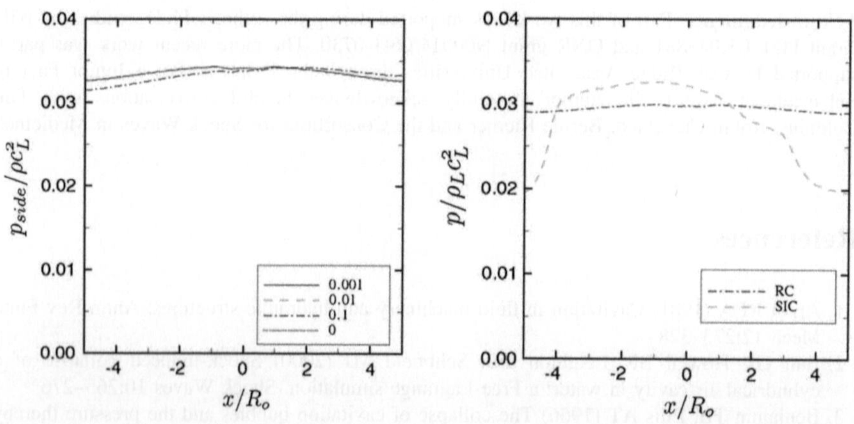

Fig. 16.12 Maximum pressure along the side wall over the course of the simulation for the collapse of a gas bubble (100 micron radius) in a cylindrical pipe. *Left* dependence on the flow velocity; *right* Rayleigh collapse versus shock-induced collapse by the same pressure ratio

pressure wave emitted during the spallation or by the collapse of neighboring bubbles. In Fig. 16.12, the shock has the same pressure ratio as in the Rayleigh collapse problem and is coming from the right, as evidenced by the sharp features on the left of the plot and the smoother distributions on the right.

16.4 Concluding Remarks

The present work shows that direct simulation of the Euler equations to study bubble dynamics and shock emission upon collapse is a powerful technique. By examining the velocity and pressure fields for Rayleigh collapse and shock-induced collapse, valuable insights can be obtained. For instance, the history of the proximal and distal side shed light on the non-spherical dynamics of shock-induced collapse, and the wall pressure provided quantitative measures of the damage potential of bubble collapse. The studies are applied to shock-wave lithotripsy, the most common treatment of kidney stones in the United States. Coupled bubble dynamics and elastic-wave propagation simulations are conducted. Two previously unexplored stone comminution mechanisms are proposed: shock-induced bubble collapse and spallation due to shocks emitted by bubble collapse. In the context of the Spallation Neutron Source, the effect of confinement on the bubble dynamics and the subsequent pressure generated by collapse are examined. As illustrated throughout this article, although many studies on cavitation erosion have focused on bubble dynamics or solid mechanics, the present results suggest that the coupling between the fluid and solid mechanics are important factors in cavitation erosion.

Acknowledgments Part of this work was supported during the author's Ph.D. studies by NIH Grant PO1 DK043881 and ONR grant N00014-06-1-0730. The more recent work was partly supported by Oak Ridge Associated Universities through the Ralph E. Powe Junior Faculty Enhancement Award. The author gratefully acknowledges helpful conversations with Tim Colonius, Robin Cleveland, Bernie Riemer and the Consortium for Shock Waves in Medicine.

References

1. Arndt REA (1981) Cavitation in fluid machinery and hydraulic structures. Annu Rev Fluid Mech 12:273–328
2. Ball GJ, Howell BP, Leighton TG, Schofield MJ (2000) Shock-induced collapse of a cylindrical air cavity in water: a Free-Lagrange simulation. Shock Waves 10:265–276
3. Benjamin TB, Ellis AT (1966) The collapse of cavitation bubbles and the pressure thereby produced against solid boundaries. Phil Trans R Soc Lond A 260:221–240
4. Best JP (1993) The formation of toroidal bubbles upon the collapse of transient cavities. J Fluid Mech 251:79–107
5. Blake JR, Taib BB, Doherty G (1986) Transient cavities near boundaries. Part 1: rigid boundary. J Fluid Mech 170:479–497
6. Bourne NK, Field JE (1992) Shock-induced collapse of single cavities in liquids. J Fluid Mech 244:225–240
7. Brennen CE (1995) Cavitation and bubble dynamics. University Press, Oxford
8. Brouillette M (2002) The Richtmyer-Meshkov instability. Annu Rev Fluid Mech 34:445–468
9. Busnaina AA, Kashkoush II, Gale GW (1995) An experimental study of megasonic cleaning of silicon wafers. J Electrochem Soc 142:2812–2817
10. Chahine GL, Duraiswami R (1992) Dynamical interactions in a multi-bubble cloud. Trans ASME J Fluids Eng 114:680–686
11. Chang CH, Liou MS (2007) A robust and accurate approach to computing compressible multiphase flow: stratified flow model and AUSM$^+$-up scheme. J Comput Phys 225:840–873
12. Chaussy C (1982) Extracorporeal shock wave lithotripsy: new aspects in the treatment of kidney stone disease. Karger, Basel
13. Cleveland RO, Sapozhnikov OA (2005) Modeling elastic wave propagation in kidney stones with application to shock wave lithotripsy. J Acoust Soc Am 118:2667–2676
14. Cohen NP, Whitfield HN (1993) Mechanical testing of urinary calculi. World J Urol 11:1–18
15. Coleman AJ, Saunders JE, Crum LA, Dyson M (1987) Acoustic cavitation generated by an extracorporeal shockwave lithotripter. Ultrasound Med Biol 13:69–76
16. Crum LA (1988) Cavitation microjets as a contributory mechanism for renal calculi disintegration in ESWL. J Urol 140:1587–1590
17. Eisenmenger W (2001) The mechanisms of stone fragmentation in ESWL. Ultrasound Med Biol 27:683–693
18. Elder SA (1959) Cavitation microstreaming. J Acoust Soc Am 31:54–64
19. Fujikawa S, Akamatsu T (1980) Effects of the non-equilibrium condensation of vapour on the pressure wave produced by the collapse of a bubble in a liquid. J Fluid Mech 97:481–512
20. Giannadakis E, Gavaises M, Arcoumanis C (2008) Modelling of cavitation in diesel injector nozzles. J Fluid Mech 629:153–193
21. Gracewski SM, Dahake G, Ding Z, Burns SJ, Everbach EC (1993) Internal stress wave measurements in solids subjected to lithotripter pulses. J Acoust Soc Am 94:652–661
22. Haas JF, Sturtevant B (1987) Interaction of weak shock waves with cylindrical and spherical gas inhomogeneities. J Fluid Mech 181:41–76
23. Haines JF, Riemer BW, Felde DK, Hunn JD, Pawel SJ, Tai CC (2005) Summary of cavitation erosion investigations for the SNS mercury target. J Nucl Mater 343:58–69
24. Hammitt FG (1980) Cavitation and multiphase flow phenomena. McGraw-Hill, New York

25. Hansson I, Kedrinskii V, Morch KA (1982) On the dynamics of cavity clusters. J Phys D Appl Phys 12:1725–1734
26. Hickling R, Plesset MS (1964) Collapse and rebound of a spherical bubble in water. Phys Fluids 7:7–14
27. Hu XY, Khoo BC, Adams NA, Huang FL (2006) A conservative interface method for compressible flows. J Comput Phys 219:553–578
28. Hua C, Johnsen E (2013) Nonlinear oscillations following the Rayleigh collapse of a gas bubble in a linear viscoelastic (tissue-like) medium. Phys Fluids 25, 083101
29. Jamaluddin AR (2005) Free-Lagrange simulations of shock-bubble interaction in extracorporeal shock wave lithotripsy. PhD thesis, University of Southampton, Southampton
30. Jiang GS, Shu CW (1996) Efficient implementation of weighted ENO schemes. J Comput Phys 126:202–228
31. Johnsen E, Colonius T (2006) Implementation of WENO schemes in compressible multicomponent flow problems. J Comput Phys 219:715–732
32. Johnsen E (2008) Numerical simulations of non-spherical bubble collapse, PhD thesis, California Institute of Technology, California
33. Johnsen E, Colonius T (2008) Shock-induced collapse of a gas bubble in shockwave lithotripsy. J Acoust Soc Am 124:2011–2020
34. Johnsen E, Colonius T (2009) Numerical simulations of non-spherical bubble collapse. J Fluid Mech 629:231–262
35. Johnsen E, Colonius T, Cleveland R (2009) Damage potential of the shock-induced collapse of a gas bubble. Proceedings of the 7th International Symposium on Cavitation, Ann Arbor, Michigan
36. Klaseboer E, Turangan C, Fong SW, Liu TG, Hung KC, Khoo BC (2006) Simulations of pressure pulse-bubble interaction using boundary element method. Comput Methods Appl Mech Eng 195:4287–4302
37. Kling CL, Hammitt FG (1972) A photographic study of spark-induced cavitation bubble collapse. Trans ASME: J Basic Eng 94:825–833
38. Knapp RT, Daily JW, Hammitt FG (1970) Cavitation. McGraw-Hill, New York
39. Kornfeld M, Suvorov L (1944) On the destructive action of cavitation. J Appl Phys 15:495–506
40. Lauterborn W, Bolle H (1975) Experimental investigations of cavitation-bubble collapse in the neighbourhood of a solid boundary. J Fluid Mech 72:391–399
41. Li SC (2000) Cavitation of hydraulic machinery. Imperial College Press, London
42. Lindau O, Lauterborn W (2003) Cinematographic observation of the collapse and rebound of a laser-produced cavitation bubble near a wall. J Fluid Mech 479:327–348
43. Lubock P, Goldsmith W (1980) Experimental cavitation studies in a model head-neck system. J Biomech 13:1041–1052
44. McClintock DA, Riemer BW, Ferguson PD, Carroll AJ, Dayton MJ (2012) Initial observations of cavitation-induced erosion of liquid metal spallation target vessels at the Spallation Neutron Source. J Nucl Mat 431(1–3):147–159 (in press)
45. Moholkar VS, Warmoeskerken MMCG, Ohl CD, Prosperetti A (2004) Mechanism of mass-transfer enhancement in textiles by ultrasound. AIChE J 60:58–64
46. Mourad PD, Roberts FA, McInnes C (2007) Synergistic use of ultrasound and sonic motion for removal of dental plaque bacteria. Compend Contin Educ. Dent 28:354–358
47. Nagrath S, Jansen K, Lahey RT Jr, Akhatov I (2006) Hydrodynamic simulation of air bubble implosion using a level set approach. J Comput Phys 215:98–132
48. Naude CF, Ellis AT (1961) On the mechanism of cavitation damage by nonhemispherical cavities in contact with a boundary. Trans ASME: J Basic Eng 83:648–656
49. Nourgaliev RR, Dinh TN, Theofanous TG (2006) Adaptive characteristics-based matching for compressible multifluid dynamics. J Comput Phys 213:500–529
50. Ohl CD, Ikink R (2003) Shock-wave-induced jetting of micron-size bubbles. Phys Rev Lett 90:214502

51. Ohl CD, Arora M, Dijkink R, Janve V, Lohse D (2006) Surface cleaning from laser-induced cavitation bubbles. Appl Phys Lett 89:074102
52. Packer M, Fishkind WJ, Fine IH, Seibel BS, Hoffman RS (2005) The physics of phaco: a review. J Cataract Refract Surg 31:424–431
53. Philipp A, Delius M, Scheffczyk C, Vogel A, Lauterborn W (1993) Interaction of lithotripter-generated shock waves with air bubbles. J Acoust Soc Am 93:2496–2509
54. Philipp A, Lauterborn W (1998) Cavitation erosion by single laser-produced bubbles. J Fluid Mech 361:75–116
55. Pishchalnikov YA, Sapozhnikov OA, Bailey MR, Williams JC, Cleveland RO, Colonius T, Crum LA, Evan AP, McAteer JA (2003) Cavitation bubble cluster activity in the breakage of kidney stones by lithotripter shockwaves. J Endourol 17:435–446
56. Plesset MS (1949) The dynamics of cavitation bubbles. Trans ASME: J Appl Mech 16:277–282
57. Plesset MS, Chapman RB (1971) Collapse of an initially spherical vapour cavity in the neighbourhood of a solid boundary. J Fluid Mech 49:283–290
58. Plesset MS, Prosperetti A (1977) Bubble dynamics and cavitation. Annu Rev Fluid Mech 9:145–185
59. Popinet S, Zaleski S (2002) Bubble collapse near a solid boundary: a numerical study of the influence of viscosity. J Fluid Mech 464:137–163
60. Rayleigh L (1917) On the pressure developed in a liquid during the collapse of a spherical cavity. Phil Mag 34:94–98
61. Riemer BW, Wendel MW, Felde DK (2010) Cavitation damage experiments for mercury spallation targets at the LANSCE—WNR in 2008. J Nucl Mater 398:207–219
62. Roberts WW, Hall TL, Ives K, Wolf JS Jr, Fowlkes JB, Cain CA (2006) Pulsed cavitational ultrasound: a non-invasive technology for controlled tissue ablation (Histotripsy) in the rabbit kidney. J Urol 175:734–738
63. Sankin GN, Simmons WN, Zhu SL, Zhong P (2005) Shock wave interaction with laser-generated single bubbles. Phys Rev Lett 95:034501
64. Suslick KS (1990) Sonochemistry. Science 247:1439–1445
65. Thornycroft J, Barnaby SW (1895) Torpedo boat destroyers. Proc Inst Of Civ Eng 122:51–103
66. Toro EF (1999) Riemann solvers and numerical methods for fluid dynamics. Springer, Berlin
67. Tomita Y, Shima A (1986) Mechanisms of impulsive pressure generation and damage pit formation by bubble collapse. J Fluid Mech 169:535–564
68. Vogel A, Lauterborn W, Timm R (1988) Acoustic transient generation by laser-produced cavitation bubbles near solid boundaries. J Acoust Soc Am 84:719–731
69. Vogel A, Lauterborn W, Timm R (1989) Optical and acoustic investigations of the dynamics of laser-produced cavitation bubbles near a solid boundary. J Fluid Mech 206:299–338
70. Walmsley AD, Laird WRE, Williams AR (1988) Dental plaque removal by cavitational activity during ultrasonic scaling. J Clin Periodontal 15:539–543
71. Walters J, Davidson JF (1962) The initial motion of a gas bubble formed in an inviscid liquid. Part 1. The two-dimensional bubble. J Fluid Mech 12:408–416
72. Wardlaw AB, Luton JA (2000) Fluid-structure interaction mechanisms for close-in explosions. Shock Vib 7:265–275
73. Yoganathan AP, He Z, Jones SC (2004) Fluid mechanics of heart valves. Annu Rev Biomed Eng 6:331–362
74. Zhang SG, Duncan JH, Chahine GL (1993) The final stage of the collapse of a cavitation bubble near a rigid wall. J Fluid Mech 257:147–181
75. Zhu S, Cocks FH, Preminger GM, Zhong P (2002) The role of stress waves and cavitation in stone comminution in shock wave lithotripsy. Ultrasound Med Biol 28:661–671

Index

K.-H. Kim et al. (eds.), *Advanced Experimental and Numerical Techniques for Cavitation Erosion Prediction*, Fluid Mechanics and Its Applications 106,
DOI: 10.1007/978-94-017-8539-6, © Springer Science+Business Media Dordrecht 2014

MIX
Papier aus verantwortungsvollen Quellen
Paper from responsible sources
FSC® C105338

If you have any concerns about our products,
you can contact us on
ProductSafety@springernature.com

In case Publisher is established outside the EU,
the EU authorized representative is:
Springer Nature Customer Service Center GmbH
Europaplatz 3, 69115 Heidelberg, Germany

Printed by Libri Plureos GmbH
in Hamburg, Germany